Environmental Footprint of Bioplastic Additives

Plastics derived from renewable sources, such as biomass or animal-based materials, offer an eco-friendly alternative to traditional petroleum-based plastics. These bioplastics, being bio-based and biodegradable, hold great promise in reducing reliance on fossil fuels and mitigating plastic pollution. However, additives play a crucial role in enhancing the properties of bioplastics, improving flexibility, durability, and overall performance. While some additives facilitate biodegradation, others – such as phthalates, bisphenol A (BPA), and brominated compounds – can release harmful substances upon breakdown, posing risks to soil and water quality.

This book offers a comprehensive understanding of bioplastic compositions, degradation processes, and the environmental impact of various additives. Through detailed life cycle analyses, it assesses how bioplastic additives compare to conventional plastic additives in terms of sustainability and ecological impact. The book also addresses the challenges of managing bioplastic waste, including recycling limitations and the lack of standardized disposal methods.

Key Features:

- In-depth analysis of bioplastic compositions, degradation processes, and additive functions
- Comparative life cycle assessment of bioplastic and conventional plastic additives
- Examination of the environmental impact of bioplastic additives on soil, wildlife, and water quality
- Discussion on innovative waste management and recycling strategies for bioplastics
- Exploration of policy frameworks, public awareness initiatives, and case studies promoting sustainable bioplastic usage

Designed to address the knowledge gap in bioplastic additive research, this book is an essential resource for researchers, policymakers, industry professionals, and environmental enthusiasts. It bridges the gap between scientific understanding and practical implementation, empowering stakeholders to make informed decisions and contribute to sustainable bioplastic waste management.

Environmental Footprint of Bioplastic Additives

Edited by
Babafemi Raphael Babaniyi, Sesan Abiodun Aransiola,
Gabriel Gbenga Babaniyi, and Naga Raju Maddela

CRC Press
Taylor & Francis Group
Boca Raton London New York

CRC Press is an imprint of the
Taylor & Francis Group, an **informa** business

Designed Cover Image: Shutterstock ID: 2482384717

First edition published 2026
by CRC Press
2385 NW Executive Center Drive, Suite 320, Boca Raton FL 33431

and by CRC Press
4 Park Square, Milton Park, Abingdon, Oxon, OX14 4RN

CRC Press is an imprint of Taylor & Francis Group, LLC

ISBN: 9781032991863 (hbk)
ISBN: 9781032991870 (pbk)
ISBN: 9781003602828 (ebk)

DOI: 10.1201/9781003602828

Typeset in Times
by Newgen Publishing UK

Contents

SECTION IV Management and Societal Impact

Foreword

Climate change is a reality and a challenge of the present

The quest for sustainable materials is increasingly imperative as the global community confronts escalating environmental challenges. Bioplastics, acknowledged as a viable alternative to conventional plastics, stand as a better innovation, environmentally friendly solution, and a commitment to social accountability. Their capacity to mitigate reliance on fossil fuels, reduce carbon emissions, and integrate seamlessly into circular economic paradigms establishes them as a cornerstone of contemporary materials science.

This book, which presents a comprehensive analysis of bioplastics and their additives, emerges at a crucial time. It evaluates various dimensions of bioplastic evolution, encompassing their fundamental characteristics and their environmental and societal repercussions. The chapters deliver an exhaustive examination of essential issues, including the advantages and challenges associated with bioplastic composites, the function of additives in performance enhancement, and pioneering recycling strategies. These discussions signify a substantial endeavor to bridge scientific inquiry with industrial applications and overarching sustainability goals. Importantly, the book accentuates environmental and societal dimensions. Chapters that investigate the implications of bioplastics on soil health, marine ecosystems, and food safety illuminate the intricate interrelations between bioplastics and ecological systems. Moreover, discussions on waste-to-energy conversion, biodegradability, and climate change mitigation illustrate the transformative potential of these materials in addressing some of the most pressing environmental issues globally.

Furthermore, this book discusses the regulatory framework, societal ramifications, and educational requirements for waste management professionals, providing a holistic perspective frequently absent in exclusively technical evaluations. As society advances toward a more sustainable future, this book serves as an indispensable resource for a diverse audience, including researchers, policymakers, educators, and industry leaders. Its judicious assessment of both the possibilities and constraints of bioplastics provides guidance for informed decision-making and responsible innovation.

I commend the authors for their comprehensive research and insightful viewpoints. This book epitomizes the efficacy of interdisciplinary collaboration and reinforces the necessity of harmonizing technological advancement with ecological and societal imperatives.

Dr. Cristóbal N. Aguilar
Full-Time Professor at BBG-DIA /UAdeC
Universidad Autónoma de Coahuila, México
December 20, 2024

Preface

The global pursuit of sustainability has spurred remarkable advancements in materials science, particularly in the development and application of bioplastics. Bioplastics, derived from renewable resources or engineered to degrade more efficiently than conventional plastics, hold immense promise for addressing critical environmental challenges. However, their adoption and integration into various sectors necessitate a comprehensive understanding of their properties, applications, and potential implications.

This book explores the complex landscape of bioplastics and their additives, providing a broad overview while examining key themes that characterize this rapidly advancing field. The chapters offer insights into the various types of bioplastic composites and the challenges they face, emphasizing innovations and their applicability across different industries. Recycling methods and their significance in promoting a circular economy highlight sustainability as a fundamental principle of bioplastic technology. Furthermore, discussions on the pros and cons of bioplastics and their additives present a balanced view of their integration into various applications. Special emphasis is given to the role of bioplastic additives such as phthalates and phenolic compounds, shedding light on their environmental implications and contributions to climate change mitigation. Cutting-edge topics such as waste-to-energy conversion, biodegradability, and the life span of bioplastics are examined to address the growing need for innovative solutions to plastic waste management.

The book also evaluates the environmental and agricultural impacts of bioplastics, including their role in soil contamination and health, the effects of additives in marine environments, and their implications for food safety and storage. Discussions on microbial degradation, waste management education, and the social aspects of bioplastic waste underscore the interconnections between science, society, and sustainability. This book is intended as a comprehensive resource for researchers, industry professionals, policymakers, and educators aiming to deepen their understanding of bioplastics and their additives. By addressing critical scientific, environmental, and societal aspects, it aims to contribute to informed decision-making and innovation in the quest for a sustainable future.

We hope this book inspires collaboration and progress, fostering solutions that harmonize material science with ecological stewardship. The chapters were contributed by 69 academicians, scientists, and researchers from eight different countries (Nigeria, Kazakhstan, China, India, the USA, South Africa, and Ecuador) across the world.

Babafemi Raphael Babaniyi
Bioresources Development Centre,
National Biotechnology Research and Development Agency (NBRDA),
Ogbomoso, Nigeria

Sesan Abiodun Aransiola, Ph.D.
Department of Microbiology, Faculty of Science,
University of Abuja,
Abuja, Nigeria

Gabriel Gbenga Babaniyi
Department of Agricultural Development and Management,
Agricultural and Rural Management Training Institute (ARMTI),
Ilorin, Nigeria

Naga Raju Maddela, Ph.D.
Departamento de Ciencias Biológicas,
Facultad de Ciencias de la Salud,
Universidad Técnica de Manabí,
Portoviejo, Ecuador

Acknowledgments

We extend our heartfelt gratitude to the contributors of this chapter for their invaluable insights and timely responses throughout the course of this project. Their enthusiasm and dedication have been instrumental in ensuring this volume was completed within the scheduled timeframe. We deeply appreciate their collaboration and unwavering support. We also express our sincere thanks to the anonymous reviewers for their constructive feedback, which significantly enhanced the quality of this book. A special note of appreciation goes to the CRC Press and production team for their exceptional guidance and unwavering support. Their expertise and commitment made it possible to complete this project within such a remarkably short period. It has been an honor and privilege to collaborate with them. Lastly, we would like to express our sincere thanks to the management and colleagues at the Biotechnology Research and Development Agency, Abuja (Nigeria), University of Abuja (Nigeria), Agricultural and Rural Management Training Institute (Nigeria), and Universidad Técnica de Manabí (Ecuador) for their unwavering support throughout the book proposal and preparation process.

Babafemi Raphael Babaniyi
Bioresources Development Centre,
National Biotechnology Research and Development Agency (NBRDA),
Ogbomoso, Nigeria

Sesan Abiodun Aransiola, Ph.D.
Department of Microbiology, Faculty of Science,
University of Abuja,
Abuja, Nigeria

Gabriel Gbenga Babaniyi
Department of Agricultural Development and Management,
Agricultural and Rural Management Training Institute (ARMTI),
Ilorin, Nigeria

Naga Raju Maddela, Ph.D.
Departamento de Ciencias Biológicas,
Facultad de Ciencias de la Salud,
Universidad Técnica de Manabí,
Portoviejo, Ecuador

About the Editors

Babafemi Raphael Babaniyi is a distinguished academic and researcher renowned for his contributions to environmental science and chemistry. He began his academic journey at Caritas University, Amorji-Nike, Emene, Enugu State, Nigeria, where he earned a Bachelor's degree in Industrial Chemistry in 2008. Driven by an insatiable passion for knowledge, he pursued a Master's degree in Environmental Chemistry at the prestigious Federal University of Technology, Akure, Ondo State, Nigeria, graduating in 2021. Currently, he is advancing his expertise as a Ph.D. candidate in Environmental Chemistry in the Department of Chemistry at the same institution. His scholarly impact extends far beyond academia. With over 40 publications to his name, including internationally recognized book chapters and articles, he has significantly enriched the global scientific community. His areas of expertise span a broad spectrum, encompassing Environmental Chemistry, Green Chemistry, Phytoremediation, Chemical Biology, Bioremediation of Soil-Contaminated Environments, Plant Protection and Animal Health, Environmental Impact Assessment, and Biodegradable Plastics. In addition to his academic achievements, he serves as Assistant Chief Research Officer at the Bioresources Development Centre, under the National Biotechnology Development Agency in Nigeria. In this role, he plays a pivotal part in advancing biotechnology research and innovation, further cementing his influence in the scientific domain. Babafemi's dedication to unraveling complex environmental challenges and his innovative approach to sustainability have established him as a trailblazer in his field. His unwavering commitment to excellence and his multifaceted contributions to science and research make him an inspiring figure and a beacon of knowledge in Environmental Chemistry.

Sesan Abiodun Aransiola is a notable Ph.D. holder in Environmental Microbiology in the Department of Microbiology, Federal University of Technology, Minna, Nigeria. He obtained his first degree (B.Tech.) and Master degree (M.Tech.) from the same department in 2009 and 2014, respectively. He has demonstrated his research expert in the production of vermicasts from vermicomposting (Vermitechnology) of organic wastes to assist plants in the remediation of polluted soil with heavy metals. Currently, he is Lecturer at the Department of Microbiology, University of Abuja, Nigeria. His expertise spans environmental health, phytoremediation, vermicomposting, marine resources, and bioremediation of pollutants soil including plastics. Aransiola's research is deeply rooted in innovative solutions for soil and water remediation, with significant contributions to microbial biotechnology and waste management. He is an award-winning researcher and has over 90 publications including book chapters, research, and review articles of good international repute with high impact factor. He has co-edited scientific books of global interest which include *Microbial Biotechnology for Bioenergy* (Elsevier); *Ecological Interplays in Microbial Enzymology*, *Prospects for Soil Regeneration and Its Impact on Environmental Protection*, *Soil Microbiome in Green Technology Sustainability*, *Marine Bioprospecting for Sustainable Blue-Bioeconomy* (Springer); *Microbial Biofilms – Applications and Control*, *Marine Greens: Environmental, Agricultural, Industrial and Biomedical Applications*, *Emerging Contaminants in Food and Food Products*, *Phytoremediation in Food Safety: Risks and Prospects*, *Biotoxins in Food: Threats and Benefits* (CRC Press) with many more in the process. Also, he has worked to the rank of an assistant chief scientific officer at Bioresources Development Centre, National Biotechnology Research and Development Agency, Nigeria where he was involved in the uses of vinasse (a by-product of ethanol) as bio-fertilizer to reclaim polluted soil for agricultural purposes. Dr. Aransiola's academic journey is complemented by a robust teaching career and active involvement in community services, where he leverages his knowledge to foster sustainability. His contributions continue to impact both academia and the broader environmental sciences community. Abiodun is a member of the Nigerian Society for Microbiology and the American Society for Microbiology among others.

Gabriel Gbenga Babaniyi is a skilled professional in animal production with a focus on physiology. Based in Nigeria, he serves as Management Development Officer at the Agricultural and Rural Management Training Institute (ARMTI) in Ilorin. His academic journey spans from a Bachelor of Agriculture from the University of Abuja to ongoing postgraduate studies in animal production at the University of Ilorin. Babaniyi's research interests include regenerative agriculture, environmental protection, and aquaculture, contributing to various published works in reputable journals. His hands-on experience in agricultural management, coupled with a strong focus on research and development, has positioned him as a key player in addressing challenges related to food security and rural development. In addition to his research endeavors, he is actively involved in training facilitation, grant writing, and agricultural consultancy. Babaniyi's academic and professional pursuits highlight his dedication to sustainable agricultural practices and rural transformation.

Naga Raju Maddela received his M.Sc. (1996–1998) and Ph.D. (2012) in Microbiology from Sri Krishnadevaraya University, Anantapur, India. During his doctoral program in the area of Environmental Microbiology, he investigated the effects of industrial effluents/insecticides on soil microorganisms and their biological activities and he has been working as Faculty in Microbiology since 1998, teaching undergraduate and postgraduate students. He worked as Prometeo Investigator (fellowship received from SENESCYT) at Universidad Estatal Amazónica, Ecuador during 2013–2015, and received "Postdoctoral Fellowship" (2016–2018) from Sun Yat-sen University, China. He also received external funding from "China Postdoctoral Science Foundation" in 2017, internal funding from "Universidad Técnica de Manabí" in 2020. He participated in national/international conferences and presented research data in China, Cuba, Ecuador, India, and Singapore. Currently, he is working as Full Professor at the Facultad de Ciencias de la Salud, Universidad Técnica de Manabí, Portoviejo, Ecuador. He has been actively publishing scientific articles, books (authored and edited), and chapters since 2007. As of now, he has published 110 articles, 25 books, and 50 book chapters.

Contributors

Muhammed Abdulazeez
Department of Microbiology
Skyline University Nigeria
Kano, Nigeria

Abdulhakeem Idris Abdulhakeem
Department of Biodiversity and Bioresources
Al-Farabi Kazakh National University
Almaty, Kazakhstan

A.O. Abdulkareem
Department of Microbiology, Faculty of Science
University of Abuja
Abuja, Nigeria

Umar Abdulrasheed
Department of Environmental Management for Agriculture
University of Hertfordshire
Hatfield, United Kingdom

Ibrahim Abdulrazaq
Department of Microbiology
University of Ilorin
Ilorin, Nigeria

Mustapha Abdulsalam
Department of Microbiology
Skyline University Nigeria
Kano, Nigeria

Peter Olabisi Abioye
Department of Microbiology
Federal University of Technology Minna
Minna, Niger State, Nigeria

Mustapha Adamu
Department of Microbiology
Federal University of Technology Minna
Minna, Niger State, Nigeria

Akintunde Adewole Adegbite
Nigerian Institute for Oceanography and Marine Research
Lagos, Nigeria

Aisha Eniola Adegbite
Department of Microbiology, Faculty of Science
University of Abuja
Abuja, Nigeria

Ojo Paul Adewoye
Bioresources Development Centre, National Biotechnology Research and Development Agency
Ogbomoso, Nigeria

Abdulmojeed Abdulsalam Adopisa
Department of Chemistry, School of Physical Sciences
Federal University of Technology Akure
Ondo, Nigeria

B.O. Akanbi
Department of Microbiology, Faculty of Science
University of Abuja
Abuja, Nigeria

Oluwatobi Sunday Akande
Department of Rural Development and Gender Issues, Agricultural & Rural Management Training Institute (ARMTI)
Ilorin, Nigeria

Joshua O. Akhigbe
Department of Microbiology, Faculty of Science
University of Abuja
Abuja, Nigeria

Feyisayo Elizabeth Akinjoko
Department of Nursing, Faculty of Nursing Science
Laadoke Akintola University of Technology Ogbomoso
Ogbomoso, Nigeria

Ulelu Jessica Akor
Department of Animal Production
University of Ilorin
Kwara State, Nigeria

Sesan Abiodun Aransiola
Department of Microbiology, Faculty of Science
University of Abuja
Abuja, Nigeria
and
Bioresources Development Centre, National Biotechnology Development Agency
Ogbomoso, Nigeria

Tolulope Timothy Arogundade
Department of Human Anatomy, Neurobiology Division, Faculty of Basic Medical Sciences
Redeemers University
Ede, Nigeria

Abraham Olufemi Asuku
Bio-Resources Development Centre, National Biotechnology Research and Development Agency
Ogbomoso, Nigeria

J.B. Austen
Department of Industrial Chemistry
Dennis Osadebay University
Asaba, Delta State, Nigeria

Helen Shnada Auta
Department of Microbiology
Federal University of Technology Minna
Minna, Niger State, Nigeria

Gabriel Gbenga Babaniyi
Department of Agricultural Development and Management, Agricultural and Rural Management Training Institute (ARMTI)
Ilorin, Nigeria

Babafemi Raphael Babaniyi
Bioresources Development Centre, National Biotechnology Development Agency
Abuja, Nigeria

Obioh Gloria Imoh Bassey
Food and Industrial Biotechnology Department, National Biotechnology Research and Development Agency
Abuja, Nigeria

Adedamola Aminat Bayo-Olugbami
Neuroscience Unit, Department of Physiology, Faculty of Basic Medical Sciences
Osun State University
Osogbo, Nigeria

Suhasini Bhatnagar
Owner, Swaroop Atomistix Pvt. Ltd.
Ghaziabad, India

Abdulmujeeb Bolaji Hamzat
Nigerian Institute for Oceanography and Marine Research, Victoria Island
Lagos, Nigeria

Maiyaki Taofiqat Damilola
Department of Food Science
University of Ilorin
Ilorin, Nigeria

P. Ebrahimbabaie
Department of Environmental Biology
Bethune-Cookman University
Daytona Beach, FL, United States
and
Department of Integrated Environmental Science
Bethune-Cookman University
Daytona Beach, FL, United States

Shehu-Alimi Elelu
Department of Chemistry
Howard University
Washington, DC
United States

P. Gnana Deepu
Center of Biotechnology
University College of Engineering Science and Technology, JNTUH
Kukatpally, Hyderabad, Telangana, India

Swapnil Godbharle
Food Evolution Research Laboratory, School of Tourism and Hospitality, College of Business and Economics
University of Johannesburg
Johannesburg, South Africa

Abisola Abiodun Ibikunle
Department of Chemistry, School of Physical Sciences
Federal University of Technology Akure
Ondo, Nigeria

N.O. Iboyi
Department of Biology and Forensic Science
Admiralty University of Nigeria
Aboh Ogwashi Ukwu, Delta State, Nigeria

Femi Ibrahim
Department of Agricultural Development and Management, Agricultural and Rural Management Training Institute (ARMTI)
Ilorin, Nigeria

Aminat Omolola Imam-Fulani
Department of Neuroscience
University of Minnesota Medical School
Minneapolis, MN, United States
and
Neuroscience Unit, Department of Physiology
University of Ilorin
Ilorin, Nigeria

A.O. Isoje
Department of Biochemistry and Molecular Biology
Dennis Osadebay University
Asaba, Delta State, Nigeria

Olumide Ayobami Iyiola
Department of Rural Development and Gender Issues, Agricultural & Rural Management Training Institute (ARMTI)
Ilorin, Nigeria

Borusu Bhargavi Lakshmi
Department of Clinical Nutrition Dietetics and Food Science, School of Allied Health Sciences
Malla Reddy University
Hyderabad, Telangana, India

Moromade Abibat Lere-Aliu
Nigerian Institute for Oceanography and Marine Research
Lagos, Nigeria

Miracle Uwa Livinus
Department of Biochemistry
Skyline University Nigeria
Kano, Nigeria

Salamatu S. Machunga-Mambula
Department of Microbiology, Faculty of Science
University of Abuja
Abuja, Nigeria

Naga Raju Maddela
Departamento de Ciencias Biológicas, Facultad la Ciencias de la Salud
Universidad Técnica de Manabí
Portoviejo, Manabí, Ecuador

Asmau M. Maude
Department of Microbiology
Federal University of Technology Minna
Minna, Niger State, Nigeria

Katherine Georgina Menon
Department of Microbiology, School of Allied Health Sciences
Malla Reddy University
Hyderabad, Telangana, India

Sheila O. Michael
Department of Biology and Forensic Science
Admiralty University of Nigeria
Aboh Ogwashi Ukwu, Delta State, Nigeria

Auwal Sagir Muhammad
School of Informatics
Xiamen University
Xiamen, China

Saheed Ibrahim Musa
Department of Biology and Forensic Science
Admiralty University of Nigeria
Aboh Ogwashi Ukwu, Delta State, Nigeria

Innocent Ojeba Musa
Department of Microbiology
Skyline University Nigeria
Kano, Nigeria

Moses Nathan
Department of Biology and Forensic Science
Admiralty University of Nigeria
Aboh Ogwashi Ukwu, Delta State, Nigeria

Pariki Nithin
Department of Clinical Nutrition Dietetics and Food Science, School of Allied Health Sciences
Malla Reddy University
Hyderabad, Telangana, India

Sarah Chioma Nwakanma
Department of Microbiology, Faculty of Science
University of Abuja
Abuja, Nigeria

Comfort Nneka Okoji
Department of Biology and Forensic Science
Admiralty University of Nigeria
Ibusa, Delta State, Nigeria

O.L. Okwute
Department of Microbiology, Faculty of Science
University of Abuja
Abuja, Nigeria

Taofeek Olawale Olabiyi
Nigerian Institute for Oceanography and Marine Research
Lagos, Nigeria

Daramola Oluwatosin Olaoluwa
Department of Chemistry, School of Physical Sciences
Federal University of Technology Akure
Ondo, Nigeria

Adunbi John Olarewaju
Bioresources Development Centre, National Biotechnology Research and Development Agency
Abuja, Nigeria

Victoria Oyinlola Olayiwola
Department of Microbiology, Faculty of Science
University of Abuja
Abuja, Nigeria

Abe Taiwo Oluwasola
Department of Chemistry, School of Physical Sciences, Federal University of Technology
Akure, Nigeria

I. Onosakponome
Department of Biochemistry and Molecular Biology
Dennis Osadebay University
Asaba, Delta State, Nigeria

Eridiong Ogbonna Onyenweaku
Food Evolution Research Laboratory, School of Tourism and Hospitality, College of Business and Economics
University of Johannesburg
Johannesburg, South Africa
and
Department of Human Nutrition and Dietetics, Faculty of Basic Medical Sciences, College of Medicine
University of Calabar
Calabar, Cross River State, Nigeria

Adewumi Yetunde Oriji
Department of Chemistry, School of Physical Sciences
Federal University of Technology
Akure, Nigeria

J.O. Otedo
Department of Biochemistry and Molecular Biology
Dennis Osadebay University
Asaba, Delta State, Nigeria

Juliana Okwena Pondei
Department of Microbiology, Faculty of Science
University of Abuja
Abuja, Nigeria

Kondakindi Venkateswar Reddy
Center of Biotechnology
University College of Engineering Science and Technology, JNTUH
Kukatpally, Hyderabad, Telangana, India

Olawoore Ayooluwa Samuel
Department of Chemistry
The Federal University of Technology
Akure, Ondo State, Nigeria

Job Oloruntoba Samuel
Department of Microbiology
Federal University of Technology Minna
Minna, Niger State, Nigeria

Tijani Abiola Tajudeen
Department of Soil Science
Federal University of Technology
Minna, Nigeria

E.C. Unoka
Department of Industrial Chemistry
Dennis Osadebay University
Asaba, Delta State, Nigeria

Fatima Adamu Yahaya
Department of Microbiology
Skyline University Nigeria
Kano, Nigeria

Amina Yahaya
Department of Microbiology, Faculty of Science
University of Abuja
Abuja, Nigeria

Abdulkareem Tajudeen Zakari
Department of Microbiology
University of Ilorin
Ilorin, Nigeria

Stella Suanu Leh-Togi Zobeashia
Food and Industrial Biotechnology Department, National Biotechnology Research and Development Agency
Abuja, Nigeria

Section I

Bioplastic Overview

1 Bio-Based Plastics and Additives

General Overview

Abe Taiwo Oluwasola and Abdulmujeeb Bolaji Hamzat

1.1 INTRODUCTION TO BIO-BASED PLASTICS AND ADDITIVES

Bio-based plastics are a rapidly emerging category of polymers derived from renewable resources, as opposed to conventional plastics that are predominantly made from petroleum-based feedstocks. These sustainable materials have garnered significant attention due to the growing concerns over the environmental impact of plastic waste and the depletion of fossil resources. Bio-based plastics are seen not only as an alternative to traditional plastics, but also as a solution to the mounting global challenges of pollution, resource scarcity, and climate change. They are designed to either biodegrade naturally, reduce carbon footprints, or be produced through more sustainable manufacturing processes (Ferreira-Filipe et al., 2021). The term "bio-based" refers to plastics made from renewable biomass sources such as plants, agricultural waste, and algae, as opposed to fossil fuels. Common bio-based plastics include polylactic acid (PLA), polyhydroxyalkanoates (PHAs), starch-based plastics, and bio-based polyethylene (Bio-PE) (Ferreira-Filipe et al., 2021). These materials are considered environmentally friendly because their raw materials can be replenished within a human timescale, reducing the reliance on finite fossil resources. However, bio-based plastics face several challenges. These include issues with cost, performance, scalability, and competition for agricultural land, which could be used for food production (Moshood et al., 2021). Furthermore, their biodegradability can be highly dependent on environmental conditions, which means that some bio-based plastics may not degrade efficiently in natural settings, such as landfills or oceans. As a result, the environmental benefits of bio-based plastics remain a subject of ongoing debate, especially when considering their full life cycle from production to disposal. Alongside bio-based plastics, the development of **additives** plays a crucial role in enhancing the functionality, performance, and processing of these materials. Additives in bio-based plastics serve to improve properties such as strength, flexibility, durability, and resistance to UV degradation (Reichert et al., 2020). These additives can include plasticizers, stabilizers, flame retardants, colorants, and impact modifiers. The growing demand for eco-friendly alternatives has encouraged innovation in the field of bio-based plastics and additives. Researchers and industry stakeholders are increasingly focusing on improving the performance of these materials, optimizing their production processes, and ensuring that their environmental impact is minimized throughout their life cycle. As technology advances, the hope is to achieve widespread adoption of bio-based plastics that offer the same performance as conventional plastics, but with far fewer environmental consequences.

DOI: 10.1201/9781003602828-2

1.2 CLASSIFICATION OF BIO-BASED PLASTICS

According to Garrison et al. (2016), bio-based plastics can be classified into two categories (source based and biodegradability).

1.2.1 Based on Source

Natural polymers are renewable materials directly extracted from natural resources with minimal processing, including starch, cellulose, and chitosan. **Starch**, derived from crops like corn, potato, and cassava, is widely used as packaging materials, biodegradable films, and food containers. Cellulose, extracted from plant cell walls, finds applications in fibers, films, and composites. **Chitosan**, obtained from chitin in crustacean exoskeletons, is utilized in medical wound dressings, water purification, and antimicrobial packaging. Chemically modified biopolymers are natural polymers altered to enhance properties like solubility, thermal stability, or mechanical strength. **Cellulose acetate**, produced by reacting cellulose with acetic acid, is used in photographic films, coatings, and cigarette filters. **Starch blends**, combining modified starch with other polymers, improve flexibility and strength and are applied in biodegradable packaging and disposable cutlery. **PLA**, synthesized from fermented sugars of corn, sugarcane, or cassava, finds uses in 3D printing, packaging, and medical implants. **Bio-polyethylene (Bio-PE)**, derived from bioethanol typically made from sugarcane, is used in bottles, films, and automotive parts.

1.2.2 Based on Biodegradability

Biodegradable bio-based plastics: They decompose into natural elements like water, carbon dioxide, and biomass under specific conditions, such as composting. **PLA** is biodegradable in industrial composting environments, while **PHAs** are produced by microorganisms through fermentation and also offers biodegradability. These plastics are commonly used in biodegradable packaging, agricultural films, and disposable tableware, helping to reduce landfill waste and address microplastic pollution.

Non-biodegradable but bio-based plastics: They are derived from renewable resources but share similar properties and durability with conventional plastics. While not designed for biodegradability, these plastics are recyclable and they also offer a lower environmental impact. **Bio-polyethylene (Bio-PE)** matches the durability and recyclability of fossil-based polyethylene (PE), while **bio-polyethylene terephthalate (Bio-PET)** is produced from bio-based ethylene glycol and fossil-based terephthalic acid. Commonly used in beverage bottles, textiles, and automotive components, these plastics help reduce reliance on fossil fuels and can be integrated into existing recycling systems.

1.3 PRIMARY TYPES OF BIO-BASED PLASTICS

Bio-based plastics are polymers derived from renewable natural resources, offering a sustainable alternative to traditional petroleum-based plastics (Singh et al., 2022). The primary types of bio-based plastics include the following:

PLA is a bio-based plastic synthesized through thefermentation of sugars from renewable resources such as corn, sugarcane, or cassava. The fermentation produces lactic acid, which is then polymerized through condensation or ring-opening polymerization to form PLA. This process is energy-efficient and results in fewer greenhouse gas emissions compared to fossil-based plastics (Rezvani et al., 2021). PLA possesses several notable properties, offers high strength and rigidity, similar to conventional plastics like polystyrene (PS), and exhibits excellent transparency, making it ideal for packaging (Farah et al., 2016). It is biodegradable under industrial composting conditions, though it has relatively low thermal resistance, limiting its use at higher temperatures. Applications of PLA are diverse, including food packaging (containers, cups, and films), 3D printing (due to its low melting temperature and ease of extrusion), and medical implants (where its biocompatibility and biodegradability are advantageous for temporary implants, sutures, and drug delivery systems) (Swetha et al., 2023).

PHAs are biodegradable plastics produced through microbial fermentation, primarily by microorganisms like *Cupriavidus necator*. These microbes synthesize PHA as an intracellular carbon and energy storage material when grown in nutrient-limited conditions. After fermentation, the polymer is extracted and purified for various applications (Koller, 2018).

PHA is biodegradable under natural environmental conditions, including marine environments, and offers high thermal resistance and mechanical strength, making it suitable for a wide range of industrial applications (Suzuki et al., 2021). Its excellent biocompatibility makes it particularly well-suited for medical and pharmaceutical uses. Common applications of PHA include agricultural films for mulching (which degrade after use), medical devices such as tissue engineering scaffolds, sutures, and biodegradable implants, as well as packaging for food and beverages, particularly when biodegradability is a key requirement (Jothiamani et al., 2019).

Starch-based plastics are produced by blending native starch with synthetic polymers or plasticizers like glycerol to enhance their mechanical properties and processability. One of the most common forms, thermoplastic starch (TPS), is created by gelatinizing starch under heat and shear (Ben et al., 2022). These plastics are inexpensive and readily available from crops such as corn, wheat, and potatoes. They are biodegradable under appropriate conditions, depending on the specific blend, but can be brittle unless modified with additives or combined with other polymers (Makhijani et al., 2015). Starch-based plastics are used in a variety of applications, including food packaging (trays, films, and bags for fresh produce), disposable items (cutlery, plates, and agricultural mulch films), and compostable products (garbage bags and carry bags that degrade in composting facilities).

Bio-Polyethylene (Bio-PE) and **Bio-Polyethylene Terephthalate (Bio-PET)** are produced from bioethanol, typically derived through the fermentation of sugarcane or corn. Bioethanol is dehydrated to form ethylene, which is polymerized to produce Bio-PE. Bio-PET is made by partially replacing fossil-based ethylene glycol with bio-based ethylene glycol, while the terephthalic acid remains fossil-based (Siracusa and Blanco, 2020). These bio-based plastics are chemically identical to their fossil-based counterparts, PE and PET, which allows them to seamlessly integrate into existing recycling streams. They are durable, lightweight, and non-biodegradable (Siracusa and Blanco, 2020). Applications of Bio-PE include flexible packaging, shrink wraps, and agricultural films, while Bio-PET is widely used for beverage bottles (e.g., soft drinks and water bottles) and in automotive parts, such as car interiors, seat fabrics, and composite materials (Reichert et al., 2020).

1.4 ADDITIVES IN BIO-BASED PLASTICS

1.4.1 Plastic Additives

Plastic additives are chemical materials incorporated into plastic for various purposes (Table 1.1) during the manufacturing process that intend to provide the maximum performance of the plastic material during the manufacturing process and usage. They are substances incorporated into bio-based plastics to modify and enhance their properties, enabling them to meet the requirements of specific applications. Additives such as plasticizers and fillers in oil composition can increase the life span, economic value, and processability of plastics, and make them cleaner (Hahladakis et al., 2018). While bio-based plastics offer sustainable alternatives, their inherent properties often need optimization to match the performance of conventional plastics. The primary purposes of additives in bio-based plastics include improving mechanical properties such as flexibility, toughness, and strength, enhancing thermal stability for high-temperature applications, and aiding in processing during manufacturing (extrusion and injection molding) (Andrew and Dhakal, 2022). Additives also extend the durability and resistance of bio-based plastics to environmental factors like UV radiation and oxidation, while introducing or improving aesthetic qualities such as color and texture (Valero et al., 2024). Plastic additive can be a liquid, powder pellet, or all three at the same time. In recent times, the most valued plastic is made from renewable sources often referred to as bio-plastic. Additives added to bio-plastic increase its value and make it more suitable for achieving a better environment when used for packaging, textile, construction materials, automotive components, and medical devices. Examples of bio-plastic-based additives include PLA, PHAs, starch-based additives, cellulose-based additives, and soy-based additives (Nanda, 2022; Rosenboom et al., 2022).

1.5 PRODUCTION PROCESSES FOR BIO-BASED PLASTICS

Plastic materials are widely regarded as essential due to their superior properties and performance compared to alternatives like metal and wood (Zhu et al., 2016; Álvarez-Chávez et al., 2012). The demand for plastics is expected to continue growing, following an upward trend that began in the 1950s (Chen et al., 2020). Given their critical role in modern society and the recognition that materials play a fundamental role in determining sustainability (Álvarez-Chávez et al., 2012), plastics remain a focus of significant attention. Replacing petroleum-based plastics with bio-based plastics is considered a promising alternative, as it can reduce reliance on fossil fuels and alleviate the burden of plastic waste in landfills. The development and commercialization of bio-based plastics for diverse applications in products and packaging are also gaining significant attention, as manufacturers seek safer and more sustainable materials to replace chemicals and substances of concern in consumer products. Bio-based materials are promising as their feedstocks are renewable, they can theoretically be composted or recycled, and their production process may be more energy-efficient than petroleum-based plastics processing.

According to Álvarez-Chávez et al. (2012), the production processes of conventional plastics and bio-based plastics have been compared to evaluate their differences and impacts.

1.5.1 Energy Consumption

Conventional Plastics: The production of conventional plastics such as PE and polypropylene (PP) is energy-intensive, primarily relying on fossil fuels like oil and natural gas. Energy consumption is typically high during polymerization and extrusion stages, with temperatures ranging from 200°C to 300°C.

Bio-Based Plastics: Bio-based plastics, particularly those made from renewable feedstocks like corn, sugarcane, and starch, generally use less energy. The fermentation process for PLA and microbial fermentation for PHA are less energy-intensive compared to the high-energy processes of conventional plastic production. However, bio-based plastics can still have higher energy consumption in some cases, particularly during feedstock cultivation and transportation.

1.5.2 Feedstock Requirements

Conventional Plastics: Petrochemical feedstocks for conventional plastics are derived from crude oil or natural gas, which are non-renewable resources that contribute to environmental degradation. While petroleum extraction does not require significant land or water, it does carry environmental costs related to extraction, refinement, and pollution.

Bio-Based Plastics: The feedstock for bio-based plastics comes from renewable agricultural sources such as corn (for PLA) or sugarcane (for Bio-PE). While these feedstocks are renewable,

TABLE 1.1
Bioplastic-Based Additives

	Types of Additives	Examples and Usage Common Bio-Based Plasticizers	
A.	**Plasticizers** are additives that increase flexibility and reduce brittleness in bio-based plastics by reducing the intermolecular forces within the polymer matrix.	• **Glycerol**, a byproduct of biodiesel production, commonly used in starch-based plastics. • **Sorbitol**, a sugar alcohol that enhances the flexibility of thermoplastic starch (TPS). • **Citric acid derivatives**, which improve the plasticity of bio-based films.	These plasticizers are primarily used in applications such as flexible films, bags, and food packaging, where enhanced flexibility and processability are required.
B.	**Stabilizers** are additives used to protect bio-based plastics from thermal degradation during processing and UV-induced degradation during use, extending the material's lifespan and performance.	• **Thermal stabilizers**, such as extracts from natural antioxidants like **tocopherols** (vitamin E) or **lignin derivatives**, which prevent thermal breakdown. • **UV stabilizers**, like bio-derived UV blockers such as **ferulic acid**, which absorb harmful UV rays to protect the plastic from degradation.	These stabilizers are commonly used in applications exposed to light and heat, such as outdoor products, agricultural films, and packaging.
C.	**Fillers and reinforcements** are additives used to enhance the mechanical properties of bio-based plastics, such as strength, rigidity, and impact resistance, while often helping to reduce material costs.	• **Natural fibers**, such as **jute**, **flax**, **hemp**, and **kenaf**, which are used as reinforcements in bio-composites. • **Bio-derived nanoparticles**, including **cellulose nanocrystals (CNCs)**, **lignin nanoparticles**, and **starch-based fillers**.	These materials are commonly used in applications like automotive parts, construction materials, and durable packaging, where enhanced mechanical properties are crucial.
D.	**Colorants and pigments** are additives used to impart or enhance color in bio-based plastics, utilizing sustainable alternatives to synthetic dyes.	• **Anthocyanins**, extracted from fruits like berries, which produce blue, red, and purple hues. • **Carotenoids**, derived from carrots, tomatoes, and algae, offering yellow, orange, and red tones. • **Chlorophyll derivatives**, which provide green pigmentation.	These natural colorants are commonly used in applications such as decorative packaging, consumer goods, and toys, where color is an important aesthetic feature.
E.	**Flame retardants** are additives used to reduce the flammability of bio-based plastics, ensuring safer use in applications such as electronics, automotive, and construction.	• **Phytic acid**, derived from plant seeds, which acts as a phosphorus-based flame retardant. • **Tannins**, natural polyphenols from plants that char when heated, creating a protective barrier. • **Intumescent additives**, bio-based materials that swell and form an insulating layer when exposed to heat.	These flame retardants are commonly used in electronics casings, automotive interiors, and furniture, where fire safety is a key concern.

they require large agricultural land areas, which can lead to land-use competition with food production, biodiversity loss, and water scarcity. The sustainability of bio-based plastics is heavily influenced by agricultural practices, including the use of fertilizers and pesticides, which can impact the environment.

1.5.3 Waste Generation

Conventional Plastics: Plastic waste is a significant environmental issue, with large volumes of non-biodegradable materials accumulating in landfills, oceans, and waterways. Many conventional plastics are not easily recyclable, resulting in high levels of waste generation and environmental pollution.

Bio-Based Plastics: Bio-based plastics can be biodegradable or compostable, helping to reduce the amount of waste that persists in the environment. However, many bio-based plastics, like Bio-PE, are still non-biodegradable, and their disposal and recycling present challenges similar to conventional plastics. Effective waste management and composting infrastructure are crucial to ensuring that bio-based plastics can deliver their potential environmental benefits.

1.6 PROCESSING TECHNIQUES

Extrusion Process: In extrusion, a polymer is heated and forced through a die to create continuous shapes such as films, sheets, or pipes. **Bio-Based Plastics**: PLA, PHA, and bio-PE are commonly processed using extrusion. However, adjustments in screw design and temperature control are required to accommodate the unique properties of bio-based plastics, such as their different melt viscosities and thermal behaviors.

Injection Molding Process: Injection molding involves injecting molten polymer into a mold to form specific shapes or parts, commonly used for manufacturing complex and high-volume plastic products. **Bio-Based Plastics**: Bio-based plastics like PLA can also be processed using injection molding, though they may require modifications in mold design, injection speeds, and cooling rates due to their different flow properties and thermal resistance compared to conventional plastics.

Blow Molding Process: In blow molding, molten polymer is inflated into a mold to form hollow objects, such as bottles. **Bio-Based Plastics**: Bio-PET and PLA are suitable for blow molding, though challenges related to their thermal and mechanical properties may require process optimization, such as adjustments in mold temperature and cooling rates.

Thermoforming Process: Thermoforming involves heating a plastic sheet until pliable, then molding it into a specific shape using vacuum or pressure. **Bio-Based Plastics**: PLA is commonly thermoformed into products like trays and packaging. However, due to PLA's tendency to deform at lower temperatures, modifications in the thermoforming process such as higher heating temperatures or different cooling techniques are necessary to achieve optimal results.

1.7 CHALLENGES IN SCALING UP BIO-BASED PLASTICS PRODUCTION

Scaling up the production of bio-based plastics poses several challenges that must be addressed to facilitate their widespread adoption (Moshood et al., 2021). These challenges include the following:

1. **Cost and Economic Viability**: Bio-based plastics often have higher production costs compared to traditional petroleum-based plastics, primarily due to the cost of raw materials (e.g., agricultural feedstocks) and the need for specialized production processes. Achieving price parity with conventional plastics is essential for large-scale adoption.
2. **Raw Material Availability**: The production of bio-based plastics relies on agricultural feedstocks like corn, sugarcane, and starch. This creates competition with food production and may lead to issues of land use, food security, and price volatility of raw materials.
3. **Performance and Material Properties**: While bio-based plastics such as PLA and PHA offer environmental benefits, they sometimes do not meet the performance standards required

for certain applications. Issues such as poor mechanical properties, low thermal stability, and limited barrier properties can limit their use in various industries.

4. **Infrastructure and Production Capacity**: The current infrastructure for bio-based plastic production is limited, and scaling up requires substantial investment in new manufacturing facilities. Many bio-based plastics also require different processing techniques from conventional plastics, requiring industry adaptation and potentially additional costs.
5. **Biodegradability and End-of-Life Management**: While many bio-based plastics are biodegradable, their degradation can depend on environmental conditions (e.g., industrial composting vs. natural settings). Additionally, the lack of consistent recycling systems and facilities for bio-based plastics can lead to improper disposal, counteracting their environmental benefits.
6. **Supply Chain and Logistics**: Scaling up production involves ensuring a stable, efficient supply chain for raw materials, as well as handling logistics for processing and distribution. These factors can be complicated by the need for regional sourcing of feedstocks and the variability of agricultural yields.
7. **Consumer and Regulatory Acceptance**: The bio-based plastics industry must navigate complex regulatory environments regarding biodegradability, safety, and sustainability claims. Furthermore, consumer trust and acceptance are essential to drive demand, and educating consumers on the benefits and limitations of bio-based plastics is crucial.

1.8 INNOVATIONS IN BIO-PLASTIC PROCESSING

Innovations in bio-plastic processing are key to overcoming the challenges of scaling up bio-based plastic production, improving material properties, and making the process more cost-effective and sustainable (Costa et al., 2023). Notable advancements in bio-plastic processing include the following:

1. **Advanced Polymerization Techniques**: Researchers are developing new, more efficient polymerization methods that reduce energy consumption and improve the overall yield and quality of bio-based plastics. Techniques such as enzyme-catalyzed polymerization are being explored to make the production process more sustainable and efficient.
2. **Blending with Conventional Plastics**: To enhance the performance and reduce production costs, bio-based plastics are being blended with conventional petroleum-based plastics. This hybrid approach combines the sustainability of bio-based materials with the strength and durability of traditional plastics, making them suitable for a broader range of applications.
3. **Biorefinery Integration**: The integration of bio-based plastic production into biorefineries, where multiple products are derived from the same biomass, is an emerging innovation. This approach optimizes the use of renewable feedstocks and allows for the efficient production of bio-plastics alongside other bioproducts, such as biofuels and chemicals.
4. **Green Solvent Processing**: The use of environmentally friendly solvents, often derived from renewable sources, is a promising innovation. Green solvents enable the production of bio-based plastics with lower environmental impact, reducing the reliance on toxic chemicals and improving the sustainability of the manufacturing process.
5. **3D Printing with Bio-Based Materials**: Bio-based plastics like PLA are being increasingly used in 3D printing, offering sustainable alternatives to conventional plastic filaments. Innovations in 3D printing technology allow for the precise and efficient use of bio-based materials, reducing waste and enabling the production of customized, eco-friendly products.
6. **Additives and Plasticizers**: Bio-based additives and plasticizers, such as those derived from plant oils or renewable fatty acids, are being developed to improve the flexibility, durability, and processing properties of bio-based plastics. These innovations help overcome the limitations of bio-plastics, such as brittleness or low thermal stability.

7. **Nanotechnology**: The integration of nanomaterials into bio-based plastics can enhance their mechanical properties, barrier resistance, and functionality. Nanocellulose, for example, is a renewable, biodegradable material that can reinforce bio-plastics, making them stronger and more versatile for different applications.
8. **Compostable and Biodegradable Blends**: Innovations in creating compostable blends and biodegradable formulations are expanding the scope of bio-plastics. Researchers are working on improving the speed and extent of biodegradation in various environmental conditions, ensuring that bio-based plastics break down efficiently in natural settings, including marine environments.
9. **Circular Economy Approaches**: Developing closed-loop systems where bio-based plastics can be reused or recycled is a key innovation. Enhanced recycling technologies, such as chemical recycling, enable the recovery of valuable monomers from bio-based plastics, reducing waste and encouraging more sustainable practices.
10. **Upcycling and Waste Utilization**: Another innovative approach is upcycling waste materials (e.g., agricultural residues or post-consumer waste) into bio-plastics. This process minimizes waste and reduces the demand for virgin feedstocks, offering an additional pathway for bio-plastic production with a lower environmental footprint.

1.9 ENVIRONMENTAL IMPACT OF BIO-BASED PLASTICS

Life cycle analysis (LCA) highlights that bio-based plastics generally have a lower carbon footprint than fossil fuel-based plastics due to their renewable feedstocks, such as corn or sugarcane, which sequester carbon during growth. For instance, PLA outperforms conventional plastics like PET and PE in emissions, though agricultural inputs and transportation can contribute significantly to its footprint. Bio-PET, while renewable, has a higher carbon footprint than PLA due to energy-intensive polymerization (Tolinski, 2011). The sustainability of bio-based plastics depends on factors like feedstock sourcing, agricultural practices, energy inputs, and end-of-life management, with sustainable farming methods playing a critical role in maximizing environmental benefits. Bio-based plastics can reduce greenhouse gas emissions by replacing fossil fuel-based feedstocks with plant-based sources that absorb CO_2 during growth, potentially achieving carbon neutrality if emissions from decomposition or combustion are offset by this absorption (Gabrielli et al., 2020). However, the extent of these reductions depends on factors such as feedstock type, agricultural practices, and transportation and processing distances, which can significantly impact the overall emissions associated with their production.

1.9.1 End-of-Life Considerations

The biodegradability and compostability of bio-based plastics like PLA and PHA depend on environmental conditions (Garrison et al., 2016). PLA can degrade into carbon dioxide, water, and organic matter within six months to a year under optimal industrial composting conditions but persists longer in marine environments due to cooler temperatures and lower microbial activity, sometimes up to two years. In contrast, PHA exhibits better biodegradability in marine settings due to its higher microbial digestibility, making it more suitable for aquatic environments (Meereboer et al., 2020). Compostability of bio-based plastics like PLA depends on the composting environment. PLA requires industrial composting facilities with high temperatures and specific microbial conditions to degrade efficiently, while it struggles to break down in home composting setups due to insufficient heat and moisture. However, limited industrial composting infrastructure in many regions poses a significant challenge to the effective management of compostable bio-based plastics. Managing bio-based plastics poses significant waste management challenges. While bio-based plastics like bio-PET integrate well into traditional recycling streams, others, such as PLA, are not compatible and can contaminate the recycling process when mixed with conventional plastics, reducing the quality of recycled materials. Advanced chemical recycling technologies for PLA and similar bio-based plastics show promise but remain in early development stages. In regions lacking proper recycling infrastructure, bio-based plastics are often sent to landfills or incineration, potentially contributing to greenhouse gas emissions.

1.10 ECOLOGICAL AND TOXICOLOGICAL IMPACT

1.10.1 Effect of Bio-Based Plastic Production on Ecosystems and Biodiversity

The production of bio-based plastics impacts ecosystems and biodiversity, particularly through land use and agriculture (Spierling et al., 2018). Increased demand for bio-based plastics can lead to habitat destruction, reduced biodiversity, and competition with food crops due to the expansion of agricultural land. Monoculture farming for feedstocks like corn and sugarcane may deplete soil nutrients, harm soil health, and increase the use of pesticides and fertilizers, negatively affecting local ecosystems (Moshood et al., 2021). However, adopting sustainable practices such as crop rotation, organic farming, agroforestry, and utilizing waste biomass can mitigate these impacts, promoting the sustainability of bio-based plastic production.

1.10.2 Potential for Toxicity and Leaching of Additives into the Environment

Bio-based plastics, like conventional plastics, often include additives such as plasticizers, flame retardants, and stabilizers to enhance performance (Hobbs, 2019). While some additives, particularly synthetic ones, may pose toxicity risks and leach into the environment during degradation, others like nontoxic flame retardants and natural dyes are more eco-friendly (Almeida et al., 2022). Additives containing heavy metals or endocrine-disrupting chemicals can harm ecosystems, highlighting the need for safer alternatives (Hamid et al., 2021). Although bio-based plastics generally have fewer toxicological concerns, their environmental impact depends on the nature of their additives. Research is advancing on bio-based and nontoxic additives, such as natural antioxidants and flame retardants, to mitigate these risks.

1.11 CONCLUSION

Bio-based plastics present promising opportunities for reducing carbon footprints and enhancing sustainability by lowering fossil fuel reliance and greenhouse gas emissions. However, their overall environmental impact is closely tied to factors such as feedstock sourcing, agricultural practices, and waste management. Key challenges include limited biodegradability in marine environments, inadequate recycling infrastructure, and potential ecological effects of large-scale production. Innovations in biotechnology, additive development, and recycling technologies will be essential to enhancing the functionality and competitiveness of bio-based plastics. Embracing a circular approach, including hybrid systems and effective waste management, is key to realizing their long-term sustainability potential. These efforts can position bio-based plastics as a cornerstone in reducing the environmental footprint of plastics and fostering a greener future.

REFERENCES

Almeida, S., Ozkan, S., Gonçalves, D., Paulo, I., Queirós, C. S., Ferreira, O., Bordado, J. and Galhano dos Santos, R. (2022). A brief evaluation of antioxidants, antistatics, and plasticizers additives from natural sources for polymers formulation. *Polymers*, 15(1), 6. https://doi.org/10.3390/polym15010006

Álvarez-Chávez, C. R., Edwards, S., Moure-Eraso, R. and Geiser, K. (2012). Sustainability of bio-based plastics: general comparative analysis and recommendations for improvement. *Journal of Cleaner Production*, 23(1), 47–56. https://doi.org/10.1016/j.jclepro.2011.10.003

Andrew, J. J. and Dhakal, H. N. (2022). Sustainable biobased composites for advanced applications: recent trends and future opportunities – a critical review. *Composites Part C: Open Access*, 7, 100220. https://doi.org/10.1016/j.jcomc.2021.100220

Ben, Z. Y., Samsudin, H. and Yhaya, M. F. (2022). Glycerol: its properties, polymer synthesis, and applications in starch-based films. *European Polymer Journal*, 175, 111377. https://doi.org/10.1016/j.eurpolymj.2022.111377

Chen, D. M. C., Bodirsky, B. L., Krueger, T., Mishra, A. and Popp, A. (2020). The world's growing municipal solid waste: trends and impacts. *Environmental Research Letters*, 15(7), 074021. https://doi.org/10.1088/1748-9326/ab8659

Costa, A., Encarnação, T., Tavares, R., Todo Bom, T. and Mateus, A. (2023). Bioplastics: innovation for green transition. *Polymers*, 15(3), 517. https://doi.org/10.3390/polym15030517

Farah, S., Anderson, D. G. and Langer, R. (2016). Physical and mechanical properties of PLA, and their functions in widespread applications—a comprehensive review. *Advanced Drug Delivery Reviews*, 107, 367–392. https://doi.org/10.1016/j.addr.2016.06.012

Ferreira-Filipe, D. A., Paço, A., Duarte, A. C., Rocha-Santos, T. and Patrício Silva, A. L. (2021). Are biobased plastics green alternatives? A critical review. *International Journal of Environmental Research and Public Health*, 18(15), 7729. https://doi.org/10.3390/ijerph18157729

Gabrielli, P., Gazzani, M. and Mazzotti, M. (2020). The role of carbon capture and utilization, carbon capture and storage, and biomass to enable a net-zero-CO_2 emissions chemical industry. *Industrial & Engineering Chemistry Research*, 59(15), 7033–7045. https://doi.org/10.1021/acs.iecr.9b06579

Garrison, T. F., Murawski, A. and Quirino, R. L. (2016). Bio-based polymers with potential for biodegradability. *Polymers*, 8(7), 262. https://doi.org/10.3390/polym8070262

Hahladakis, J. N., Velis, C. A., Weber, R., Iacovidou, E. and Purnell, P. (2018). An overview of chemical additives present in plastics: migration, release, fate and environmental impact during their use, disposal and recycling. *Journal of Hazardous Materials*, 344, 179–199. https://doi.org/10.1016/j.jhazmat.2017.10.014

Hamid, N., Junaid, M. and Pei, D. S. (2021). Combined toxicity of endocrine-disrupting chemicals: a review. *Ecotoxicology and Environmental Safety*, 215, 112136. https://doi.org/10.1016/j.ecoenv.2021.112136

Hobbs, C. E. (2019). Recent advances in bio-based flame retardant additives for synthetic polymeric materials. *Polymers*, 11(2), 224. https://doi.org/10.3390/polym11020224

Jothimani, B., Venkatachalapathy, B., Karthikeyan, N. S. and Ravichandran, C. (2019). A review on versatile applications of degradable polymers. *Green Biopolymers and their Nanocomposites*, 403–422. https://doi.org/10.1007/978-3-030-04324-7_16

Koller, M. (2018). A review on established and emerging fermentation schemes for microbial production of polyhydroxyalkanoate (PHA) biopolyesters. *Fermentation*, 4(2), 30. https://doi.org/10.3390/fermentation4020030

Makhijani, K., Kumar, R. and Sharma, S. K. (2015). Biodegradability of blended polymers: a comparison of various properties. *Critical Reviews in Environmental Science and Technology*, 45(16), 1801–1825. https://doi.org/10.1080/10643389.2014.955627

Meereboer, K. W., Misra, M. and Mohanty, A. K. (2020). Review of recent advances in the biodegradability of polyhydroxyalkanoate (PHA) bioplastics and their composites. *Green Chemistry*, 22(17), 5519–5558. https://doi.org/10.1039/D0GC01647K

Moshood, T. D., Nawanir, G., Mahmud, F., Mohamad, F., Ahmad, M. H. and Abdul Ghani, A. (2021). Expanding policy for biodegradable plastic products and market dynamics of bio-based plastics: challenges and opportunities. *Sustainability*, 13(11), 6170. https://doi.org/10.3390/su13116170

Nanda, S., Patra, B. R., Patel, R., Bakos, J. and Dalai, A. K. (2022). Innovations in applications and prospects of bioplastics and biopolymers: a review. *Environmental Chemistry Letters*, 20(1), 379–395. https://doi.org/10.1007/s10311-021-01243-6

Reichert, C. L., Bugnicourt, E., Coltelli, M. B., Cinelli, P., Lazzeri, A., Canesi, I., Braca, F., Martínez, B. M., Alonso, R., Agostinis, L. and Verstichel, S. (2020). Bio-based packaging: materials, modifications, industrial applications and sustainability. *Polymers*, 12(7), 1558. https://doi.org/10.3390/polym12071558

Rezvani Ghomi, E., Khosravi, F., Saedi Ardahaei, A. and Dai, Y. (2021). The life cycle assessment for polylactic acid (PLA) to make it a low-carbon material. *Polymers,* 13(11), 1854. https://doi.org/10.3390/polym13111854

Rosenboom, J. G., Langer, R. and Traverso, G. (2022). Bioplastics for a circular economy. *Nature Reviews Materials*, 7(2), 117–137. https://doi.org/10.1038/s41578-021-00407-8

Singh, N., Ogunseitan, O. A., Wong, M. H. and Tang, Y. (2022). Sustainable materials alternative to petrochemical plastics pollution: a review analysis. *Sustainable Horizons*, 2, 100016. https://doi.org/10.1016/j.horiz.2022.100016

Siracusa, V. and Blanco, I. (2020). Bio-polyethylene (Bio-PE), bio-polypropylene (Bio-PP) and bio-poly(ethylene terephthalate) (Bio-PET): recent developments in bio-based polymers analogous to petroleum-derived ones for packaging and engineering applications. *Polymers*, 12(8), 1641. https://doi.org/10.3390/polym12081641

Spierling, S., Knüpffer, E., Behnsen, H., Mudersbach, M., Krieg, H., Springer, S., Albrecht, S., Herrmann, C. and Endres, H. J. (2018). Bio-based plastics – a review of environmental, social and economic impact assessments. *Journal of Cleaner Production*, 185, 476–491. https://doi.org/10.1016/j.jclepro.2018.03.014

Suzuki, M., Tachibana, Y. and Kasuya, K. (2021). Biodegradability of poly(3-hydroxyalkanoate) and poly(ε-caprolactone) via biological carbon cycles in marine environments. *Polymer Journal*, 53, 47–66. https://doi.org/10.1038/s41428-020-00396-5

Swetha, T. A., Bora, A., Mohanrasu, K., Balaji, P., Raja, R., Ponnuchamy, K., Muthusamy, G. and Arun, A. (2023). A comprehensive review on polylactic acid (PLA) – synthesis, processing and application in food packaging. *International Journal of Biological Macromolecules*, 234, 123715. https://doi.org/10.1016/j.ijbiomac.2023.123715

Tolinski, M. (2011). *Plastics and sustainability: towards a peaceful coexistence between bio-based and fossil fuel-based plastics*. John Wiley & Sons.

Valero, L., Gainche, M., Esparcieux, C., Delor-Jestin, F. and Askanian, H. (2024). Vegetal polyphenol extracts as antioxidants for the stabilization of PLA: Toward fully biobased polymer formulation. *ACS Omega*, 9(7), 7725–7736. https://doi.org/10.1021/acsomega.3c07236

Zhu, H., Luo, W., Ciesielski, P. N., Fang, Z., Zhu, J. Y., Henriksson, G., Himmel, M. E. and Hu, L. (2016). Wood-derived materials for green electronics, biological devices, and energy applications. *Chemical Reviews*, 116(16), 9305–9374. https://doi.org/10.1021/acs.chemrev.6b00225

2 Bioplastics Economy Sustainability

P. Ebrahimbabaie, Saheed Ibrahim Musa, N.O. Iboyi, Moses Nathan, and Comfort Nneka Okoji

2.1 INTRODUCTION TO BIOPLASTICS AND SUSTAINABILITY

2.1.1 Bioplastics in the Context of Environmental Sustainability

The environmental sustainability of bioplastics has gained increasing attention as the world seeks alternatives to conventional petroleum-based plastics. Bioplastics, derived from renewable biological resources such as corn, sugarcane, and cellulose, offer potential advantages in reducing reliance on fossil fuels and lowering greenhouse gas (GHG) emissions during production (Gourmelon, 2015). However, their environmental benefits must be assessed comprehensively across the product life cycle to understand their full impact on sustainability.

One of the key arguments in favor of bioplastics is their potential for carbon neutrality. Since the carbon dioxide (CO_2) absorbed during the growth of biomass is released upon degradation or combustion, bioplastics can have a closed carbon loop. Studies show that certain bioplastics, like polylactic acid (PLA) and polyhydroxyalkanoates (PHAs), have lower carbon footprints compared to petroleum-based plastics (Philp *et al.*, 2013). Additionally, using renewable feedstocks can help mitigate the depletion of non-renewable resources, which is a critical component of environmental sustainability.

However, the environmental benefits of bioplastics are not without challenges. One major concern is the agricultural land use required to grow the biomass needed for production. Large-scale bioplastics production can contribute to deforestation, loss of biodiversity, and increased water consumption, especially when feedstocks like corn and sugarcane are grown in regions with limited natural resources (Shen *et al.*, 2010). Furthermore, not all bioplastics are biodegradable, which undermines their role in addressing plastic pollution. Bioplastics like PLA, for instance, often require specific industrial composting conditions for degradation, which are not widely available (European Bioplastics, 2022).

To enhance the sustainability of bioplastics, innovations such as the use of non-food biomass and the development of truly biodegradable bioplastics are essential. Recent research has focused on using agricultural waste or algae as raw materials, which can reduce the environmental pressures associated with crop cultivation (Narancic and O'Connor, 2019). Furthermore, improving waste management infrastructure and ensuring that bioplastics are appropriately sorted and processed are key factors in maximizing their environmental benefits.

Bioplastics present a promising but complex solution to the environmental challenges posed by traditional plastics. While they offer potential advantages in reducing carbon emissions and reliance on fossil fuels, a holistic approach is needed to address issues related to land use, resource consumption, and biodegradability. As the technology advances, sustainable production methods and improved waste management will be crucial in realizing the full environmental potential of bioplastics.

 DOI: 10.1201/9781003602828-3

2.1.2 Importance of Economic Sustainability for Bioplastics

Sustainability in bioplastics extends beyond their use of renewable resources; it also depends on their production processes, recyclability, and environmental impact. Bioplastics have the potential to drive industries toward a circular economy, with many bio-based alternatives available for fossil-based applications. However, issues such as high costs, limited recycling infrastructure, and inadequate end-of-life management present challenges. Effective sorting technology, clear recycling definitions, and government incentives are crucial for improving circularity. While biodegradation alone isn't a complete solution, advancements in recycling technologies and efficient biorefinery processes are needed to enhance the economic and environmental sustainability of bioplastics. Ongoing research, policy support, and improved product labeling will be key in addressing these challenges and achieving true sustainability in bioplastics (Rosenboom *et al.*, 2022).

2.1.3 Market Growth and Economic Impact of Bioplastics

The bioplastics market has witnessed significant growth over the past decade, driven by increasing environmental awareness, government regulations, and a shift toward sustainable materials. In 2022, the global bioplastics market was valued at approximately USD 10.7 billion, and it is expected to reach USD 27.9 billion by 2028, growing at a compound annual growth rate (CAGR) of around 16.1% (Markets and Markets, 2023). This surge is largely due to the demand for sustainable packaging, which constitutes the largest segment in the bioplastics industry. Companies in the food, beverage, and cosmetics industries are increasingly adopting bioplastics to meet consumer demand for eco-friendly products, fueling market expansion.

The economic impact of the bioplastics industry extends beyond the environment. Bioplastics production creates jobs in research, development, and manufacturing sectors. The industry's growth is expected to generate more than 300,000 new jobs globally by 2030 (European Bioplastics, 2021). Additionally, as bioplastics rely on renewable resources such as corn, sugarcane, and agricultural waste, the industry contributes to the agricultural economy. This interconnection supports rural development, particularly in regions where biomass feedstocks are cultivated. Despite these positive trends, the economic impact of bioplastics is influenced by several challenges. One of the key issues is production cost. Bioplastics are currently more expensive to produce than traditional petroleum-based plastics, which limits their broader market penetration (Shen *et al.*, 2020). However, as technology improves and economies of scale are realized, production costs are expected to decrease, making bioplastics more competitive in the global market.

Government policies and regulations promoting the use of renewable materials and reducing plastic waste are also pivotal in shaping the economic landscape of bioplastics. Many countries have introduced incentives, tax benefits, and subsidies for bioplastic manufacturers, further accelerating market growth. Moreover, the rising cost of crude oil and concerns over plastic pollution have motivated businesses to invest in bioplastic solutions, creating new economic opportunities.

2.2 PRODUCTION COSTS AND SCALABILITY CHALLENGES

Plastics are a diverse class of polymers with a wide range of properties and attributes that are primarily produced from fossil fuels. Approximately 90% of the overall production is made up of plastics derived from fossil fuel. Overall, 7% out of the 34 million tons of the waste plastic produced annually by humans are reprocessed. Overall, 93% of the remain is thrown away in recycling station or poured out into waterways. Currently, 4–8% of the oil used worldwide is used in the plastics industry. But by 2050, it's expected that this figure would increase to 20%. The plastics business seemed to help people in its early stages. But as it has been used more frequently over time, it has hurt society. The success and spread of plastic can be attributed to its affordability, toughness, strength-to-weight ratios, and contributions to everyday convenience. Plastic products, which are used in many aspects of our lives, are causing significant environmental damage. Each year, millions of tons of these non-biodegradable polymers end up in the environment. Global concern over persistent plastic waste is growing since there are not enough effective waste management options in place at the moment. The first difficulty with too much plastic in the environment is the "landfill problem." Growing amounts of plastic garbage have become an issue in many regions of the world as a result of shrinking disposal capacity and rising landfill fees worldwide. The "accumulation of plastics in water bodies is one of the key environmental issues" is one of the worries. For example, long-term research in the North Atlantic found 580,000 plastic particles per square kilometer in one saline sample. Burning plastic releases CO_2 and methane (Baheri, 2007). These gases are referred to as GHGs when they are emitted into the atmosphere. Additionally, "durability"—that is, non-biodegradability—is the significant issue. Plastic will be in the environment for hundreds of years and cannot be recycled. "Competition for crude oil and energy security" is the economic problem. Solutions to these problems are within reach as a result of rising public awareness of the detrimental environmental consequences of plastic polymers derived from petrochemicals. Many studies have been carried out in an effort to handle plastic waste in an environmentally responsible way. Unlike fossil fuels, bioplastics are synthetic or biodegradable polymers made from microorganisms or plants. The appearance and uses of bioplastics are similar to those of traditional plastics. The term "bio," which is used interchangeably but refers to biodegradable or bio-based bioplastics, is the only exception. All bio-based plastics are biodegradable, but not all biodegradable plastics are bio-based (Singh, 2020). Biodegradable polymers decompose and lose their biophysical properties when exposed to microbes, releasing CO_2 in aerobic conditions and methane in anaerobic ones after a certain period of time. How long something takes to decompose depends on a number of environmental elements, including temperature, the substance, humidity, and the location of the decomposition. Microorganisms can break down biodegradable plastics into humus, which is nutrient-rich and free of hazardous metals. The guidelines must be adhered to by the biodegradable polymers. Since particles must also decompose into residues smaller than 1 mm in diameter during this time, not all biodegradable polymers can be composted. Bio-based plastic, a different kind of bioplastic, is made from a variety of essential plant-based materials that are not always biodegradable. Bioplastics are made from organic polymeric components, which are present in bacteria, plants, and mammals. The main elements for bioplastics are biological raw materials like starch, cellulose, and lactic acid that are chemically generated, or they can be taken from biological precursors made by bacteria, plants, and animals (Funk *et al.*, 2019). Large-scale production and usage of bioplastics could help overcome the problems connected with fossil fuels and conserve their limited supply.

Reduced carbon emissions and more possibilities for waste disposal are two other advantages of recycling chemicals and biological materials. Biodegradability would solve the world's plastic pollution problems. Because bioplastics are compostable—they just rot and decompose—they may be applied to soil without damaging it. Nevertheless, certain bioplastics could leave behind dangerous residues in the form of fragments of fragmented soil. For example, because the maritime environment does not provide a suitable environment for the breakdown of biodegradable plastics, certain bioplastics might only decompose at high temperatures in a specialized composter. On the other hand, effective methods for collecting, sorting, and recycling could lead to a greater recovery of bioplastic recycling resources. Furthermore, in terms of mechanical strength and thermal stability, bioplastics products perform better than conventional virgin plastic. Totes, super-absorbent diapers, sewage treatment, a variety of packaging applications, health and prosthetics, catering supplies, and agricultural composting are just a few of the possibilities for bioplastics (Kalra, 2002). Bioplastics are more expensive than conventional plastics. Bioplastics have become more useful and accessible due to a number of factors, including the desire to lessen dependency on fossil fuels, the growing interest in sustainable development, and changes in waste management attitudes and practices.

2.3 ADVANTAGES AND DISADVANTAGES OF BIOPLASTICS

The use of biodegradable polymers is quite promising. The following are a few benefits of bioplastics:

1. One can reduce their carbon impact considerably. However, it is crucial to remember that whether a bioplastic retains carbon over time from the developing plant or from the atmosphere has a big impact on the plastic's carbon footprint. The CO_2 that the plant absorbs during photosynthesis is sequestered by plastic made from biological sources. Therefore, this sequestration is reversed when bioplastic decomposes into CO_2 and water. On the other hand, CO_2 can be permanently stored in a biodegradable plastic that resembles polyethylene (PE) or other common polymers. The material retains its ability to hold CO_2 even after it has been recycled multiple times. Renewable resources can be used to make bioplastics instead of fossil fuels.
2. GHG emissions are reduced by the decreased carbon footprint of renewable resources. About 80% less CO_2 is released during the production of bioplastics than during the production of petrochemical plastics. Additionally, compared to petrochemical plastics, bioplastics use 65% less energy to produce. Other benefits include reduced carbon emissions, energy efficiency, and environmental safety. Bioplastics, for example, are more financially viable and can increase a product's value-add through green marketing campaigns. Since "80% of European buyers desire to buy things with a low environmental impact," bioplastics make a strong argument for marketers. PLA and PHB, two starch-based bioplastics, are harmless and do not cause harm to humans. Because they do not taste like food and do not absorb pollutants like bisphenol (BPAs) like some oil-based plastics do, they are great for food packing. The "plastic spectrum" is one tool that helps with the selection of plastics. Since they are formed from naturally existing materials that can be replenished and are positioned to the right of the pyramid, bio-based bioplastics are the most prominent in this category (Figure 2.1).

Plastic spectrum

Polyvinyl chloride is referred to as PVC, polyurethane as PU, polystyrene as PS, and acrylonitrile butadiene styrene as ABS. The letters PC, PET, PE, and PP stand for polycarbonate, polyethylene terephthalate, polyethylene, and polypropylene, respectively. It has been shown that there are several applications for biodegradable bioplastics. Drugs and medication are recognized as some of the original uses. Some of the disadvantages of the plastic spectrum are based on the saying that bioplastics offer many significant benefits over petrochemical plastics. However, they also have some disadvantages that need to be taken into account. A few of them are covered in the section below. It is estimated that the cost of bioplastics is twice that of regular plastics. There will need to be cost-cutting measures put in place as the number of significant bioplastics production facilities rises. Trash containing bioplastics should be disposed of properly to prevent littering and pollution of the land and water. Bioplastics have the potential to pollute the recycling procedure if they are not isolated from regular plastics. For instance, infrared radiation cannot be used in a waste separation system to separate bioplastics. Like ordinary plastic, bioplastic waste littering has a major negative impact on wildlife. Landfill disposal of bioplastic waste has the potential to increase GHG emissions. Due to mechanical

FIGURE 2.1 Spectrum based on the different plastic.

characteristics like greater water vapor permeability than ordinary plastic, ease of shredding like tissue paper, or extreme brittleness, certain Compared to oil-based polymers, bioplastics have a shorter lifespan (Po-Keung and Man, 2013). For instance, certain bioplastics derived from algae decompose in the water within hours, rendering them both delicate and biodegradable. Crop-based bioplastics require fertile land, water, and fertilizer and are weather-dependent. Drought and other natural calamities may jeopardize the supply of raw materials needed to make bioplastics. It's easy to become perplexed when bioplastic is referred to as compostable because of a terminology error. Just as organic food waste cannot be composted at home, bioplastics must be handled in an industrial composting facility, which isn't always available. The names "bioplastics" and "similar terminology" are also being misused by other producers in an attempt to deceive consumers into buying their goods. Buzzwords such as "non-toxic," "environmentally friendly," and "biodegradable" are employed by manufacturers to mislead consumers who are overworked and ignorant. The production of bioplastics is not subject to any laws. By 2019, the manufacturing of bioplastics had increased to almost 6.7 million tons. Nonetheless, a lot of nations still lack regulations controlling the production, utilization, or removal of waste.

2.4 TYPES OF BIOPLASTICS

Bioplastics are available in a range of sizes and forms, each with unique properties and uses, as well as different manufacturing processes and raw materials. Currently, bioplastics are divided into the following categories: Starch-based bioplastic EverCorn and Nature Works were the first companies to produce bioplastics using starch replaced from corn. These plastics were created by combining biodegradable starch polymeric components with petrochemical plastic polymer. Nowadays, starch-based polymers are made from potatoes, corn, wheat, and tapioca. When these bioplastics degrade, bacteria target the polymer's starch molecules, which leads to the disintegration of the plastic polymer. However, these starch-substituted bioplastics are not acceptable for industrial usage due to their physical and chemical properties. Furthermore, the presence of non-biodegradable plastic waste in soil and water could potentially lead to disruptions in the environment. Additionally, Spence *et al.* (1996) stated that starch is employed in the microbial fermentation methods used to make these bioplastics.

PLA and PHAs are produced by microorganisms digesting starch, as an example. In addition, bioplastics like partially bio-based polyethylene terephthalate (PET) and bio-based PE and PVC are produced using starch and other carbohydrates as their source materials (Lopez-Gil *et al.*, 2014).

2.5 BIOPLASTICS BASED ON CELLULOSE

The glucose units in cellulose, a glucose polymer, are connected by 1,4-glycosidic bonds. All prominent plants, green algae, and the majority of fungi's membranes include cellulose. Natural cellulose is extracted or chemically modified to produce cellulosic polymers. The two most common sources of cellulosic plastic are wood pulp and cotton fibers. The two types of cellulosic polymers are regenerated cellulose and organic cellulose esters. Organic cellulose esters utilize almost 20% of all chemical-grade pulp produced worldwide.

When cellulose is esterified with organic acids, organic cellulose esters are produced. The three most often utilized in the industry are cellulose acetate (CA), cellulose acetate propionate (CAP), and cellulose acetate butyrate (CAB) (Ach, 1993). Organic cellulose esters are used in textile fibers, pharmaceuticals, packaging film, cigarette filters, and a few other specialized industrial uses. Cellulose that has been chemically dissolved and then reassembled into fibers or films is used to make cellulose regeneration. Presently, more than 60% of the chemical-grade pulp produced worldwide is made from cellulose regenerated. Regenerated cellulose can be found in rayon, Lyocell, viscose, and viscose silk, among other materials.

2.6 POLYLACTIC ACID BIOPLASTICS

Nowadays, PLA and polylactide polymers are the two main bioplastics available on the market. The primary source of lactic acid (LA), a lactic acid-based product, is bacterially fermented starch from maize, cassava, potatoes, sugarcane, and sugar beetroot. These PLA bioplastics can address the issues with bioplastics that substitute starch. Bacteria use plant starch as a monomer to convert it to lactic acid, which is chemically treated to join the molecules into lengthy many metric chains. The PLA substance resembles traditional petrochemical plastics and is biodegradable. With modern plastics manufacturing equipment, PLA plastics offer great stiffness, durability, transparency, thermoplasticity, and increased efficiency.

Plastics such as PLA are widely utilized in packaging materials because they have similar properties to those of PE and PP. Moreover, the production of PLA plastic conserves two thirds of the resources needed to make polymers derived from fossil fuels. When PLA polymers decompose in landfills, they emit around 70% less GHG than conventional plastics. Granulated PLA and PLA-blends are offered in several scores for usage in film, cups, bottles, beverage containers, molded parts, and other everyday items. Plastics based on PLA are being utilized more often in a variety of industries, such as household goods, textiles, cosmetics, and pharmaceuticals. In the automobile industry, PLA-based plastics are also used to make dashboards, door tread plates, and other components.

2.7 BIOPLASTICS BASED ON POLYHYDROXYALKANOATES

Plant-based starch is converted by microbial fermentation into PHA-based polymers. These PHA-based polymers share several physiochemical characteristics with polyesters, PE, and PP. Intercellular PHA particles are produced by a variety of biological microorganisms when there are abundant carbon sources available. The most popular PHAs utilized in the manufacturing of plastics are polyhydroxy butyric acid (PHB), polyhydroxybutyrate (PHB), polyhydroxy valerate (PHV), and poly-3-hydroxybutyrate-valerate (PHBV). Two well-known PHA-based polymers that are used extensively are Bonello and Bippolo. Bippolo is a poly 3-hydroxybutyrate-co-valerate (PHBV) copolymer of polyhydroxybutyrate (PHB), whereas Bonello is a chemical compound of PHA-based biodegradable plastic. Prokaryotes and archaea belonging to over 100 genera are capable of producing PHA intracellularly. This includes bacteria like *Pseudomonas* species, *Alcaligenes* species, and many lamentous taxa including *Nocardia* spp.

The phase gene contributes to the synthesis and build-up of PHA in bacteria. Certain bacteria have the ability to store up to 80% of their body weight as PHA polymer granules in the cytoplasm of cells when they are under restricted nutritional conditions, thanks to novel transgenic carbohydrate fermentation (Emadian *et al.*, 2017). By breaking the cell wall and extracting the PHA polymer using organic solvents in an aqueous-based extraction process, the polymer is purified. After refinement, the PHA biopolymer is formed into lattices for use in the production of plastics. Through genetic engineering, scientists have produced a large number of genetically modified plants that produce PHA from the phase gene.

Switchgrass, or *Panicum virgatum*, has been employed as a genetically modified host to generate PHA. Subsequently, researchers created genetically altered maize to generate PHA, and the plants were effectively cultivated as a plastic crop. Using multi-gene expression approaches, Metabolic, a European bioscience engineering company, reported the successful launch of a bio-based PHA production program in switchgrass, camelina, and sugarcane. Recent years have seen the production of biopolymers from aerobic wastewater treatment sludge and anaerobic digesters, including PHAs and alginates. One of the many advantages of using inexpensive waste sludge to create biopolymers from sludge is the low cost of manufacture. PHA-based plastics are genuinely biodegradable; in aerobic and anaerobic circumstances, a variety of bacteria can break them down entirely in less than a year, generating water and CO_2 in the former case. Because it is a natural and non-toxic polymer, it is used in a variety of industries, including food packaging, agriculture, and pharmaceuticals.

2.8 DROP-IN BIOPLASTICS

Drop-ins are a kind of bioplastic that can come from renewable resources all together or in part, but they don't break down naturally. For drop-in polymers, corn, sugarcane, and sugar beets are the most popular feedstock. This kind of plastic combines organic and petrochemical polymers. Drop-ins and typical plastics differ only in that they can have a foundation made entirely or partially of renewable bio-based input materials. The current trend is an increase in the use and production of drop-in bio-based products. These bioplastics, which are derived from bioethanol, a renewable resource, include bio-PE, bio-PP, and bio-PET, as well as conventional polymers like PE, PP, and PVC (Ashok *et al.*, 2016).

One kind of bioplastic known as "drop-ins" is derived fully or primarily from renewable resources. They are not biodegradable, though. Films and bottles are made of bio-low-density polyethylene (LDPE), while production processes often use bio-PET polymers. The main advantage is that since bioplastics are made from renewable resources, they are biodegradable and have a lower carbon footprint. Production costs can be reduced by using existing facilities for the processing, production, and recycling of traditional plastics.

2.9 BIOPLASTIC WASTE MANAGEMENT OPTIONS

Garbage bioplastics are recovered and handled similarly to regular plastic garbage, depending on the product, composition, volumes, and recovery technology available. Avoidance and management: At this stage of the waste hierarchy, production methods and materials must be used to minimize resource usage while optimizing product functionality. Plastics have proven their usefulness time and time again, making products stronger, lighter, and thinner as a result of their application. The large number of bioplastic goods already on the market and their rapid demand are proof of their success (Katiyer, 2020).

2.10 RECYCLING OPTIONS

2.10.1 Reuse

Numerous multipurpose plastic products are available on the market. PET bottles and PET bottles made of biomaterials can be cleaned and then recycled within recycling systems. Bio-based PE or starch carrier bags can be reused multiple times before the material expires. Numerous reusable products can be made with bioplastics. Recycling is the most practical option for the long-term management of bioplastics, and it is superior to landfilling when it comes to incineration with energy recovery. All recycling technologies, including material, chemical, and organics recycling, are compatible with the management of bioplastics. Because bioplastic garbage degrades naturally, it presents a challenge to remove plastics from municipal waste due to the differences in fundamental elements between bioplastics and conventional plastics (Katiyer, 2020).

Additionally, it can be utilized as a composting substrate. As a result, a product containing valuable organic elements can be recovered. Washing, density separation, and compounding techniques are used by mechanical recycling companies to turn these separated materials into recyclables. Bioplastics can easily be recycled with standard plastics, depending on the basic ingredients. In contrast to bio-based PET, which is recycled in the PET stream, bio-based PE is recycled in the PE stream. Bioplastics improve the recycling rate in this method. Waste PLA polymer can be mechanically recycled multiple times without degrading its characteristics. After recycling, the materials can be used to make pallets, garden furniture, pipework, and plastic lumber (Ghosh and Jones, 2021).

Chemical recycling is an additional technique for recycling bioplastics. It entails remitting and degranulating waste biopolymer to produce a new product. Waste biopolymers can sometimes be converted into chemical building blocks like monomers, which can be used to create additional biopolymers. This type of recycling is also known as feedstock recycling. This process can be used to recycle PLA-based bioplastics, which are then recycled into lactic acid and made into new PLA-based products in the United States and Belgium. Another chemical recycling technique is the Loopla method, which hydrolyzes PLA to produce lactic acid. Many parts of the world use composting, a technique for turning biodegradable plastic trash into nutrient-rich soil additions.

These biodegradable polymer wastes could be broken down into simpler molecules by natural microbes. Next, in aerobic environments, they are broken down into CO_2 and water, and in anaerobic environments, into methane and CO_2. Composting, a type of material recycling that includes organic recycling, is regarded as an environmentally sound and sustainable approach to managing bioplastic waste. Bioplastics don't release any dangerous substances throughout the composting process, according to International Solid Waste Association (ISWA). Like other organic manures, compost made from biodegradable plastic trash improves plant growth indices and soil quality.

2.10.2 Mechanical Recycling

Big surface films (like carrying bags), large hollow containers, and construction materials are examples of easily recyclable plastic products. Recyclability of plastic products is determined by the design of the product, the makeup of the materials, and how economical the recycling procedure is. To make work easier, products that don't consist of complex material mixtures can be simply separated into recyclable materials. Bio-based versions of PE, PET, and (coming) PP are chemically and physically comparable to their fossil fuel-derived counterparts, meaning they can be recycled in the current systems. When commercial quantities and revenues have increased to the point that the required investments can be made, it will be viable to recycle other bioplastics for which there is now no obvious recycling method. Soon, new distinct streams (such as those with PLA) will be available (Ghosh and Jones, 2021).

2.10.3 Organic Recycling/Composting

Plastic items that are recyclable and biodegradable, such as bags, packaging, and cutlery, can expand the range of options for treating garbage. In addition to energy recovery and mechanical recycling, industrial composting, also known as organic recovery or recycling, is a practical waste management option. Under these circumstances, neither the plastics nor the biowaste can be mechanically recycled when they are combined with plastic products. Because the combined waste contains biodegradable polymers, it can be recycled organically, moving it up the waste hierarchy from recovery to recycling (a higher-ranking treatment option). With this approach, biowaste is kept out of landfills and other recycling streams, making composting more beneficial and separation easier.

2.10.4 Recovering Feedstock

When mechanical recycling presents technological challenges or is not financially viable, feedstock recovery is recommended. This process works with a variety of materials and is less prone to contamination. Trash made of plastic can be recycled and used in place of other materials to make steel and concrete. Gasification of plastics—both bio-based and non-bio-based—as well as municipal garbage can result in the production of syngas. Based on the amount of waste organic materials and bio-based polymers, the generated syngas has a high bio-based component. Syngas can be used as fuel, energy source, or heat source. It can also be used to produce methanol and ethanol, as well as petrol (synthetic petroleum) (Guilbert *et al.*, 2006).

2.10.5 Landfill

Utilizing resources inefficiently is landfilling. It is being phased away gradually, even though it remains one of the most popular disposal options in many European countries. Landfills disposed of about 31% of plastic waste in 2014, down from 47% in 2012. European Bioplastics advocates for improved plastic waste recovery and recycling as well as a prohibition on the disposal of plastic items in landfills throughout the continent.

2.10.6 Energy Recuperation

Various methods, including pyrolysis, incineration, and anaerobic digestion, are employed to produce energy from waste bioplastics. It can only be executed, though, when every recyclable part of the bioplastic waste has been retrieved. Bioplastics can be burned in conventional plastic waste incinerators to produce energy because of their high caloric value. Despite this, some bioplastics have lower gross caloric values (GCV), such as natural cellulose fiber and starch. Because they resemble wood, it is possible to recover energy from bioplastic wastes by incinerating them (Yang, 2018).

Burning bioplastic waste, which is recovered and used to make a new product of bio-based, will produce a significant amount of CO_2. Consequently, burning garbage made of bioplastics will turn into a viable option. Anaerobic digestion is a useful technique for managing bioplastic wastes from different garbage collection programs and the mechanical processing of bioplastic wastes. Anaerobic conditions cause bioplastic waste to break down into CO_2 and methane. The byproducts of anaerobic digestion, biogas and digestate, are useful as organic manures and renewable energy sources, respectively. Song *et al.* claim that they are appropriate for anaerobic digestion, which produces methane from waste biopolymers that can be used to generate electricity (Volova *et al.*, 2017).

2.10.7 Bioplastics Market Demand

There has been debate over the usefulness of bioplastics for the last 20 years. Conversely, bioplastics have only lately become a big part of our everyday life. The usage of bioplastics in the general/open market has skyrocketed as a result of ongoing development and research on bioplastics-related activities raising environmental consciousness. Strong legal changes aimed at reducing the use of plastic have also increased demand in a number of nations for bioplastics and recycled plastic. For now, bioplastics only make a small difference. Currently, 1% of the plastic market is made up of bioplastics. At the 12th conference of European Bioplastics held in Berlin, European Bioplastics presented its annual market data update, which reaffirmed the continued increase of the world bioplastics industry (Musa *et al.*, 2024).

Global production of bioplastics amounts to 2.05 million tons. Bioplastics are increased at a rate of 20–25% annually by 2022, whereas regular plastics will grow at a rate of about 49% annually to 2.44 million tons (Gloria *et al.*, 2024). The global bioplastics market is expected to develop at a rate of more than 20% annually, according to the European bioplastics market. The sole disadvantage facing the global bioplastics market is its higher production costs compared to conventional plastics. On the other side, technological progress can defeat it. In addition, the production of petroleum-based bioplastics has benefited from the growing price of crude oil compared to petroleum-based plastics (Song *et al.*, 2009; Musa *et al.*, 2023).

Roughly 56% of all bioplastics are produced using drop-in bio-based ingredients. Bio-PET, bio-PA, and bio-PE are examples of bioplastics. Drop-in bio-based PET leads the market in individual varieties when compared to other bioplastics, like those derived from starch. The bioplastics industry anticipates that PLA and PHA-based plastic manufacture will expand due to their enhanced qualities and feasible end-of-life options. The global market for biodegradable plastics is divided into a number of areas, including packaging, textile, agriculture, injection, electrical, medical, building construction, and molding. Their largest market is the packaging sector, which produces 60% of all bioplastics worldwide (Moses *et al.*, 2023).

Because functional polymers are becoming more and more capable, prospects are developing for industries including building and construction, automotive, and transportation. With 48% of the global bioplastics market in 2022 (million tons), however, PAC is now the largest market.

The automobile, packaging, and electronics industries are three businesses that could profit from the use of bioplastics: The automotive sector contributes significantly to commercial trash, both during the product's life cycle and during disposal. The inside of cars produces an enormous amount of plastic garbage, most of which is disposed of in landfills and pollutes the environment (Musa *et al.*, 2022c).

Using polymeric compounds originating from biological sources is an approach that reduces waste management issues and environmental pollution while producing a low carbon footprint. It is extremely simple to get rid of. The overall weight of the vehicle is decreased when plastics are used in place of metals, increasing fuel efficiency by 3–7%. While still meeting the industry standard for safety in the transportation sector, BBPs can include robust dashboard components and strong exterior and interior elements. Items include steering wheels and covers for airbags and seats. Specialized bio-polyamides can be utilized to fabricate brake noses, tubing, gasoline lines, connectors, and brake lines (Musa *et al.*, 2022b).

Government support, feedstock capabilities, and environmental awareness campaigns are helping bioplastics enterprises in India; however, more effort is needed in the areas of production, raw materials, and technology development (Shamsuddin *et al.*, 2017). Promoting the long-term benefits of bioplastic and increasing environmental awareness are the initial steps toward accomplishing this shift. Researchers in India are creating bioplastics. IIT-Guwahati is the source of the latest bioplastics development, which is currently undergoing commercialization. BioGreen is India's first biotechnology firm producing biodegradable products. EnviGreen, Pastorage, Eco Life, and TrueGreen are already making bioplastics in India (Musa *et al.*, 2022a).

A significant advancement that may provide long-term, environmentally beneficial answers for the plastic pollution problem is bioplastic. The best course of action is to employ them for industrial purposes utilizing the most economical method—bacterial fermentation—in an environmentally sustainable way. Biofuel, biogas, and bio-oil can all be produced from biomass. Due to its long-term viability and potential to create bio-oil in an environmentally sustainable manner through mutagenesis, the market for biopolymers is anticipated to rise significantly in the next years. The biotechnology of microbes has rekindled optimism for the production of bioplastics, which might have a significant environmental impact. However, in order to overcome current obstacles, output needs to be raised (Musa and Ikhajiagbe, 2023a).

Globally, bioplastics research has become a cutting-edge field of study for scientists. The goal of finding environmentally beneficial substitutes for materials sourced from fossil fuels has driven this continual advancement. Excellent biodegradability is a possible benefit of bioplastics, which could aid in addressing global environmental problems such as trash, especially in urban, river, and marine environments. Further research and development on this topic might be possible with focused, ongoing investigation. Bioplastics break down into CO_2; they are environmentally beneficial. Consequently, there is a rapidly growing need for bioplastic applications (Ikhajiagbe *et al.*, 2022b). Commercialization of bioplastics products is warranted due to its renewable nature, biodegradability, composability, and ecological benefits. There is a bioplastic on the market with similar qualities to conventional plastic materials and applications; therefore, bioplastics are currently widely employed in many items that we use on a daily basis. They may also provide additional benefits. The range of applications keeps expanding, encompassing a variety of worthwhile industries like electronics, food, toys, and textiles (Ikhajiagbe *et al.*, 2022a).

2.11 BIOPLASTICS ECONOMY SUSTAINABILITY

The realization that environmental issues and economic development are inextricably linked is spreading among national governments and multilateral organizations. Degradation of the environment can obstruct economic growth, and many development projects exhaust the environmental resources that are their basis. According to Gro Harlem and Oslo (1987), there is no universally accepted collection of indicators needed to perform measurements, no ideal tool for evaluating sustainability, and no comprehensive or uniform concept of sustainability. Notwithstanding their drawbacks, life cycle studies seem to be the sensible place to start. It appears possible that agreement might be obtained on the necessary actions for environmental sustainability, with subsequent effects on the economy and society. Essentially, "development that meets the needs of the present without compromising the ability of future generations to meet their own needs" defines sustainable development. Nonetheless, this straightforwardness allows for nearly limitless flexibility. At least two definitions of sustainability, one somewhat close to the one above, have been provided by the Organisation for Economic Co-operation and Development (OECD). "Non-declining trends of economic growth and development that might be impaired by natural resource depletion and environmental degradation" is another, to put it in more technical terms. Sustainability encompasses social and economic effects in addition to environmental ones. A growing number of businesses across all industrial sectors have implemented comprehensive sustainability objectives that cover these effects due to increased customer accountability and consciousness of their own role (Blanch *et al.*, 2010).

Many businesses view implementing a sustainability strategy as a means of outperforming their rivals in the marketplace. However, because it might be hard to find trustworthy indicators, assessing the effects on society is still quite challenging. An OECD article was based on the Building for Environmental and Economic Sustainability (BEES) model developed by the US Department of Commerce's National Institute of Standards and Technology (NIST) to assess a product's overall environmental and health attributes. More recently, a collection of initiatives for social, economic, and environmental well-being has been developed by the Global Bioenergy Partnership (GBEP). According to Isles and Martin, certain significant bioplastics chemical businesses are creating business strategies that address sustainability and present the argument for using bioplastics. Not surprisingly, the fastest-growing product line in the bio-based products business is bioplastics. Making bioplastics valuable from an ecological and social perspective in order to make money, collaborating with societal actors to establish and monitor this value, and collaborating with suppliers and customers to create value that benefits all stakeholders are the components of a company's sustainability task.

2.12 LIFE CYCLE ASSESSMENT (LCA) OF BIOPLASTICS

The life cycle of a bioplastic is being examined. Because bioplastics have the potential to replace petro-plastics in some applications and are anticipated to have positive environmental benefits, the life cycle of a bioplastic is being studied. Biopolymers like polybutylene adipate terepthalate (PBAT) and polybutylene succinate (PBS) are made from blended starches, whereas PLA and PHA are made from land. While petrochemical basic materials are not biodegradable, nearly all standard PCL plastics, including PE, PP, and PET, are biodegradable and made from renewable raw resources. Lastly, because final disposal options have important environmental implications for the usage of bioplastics, including the capacity to ameliorate climate change, they are taken into consideration. The life cycle of a bioplastic includes the following: The development of the biomass source, the harvesting of raw materials, the processing of biomass prior to bio-refining, fermentation and downstream processing to purify the plastic, subsequent injection or blow molding to create products, sale, use, and end-of-life options, and transportation at different stages of the life cycle. Plastics can be recycled in two primary ways:

- Reusing the finished product once it has been collected, cleaned, and re-distributed.
- Overall, 80% of the plastics business is made up of thermoplastics, which may be remelted and remolded into the same or a different product after being collected. Although recycling offers a significant solution to some of the environmental problems caused by plastics, the process is not always simple. Compared to, say, glass and metals, plastics require more processing. Because different plastics have a tendency to phase-separate when melted together, functioning more like water and oil and creating these layers, it is necessary to separate them. The resulting material's structural fragility restricts the amount of recycling that can be done with these kinds of plastic blends when coloring pigments or other additives are added and cause problems with recycling. Furthermore, the recovered plastic may not be the same when biodegradable and petro-plastics are mixed. Recycling plastics is a complicated process even though it may appear simple at first.

When the OECD thoroughly examined biodegradability in the past, they discovered that there was no universally accepted standard. Instead, a number of assays for biodegradability have been created. Below are the main typical tests. The requirements for biodegradability and composability are as follows:

- Ready biodegradability: A high concentration of the test material is utilized, and non-specific characteristics such as CO_2 production are used to quantify biodegradation. The test is carried out under aerobic conditions.
- Anaerobic biodegradability: In a screening test for anaerobic biodegradability, possible biodegradability under extremely low oxygen (anoxic) conditions may be investigated. The anaerobic conditions of the waste are one significance of this.
- Simulation tests: When a chemical is present in the environment at low concentrations, it may degrade quickly even if it does not meet the requirements for easy biodegradability or even inherent biodegradability. We run simulation experiments to see if this is the case.
- Ultimate biodegradability: The test chemical or polymer's ability to biodegrade into CO_2, biomass, H_2O, and other inorganic compounds such as NH_3. Therefore, the biodegradability assessment is not the outcome of a single test. It is possible for test results to be misinterpreted or for there to be insufficient information, which could lead to mistakes in things like plastic product labeling. Another concern is the extent of biodegradation of bioplastics. It's well knowledge that biodegradation leads to a bioplastic's complete elimination from the environment by mineralization to CO_2 and H_2O, but oxo-biodegradable plastics contain metal salts that act to catalyze degradation, and this degradation can result in residual micro-fragments of plastic and metals that remain in the environment without being a visible contaminant.

Although the effects of tiny plastic fragments produced during the degradation process on soil, water, flora, and fauna are still poorly understood, it is known that these fragments are easily carried to waterways and readily re-distributed throughout the environment. Once in waterways, they may have a variety of effects, including the potential to enter food chains, and their accumulation in soil may harm ecosystems. Because complete biodegradability and composability cannot be assumed, business organizations are beginning to take action in the lack of solid scientific proof on the degree of biodegradability. For example, a prominent UK grocery chain, having delivered over 2 billion oxo-biodegradable bags to customers, has stopped using the bags over uncertainties about their environmental benefits (Aylott, 2011). When a solid material completely converts to minerals and biomass in a given amount of time under specific conditions in a composting system, it is said to be compostable. As a result of its decomposition, compost is produced, which is a naturally occurring fertilizer that can aid in reestablishing soil fertility, managing weed growth, preserving groundwater, and minimizing soil erosion.

Requirements for land for raw materials used in bioplastics production Standard Life Cycle Assessments (LCAs) include land use, and discussions of sustainability and biodiversity frequently touch on this topic. Discussions about the rivalry for land use between plants grown solely as food crops and biomass designated for the manufacture of biofuels have been centered on how biofuels affect patterns of land use. However, as 92% of all arable land worldwide is utilized for the production of food and animal feed, 6% is used for industrial materials, and only 2% is used for biofuels, biofuels have very little effect on patterns of land use. Despite this, there has been a lot of talk about how recent spikes in food prices are related to the usage of land for biofuels. This debate over "food versus fuel" is also taking place around bioplastics (Zhang *et al.*, 2010) albeit to a lesser extent than for biofuels so far. According to estimates (Carus and Piotrowski, 2009), bioplastics had about 250 times smaller impact on food markets, agricultural prices, and land competitiveness in 2008 than biofuels did. Though more research is needed on the subject, it is evident that producing bio-based energy requires a lot more area than producing bioplastics. The creation and application of bioplastics, technical advancements, and consumer demand have had an impact on how bioplastics have evolved over time. Three generations can be used to describe the evolution of bioplastics and their uses; these generations now coexist in parallel and differ in certain ways. The initial wave of bioplastics was distinguished by its exclusive concentration on biodegradable and/or compostable plastics, which were mainly meant for straightforward packaging uses. These biodegradable PHAs are produced on an industrial scale through direct fermentation and were first produced in the 1980s and 1990s. They have found it difficult to compete with the market-dominating petrochemical thermoplastics on price and performance. All varieties of bioplastics, though, are probably going to endure in some applications as they have a niche that makes it difficult for other materials to replace them. The performance of the second generation was better than that of the first. They are more competitive in several applications, particularly packaging, and offer superior processing and performance qualities. In fact, the packaging, gardening, and agricultural industries expressly designed these second-generation materials to be compostable and biodegradable. Production of second-generation bioplastics has scaled up to an industrial level. Recently, some have begun to spread into other fields of application, such as engineering applications, and other industries, including the automotive and consumer electronics. Bioplastics of the third generation are significantly more resilient than the first, moving from biodegradability to resilience. Bio-based versions of the three main thermoplastics that control the market, PE, PP, and PET, are typically available. PVC should be bio-based by 2015, but bio-based PE and PP have only recently been recognized as promising substitutes for the main thermoplastics; bio-PE and bio-PP are chemically produced from fermentation-generated monomers rather than being directly fermented. Given that their carbon content is derived from renewable resources, they can be classified as bioplastics and may help reduce GHG emissions. Predictions indicate a dramatic shift in the global trend of bioplastics production, with durable bio-based materials expected to take the lead. Nonetheless, a noteworthy update comes from the industry association European Bioplastics. Production volumes in the market, which was estimated to be 1.2 million tons in 2011, might expand five times to over 6 million tons by 2016. The product that is anticipated to have the biggest impact on this expansion is bio-based PET (used in plastic bottles), which currently makes up around 40% of the global bioplastics production capacity due to demand from major producers of carbonated drinks, the present production volume is anticipated to increase to more than 4.6 million tons by 2016. The nova-Institute forecast early in 2013 that drop-in polymers, specifically bio-PET13, would be the main cause of the projected 12 million ton increase in bioplastics output by 2020. Bio-based proportion should rise from 1.5% in 2011 to 3% in 2020, with an estimated 400 million tons of polymers produced worldwide in 2020. Synthetic biology methods are already being used, at least in lab settings, to produce bioplastics as a next-generation advancement. This advancement could lead to more cost-effective, scalable, and controlled industrial scale production. Furthermore, studies are investigating the production of bioplastics using fermentation techniques and plants. PHAs are the primary polymers produced directly by fermentation. To match the well-studied and produced high performance of petro-plastics, research has focused on altering the structure of PHAs to create bioplastics with enhanced performance.

Currently included in commercial toolset of synthetic biology is metabolic engineering in conjunction with minimal genome methodologies (Briassoulis and Dejean, 2010), which determine the least number of genes needed to support basic life. With its shown ability to reduce GHG emissions, PLA is a thermoplastic generated from biomass and a promising alternative to certain petro-plastics. Nowadays, ring opening polymerization (ROP) of lactide, a cyclic dimer of lactate, is used to create commercial PLA products instead of direct fermentation (Metz *et al.*, 2012). Because of this, PLA can be considered a bio-based plastic even if the polymer is chemically synthesized; its monomers are generated through fermentation. In order to produce PLA and lactate-containing copolymers efficiently by direct fermentation without the need for a difficult chemical process, the metabolic pathways of *Escherichia coli* have recently been modified (Barker, 2007). This allays worries that leftover heavy metal catalysts from polymerization could jeopardize the chemical process in applications involving humans and food. Given the benefits of fermentative PLA synthesis, which may address issues with lactide-based chemical PLA synthesis, the development of an effective method for producing PLA using engineered PHA biosynthesis systems could have a substantial commercial impact. This is a path toward using synthetic biology to produce polyesters that are customized and unnatural. It is reasonable to anticipate that similar production techniques will be used frequently in the future to produce bioplastics as well as numerous other products, including fine chemicals and pharmaceuticals. The synthesis of bioplastics in plants as opposed to fermentation from microbes is a different area of study. According to Kurdikar *et al.* (2001), LCA has indicated that producing biopolymers in plants could result in more significant reductions in GHG emissions compared to fermentation techniques, provided that the yield is high. According to van Beilen and Poirier (2012), the direct synthesis of biopolymers in plants is also potentially less labor-intensive and more energy-efficient. But fermentation is more effective in terms of process control and product yields. Another significant limitation is seasonal harvesting; however, as Somleva *et al.* (2008) showed, switchgrass can be used to produce polyhydroxybutyrate (PHB) at a lower cost than it does now. This non-food crop has shown itself to be susceptible to the intricate metabolic engineering required to produce high-value biomaterials with co-products that are biofuels produced from lignocellulose. For all future generations of bioplastics, recycling and disposal will be crucial, regardless of how bioplastics are made in the future. These and other bioplastic end-of-life possibilities are discussed in the next section (Kingsley *et al.*, 2024b). Options for the end of life of bioplastics

> While PLA has a lot of potential to be a successful substitute for traditional plastics in the US market, the end-of-life issue is still a big challenge that needs to be resolved before its actualtrue cradle-to-cradle nature can be realised.

One of the most important factors for bioplastics is end-of-life alternatives. According to Hermann *et al.*, a "cradle-to-factory gate" LCA overlooks the energy, financial, and environmental effects of ultimate disposal; instead, a "cradle-to-grave" evaluation evaluates a material's full environmental impact. Given that many plastics are designed to be used only once, this is an important factor to take into account (Kingsley *et al.*, 2024a). A cradle-to-grave LCA is likely to yield quite different results. Regardless of the end-of-life option, the plastics must also be gathered and transported to the appropriate location, which necessitates additional energy and other transportation-related inputs. Depending on the kind of plastic, there are several end-of-life options:

- Recycling;
- Anaerobic and aerobic composting;
- Incineration (with or without energy recovery);
- Recovery of monomers.

End-of-life effects rely significantly on local infrastructure. To assess the costs and advantages to society of any end-of-life choice, information on the inputs and outputs of each process is required. Obtaining the data for these inputs and outputs may occasionally be challenging from a technical and political standpoint. A prime example of the wide range of actual disposal regimes that exist worldwide is the situation in the European Union (EU), where the use of landfills as a way of disposing of municipal solid waste varies from 0% (in three of the 27 EU countries) to 100% (in three other EU countries). Furthermore, a bioplastic may not be suitable for a certain end-of-life option. Food packaging is one potential large-scale use for biodegradable polymers, which may seem best suited for recycling and then disposal. However, the majority of this material is now disposed of in landfills since a normal recycling facility is ill-equipped to handle food-contaminated packaging.

Municipal garbage amounts are landfilled, burned, recycled, and composted as a fraction of the total amounts processed, as indicated by the percentage on the x-axis. Both anaerobic and aerobic composting have not been utilized much as a solid waste treatment solution, despite its seeming appeal. For many years, it has been known that low-quality compost can have many health risks, including elevated levels of heavy metals and other pollutants like glass (De Wilde and Boelens, 1998, for example). Separated garbage collection has been implemented in numerous nations in the recent past. Composting is becoming more cost-competitive because of rising landfill charges. The disposal of putrescible waste in landfills is prohibited in several nations. Foul air is produced during aerobic composting, and it should not be allowed to leave the site. Odor control and substantially greater process control are two benefits of the development of in-vessel composting. Large-scale composting necessitates collection, separation, and transportation to an industrial composting facility, all of which are costly and could put the process at a disadvantage compared to other end-of-life options. The amount of CO_2 released during incineration, for instance, can be more alluring than that of composting, provided that power and heat are recovered during the process. Because of their large molecular weight, bioplastics may be suitable candidates for anaerobic biological processing. Although there is little data based on real-world experience, the potential yields of biogas seem promising. A recent investigation under anaerobic conditions found no evidence of direct biological breakdown of a commercially available PLA. Analogous to aerobic composting, it's possible that PLA decomposition in a landfill would require a chemical hydrolysis stage before any biodegradation could occur. This process is expected to take more than a century at 20°C, and the PLA would degrade very slowly in that environment. The ideal bioplastic is one that breaks down aerobically and doesn't release methane into the atmosphere when dumped in a landfill. Reusing because it has the same difficulties as recycling petro-plastics, thermo-mechanical recycling of bioplastics is now very unpopular. Down-cycling refers to the process of converting. Because biopolymers often have worse thermo-mechanical qualities than petro-plastics, depolymerization during thermal load—which is made worse by hydrolysis in the presence of moisture—is likely to cause materials into lower-quality and less affordable functionality. The presence of only one type of plastic in a waste stream is highly preferred by conventional petro-plastics recycling plants, which makes sorting necessary and costly. This also applies to bioplastics, where considerable quality loss in the recycled material is to be expected. The use of PLA bottles in PET recycling procedures has already drawn criticism because of the mixing of polymer types (Verespej, 2009). On the other hand, bioplastic detection systems utilizing near-infrared (NIR) sensing are undergoing testing. After bioplastics have reached significant product and recovery volumes, mono-material mechanical recycling could be a feasible and sustainable end-of-life solution. It is also feasible to chemically recycle PLA back to lactic acid as a monomer-by-monomer recovery. Approximately 99% of sorted PLA waste can be hydrolyzed back into lactic acid after being cleaned, shredded, and processed (European Bioplastics, 2010). Once more, in order for hydrolysis plants to be profitable, a sufficient amount of PLA waste streams from industry and consumers must be produced. Burning with or without recovery of energy, petro-plastics have a high-calorie density per unit volume, making them suitable fuels. Petro-plastics typically have larger calorie values (about 25 GJ per ton) than bioplastics (about 10 GJ per ton), with the exception of the bio-based alternatives to traditional plastics, which share the same molecules and hence have the same calorie counts. Calorie counts of trash, fuels, and plastics. Burning plastic waste often reduces its volume by 90–99%, which greatly lowers the amount of landfill space needed to get rid of the waste. High moisture content trash, such as food, reduce incineration's efficacy. However, this is a preferred end-of-life option for bio-based plastics in countries with advanced infrastructure for energy recovery from incineration. Regardless of the raw materials used in their production, as well as their biodegradability and composability, all bioplastics and their composites can benefit from energy recovery from incineration. For example, 10 MT of plastic waste are disposed of year in Japan, of which 15 MT are produced and 15 MT are produced, of which about half are burned. In most cases, incineration is combined with heat recycling and power generation (Comfort *et al.*, 2023).

One issue raised is the potential rivalry between agricultural output and the extraction of raw materials for bio-based sector. Producing bioplastics won't compete with food production in the near future, according to European Bioplastics, which estimates that it will occupy less than 0.02% of the world's agricultural land. Conversely, a Greenpeace report stated that location and whether the land is concentrated in a few areas are crucial factors to take into account (Jia, 2020). The paper also included a calculation of its own regarding the impact of switching from plastic to PLA packaging. According to the computation, this would require 32% of the world's yearly maize production to be transferred to PLA, or 1% of total corn production (Korol *et al.*, 2019).

Additionally, the growing of crops, such as maize suited for PLA manufacturing, necessitates a significant amount of fresh water to produce bioplastics. Though some concerning findings indicate that substituting bioplastics for water footprint analysis of the manufacture of bio-based plastics is not commonly carried out in LCA studies, and the production of packaging in Europe would increase the associated water use to over one fifth of the EU's total freshwater use.

2.13 POLICIES PROMOTING BIOPLASTICS DEVELOPMENT AND USE

Bioplastics, which are derived from renewable biomass sources such as plants, algae, or microorganisms, offer a sustainable alternative to traditional fossil fuel-based plastics. Governments and organizations worldwide recognize the potential of bioplastics to mitigate climate change, reduce GHG emissions, and promote a circular economy. Since these bioplastics have proven to be effective in mitigating environmental and climate-related issues, it is important that the government and relevant stakeholders design some policies that will promote the development, continued use and research in bioplastics. To this end, there are several policies that have been approved and implemented in different parts of the world regarding bioplastics (Singh, 2020). Some of them include the tax incentives. Through tax incentives, governments, for example in the United States (Funk *et al.*, 2019), have offered tax breaks and sometimes credits to companies investing in bioplastics research, development, and production. Italy also comes up with its Italy's Biodegradable Plastic Tax which imposes a tax on non-biodegradable plastics, encouraging bioplastics use. In addition, government in some countries have considered giving grants and funding, including other financial supports to companies with high production of bioplastics, as well as companies that are engaging research in bioplastics, companies using bioplastics for their pilot projects, and companies on commercial-scale production of bioplastics.

In addition, the government has introduced some regulatory frameworks that establish standards, certifications, and labeling requirements for bioplastics and its production processes. The aim of this policy is also to encourage and promote bioplastics use and development all over the world (Karla *et al.*, 2002). The government of most countries, especially in Africa, has prioritized bioplastics in public procurement policies, making bioplastics the first and most preferred in line of its procurement. This has boosted the use and development of bioplastics in these regions. For example, Nigeria has witnessed about 15 new companies that are developing road asphalt and compound flooring using bioplastics. In addition, Ghana has also witnessed about 30% increase in plastic processing industries since 2020. In Abuja and Lagos, Nigeria, bioplastic waste disposal basket has been designed to improve the use and recycling of bioplastic in the country (Po-Keung *et al.*, 2013). Governments in Germany and Russia have implemented sustainable waste management policies that encourages bioplastics recycling and composting, through the Germany's National Bioeconomy Strategy that aims to increase bioplastics production and use through research funding and regulatory support.

Furthermore, some countries such as Japan introduced Japan's Bioplastics Promotion Policy, which is a sustainable policy that supports bioplastics development through research funding and industry partnerships. These could serve as some way of giving economic incentives to bioplastics companies. There are some international organizations that have been assisting in ensuring a proper policy on bioplastics use and development. These companies are as follows:

1. United Nations Environment Programme (UNEP): UNEP promotes bioplastics through its Global Programme of Action for the Protection of the Marine Environment from Land-based Activities.
2. EU's Bioeconomy Strategy: The EU supports bioplastics development through research funding, regulatory frameworks, and market incentives.
3. International Organization for Standardization (ISO): ISO develops standards for bioplastics, ensuring consistency and quality.

2.14 ECONOMIC INCENTIVES AND SUBSIDIES FOR BIOPLASTICS INDUSTRIES

The bioplastics industry has experienced significant growth in recent years, driven by increasing demand for sustainable and environmentally friendly products. Governments and organizations worldwide recognize the potential of bioplastics to reduce GHG emissions, mitigate climate change, and promote a circular economy. To support the development and growth of the bioplastics industry, various economic incentives and subsidies have been implemented. Some of which include the following:

1. Tax Credits: Direct tax credits or rebates for bioplastics production, research, and development.
2. Grants: Financial support for bioplastics research, pilot projects, and commercial-scale production.
3. Low-Interest Loans: Access to affordable financing for bioplastics companies.
4. Feed-in Tariffs: Guaranteed prices for bioplastics producers selling renewable energy.
5. Carbon Credits: Credits for bioplastics producers reducing GHG emissions.
6. Production Subsidies: Direct financial support for bioplastics production.
7. Import/Export Tariffs: Reduced or eliminated tariffs on bioplastics imports/exports.
8. Green Bonds: Specialized bonds financing bioplastics projects.

In addition, government from various countries have also introduced their respective initiatives and subsidies for companies that are into production and development of bioplastics products. For example, the United States have developed the Biodegradable Products Institute (BPI) certification tax credit, USDA BioPreferred program, Department of Energy (DOE) grants for bioplastics R&D. The EU came up with Bio-based Industries Consortium (BIC) funding, Horizon 2020 research and innovation program, EU Emissions Trading System (EU ETS) carbon credits. While China introduced the 13th Five-Year Plan for Biodegradable Materials, National Development and Reform Commission (NDRC) subsidies, Ministry of Science and Technology (MOST) grants. India on the other hand introduced the National Bioplastics Policy, Department of Biotechnology (DBT) funding, and Ministry of New and Renewable Energy (MNRE) subsidies.

2.15 INTERNATIONAL REGULATIONS AND THEIR ECONOMIC IMPACT ON BIOPLASTICS ADOPTION

The bioplastics industry has experienced significant growth due to increasing demand for sustainable and environmentally friendly products. International regulations play a crucial role in promoting bioplastics adoption, mitigating climate change, and reducing plastic waste. This write-up examines key international regulations, their economic impact, and the future outlook for the bioplastics industry.

2.16 KEY INTERNATIONAL REGULATIONS

1. EU's Single-Use Plastics Directive: Bans single-use plastics and promotes biodegradable alternatives.
2. UNEP Global Partnership on Marine Litter: Encourages countries to reduce marine litter, including plastic waste.
3. ISO 14855: Establishes standards for biodegradable plastics.
4. REACH (Registration, Evaluation, Authorization, and Restriction of Chemicals): Regulates chemical use in bioplastics production.
5. G7 Ocean Plastics Charter: Aims to reduce plastic waste, increase recycling, and promote sustainable alternatives.

2.17 ECONOMIC IMPACT

1. Increased Demand: Regulations drive demand for bioplastics, stimulating market growth.
2. Investment Incentives: Governments offer incentives for bioplastics production, research, and development.
3. Job Creation: Bioplastics industry expansion creates new employment opportunities.
4. Reduced GHG Emissions: Bioplastics production generates fewer emissions than traditional plastics.
5. Waste Reduction: Biodegradable bioplastics reduce plastic waste, mitigating environmental impacts.

2.18 CONCLUSION

The bioplastics economy has emerged as a promising solution to mitigate the environmental impacts associated with traditional plastics. The production of bioplastics from renewable biomass sources, such as corn starch, sugarcane, or potato starch, offers a more sustainable alternative to fossil fuel-based plastics. Bioplastics have the potential to reduce GHG emissions, conserve non-renewable resources, and decrease the amount of plastic waste in oceans and landfills. However, the bioplastics economy is not without its challenges. The production of bioplastics requires large amounts of land, water, and energy, which can lead to competition with food crops and other land uses. Additionally, the end-of-life management of bioplastics is still in its infancy, and the lack of standardized recycling protocols and infrastructure can hinder the development of a circular bioplastics economy.

Despite these challenges, the bioplastics economy has the potential to play a critical role in the transition toward a more sustainable and circular economy. Governments, industries, and consumers must work together to create an enabling environment for the development of the bioplastics economy. This can be achieved through the implementation of policies and regulations that support the production and use of bioplastics, investments in research and development to improve the performance and sustainability of bioplastics, and education and awareness-raising campaigns to promote the benefits of bioplastics. Ultimately, the success of the bioplastics economy will depend on its ability to balance economic, social, and environmental sustainability. By adopting a holistic approach that considers the entire value chain of bioplastics, from production to end-of-life management, we can create a more sustainable and circular bioplastics economy that benefits both people and the planet.

REFERENCES

Ach, A. (1993). Structure-directed synthesis of cyclacene and polyacene derivatives. *Journal of Macromolecular Science, Part A: Pure and Applied Chemistry*. 30: 733–740.

Ashok, A., Rejeesh, C.R., and Renjith, R. (2016). Biodegradable polymers for sustainable packaging applications: A review. *International Journal of Bionics and Biomaterials*. 2: 14–24.

Aylott, M. (2011). Role of bioplastic in nation building. *International Journal of Sustainable Research*. 2(1): 23–31.

Bagheri, A. (2007). Fate of so-called biodegradable polymers in seawater and freshwater. *Global Challenges*. 1: 17–48.

Barker, T, Bashmakov, I, Berstein, L., Bogner, J.E., Bosch, P., Dave, R., Davidson, O., Fisher, B.S., and Gupta, S. (2007). Technical summary. In: Climate Change (2007): Mitigation. Contribution of Working Group III to the Fourth Assessment, Report of the Intergovernmental Panel on Climate Change.

Blanch, H.W., Simmons, B.A., Oleskowicz-Popiel, P., and Klein-Marcuschamer, D. (2010). Technoeconomic analysis of biofuels: A wiki-based platform for lignocellulosic biorefineries. *Biomass Bioenergy*. 34: 1914–1921.

Brizga, J., Hubacek, K., and Feng, K. (2020). The unintended side effects of bioplastics: Carbon, land, and water footprints. *One Earth*. 3: 45–53.

Carus, M., and Piotrowski, S. (2009). Land use for bioplastics. *Bioplastics Magazine*. 4: 46–49.

Comfort, N.O., Okoji, A. I., Ibrahim, M.S., and Obinna, O. (2023). Comparative analysis of adaptive neuro-fuzzy inference system (ANFIS) and RSRM models to predict DBP (trihalomethanes) levels in the water treatment plant. *Arabian Journal of Chemistry*. 15(10): 37–94.

De Wilde, B., and Boelens, J. (1998). Prerequisites for biodegradable plastic materials for acceptance in real-life composting plants and technical aspects. *Polymer Degradation and Stability*. 59: 7–12.

Emadian, S.M., Onay, T.T., and Demirel, B. (2017). Biodegradation of bioplastics in natural environments. *Waste Management*. 59: 526–534.

Enerijiofi, K.E., Ekene, C., Musa, S.I., Okoji, C., Moses, N., Ebrahimbabaie, P., Onuche, I.N., and Adetunji, C.O. (2024a). Environmental pollution, causes types, fates, effects and control strategies. *Phytoremediation in Food Safety*. Taylor and Francis Ltd.

Enerijiofi, K.E., Musa, S.I., Igiebor, F.A., Lawani, M., Chuka, N.E., Odozi, E.B., and Ikhajiagbe, B. (2024b). Nanotechnological approaches for enhanced microbial activities, sustainable agriculture, and the environment. *Sustainable Agriculture: Nanotechnology and Biotechnology for Crop Production and Protection*. De Gruyter, pp. 153–184.

European Bioplastics (2010). Feedstock recovery of post-industrial and post-consumer polylactide bioplastics. *European Bioplastics Fact Sheet*, p. 40.

Funk, C., Grabert, M., Oey, M., and Hankamer, B. (2019). Green bioplastics as part of a circular bioeconomy. *Trends in Plant Science*. 24: 237–241.

Ghosh, K., and Jones, B.H. (2021). Roadmap to biodegradable plastics—Current state and research needs. *ACS Sustainable Chemistry and Engineering*. 9: 61–70.

Gourmelon, G. (2015). Global plastic production rises, recycling lags. *Journal of Vital Signs*. 3(2): 23–34.

Guilbert, S., Morel, M.H., Gontard, N., and Cuq, B. (2006). Signing new materials from wheat protein. *ACS Symposium Series*. 921: 33–39.

Ikhajiagbe, B., Anoliefo, G.O., Igiebor, F.A., and Musa, S.I. (2022a). Enhancement of bioprocesses for the reclamation of impacted environment. *Bioenergy and Environmental Biotechnology for Sustainable Development*. CRC Press. 1st Edition, p. 24.

Ikhajiagbe, B., Anoliefo, G.O., Olise, O.F., Adekunle, I.J., and Musa, S.I. (2022b). Environmental reclamation in Nigeria: Past, present and future. Biotechnology society of Nigeria. *Bioenergy and Environmental Biotechnology for Sustainable Development*. CRC Press. 1st Edition, p. 38.

Jia, Z. (2020). Biodegradable Plastics: Breaking Down the Facts—Production, Composition and Environmental Impact. *Greenpeace East Asia*. Hong Kong, China, p. 23.

Kalra, B., and Gross, R. (2002). Biodegradable polymers for environments. *Science*. 297: 803–815.

Katiyer, V. (2020). *Sustainable Polymers for Food Packaging: An Introduction*. De Gruyter, p. 200.

Korol, J., Hejna, A., Burchart-Korol, D., Chmielnicki, B., and Wypiór, K. (2019). Water footprint assessment of selected polymers, polymer blends, composites, and biocomposites for industrial application. *Polymers*. 11: 17–91.

Kurdikar, G.H. Dennis, J.S., von Blottnitz, H., and Harrison, S.T.L. (2001). Environmental analysis of plastic production processes: Comparing petroleum-based polypropylene and polyethylene with biologically-based poly-β-hydroxybutyric acid using life cycle analysis. *Journal of Environmental Sciences*. 23(1): 23–43.

Lopez-Gil, A., Rodriguez-Perez, M.A., De Saja, J.A., Bellucci, F.S., and Ardanuy, M. (2014). Strategies to improve the mechanical properties of starch-based materials: Plasticization and natural fibers reinforcement. *Polimeros*. 24: 36–39.

Markets and Markets. (2023). Bioplastics market by type and region – Global Forecast to 2028. *National Development Policy*. 2: 56–98.

Metz, B., Davidson, O.R., Bosch, P.R., Dave, R., and Meyer, L.A. (2012). Cambridge University Press, Cambridge, and New York, NY. *BC Bio-Economy British Columbia Committee on Bioeconomy*. Handbook, p. 230.

Musa, S.I. and Ikhajiagbe, B. (2023). Seed bio-priming with phosphate solubilizing bacteria strains to improve rice (*Oryza sativa* L. var. FARO 44) growth under ferruginous ultisol conditions. *Journal of Biotechnology, Computational Biology and Biotechnology*. 104(1): 33–51

Musa, S.I. and Ikhajiagbe, B. (2024). Effects of plant growth promoting bacteria (PGPB) rhizo-inoculation on soil physico-chemical, bacterial community structure and root colonization of rice (*Oryza sativa* L. var. FARO 44) grown in ferruginous ultisol conditions. *Science World Journal*. 19(3): 894–899.

Musa, S.I., Isitua, C.C., Moses, N., and Dioha, I.J. (2022b). Challenges and future scope of bionanotechnology in algal biofuel and life cycle analysis. *Bionanotechnology towards Green Energy*. Taylor and Francis, CRC Press, p. 567.

Musa, S.I., Moses, N. and Ikhajiagbe, B. (2022c). Seed priming with phytohormones. *Plant Hormones: Recent Advances, New Perspectives and Applications*. IntechOpen, p. 456.

Musa, S.I., Ogala, H., Okoji, C., Moses, N., Nathaniel, I., and Ojei, S. (2022a). Applications of bio-nanoenabled materials and coatings for affordable, reliable and sustainable green energy production. *Bionanotechnology towards Green Energy*. Taylor and Francis, CRC Press, p. 567.

Musa, S.I., Shehu, M.S., Ogala, H., Onukwu, I.S., Ebrahimbabaie, P., Okoji, C., Chodize, E., Moses, N., and Ezinne, O.M. (2023). Global advances in bioenergy production technologies. *Microbial Biotechnology for Bioenergy*. Elsevier.

Narancic, T., and O'Connor, K.E. (2019). Plastic waste as a global challenge: Are biodegradable plastics the answer to the plastic waste problem. *Microbiology*. 165(2): 129–137.

Nathan, M., Ezenwata, I.S., and Musa, S.I. (2023). Some physicochemical properties and heavy metals in water from Oboshi river, Ibusa in Delta State, Nigeria. *Journal of Applied Sciences and Environmental Management*. 27(6): 1291–1294.

Okoji, C., Musa, S.I., Moses, N., Ogala, H., Onukwu, I.S., Chidozie, E., Ezinne, Omoregie, G.O., Igiebor, F.A., Mshelmbula, B.P., Musa, S.I., Osagie, P., Adebiyi, M., Igbinosun, C.E., and Ikhajiagbe, B. (2024).

Nanobioherbicide and photosynthetic pigment synthesis. Charles Oluwaseun Adetunji and Julius Kola Oloke (eds.). *Handbook of Agricultural Biotechnology.* Volume II Nanobioherbicides (97–124). Scrivener Publishing LLC.

Philp, J.C., Ritchie, R.J., and Guy, K. (2013). Biobased plastics in a bioeconomy. *Trends in Biotechnology.* 31(2): 65–67.

Po-Keung, W., and Man, C.L. (2013). Development and sustainability of bioplastics: A review. *World Scientific.* 2: 53–59.

Rosenboom, J.G., Langer, R., and Traverso, G. (2022). Bioplastics for a circular economy. *Nature Reviews Materials.* 7: 117–137.

Shamsuddin, I.M., Jafar, J.A., Abdulrahman, S.A.S., Yusuf, S., Lateefah, M., and Aminu, I. (2017). *Advances in Biosciences and Bioengineering.* 5: 63–70.

Shen, L., Haufe, J., and Patel, M.K. (2010). Product overview and market projection of emerging bio-based plastics. *PRO-BIP.* 1: 1–243.

Shen, L., Haufe, J., and Patel, M.K. (2020). Product overview and market projection of emerging bio-based plastics. *PRO-BIP.* 1, 1–243.

Singh, R. (2020). Industrial biotechnology: An Indian perspective. *Industrial Biotechnology.* 16: 21–25.

Somleva, M., Snell, K., Beaulieu, J., Peoples, O., Garrison, B., and Patterson, N. (2008). Production of polyhydroxybutyrate in switchgrass, a value-added co-product in an important lignocellulosic biomass crop. *Plant Biotechnology Journal.* 6: 663–678.

Song, J.H., Murphy, R.J., Narayan, R., and Davies, G.B.H. (2009). Plant bio extract. *Philosophical Transactions of the Royal Society of Biological Sciences.* 3(6): 23–29.

Spence, K.E., Allen, A.L., Wang, S., and Jane, J. (1996). Soil and marine biodegradation of protein starch plastics. *ACS Symposium Series.* 627: 149–157.

Van Beilen, J.B., and Poirier, Y. (2012). Plants as factories for bioplastics and other novel biomaterials. Chapter 30. Gautam Altman and P.M. Hasegawa (eds.). *Plant Biotechnology and Agriculture: Prospects for the 21st Century.* Academic Press. ISBN 978-0-12-381466-1.

Verespej, K. (2009). Casein core-polysaccharide shell nanocomplexes stable at pH 4.5 enabled by chelating and complexation properties of dextran sulfate. 105723.

Volova, T.G., Vinnik, Y.S., Shishatskaya, E.I., Markelova, N.M., and Zaikov, G.E. (2017). Application of polymers. *Natural Based Polymers for Biomedical Applications.* 2(1): 46–50.

Yang, G. (2018). Food science and nutrition. *John Wiley and Sons.* 3(2): 34–39.

Zhang, S.Z., Penner, G.B., Yang, W.Z., and Oba, M. (2010). Effects of partially replacing barley silage or barley grain with dried distillers' grains with soluble on rumen fermentation and milk production of lactating dairy cows. *Journal of Dairy Science.* 93(7): 3231–3242.

3 Advances in Bioplastic Additives

Katherine Georgina Menon, Pariki Nithin, Borusu Bhargavi Lakshmi, P. Gnana Deepu, and Kondakindi Venkateswar Reddy

3.1 INTRODUCTION

The most widely accepted definition of a bioplastic is a polymer that is either biobased, biodegradable, or both (Costa et al., 2023). This highlights the crucial point that not all bioplastics are biodegradable, that is, being biobased does not automatically make a material biodegradable and that some non-biobased plastics, like poly(butylene adipate-co-terephthalate) (PBAT), are biodegradable (Chan et al., 2022). Biodegradability and renewability are two sustainability concepts that apply to biobased polymers, or bioplastics. They can be utilized to limit the harm caused by plastic waste that can take hundreds of years to get decomposed. In a variety of applications, including food packaging, agriculture, and hygiene, bioplastics are used with their own benefits (Kumar et al., 2023).

Bioplastics are materials that come from renewable sources and can be utilized to decrease the issue of plastic squander that is choking the planet and sullying the environment (Costa et al., 2023). Plastic is the third most utilized petroleum subsidiary in the world; each year 200 million tons of plastic are devoured on the planet. It comes from a non-renewable source (petroleum); it is sullying and non-biodegradable (Lackner, 2000).

Additives are usually substances added to a process to improve or increase the overall quality or specific properties of a material. Additives in bioplastic production are used to increase biodegradability of the plastic compound (Deanin, 1975).

Bioplastics have been brought into focus as a potential substitute for the traditional plastics that are generally sourced from renewable biomass against the backdrop of rising demand for sustainable materials (Ali et al., 2023). Bioplastics have inherent merits in terms of environment performance and applicable use and may be highly improved with various additives. These additives play the key role in the ability of bioplastics to alter their physical, mechanical, and aesthetic properties to be applied in various fields ranging from packaging to automotive components (Folino et al., 2023). Additives can be broadly categorized into several types, namely plasticizers, stabilizers, fillers, colourants, and processing aids (Marturano et al., 2017; Safitri et al., 2022). Each has its rationale: plasticizers make the material more flexible (Eslami et al., 2023), stabilizers increase the durability of the material (Jia et al., 2018), fillers give some strength to the material (Hallstein et al., 2024), colourants add beauty (Mariano et al., 2014), and processing aids assist in the manufacturing (Reichert et al., 2020).

DOI: 10.1201/9781003602828-4

In the incorporation of these additives, the precision is high and it allows for the optimization of the performance of bioplastics without losing their biodegradability and sustainability. The major challenges lie in the sourcing and cost-effectiveness aspects and require full evaluation of environmental impact for these bioplastic formulations (Boey et al., 2022; Rahmati et al., 2024).

3.1.1 Bioplastic Production Process

Bioplastics can be produced from primary raw materials like rice, sorghum, soy, beet, corn, palm, barley, sugar cane, wheat, and potatoes. Secondary materials such as biowaste, spent cooking oil, and petroleum-based feed are also being used to produce bioplastics. Potato peel waste has the potential to be used as a feedstock for high-value bioproducts like biochemicals, biofuels, and bioplastics (Krishnamurthy et al., 2019). *Cupriavidus necator*, a recombinant amylolytic strain, produced polyhydroxyalkanoates (PHA) from purple sweet potato waste, and the PHA titre increased to 3.65 g L^{-1}. A variety of bioplastics, including lignin- and starch-based polymers, as well as active compounds, such as polyphenols, fatty acids, and polysterol, can be made from lignocellulosic biomass (Thomas et al., 2023). The production of bioplastics can also be achieved by using microalgae to treat food wastes and extract different types of polymers for use as building blocks. Using biorefining techniques, microalgal biomass can be converted into a bioplastic polymer like PHA, which could serve as a feedstock for future advancements in the bioplastics sector. To produce bioplastics, the downstream processing of microalgae to extract such bioactive compounds (cellulose and lipids) may show promising results. Bioplastics produced from microorganisms are becoming more and more popular as a viable solution for environmental sustainability. Biomass waste from corn cobs was treated to produce fermentable sugars and, because of the bacterial activity, was able to produce astaxanthin and PHA (Khomlaem et al., 2023).

Organic wastes can be used to produce PHA and astaxanthin, at a reasonable cost and it is a renewable resource. Thus, the utilization of organic wastes as raw materials is crucial for resolving issues related to waste management. It is also widely agreed upon that treating waste, as a whole, reduces greenhouse gas emissions because waste recovery and prevention also reduce emissions in other economic sectors.

3.2 TYPES OF ADDITIVES IN BIOPLASTIC PRODUCTION

Bioplastics present new alternative options for conventional plastics within the backdrop of increased demand for sustainable materials that carry a reduced environmental impact and are mainly derived from renewable resources (Islam et al., 2024; Atiwesh et al., 2021). The performance and functionality of bioplastics are improved through additive addition during their production process (Abu et al., 2024). Some additives improve mechanical properties, increase flexibility, enhance durability, and provide aesthetic appearance (Crawford, 1998). The most common groups of additives are plasticizers, stabilizers, fillers, colourants, antimicrobials, compatibility agents, foaming agents, impact modifiers, flame retardants, and biodegradable additives. Every type has a crucial role in ensuring that bioplastics meet the specific requirements of different applications without losing the benefits associated with their environmental comfort. Additives types are illustrated in Figure 3.1.

Plasticizers: Plasticizers are compounds added to polymers to increase their flexibility, workability, and durability. In the context of bioplastics, which can often be brittle and rigid, plasticizers play an essential role in improving the mechanical properties and processing capabilities of these materials. By modifying the physical characteristics of bioplastics, plasticizers facilitate their use in various applications, such as packaging, automotive components, and consumer goods. As the demand for environmentally friendly materials grows, the choice of plasticizers has become increasingly important, leading to the exploration of safer and more sustainable options.

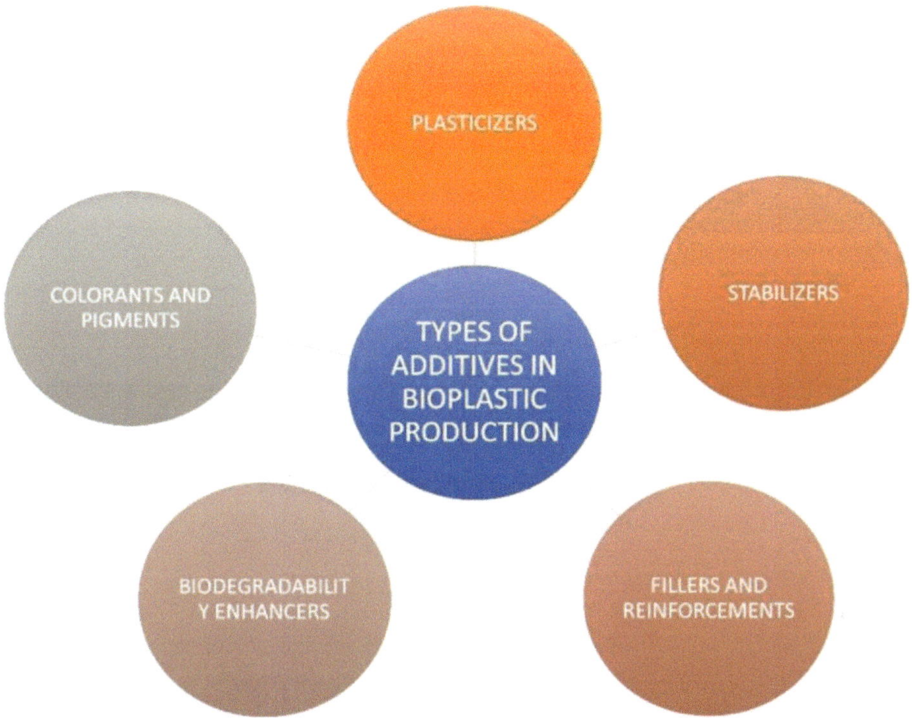

FIGURE 3.1 Types of bioplastic additives.

3.2.1 Commonly Used Plasticizers for Bioplastics

Glycerol: This is the most commonly found natural plasticizer, obtained through the saponification of fats and oils. Glycerol is a non-toxic agent and is said to be well compatible with starch-based bioplastics. Glycerol enhances flexibility and diminishes brittleness in the biodegradable plastics. Moreover, it is linked to the biodegradability of bioplastics towards emerging trends in sustainability.

Sorbitol: Sorbitol is a sugar alcohol extracted from fruits and similar sources. Sorbitol, like glycerol, can impart mechanical properties in starch-based bioplastics. Its potential to add flexibility and reduce moisture pickup makes it also a value-added plasticizer in many applications of bioplastics like food packaging and disposable items (González et al., 2021).

Acetyl tributyl citrate (ATBC): ATBC is a biodegradable plasticizer with its origin from citric acid. It is non-toxic and can be used for all sorts of applications that involve direct contact with food. ATBC improves flexibility and processability of bioplastics, maintaining the biodegradability. ATBC is very widely used in movies, coatings, and processing technologies for flexible packaging material (Tsou et al., 2014).

Polyethylene glycol (PEG): PEG is an industrial polymer derived from renewable feedstocks. PEG acts as a general-purpose plasticizer and enhances the flexibility and processing properties of most bioplastics. Most important, it acts as a viscosity-reducing agent that helps to decrease the viscosity of the formulation for ease in shaping and moulding the material during manufacturing (Yu et al., 2017).

Dioctyl terephthalate (DOTP): DOTP is a non-phthalate alternative to traditional plasticizers which are becoming increasingly available in the market of bioplastics. DOTP does not contain the environmental and health problems caused by phthalates, but it offers excellent flexibility and durability properties. DOTP plasticizer is generally the first choice for applications where long lifespan and flexibility in the final materials are important, such as flooring and automotive parts.

3.2.2 Stabilizers

Stabilizers are those chemical additives that one adds to bioplastics to enhance the performance and stability of a material. These additives help to protect the material from environmental factors like heat, light, and moisture from degrading its properties. In other words, stabilizers can affect several physical properties of bioplastics, namely thermal stability, flexibility, mechanical strength, and resistance to oxidation and UV light. The above challenges are what stabilizers help resolve so that bioplastics can retain integrity and functionality throughout an intended lifespan, more suitable for a wide array of applications (Marturano et al., 2023).

3.2.2.1 Commonly Used Stabilizers For Bioplastics

Antioxidants: These play the most crucial role in preventing oxidative degradation during processing and application. Bioplastics are often prone to oxidation, and this can make free radicals form which may degrade their mechanical and thermal properties (Ismayati et al., 2024). Common antioxidants are tocopherols-vitamin E, ascorbic acid-vitamin C, and synthetic compounds, such as BHT—-butylated hydroxytoluene (Almeida et al., 2022). Adding these antioxidants ensures the integrity of bioplastics at all stages, from production to end application.

UV stabilizers: UV radiation causes severe degradation of bioplastics, particularly when used outdoor. Benzotriazole and HALS are known UV stabilizers, where they absorb or reflect harmful UV rays to prolong the lifespan of bioplastic products. The addition of these stabilizers enhances the weatherability and aesthetic characteristics of bioplastics, thus making them acceptable for an even broader spectrum of applications.

Thermal stabilizers: Thermal stabilizers are key additives in bioplastic production, enhancing the thermomechanical stability of polymers during their processing and use. These additives resist degradation because of high temperature; otherwise, the mechanical properties and aesthetics of bioplastics could degrade because of the said process. Most of them used are metal salts, phenolic antioxidants, and phosphorus-based compounds that work through scavenging the free radicals and interfering with the pathways of degradation, and effective heat distribution. The presence of thermal stabilizers will ensure better processing efficiency and hence longer longevity, hence creating room for bioplastics to be competitive against their traditional counterparts.

Antimicrobial agents: Antimicrobial agents are meant to inhibit the growth of bacteria, fungi, and other microorganisms—to enhance the hygiene and safety of bioplastics (Abang et al., 2023). There are various sources of natural agents, like chitosan or essential oils; others are synthetic, such as silver nanoparticles, which are reported to act most powerfully against microorganisms (Suryanegara et al., 2021). Bioplastics can be developed for specific applications in food packaging and medical devices (Nanda et al., 2022). Specific antimicrobial agents would help prevent contamination of the product during storage and maintain product cleanliness, thus extending the shelf life of the products and further building consumer confidence in their use of bioplastic materials.

3.2.3 Fillers and Reinforcements

Fillers that are introduced to the bioplastics include the addition of calcium carbonate, talc, and cellulose. These contribute to mechanical strength and resistance against thermal stability. The inclusion of such a filler reduces costs in the production process and complements the overall performance of bioplastics. Natural fibre reinforcements, for example, flax and hemp, have strength and also have renewability as a characteristic, which is a compliment for this kind of production (Mohammed et al., 2023).

- Mineral fillers include calcium carbonate and talc, considered relatively inexpensive reinforcements that are used to enhance stiffness as well as impact resistance.
- Clay-based filler can be chosen for any improvement in barrier or processing properties of the polymer (Helanto et al., 2021).
- Organic fillers such as wood flour and natural fibres, like flax, hemp, and jute, provide reinforcement to the polymer matrix and enhance its overall mechanical properties, derived from renewable resources.
- Synthetic fillers, such as glass fibres, tend to increase the strength and stiffness quite remarkably, even if they are not biodegradable.
- Natural fibre-based reinforcements, such as hemp and bamboo fibres, are known to exhibit a high strength-to-weight ratio, and, therefore, have applications in automotive materials, construction materials, and many others (Azman et al., 2021).
- Also, a mixture of two or more fillers and reinforcements in hybrid composites can be used to create materials where each component can actually benefit from the properties brought about by its neighbour, enhancing overall performance in each application. The various bioplastics can be carefully selected and combined to be optimized towards the goal of a particular application while still keeping all aligned towards sustainability (Khalid et al., 2021).

3.2.4 Biodegradability Enhancers

Biodegradability enhancers are critical additives in the manufacture of bioplastics, which could allow them to easily decompose naturally via biological means. Among available biodegradability enhancers include natural additives like starch and cellulose, which are sources of nutrients available for microorganisms to degrade the plastic matrix (Marichelvam et al., 2019). Addition of amylases and cellulases accelerates the breakdown process because this breaks the larger polymer chains into smaller entities, easier for microbes to assimilate. Microbial inoculants also introduce strains of bacteria or fungi which actively degrade the bioplastics and cause a faster decomposition rate. Biodegradable plastics may also be conditioned favourably towards microbial action by incorporating other plasticizers, such as citrates and glycerol, which favourably enhance flexibility and, consequently, biodegradability. Additives with controlled release can themselves degrade in a controlled manner, like aliphatic polyesters, and be used for developing materials that degrade at predictable rates for specific applications (Ahsan et al., 2023). Enhancers are useful for microbial colonization, enzyme activity, and nutrient availability; all these will enhance the biodegradation of bioplastics. The addition of such enhancers is crucial because bioplastics can meet standards on compostability and open further opportunities for packaging, agriculture, and more (Abe et al., 2021). Nevertheless, the integration of biodegradation enhancers along with maintaining mechanical properties and processing characteristics—fundamental functionality and attractiveness in the market—is very important.

3.2.5 Colourants and Pigments

Colourants and pigments are two of the primary factors that give bioplastics aesthetic and functional appeal (Niaounakis et al., 2015). However, with this growing demand in the market, manufacturers learn the hard way that colourants not only aesthetically appeal to them but also perform purposeful functions, such as branding, identification, and UV protection.

- Organic colourants and pigments fall into two large categories. Organic colourants could be of natural origin, such as extracted from plants or insects, or synthetic organic compounds. Among the most widely used natural colourants are beetroot extract, turmeric, and indigo. Organic colourants offer an accessible range of shades as well as compatibility with a bioplastic matrix. However, they might lack in terms of lightfastness and stability, which can influence the final colour life product.

- Inorganic pigments are mineral-based and known for their lightfastness and stability characteristics; however, there may be some exceptions. These include titanium dioxide or white, iron oxide used in red, yellow, and brown, and ultramarine blue. These pigments are very useful for durable applications such as extrusion products and packaging materials. As inert, they do not react with the bioplastic matrix thus keeping the integrity of the material but simultaneously imparting rich colours (Charvat, 2004).

Some factors that determine the choice of colourants and pigments are based on the intended use of the bioplastic, type of application method, as well as end-use. For example, products being used for food packaging should have non-toxic colourants that can stand in line with the regulatory guidelines provided by food safety organizations. Bioplastics used outdoors might require UV-stabilized pigments to increase their sunlight resistance and, therefore, extend the life span of the material. Besides aesthetic purposes, colourants and pigments functionally contribute to the bioplastic. Some examples of this are that certain pigments provide UV protection and shield the bioplastic from UV-related degradation due to exposure to sunlight.

3.3 INNOVATIONS IN BIOPLASTIC PRODUCTION

Additives for bioplastics are now at the forefront of sustainable material science evolution, offering great enhancements in terms of performance, functionality, and environmental compatibility. Given the now ever-increasing demand for more environmentally friendly alternatives to the rather environmentally unfriendly traditional plastics, a wide array of additives for bioplastics is currently under investigation for increasing the versatility of bioplastics in various applications (Cruz et al., 2022). Promising products include biodegradation accelerators that ensure easy biodegradability, functional additives for enhanced barrier and antimicrobial properties, and sustainable colourants that achieve aesthetics without environmental loss (Perera et al., 2023). Novel smart additives provide the bioplastics with the ability to react to changes in their environment, and processing aids offer improved manufacturability that partially bridges the gap to production shortcomings of bioplastics. Overall, these innovations expand the application of bioplastics while directing them to meet broader sustainability goals, thus setting out a more sustainable future for material science.

3.3.1 Biobased Additives

Biobased additives are the elements playing a very significant role in developing eco-friendly bioplastics sourced from renewable biomass resources such as plant material and agricultural byproducts. Additives enhance toughness, flexibility, and biodegradation of bioplastics, and as green products become the demand today, biobased additives are an excellent opportunity that not only makes the properties of bioplastics better but also aligns with ecological objectives. Such bioplastics are effective and sustainable, consisting of materials including starch, cellulose, lignin, and natural fibres, which contribute towards a more circular economy-possibly with much less fossil-fuel dependency (Rosenboom et al., 2022).

Among the biobased additives, one of the most abundant in bioplastics is starch. It is obtained from crops like corn, potatoes, and tapioca. When used blended with biopolymers like PLA (polylactic acid), starch enhances biodegradability while reducing the production cost with better mechanical properties suitable for packaging applications. Currently, cellulose nanocrystals (CNCs) recovered from the plant cell wall acquire significance because of their ultra-high strength and sturdiness.

Such properties have the capacity for significantly enhancing the mechanical and thermal stability of bioplastics in a way that can be suitable for light-weighted applications with maintaining biodegradability (Poulose et al., 2022).

Lignin, being of natural origin and occurring within plant cell walls, is fast becoming a more common component in bioplastic production as a way of bioplastic reinforcement and imparting UV stability (Mariana et al., 2021; Yang et al., 2019). Its chemical composition enables it to withstand degradation while at the same time taking part in the process that makes them biodegradable and also supports the interest of waste reduction associated with agriculture as byproducts of other industries. Chitosan, derived from shells of certain crustaceans, could potentially introduce biodegradability and antimicrobial activity into packaging materials, especially foodstuffs, thus ensuring extended shelf life while also inhibiting microbial growth (Barik et al., 2024; Zhang et al., 2023). Biobased plasticizers can replace traditional ones such as petroleum-based, like citrates and epoxidized vegetable oils that may make it more flexible yet processable without losing its biodegradable ability (Vieira et al., 2011). Other plant-based natural fibres such as hemp, jute, or flax can also be used to introduce more mechanical strength as well as reduce weight while making the sustainable composite material applicable in different fields, including automotive and consumer goods products.

Fillers and other adhesive properties can be added to plant-based proteins like soy and wheat gluten. This may give bioplastics not only a better structure but also biodegradability. Natural preservatives through essential oils with antimicrobial properties can be incorporated for enhanced food packaging functionalities (Chotiradsirikun et al., 2023).

The challenges in the production of bioplastics constitute a major aspect against the promise of a new sustainable future for traditional petroleum-based plastics (Serrano-Aguirre & Prieto, 2024). The use of additives entails a series of challenges related to compatibility, costs, performance gaps, sourcing, and sometimes even regulation.

3.3.2 Compatibility

Compatibility between biobased additives and base polymers is the major challenge during the preparation of bioplastics. Most renewable resource-based bioplastics have different properties related to chemical and physical behaviour from those of conventional petroleum-based plastics. Biobased plasticizers and fillers are known to increase the chance of phase separation when added. Incompatibility can lead to reasons like decreased mechanical strength, brittleness, or poor processing behaviour. A considerable amount of work in formulating needs to be done for effective integration through modification of the additives' surface properties or the use of compatibilizers which enhance the interaction between the dissimilar materials, and otherwise, the final product will not meet the expected performance requirements of the intended application (Imre et al., 2013).

3.3.3 Regulatory Compliance

Biobased materials are no easy task to navigate through the regulatory landscape. Standards and certifications vary from region to region and country to country, often due to environmental policies, safety standards, and market conditions related to bioplastics (Jayakumar et al., 2023). All these fragmentation leads to confusion and higher compliance costs since manufacturers will have to undertake extensive testing and documentation to comply with myriad requirements. The whole process of certification is also protracted, thereby incurring delays in product development and market entry. To encourage innovation and ease the adoption of biobased additives, a more harmonized framework could be implemented as needed, along with continued communication between manufacturers and regulatory authorities.

3.3.4 Sourcing and Supply Chains

Sourcing and supply chains pose different challenges for biobased materials. Good-quality additives made from biobased materials may not be always available because of market demand for raw materials, among other reasons related to agricultural variability or certain climate-related conditions (Dams et al., 2023). Renewable sourcing would have to ensure responsible food production and avoid causing deforestation. There is a need for real collaboration with farmers, manufacturers, and appropriate regulatory bodies in building a reliable chain of supply that is sustainable. Building such infrastructure requires time and investment and can further limit the fast adoption of biobased materials in the industry.

3.4 NANOTECHNOLOGY IN BIOPLASTIC ADDITIVES

Common nanotechnology applied for bioplastic additives are different types of nanomaterials to ensure improvement in properties and performance of bioplastics (Abioye et al., 2021). The most commonly used nano clays are montmorillonite and kaolin that enhance mechanical strength, barrier properties, and thermal stability (Bangar et al., 2023). Clay nanoparticles tend to build a layered structure in the polymer matrix while increasing rigidity and reducing gas permeability that make them particularly efficient in food packaging applications (Perera et al., 2024).

Carbon nanotubes (CNTs) present excellent mechanical properties and conductivity and, hence, enhance tensile strength as well as impact resistance when added to bioplastics (Ramontja et al., 2009). They also enhance thermal conductivity making bioplastics versatile for application in electronics as well as automobile components (Lagaron et al., 2011). Silica nanoparticles are another option that improves both mechanical as well as thermal properties besides UV resistance to increase application range (Saleh et al., 2022). Silica has low toxicity and the potential for enhancing the biodegradability, and such an attractive choice for green formulations is considered (Kong et al., 2023; de Azêvedo et al., 2021).

Bioderived CNCs are biodegradable and can reinforce bioplastics without compromising their eco-advantages. Tensile strength, flexibility, and thermal stability of these CNCs ensure that all the features of the final products stay intact with their eco-friendly nature (Muhammad et al., 2019; Rismawati et al., 2023). Metal nanoparticles such as Ag and ZnO were added for their antimicrobial activity against microbial growth, thus fixing their place as future promising candidates in food packaging and medicinal applications (Navaf et al., 2023; Zaim et al., 2023). Their properties can improve the storage lifetime of the material and relate to sustainability objectives. Finally, graphene oxide possesses better mechanical, thermal, and electrical properties. This makes it potentially stronger than the original bioplastics and enhances the barrier properties (Ajaj et al., 2024). Functionalization is also possible to create selective properties, such as increased hydrophobicity or electrical conductivity. These nanomaterials together enable the manufacturing of high-performance bioplastics but also open opportunities for innovative applications and can contribute to more sustainable development of high-performance materials (Islam et al., 2024; Atiwesh et al., 2021).

Adding bioplastics with sustainable, renewable additive sources is an important step forward in trying to mitigate some of the environmental impacts associated with traditional plastics. Materials applied include natural fibres like hemp and flax, crop-based starch, cellulose from plants, and proteins based on agricultural byproducts (Oluwasina et al., 2021). All these materials have enhanced the mechanical properties, thermal stability, and biodegradability of bioplastics (Elfaleh et al., 2023). As the volumes of renewable additives grow, so does the reliance lessened on fossil fuels, and it benefits more directly the local economies. However, challenges such as these are the quality and performance of these natural materials have to be maintained constant, and their use could only go wider if manufactured by bigger industrial scales, and constantly changing regulatory requirements will pose a challenge (Zhao et al., 2023). Even with the above-mentioned challenges, the inclusion of sustainable additives marks a great milestone towards green materials in a circular economy and is finally addressing a global concern regarding pollution in plastic (Acharjee et al., 2023; Mastrolia et al., 2022).

3.5 PERFORMANCE ENHANCEMENT THROUGH ADDITIVES

3.5.1 Mechanical Property Improvements

The inclusion of biobased additives greatly enhances the mechanical properties of bioplastics, and therefore, they compete with conventional plastics. Some of the best reinforcements are natural fibres-hemp, flax, jute, etc. (Awal et al., 2015). Their addition can increase tensile strength by 30–50%, which generally increases overall durability and stiffness for applications requiring robust materials. The starch, which is often used as a plasticizer, has so far given bioplastics higher flexibility and elongation at break, so they better resist the mechanical stress without cracking or breaking. Therefore, they are better applied in applications with dynamics, namely flexibility, particularly.

CNCs enhance the mechanical performance notably by increasing tensile strength and thermal stability considerably. The nanoscale dimensions of CNCs allow for uniform dispersion within the polymer matrix and thus grant greater mechanical properties without sacrificing biodegradability. In addition to that, protein-based additives sourced from soy and wheat also add a bit to the improved mechanical performance (Bhaskar et al., 2023). They do increase strength and add that with even more functionalities, such as antimicrobial properties.

Overall, strategic applications of biobased additives significantly improve the mechanical properties of bioplastics, thus allowing further expansion of the range of their potential applications and supporting a shift towards more sustainable materials.

3.5.2 Thermal Stability and Processing Benefits

For high temperature usage applications such as automotive parts, sterilized packaging, thermal stability is important. Apart from providing better mechanical properties, the addition of biobased additives like flax or hemp fibres also has a positive impact on heat deflection temperatures, thus enhancing the ability of bioplastics to operate at elevated temperatures. Other nanomaterials would include silica nanoparticles and CNCs, which can help in dispersing heat through the polymer matrix for minimizing thermal degradation risks. Thermally stressed bioplastics may also be preserved through oxidative degradation by using natural antioxidants as sources of biobased thermal stabilizers.

Biobased additives significantly advance the ability to manufacture. Other natural plasticizers, such as glycerol and sorbitol, make it easier for bioplastics to be stretched and lower their viscosity to be easier for processing and shaping in the production (Benitez et al., 2024). Biobased compatibilizers improve the dispersion and adhesion of materials that are part of the polymer matrix which leads to a more homogeneous final product and smoother workflow in the process (Chotiradsirikun et al., 2023). These improvements not only enhance the performance of bioplastics but also bring wider implications in terms of environmental and economic changes. Increased applicability of bioplastics cuts across several sectors, which contributes to reduced fossil-fuel reliance and lower emissions of greenhouse gases. More efficient processing can also lead to a decrease in production cost, therefore making the bioplastics competitive with the usual plastics. As the research and innovation in this area continue to advance, bioplastics are promising high-performance, more environmentally friendly alternatives to conventional plastics: building towards an increasingly sustainable materials future.

3.5.3 Barrier Properties and Applications in Packaging

Barrier properties refer to the prevention of gas, moisture, and other substances' permeability through the material (Marano et al., 2022). In the context of packaging, it is critical to avoid spoilage as well as maintain product integrity. In this regard, natural fibres such as cellulose and chitosan can promote those properties by creating a denser network within the polymer matrix, thus reducing the free volume available for gas and moisture transmission (Wu et al., 2021). Other nanomaterials, like nano clays and silica nanoparticles, open original possibilities: these nano-objects have a very high surface area, which reduces the permeabilities in gases and makes them excellent candidates for nanocomposites (Homaeigohar et al., 2020). The generally used plasticizers, for flexibility adjustment, may be a two-edged sword: if used reasonably, they allow an overall improvement of the polymer matrix structure, which should translate into better barrier properties. Additional moisture and oxygen barrier capability may be added by copolymerizing bioplastics with other polymers, such as polyvinyl alcohol or PLA (Ilyas et al., 2022).

Improvements in barrier properties allow for more functionality and wider application possibilities of bioplastics in food packing, medical applications, electronics, etc. in order to minimize the dependency on traditional plastics and contribute to the agenda towards sustainability. Improved barrier properties can extend shelf life and avoid waste through reduced product spoilage. However, variability in the performance of natural additives is still a challenge, together with the complexity in optimizing enhanced barrier properties within a background of overall mechanical and thermal properties. Further research should thus optimize formulations and exploit innovation in nanotechnology and material science to produce novel additives enhancing barrier properties while maintaining the sustainability of bioplastics. As the industry keeps innovating in response to these challenges, the promise of bioplastics becomes increasingly promising for providing effective alternatives to traditional plastics, forking a trail to the future of an environmentally friendly society (Ghosh et al., 2021).

3.6 ENVIRONMENTAL IMPACT AND BIODEGRADABILITY OF ADDITIVES

Bioplastics are very often derived from renewable feedstocks such as cornstarch or sugarcane and are bio-engineered to offer a more sustainable alternative to traditional petroleum-based plastics (Pooja et al., 2023). However, additives in these materials may be hugely influencing factors on their overall environmental footprint. This may negate any sustainability from the bioplastic itself. Such production processes may be highly energy intensive and hence damaging to the environment, thereby offsetting the advantages of bioplastic as the base material (Folino et al., 2020).

The end-of-life behaviour of bioplastic additives also holds a vital place in its ecological footprint. Though several additives are developed with full biodegradability potential, the other ones decompose poorly, which in return causes pollution and aggravates plastic waste (Folino et al., 2023). It is highly scary when the additive is of non-biodegradable kinds because these persist in nature, damage the ecosystems as well as human health (Gilani et al., 2023). These also pose risks to both human and environmental well-being due to the release of toxic substances during the production, usage, or disposal phases of bioplastics.

The chemical nature of the bioplastic additive and compatibility with the bioplastic matrix have a significant influence on their biodegradability. The general biodegradation of the composite material is enhanced through easier decomposition in suitable environments by natural polymers among others. On the other hand, synthetic additives impede the process by complicating the biodegradation pathways. Environmental conditions like temperature, moisture, and microbial activity contribute to a great extent to the speed of biodegradation for bioplastics and additives (Agboola et al., 2017). Wrong disposal contributes to the inhibition of the biodegradation process. As a result, this material will have an extended existence in the landfills or natural environment.

The other factors considered by manufacturers to determine the compostability of bioplastics and their additives include regulatory standards like ASTM D6400 and EN 13432 (Yu et al., 2024). Complying with such standards would ensure the material degrades well in the conditions given by industrial composting, thereby reducing their impacts on the environment. In a nutshell, bioplastic additives can indeed upgrade the properties of bioplastics, but environmental impacts and biodegradability have to be seriously taken into account. Bioplastics can maximize the ecological benefits and minimize adverse consequences associated with their use by choosing sustainable additives and innovative practices during their development and disposal.

3.7 BIODEGRADABILITY AND COMPOSTABILITY: THE ROLE OF ADDITIVES

Biodegradability and compostability are the critical attributes that define the level of environmental effect bioplastics have and their sustainability level. Additives are important since they will enhance these features, making the bioplastics more functional and environmentally friendly (Moshood et al., 2022). Biodegradability is the ability of a material to break down into natural substances such as water, carbon dioxide, and biomass through microorganisms. In this case, compostability is specifically related to a material's capability to decompose within the composting environment with higher temperatures, moisture, and microbial activity (Samir et al., 2022).

Additives play a significantly vital role in determining the degradation rate and efficiency of the process. Natural polymers are excellent additives that ensure an adequate food supply for microbes that facilitates microbial activity (Emadian et al., 2017). That breaks down faster, eliminating chances of bioplastics' presence in the environment for a long time. There are specific additives engineered specifically to enhance composting capability by facilitating decomposition through the action of composting organisms (Crema et al., 2024). Specific enzymes may thus be added to increase the rate of decomposition, hence supports good quality (Rahmati et al., 2024).

While improving the biodegradability and compostability of bioplastics, the additives must ensure that they do not sacrifice or even improve the mechanical and thermal properties of bioplastics (Haider et al., 2019). Good additives would exhibit flexibility, strength, and toughness, which would then make the bioplastics suitable for a wide array of applications, from packaging to consumer products. It is a great balance which would be necessary to render bioplastics viable in the marketplace. Other environmentally friendly plasticizers from renewable sources enhance the flexibility and processing characteristics of bioplastics.

3.8 TOXICITY OF ADDITIVES

The more bioplastics and other alternative materials grow, the more important it is to discuss environmental concerns: the safety and sustainability of bioplastics. The two major problems are toxicity from additives and the leaching of these materials into the environment. This understanding will be important in any attempt to assess the full impact of bioplastics (Boisseaux et al., 2023).

Many of the bioplastics are formulated with additives that provide some further improvements in flexibility, durability, and resistance to degradation. However, these additives can be toxic to human beings and the environment. For instance, certain synthetic plasticizers, stabilizers, and colourants emit toxins at stages in production, usage, or disposal. These may find their ways into the environment and affect water and soil quality, killing plants and other animal species. Toxicity concerns are only applicable in food packaging applications wherein harmful additives may leach through into food products, which could poison humans (Gunaalan et al., 2020) (Shaikh et al., 2021). Regulatory bodies usually provide very limited guidelines on the types of additives permissible in food-contact materials; however, the complexity involved in chemical interactions makes assessment quite challenging.

3.8.1 Leaching of Additives

Leaching is the process in which additives are leached out of the material into their surroundings through chemical agents, moisture, or heat. This is more alarming for biodegradable plastics since their aim is to create materials with no residual harm that can be harmful in nature (Nayanathara et al., 2024). When additives leach into soil or water systems, they can cause ecosystem disruptions and harm organisms. For instance, toxic substances can change aquatic life by causing reduced biodiversity and changes in food chains. Leached additives in agricultural settings negatively affect soil health and crop growth, complicating the issue of sustainability further (Maddela et al., 2023).

3.8.2 Environmental Risk Mitigation

Addressing the above concerns would require the development of safer additives, which are less harmful to the environment and human health (Saharan et al., 2022). Scientists have been exploring biodegradable, naturally occurring additives that could be used to enhance the properties of bioplastics without subjecting some harmful substances to the environment (Cruz et al., 2022). Testing for toxicity and leaching potential should also now serve as the foundation for new formulations of bioplastics (Okoye et al., 2022).

Regulatory frameworks also have an important role in controlling the application of additives. For instance, strict regulations can ensure that only safe materials are used on bioplastics, especially when applying to foods or other sensitive applications. More transparency in supply chains and stricter tests for leaching and toxicity might increase consumers' confidence and promote even more responsible practices.

3.9 APPLICATIONS OF BIOPLASTIC ADDITIVES

Bioplastic additives are critically essential for various applications in enhancing the performances and sustainability of bioplastics (Figure 3.2). Such additives enhance the barrier properties, flexibility, and sealability of food packaging, so that the materials maintain freshness and safety (Harnkarnsujarit et al., 2021; Siddiqui et al., 2024). Natural additives, such as starch and cellulose, help to enhance the biodegradability, and antimicrobial agents could limit bacterial growth to enhance the shelf life of perishable products. Bioplastic additives have many applications in agriculture to make films for mulching, crop protection, and greenhouse coverings biodegradable by including natural waxes and plasticizers that improve the mechanical properties and performance.

3.9.1 Packaging Industry

Bioplastic additives have been integral, especially in food and beverage applications, in order to improve performance, safety, and sustainability in the packaging sector (Hussain et al., 2024). Their functions include improving barrier properties to water, oxygen, and light, the greatest demand to extend the shelf-life of susceptible products (Rahman et al., 2019). Thus, nano clays, waxes, and other biopolymers were added to bioplastics for forming effective barriers that decrease the spoilage (Fletcher et al., 2024). Many bioplastics are also designed to be biodegradable: with starch and cellulose additives, they safely break down after being thrown away. This characteristic has great appeal in environmentalist circles and for companies committed to reducing plastic waste in general (Zhao et al., 2024).

Some additives impart packaging with antimicrobial properties for the safety of foods. Natural agents including essential oils, silver nanoparticles, and plant extracts have been used to avoid

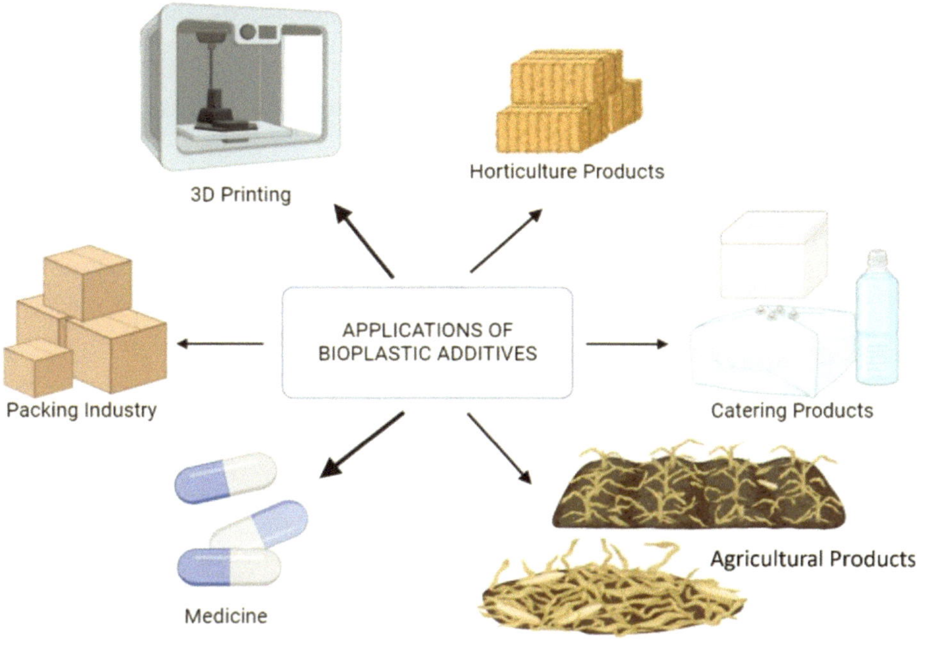

FIGURE 3.2 Applications of bioplastic additives.

growth of bacteria and fungi and are highly useful in packing perishable products like meat and dairy products (Mishra et al., 2018). Other plasticizers, like citrate, which is a biodegradable one used for flexible and processable packaging, could improve the mechanical strength of bioplastics while it withstands temperature fluctuation during storage and transportation (Mohammed et al., 2022; Narancic et al., 2020).

Additive technologies further enhance aesthetics in bioplastic formulations (Hussain et al., 2024). Packaging designs often play an imperative role in consumer preferences for their packaging. Thus, additives are useful in adding colours, gloss, and/or clarity to bioplastics for enhancing the product presentation. It makes them not only visually appealing at shop displays but also as brand identities emphasizing sustainability (Donkor et al., 2023; Zhao et al., 2023). Consumers will pay more for this product if packaged with attractive environment-friendly materials. Bioplastic additives are selected and compounded to the highest food safety standards so that they do not leach harmful substances into food products. This is critical in consumer confidence building, such that packaging fulfils its basic function of protecting food without undermining its safety.

3.9.2 Medical Applications: Additives for Biocompatibility

Bioplastic additives are one of the most innovative and critical components in medical applications, which provide creative solutions for improving the performance and sustainability of medical devices, packaging, and healthcare products (Arif et al., 2023) (Ibrahim et al., 2021). One of the prime requirements for medical applications is biocompatibility, and it is one of the aspects that these additives improve greatly in bioplastics, such as PLA and PHA, ensuring that these materials do not provoke negative biological reactions (Kong et al., 2023).

Other added plasticizers and impact modifiers can increase the tensile strength and elasticity of mechanical properties to enhance the material so that the bioplastic can be used as flexible tubing or sutures. In addition, an added antimicrobial agent prevents bacterial and fungal growth; hence, it will be suitable for wound dressing and medical instrument applications wherein sterility maintenance is vital (DeStefano et al., 2020).

Bioplastic additives also can make it possible for a design of drug delivery systems having controlled drug release and therefore will be in the capacity to gradually as well as constantly release the therapeutic agent with time (Kumar et al., 2023). Such kind of design may even come particularly helpful in the area of oncology (Gheorghita et al., 2021). In the midst of these benefits, however, are issues such as the very complex regulation for medical devices and how bioplastics production and degradation processes can be more environmental-friendly (Jabeen et al., 2024). Bioplastic additives have been an important breakthrough in medical applications and set the course to a safer and more sustainable solutions in healthcare, as also in response to the increasingly urgent emphasis of the industry towards environmental responsibility (Tsung et al., 2023).

3.9.3 Agricultural Applications: Additives for Controlled Degradation

Bioplastic additives increasingly appear to be recognized with a future in agricultural uses. They may provide agricultural enterprises with innovative solutions whereby it can be more environmentally friendly, increase crop production, and have a lesser adverse effect on the environment (Rosseto et al., 2019; Park et al., 2024). Biodegradable mulch films are some of the key applications. They help in covering the soil to achieve some weed suppression, maintain moisture, and regulate the temperatures (Campanale et al., 2024). These films get the benefits from bioplastic additives. They will enhance the properties of such films, helping them in effective performance. They can naturally degrade inside the soil; thus, they reduce plastic wastes and improve health in soils (Huang et al., 2023).

Bioplastic additives are also added to the seed coatings so that there is an increased germination and growth from the side of seeds while providing necessary nutrients, fighting pests, and gradually releasing the water retaining property (Lakhiar et al., 2024). The use of bioplastic additives in controlled release fertilizers makes fertilization more environmentally friendly by synchronizing nutrient availability to the needs of plants while minimizing environmental runoff. Other environment-friendly alternatives of using plastic pots and containers include making biodegradable pots and containers from bioplastics. They will break down in the soil and gradually fertilize it as they disintegrate. Bioplastic additives also enhance crop protection products such as biodegradable nets and covers that shield crops from pests but are far less damaging to the environment than regular plastics (Hofmann et al., 2023). More uses involve the integration of bioplastics into the soil amendment, which endows it with a well-structured soil and facilitates a proper population of desirable microbes. In water-limited regions, these could be employed in creating an absorbing gel that holds soil moisture and enhances the crop's yield during dry weather conditions (Patra et al., 2022).

As mentioned above, bioplastic materials are used extensively in agricultural practices in the form of mulching films, seedling trays, and irrigation tubing (Chia et al., 2020). However, persistence of plastic waste poses severe environmental risks such as soil degradation, harm to wildlife, and disruption of agricultural ecosystems. Thus, controlled degradation additives can help in these issues by allowing bioplastics to degrade at an appropriate rate, aligning with crop cycles and avoiding soil contamination.

Additional surfaces for microbial colonization by starch and cellulose fillers lead to fast degradation of the polymer matrix. Enzymatic additive acts as biological catalyst by specifically breaking polymer chains with microorganisms into smaller fragments that result in enormous shortening of bioplastic life expectancy in well-suited environments (Islam et al., 2024).

The use of these additives also offers some advantages including environmental sustainability since the product degrades on time, enrichment of soil with organic matter, and easier removal processes in agricultural fields since they don't necessarily absorb much labour.

3.10 CHALLENGES AND FUTURE DIRECTIONS

3.10.1 Limitations of Additives in Bioplastics

Incorporation of additives, in order to enhance the performance and degradation of bioplastics, poses numerous challenges to the effectiveness of bioplastics (Negrete-Bolagay et al., 2024). For example, using biodegradable fillers like starch to enhance the biodegradability reduces the mechanical strength, thermal stability, and barrier properties. Such trade-off might limit the use of bioplastics in industries with persistent material applications, such as packaging (González et al., 2023) and construction. Performance of enzymatic additives could also be less predictable, as such performance is a function of how the environment affects such additive performance-temperature and moisture levels, among others (Kumari et al., 2023). The core problem, then, is one of striking a balance between enhanced biodegradability without diminishing desirable physical properties.

Another drawback that is associated with bioplastic additives is that their degradation rates can be variable (Imre et al., 2013). Many factors of the environment are associated with a change in degradation rate. For example, temperature, humidity, and microbial activity are some few factors of the environment which are involved in this aspect. In this respect, for instance, pro-oxidant additives may degrade slowly because of the design requirement of fast degradation under many conditions. This is due to unpredictable breakdowns of such additives. Bioplastics, therefore, could not be used for all agricultural practices, where their use will have to match degradation patterns to crop cycles for easy removal. Without reliable degradation performance, the intended environmental benefit of bioplastics can be undermined, creating problems with lingering plastic waste if the materials do not break down (Ali et al., 2023).

Yet another environmental concern regarding additives is their impact. Many of them are engineered to be more biodegradable, but the fuller environmental implications are not always well understood (Chamas et al., 2020). For instance, chemical additives could break down into dangerous byproducts that could be transferred into soil and water and have a deleterious impact on ecosystems and human health (Adetunji et al., 2023).

Economic factors are another big limitation for bioplastic additives. The use of good-quality additives, including specific enzymes or advanced pro-oxidants, would normally raise the cost of production in such a way that the bioplastics end up being less competitive than their conventional plastics counterparts based on price-sensitive markets (Moshood et al., 2022). Specialized production techniques are also required to introduce and distribute additives in the right manner to increase the complexity of the manufacturing process and, accordingly, the burden of the cost.

Lastly, regulatory challenges also pose significant barriers for the industry of bioplastics. Many regions lack a well-defined regulatory framework governing the usage of bioplastics and their additives, thereby it may hamper the innovation and commercial viability of a bioplastic product. The manufacturer may not have known which additives are allowed, and which testing protocols apply or which ones do not apply. This creates consumer confusion regarding the environmental claims tied to these products. The appropriate response to these regulatory challenges, while there is growth and assurance towards the consumers, would be continuing developments of such an industry (Fletcher et al., 2021).

3.10.2 Future Trends in Bioplastics

Bioplastics can be considered as an environmentally friendly alternative to traditional, petroleum-based plastics and are produced from renewable resources (Serrano et al., 2024). Bioplastic additives improve performance and biodegradability, and several trends are emerging in the near future. Advanced biodegradable materials become one of the most impressive trends in this area. The different natural and synthetic biodegradable additives, including biopolymers, starches, and cellulose, which improve the mechanical properties of bioplastics, also degrade at higher rates in different environments (Samir et al., 2022). More effective biodegradable additives, that break down fast under certain conditions and hence reduce the environmental impact, are to be found in the new innovations in blending techniques and formulation science (Atiwesh et al., 2021).

Another exciting area of focus can be the use of enzymatic additives in bioplastics. Embedding particular enzymes in a material may allow manufacturers to enable degradation at a controlled rate that has been determined, enabling materials to break down as needed. Future enzyme engineering might make possible more stable, efficient enzymes that work under broader ranges of environmental conditions. A growth trend in the use of biobased additives from agricultural byproducts and waste materials also adds value to the waste streams through the circular economy, reducing dependence on fossil-based energy sources.

Nanotechnology is going to play a critical role in transforming bioplastic additives. It is going to upgrade the physical and mechanical properties of bioplastics. Nano-fillers, for example, nano clay or CNCs, show promise in providing strength, thermal stability, and barrier properties without raising weight too much (Ferreira et al., 2018). Future studies will most likely focus on the optimum dispersion of these nano-additives in bioplastics so that the properties are homogeneously upgraded. Nanotechnology will also combine with biodegradable material to produce high-performance bioplastics, which can be eco-friendly (Gupta et al., 2024).

Increased awareness on the side of the consumers with regard to environmental issues creates pressure on manufacturers to come up with sustainable materials. This should further boost the regulation development of clearer labelling and standards of bioplastics and additives (de Sousa et al., 2023). Regulations on the need for biodegradability and environmental impacts of additives are also anticipated to make a stronger stand in terms of their demand for sustainable practices. Manufacturers will likely invest in developing innovative bioplastic additives that will meet regulatory standards and align with customer expectations for sustainability and safety. Another characteristic trend would be the integration of bioplastics in circular economy. This simply means managing and designing products as well as materials along the length of their life cycles using some principles like reuse, recycling, and composting (Tennakoon et al., 2023). Such additives that enhance recyclability or compostability for bioplastics will be crucial, and future innovations may target additive developments in helping sort bioplastics through the recycling process at the point of waste management and keep such materials from becoming waste products within landfills.

3.11 REGULATORY AND MARKET CONSIDERATIONS

Important aspects of regulations deal with safety and environmental protection towards the environment at hand while building consumer confidence. Based on this perspective, many regulatory bodies all over the world, of them, U.S. Environmental Protection Agency (EPA), European Food Safety Authority (EFSA) regulates acceptability of additives in bioplastics, mainly focusing on utilization, along with safety, assessment regarding the environment, and the standards of biodegradable properties. For instance, the European Union has under it Regulation (EU) 2019/1020 that addresses all plastic products provided for sale so that their safety and environmental compatibility can be assured. Similarly, for the United States, their federal law regulates bioplastic and additives used into the product especially when utilized as food contact. As structures of future policy are designed, they may focus more and more on transparency, compelling manufacturers to provide safety and environmental impact information to users.

Consumer awareness and perception also form a very significant part of the dynamics of the bioplastics market (EFSA, 2023). More and more consumers are being made aware of the impacts plastics have on the environment; they tend to have little information regarding bioplastics and its additives. This will make them question the legitimacy of some of the producer claims on biodegradability and sustainability (Weinrich., 2023). All these can be achieved through clear labelling and transparency in the materials used, with bioplastic additives. Both regulators and market participants will have trust in sustainable products. As manufacturers try to create products with environmental responsibility, the additive itself needs to be sustainable and biodegradable. There is a revived interest in the focus by the regulatory agencies on the life cycle impacts of bioplastics and its additives. This is much further than sourcing, production, and end-of-life options.

Equally relevant to this topic is the biodegradability performance of the additives. While it would add biodegradability without impacting the ultimate strength, durability, and functionalities for performance of the end product, sustainability requirements on the one hand conflict with market aspirations, potentially making such pressures forces for innovative solutions that will balance out two seemingly conflicting requirements (Negrete et al., 2024). Thus, this is going to demand regulatory considerations as well as market concerns related to incorporating bioplastic additives into general use. Where the demand for sustainable material is rising, so do the regulations of safety and environmental concerns with transparency (Silva et al., 2023). The manufacturer has to innovate, be transparent, and look forward to marketing approaches that overcome the challenges effectively. It supports a future that is sustainable while responding to consumer demand and being in line with regulatory expectations (Geissdoerfer et al., 2018).

3.12 CONCLUSION

The field of bioplastics has witnessed significant advancements in recent years, particularly in the development and incorporation of innovative additives. These advancements aim to enhance the performance, durability, and environmental benefits of bioplastics, bridging the gap between sustainability and functionality. Bioplastic additives, including plasticizers, fillers, stabilizers, and compatibilizers, are being tailored to address key challenges such as mechanical strength, thermal stability, and biodegradability. One of the most notable advancements is the use of natural and biobased additives derived from renewable resources. These additives not only improve the properties of bioplastics but also align with the overarching goal of reducing dependence on fossil fuels and minimizing environmental impact. For example, plant-based plasticizers have been developed to enhance flexibility, while mineral fillers improve mechanical strength without compromising biodegradability. Additionally, nanotechnology has introduced nanoscale additives that significantly enhance barrier properties and thermal stability, expanding the application scope of bioplastics in packaging, automotive, and biomedical industries.

Moreover, the integration of functional additives, such as antimicrobial agents and antioxidants, has further elevated the utility of bioplastics, particularly in food packaging and medical applications. These multifunctional additives ensure product safety and longevity, meeting the demands of diverse industries. Despite these advancements, challenges remain, including the cost of production, compatibility issues with conventional plastics, and limited scalability. Ongoing research aims to optimize additive formulations to address these barriers, paving the way for wider adoption of bioplastics. The progress in bioplastic additives underscores a crucial shift towards sustainable materials without compromising performance. By harnessing innovative additives, the bioplastics industry is well-positioned to contribute significantly to global sustainability efforts while meeting the demands of modern applications.

REFERENCES

Abang, S., Wong, F., Sarbatly, R., Sariau, J., Baini, R., & Besar, N. A. (2023). Bioplastic classifications and innovations in antibacterial, antifungal, and antioxidant applications. *Journal of Bioresources and Bioproducts*, 8(4), 361–387.

Abe, M. M., Branciforti, M. C., & Brienzo, M. (2021). Biodegradation of hemicellulose-cellulose-starch-based bioplastics and microbial polyesters. *Recycling*, 6(1), 22.

Abioye, A. A., Afolalu, S. A., Yusuf, O. O., Emetere, M. E., Ongbali, S. O., Mfon, U., Samuel, O. D., & Olayinka, A. (2021). Nanotechnology applications in synthesis of bioplastic. *IOP Conference Series: Materials Science and Engineering*, 1107(1), 012190.

Abu Ubaidah, A. S., Mohamad Puad, N. I., Hamzah, F., Azmi, A. S., & Ahmad Nor, Y. (2024). Synthesis of bioplastics and the effects of additives on the mechanical, thermal and biodegradable properties. *Polymers from Renewable Resources*, 15(4), 430–490.

Acharjee, S. A., Bharali, P., Gogoi, B., Sorhie, V., Walling, B., & Alemtoshi. (2023). PHA-based bioplastic: A potential alternative to address microplastic pollution. *Water, Air, & Soil Pollution*, 234(1), 21.

Adetunji, A. I., Oberholster, P. J., & Erasmus, M. (2023). From garbage to treasure: A review on biorefinery of organic solid wastes into valuable biobased products. *Bioresource Technology Reports*, 24, 101610.

Agboola, O., Sadiku, R., Mokrani, T., Amer, I., & Imoru, O. (2017). Polyolefins and the environment. In *Polyolefin Fibres* (pp. 89–133). Elsevier.

Ahsan, W. A., Hussain, A., Lin, C., & Nguyen, M. K. (2023). Biodegradation of different types of bioplastics through composting—A recent trend in green recycling. *Catalysts*, 13(2), 294.

Ajaj, Y., AL-Salman, H. N. K., Hussein, A. M., Khaleel Jamee, M., Abdullaev, S., Omran, A. A., Morad Karim, M., Abdulwahid, A. S., Mahmoud, Z. H., & kianfar, E. (2024). Effect and investigating of graphene nanoparticles on mechanical, physical properties of polylactic acid polymer. *Case Studies in Chemical and Environmental Engineering*, 9, 100612.

Ali, S., Isha, & Chang, Y.-C. (2023). Ecotoxicological Impact of Bioplastics Biodegradation: A Comprehensive Review. *Processes*, 11(12), 3445.

Almeida, S., Ozkan, S., Gonçalves, D., Paulo, I., Queirós, C. S. G. P., Ferreira, O., Bordado, J., & Galhano dos Santos, R. (2022). A brief evaluation of antioxidants, antistatics, and plasticizers additives from natural sources for polymers formulation. *Polymers*, 15(1), 6.

Arif, Z. U., Khalid, M. Y., Noroozi, R., Hossain, M., Shi, H. H., Tariq, A., Ramakrishna, S., & Umer, R. (2023). Additive manufacturing of sustainable biomaterials for biomedical applications. *Asian Journal of Pharmaceutical Sciences*, 18(3), 100812.

Atiwesh, G., Mikhael, A., Parrish, C. C., Banoub, J., & Le, T.-A. T. (2021). Environmental impact of bioplastic use: A review. *Heliyon*, 7(9), e07918.

Awal, A., Rana, M., & Sain, M. (2015). Thermorheological and mechanical properties of cellulose reinforced PLA bio-composites. *Mechanics of Materials*, 80, 87–95.

Azman, M. A., Asyraf, M. R. M., Khalina, A., Petrů, M., Ruzaidi, C. M., Sapuan, S. M., Wan Nik, W. B., Ishak, M. R., Ilyas, R. A., & Suriani, M. J. (2021). Natural fiber reinforced composite material for product design: A short review. *Polymers*, 13(12), 1917.

Bangar, P. S., Ilyas, R. A., Chaudhary, N. *et al.* (2023). Plant-based natural fibers for food packaging: A green approach to the reinforcement of biopolymers. *Journal of Polymers and the Environment,* **31**, 5029–5049. https://doi.org/10.1007/s10924-023-02849-3

Barik, M., BhagyaRaj, G. V. S., Dash, K. K., & Shams, R. (2024). A thorough evaluation of chitosan-based packaging film and coating for food product shelf-life extension. *Journal of Agriculture and Food Research*, 16, 101164.

Benitez, J. J., Florido-Moreno, P., Porras-Vázquez, J. M., Tedeschi, G., Athanassiou, A., Heredia-Guerrero, J. A., & Guzman-Puyol, S. (2024). Transparent, plasticized cellulose-glycerol bioplastics for food packaging applications. *International Journal of Biological Macromolecules*, 273, 132956.

Bhaskar, R., Zo, S. M., Narayanan, K. B., Purohit, S. D., Gupta, M. K., & Han, S. S. (2023). Recent development of protein-based biopolymers in food packaging applications: A review. *Polymer Testing*, 124, 108097.

Boey, J. Y., Lee, C. K., & Tay, G. S. (2022). Factors affecting mechanical properties of reinforced bioplastics: A Review. *Polymers*, 14(18), 3737.

Boisseaux, P., Hopkinson, P., Santillo, D., Smith, C., Garmulewicz, A., Powell, Z., & Galloway, T. (2023). Environmental safety of second and third generation bioplastics in the context of the circular economy. *Ecotoxicology and Environmental Safety*, 256, 114835.

Campanale, C., Galafassi, S., di Pippo, F., Pojar, I., Massarelli, C., & Uricchio, V. F. (2024). A critical review of biodegradable plastic mulch films in agriculture: Definitions, scientific background and potential impacts. *TrAC Trends in Analytical Chemistry*, 170, 117391.

Chamas, A., Moon, H., Zheng, J., Qiu, Y., Tabassum, T., Jang, J. H., Abu-Omar, M., Scott, S. L., & Suh, S. (2020). Degradation rates of plastics in the environment. *ACS Sustainable Chemistry & Engineering*, 8(9), 3494–3511.

Chan, C. M., Lyons, R., Dennis, P. G., Lant, P., Pratt, S., & Laycock, B. (2022). Effect of toxic phthalate-based plasticizer on the biodegradability of polyhydroxyalkanoate. *Environmental Science & Technology*, 56(24), 17732–17742.

Charvat, R. A. (2004). Colorants for plastics. In *Kirk-Othmer Encyclopedia of Chemical Technology*. Wiley.

Chia, W. Y., Ying Tang, D. Y., Khoo, K. S., Kay Lup, A. N., & Chew, K. W. (2020). Nature's fight against plastic pollution: Algae for plastic biodegradation and bioplastics production. *Environmental Science and Ecotechnology*, 4, 100065.

Chotiradsirikun, S., Than-ardna, B., Guo, R., Bhalla, A. S., & Manuspiya, H. (2023). Improvement of compatibility, mechanical, thermal and dielectric properties of poly(lactic acid) and poly(butylene adipate-co-terephthalate) blends and their composites with porous clay heterostructures from mixed surfactant systems. *Polymers for Advanced Technologies*, 34(2), 702–716.

Costa, A., Encarnação, T., Tavares, R., Todo Bom, T., & Mateus, A. (2023). Bioplastics: Innovation for green transition. *Polymers*, 15(3), 517.

Crawford, R. J. (1998). General properties of plastics. In *Plastics Engineering* (pp. 1–40). Elsevier.

Crema, A., Dinelli, E., Fabbri, E., Galletti, P., Greggio, N., Lastella, V., Parodi, A., Pasteris, A., Pedrizzi, M., & Samorì, C. (2024). Additives in bioplastics: Chemical characterization, migration in water and effects on photosynthetic organisms. *Science of the Total Environment*, 955, 177205.

Cruz, R. M. S., Krauter, V., Krauter, S., Agriopoulou, S., Weinrich, R., Herbes, C., Scholten, P. B. v., Uysal-Unalan, I., Sogut, E., Kopacic, S., Lahti, J., Rutkaite, R., & Varzakas, T. (2022). Bioplastics for Food Packaging: Environmental Impact, Trends and Regulatory Aspects. *Foods*, 11(19), 3087.

Dams, B., Maskell, D., Shea, A., Allen, S., Cascione, V., & Walker, P. (2023). Upscaling bio-based construction: challenges and opportunities. *Building Research & Information*, 51(7), 764–782.

de Azêvedo, L. C., Rovani, S., Santos, J. J., Dias, D. B., Nascimento, S. S., Oliveira, F. F., Silva, L. G. A., & Fungaro, D. A. (2021). Study of renewable silica powder influence in the preparation of bioplastics from corn and potato starch. *Journal of Polymers and the Environment*, 29(3), 707–720.

de Sousa, F. D. B. (2023). Consumer awareness of plastic: An overview of different research areas. *Circular Economy and Sustainability*, 3(4), 2083–2107.

Deanin, R. D. (1975). Additives in plastics. *Environmental Health Perspectives*, 11, 35–39.

DeStefano, V., Khan, S., & Tabada, A. (2020). Applications of PLA in modern medicine. *Engineered Regeneration*, 1, 76–87.

Donkor, L., Kontoh, G., Yaya, A., Bediako, J. K., & Apalangya, V. (2023). Bio-based and sustainable food packaging systems: relevance, challenges, and prospects. *Applied Food Research*, 3(2), 100356.

EFSA Scientific Committee (SC), Bampidis, V., Benford, D., Bragard, C., Hernández-Jerez, A., Bennekou, S. H., Koutsoumanis, K. P., Lambré, C., Machera, K., & Mullins, E. (2023, October 21). Guidance on protocol development for EFSA generic scientific assessments. *EFSA* Journal, 21(10), e08312.

Elfaleh, I., Abbassi, F., Habibi, M., Ahmad, F., Guedri, M., Nasri, M., & Garnier, C. (2023). A comprehensive review of natural fibers and their composites: An eco-friendly alternative to conventional materials. *Results in Engineering*, 19, 101271.

Emadian, S. M., Onay, T. T., & Demirel, B. (2017). Biodegradation of bioplastics in natural environments. *Waste Management*, 59, 526–536.

Eslami, Z., Elkoun, S., Robert, M., & Adjallé, K. (2023). A review of the effect of plasticizers on the physical and mechanical properties of alginate-based films. *Molecules*, 28(18), 6637.

Ferreira, F. V., Dufresne, A., Pinheiro, I. F., Souza, D. H. S., Gouveia, R. F., Mei, L. H. I., & Lona, L. M. F. (2018). How do cellulose nanocrystals affect the overall properties of biodegradable polymer nanocomposites: A comprehensive review. *European Polymer Journal*, 108, 274–285.

Fletcher, C. A., Aureli, S., Foschi, E., Filho, W. L., Barbir, J., Beltrán, F. R., Lehtinen, L., & Banks, C. E. (2024). Implications of consumer orientation towards environmental sustainability on the uptake of bio-based and biodegradable plastics. *Current Research in Environmental Sustainability*, 7, 100246.

Fletcher, C. A., Niemenoja, K., Hunt, R., Adams, J., Dempsey, A., & Banks, C. E. (2021, September 23). Addressing stakeholder concerns regarding the effective use of bio-based and biodegradable plastics. *Resources*, 10(10), 95.

Folino, A., Karageorgiou, A., Calabrò, P. S., & Komilis, D. (2020). Biodegradation of wasted bioplastics in natural and industrial environments: A Review. *Sustainability*, 12(15), 6030.

Folino, A., Pangallo, D., & Calabrò, P. S. (2023). Assessing bioplastics biodegradability by standard and research methods: Current trends and open issues. *Journal of Environmental Chemical Engineering*, 11(2), 109424.

Geissdoerfer, M., Vladimirova, D., & Evans, S. (2018). Sustainable business model innovation: A review. *Journal of Cleaner Production*, 198, 401–416.

Gheorghita, R., Anchidin-Norocel, L., Filip, R., Dimian, M., & Covasa, M. (2021). Applications of biopolymers for drugs and probiotics delivery. *Polymers*, 13(16), 2729.

Ghosh, K., & Jones, B. H. (2021). Roadmap to biodegradable plastics—Current state and research needs. *ACS Sustainable Chemistry & Engineering*, 9(18), 6170–6187.

Gilani, I. E., Sayadi, S., Zouari, N., & Al-Ghouti, M. A. (2023). Plastic waste impact and biotechnology: Exploring polymer degradation, microbial role, and sustainable development implications. *Bioresource Technology Reports*, 24, 101606.

González-López, M. E., Calva-Estrada, S. de J., Gradilla-Hernández, M. S., & Barajas-Álvarez, P. (2023). Current trends in biopolymers for food packaging: a review. *Frontiers in Sustainable Food Systems*, 7, 1225371.

González-Torres, B., Robles-García, M. Á., Gutiérrez-Lomelí, M., Padilla-Frausto, J. J., Navarro-Villarruel, C. L., Del-Toro-Sánchez, C. L., Rodríguez-Félix, F., Barrera-Rodríguez, A., Reyna-Villela, M. Z., Avila-Novoa, M. G., & Reynoso-Marín, F. J. (2021). Combination of sorbitol and glycerol, as plasticizers, and oxidized starch improves the physicochemical characteristics of films for food preservation. *Polymers*, 13(19), 3356.

Gunaalan, K., Fabbri, E., & Capolupo, M. (2020). The hidden threat of plastic leachates: A critical review on their impacts on aquatic organisms. *Water Research*, 184, 116170.

Gupta, R. K., Gawad, F. A. el, Ali, E. A. E., Karunanithi, S., Yugiani, P., & Srivastav, P. P. (2024). Nanotechnology: Current applications and future scope in food packaging systems. *Measurement: Food*, 13, 100131.

Haider, T. P., Völker, C., Kramm, J., Landfester, K., & Wurm, F. R. (2019). Plastics of the future? The impact of biodegradable polymers on the environment and on society. *Angewandte Chemie International Edition*, 58(1), 50–62.

Hallstein, J., Metzsch-Zilligen, E., & Pfaendner, R. (2024). Long-term thermal stabilization of poly(lactic acid). *Materials*, 17(11), 2761.

Harnkarnsujarit, N., Wongphan, P., Chatkitanan, T., Laorenza, Y., & Srisa, A. (2021). Bioplastic for sustainable food packaging. In *Sustainable Food Processing and Engineering Challenges* (pp. 203–277). Elsevier.

Helanto, K., Talja, R., & Rojas, O. J. (2021). Effects of talc, kaolin and calcium carbonate as fillers in biopolymer packaging materials. *Journal of Polymer Engineering*, 41(9), 746–758.

Hofmann, T., Ghoshal, S., Tufenkji, N., Adamowski, J. F., Bayen, S., Chen, Q., Demokritou, P., Flury, M., Hüffer, T., Ivleva, N. P., Ji, R., Leask, R. L., Maric, M., Mitrano, D. M., Sander, M., Pahl, S., Rillig, M. C., Walker, T. R., White, J. C., & Wilkinson, K. J. (2023). Plastics can be used more sustainably in agriculture. *Communications Earth & Environment*, 4(1), 332.

Homaeigohar, S., & Boccaccini, A. R. (2020). Antibacterial biohybrid nanofibers for wound dressings. *Acta Biomaterialia*, 107, 25–49.

Huang, F., Zhang, Q., Wang, L., Zhang, C., & Zhang, Y. (2023). Are biodegradable mulch films a sustainable solution to microplastic mulch film pollution? A biogeochemical perspective. *Journal of Hazardous Materials*, 459, 132024.

Hussain, S., Akhter, R., & Maktedar, S. S. (2024). Advancements in sustainable food packaging: from eco-friendly materials to innovative technologies. *Sustainable Food Technology*, 2(5), 1297–1364.

Ibrahim, N. I., Shahar, F. S., Sultan, M. T. H., Shah, A. U. M., Safri, S. N. A., & Mat Yazik, M. H. (2021). Overview of bioplastic introduction and its applications in product packaging. *Coatings*, 11(11), 1423.

Imre, B., & Pukánszky, B. (2013). Compatibilization in bio-based and biodegradable polymer blends. *European Polymer Journal*, 49(6), 1215–1233.

Islam, M., Xayachak, T., Haque, N., Lau, D., Bhuiyan, M., & Pramanik, B. K. (2024). Impact of bioplastics on environment from its production to end-of-life. *Process Safety and Environmental Protection*, 188, 151–166.

Ismayati, M., Fatah, N. A. N., Ernawati, E. E., Juliandri, Kusumaningrum, W. B., Lubis, M. A. R., Fatriasari, W., Solihat, N. N., Sari, F. P., Halim, A., Cholilie, I. A., & Tobimatsu, Y. (2024). Antioxidant and UV-blocking activity of PVA/tannin-based bioplastics in food packaging application. *International Journal of Biological Macromolecules*, 257, 128332.

Jabeen, N., & Atif, M. (2024). Polysaccharides based biopolymers for biomedical applications: A review. *Polymers for Advanced Technologies*, 35(1).

Jayakumar, A., Radoor, S., Siengchin, S., Shin, G. H., & Kim, J. T. (2023). Recent progress of bioplastics in their properties, standards, certifications and regulations: A review. *Science of the Total Environment*, 878, 163156.

Jia, P., Xia, H., Tang, K., & Zhou, Y. (2018). Plasticizers derived from biomass resources: A short review. *Polymers*, 10(12), 1303.

Khalid, M. Y., al Rashid, A., Arif, Z. U., Ahmed, W., Arshad, H., & Zaidi, A. A. (2021). Natural fiber reinforced composites: Sustainable materials for emerging applications. *Results in Engineering*, 11, 100263.

Khomlaem, C., Aloui, H., Singhvi, M., & Kim, B. S. (2023). Production of polyhydroxyalkanoates and astaxanthin from lignocellulosic biomass in high cell density membrane bioreactor. *Chemical Engineering Journal*, 451, 138641.

Kong, U., Mohammad Rawi, N. F., & Tay, G. S. (2023). The potential applications of reinforced bioplastics in various industries: A Review. *Polymers*, 15(10), 2399.

Krishnamurthy, A., & Amritkumar, P. (2019). Synthesis and characterization of eco-friendly bioplastic from low-cost plant resources. *SN Applied Sciences*, 1(11), 1432.

Kumar, S., Dhiman, R., Prudencio, C. R., da Costa, A. C., Vibhuti, A., Leal, E., Chang, C.-M., Raj, V. S., & Pandey, R. P. (2023). Chitosan: Applications in drug delivery system. *Mini-Reviews in Medicinal Chemistry*, 23(2), 187–191.

Kumari, R., Singha, L. P., & Shukla, P. (2023). Biotechnological potential of microbial bio-surfactants, their significance, and diverse applications. *FEMS Microbes*, 4. https://doi.org/10.1093/femsmc/xtad015

Lackner M. Bioplastics. Kirk-Othmer Encyclopedia of Chemical Technology. 2000 Dec;4:1–41

Lackner, M. (2015). Bioplastics. In *Kirk-Othmer Encyclopedia of Chemical Technology* (pp. 1–41). Wiley.

Lagaron, J. M., & Lopez-Rubio, A. (2011). Nanotechnology for bioplastics: Opportunities, challenges and strategies. *Trends in Food Science & Technology*, 22(11), 611–617.

Lakhiar, I. A., Yan, H., Zhang, J., Wang, G., Deng, S., Bao, R., Zhang, C., Syed, T. N., Wang, B., Zhou, R., & Wang, X. (2024). Plastic pollution in agriculture as a threat to food security, the ecosystem, and the environment: An overview. *Agronomy*, 14(3), 548.

Maddela, N. R., Kakarla, D., Venkateswarlu, K., & Megharaj, M. (2023). Additives of plastics: Entry into the environment and potential risks to human and ecological health. *Journal of Environmental Management*, 348, 119364.

Marano, S., Laudadio, E., Minnelli, C., & Stipa, P. (2022). Tailoring the barrier properties of PLA: A state-of-the-art review for food packaging applications. *Polymers*, 14(8), 1626.

Mariana, M., Alfatah, T., H.P.S., A. K., Yahya, E. B., Olaiya, N. G., Nuryawan, A., Mistar, E. M., Abdullah, C. K., Abdulmadjid, S. N., & Ismail, H. (2021). A current advancement on the role of lignin as sustainable reinforcement material in biopolymeric blends. *Journal of Materials Research and Technology*, 15, 2287–2316.

Mariano, M., el Kissi, N., & Dufresne, A. (2014). Cellulose nanocrystals and related nanocomposites: Review of some properties and challenges. *Journal of Polymer Science Part B: Polymer Physics*, 52(12), 791–806.

Marichelvam, M. K., Jawaid, M., & Asim, M. (2019). Corn and rice starch-based bio-plastics as alternative packaging materials. *Fibers*, 7(4), 32.

Marturano, V., Cerruti, P., & Ambrogi, V. (2017). Polymer additives. *Physical Sciences Reviews*, 2(6). https://doi.org/10.1515/psr-2016-0130

Marturano, V., Marotta, A., Salazar, S. A., Ambrogi, V., & Cerruti, P. (2023). Recent advances in bio-based functional additives for polymers. *Progress in Materials Science*, 139, 101186.

Mastrolia, C., Giaquinto, D., Gatz, C., Pervez, Md., Hasan, S., Zarra, T., Li, C.-W., Belgiorno, V., & Naddeo, V. (2022). Plastic pollution: Are bioplastics the right solution? *Water*, 14(22), 3596.

Mishra, R. K., Ha, S. K., Verma, K., & Tiwari, S. K. (2018). Recent progress in selected bio-nanomaterials and their engineering applications: An overview. *Journal of Science: Advanced Materials and Devices*, 3(3), 263–288.

Mohammed, A. A. B. A., Hasan, Z., Omran, A. A. B., Elfaghi, A. M., Khattak, M. A., Ilyas, R. A., & Sapuan, S. M. (2022). Effect of various plasticizers in different concentrations on physical, thermal, mechanical, and structural properties of wheat starch-based films. *Polymers*, 15(1), 63.

Mohammed, M., Jawad, A. J. M., Mohammed, A. M., Oleiwi, J. K., Adam, T., Osman, A. F., Dahham, O. S., Betar, B. O., Gopinath, S. C. B., & Jaafar, M. (2023). Challenges and advancement in water absorption of natural fiber-reinforced polymer composites. *Polymer Testing*, 124, 108083.

Moshood, T. D., Nawanir, G., Mahmud, F., Mohamad, F., Ahmad, M. H., & AbdulGhani, A. (2022). Sustainability of biodegradable plastics: New problem or solution to solve the global plastic pollution? *Current Research in Green and Sustainable Chemistry*, 5, 100273.

Muhammad, A., Roslan, A., Sanusi, S. N. A., Shahimi, M. Q., & Nazari, N. Z. (2019). Mechanical properties of bioplastic form cellulose nanocrystal (CNC) mangosteen peel using glycerol as plasticizer. *Journal of Physics: Conference Series*, 1349(1), 012099.

Nanda, S., Patra, B. R., Patel, R., Bakos, J., & Dalai, A. K. (2022). Innovations in applications and prospects of bioplastics and biopolymers: a review. *Environmental Chemistry Letters*, 20(1), 379–395.

Narancic, T., Cerrone, F., Beagan, N., & O'Connor, K. E. (2020). Recent advances in bioplastics: Application and biodegradation. *Polymers*, 12(4), 920.

Navaf, M., Sunooj, K. V., Aaliya, B., Akhila, P. P., Sudheesh, C., Mir, S. A., & George, J. (2023). Impact of metal and metal oxide nanoparticles on functional and antimicrobial activity of starch nanocomposite film; a review. *Measurement: Food*, 11, 100099.

Nayanathara Thathsarani Pilapitiya, P. G. C., & Ratnayake, A. S. (2024). The world of plastic waste: A review. *Cleaner Materials*, 11, 100220.

Negrete-Bolagay, D., & Guerrero, V. H. (2024). Opportunities and challenges in the application of bioplastics: Perspectives from formulation, processing, and performance. *Polymers*, 16(18), 2561.

Niaounakis, M. "Medical, dental, and pharmaceutical applications." Biopolymers: applications and trends (2015): 291–405.

Okoye, C. O., Addey, C. I., Oderinde, O., Okoro, J. O., Uwamungu, J. Y., Ikechukwu, C. K., Okeke, E. S., Ejeromedoghene, O., & Odii, E. C. (2022). Toxic chemicals and persistent organic pollutants associated with micro-and nanoplastics pollution. *Chemical Engineering Journal Advances*, 11, 100310.

Oluwasina, O. O., Akinyele, B. P., Olusegun, S. J., Oluwasina, O. O., & Mohallem, N. D. S. (2021). Evaluation of the effects of additives on the properties of starch-based bioplastic film. *SN Applied Sciences*, 3(4), 421.

Park, H., He, H., Yan, X., Liu, X., Scrutton, N. S., & Chen, G.-Q. (2024). PHA is not just a bioplastic! *Biotechnology Advances*, 71, 108320.

Patra, S. K., Poddar, R., Brestic, M., Acharjee, P. U., Bhattacharya, P., Sengupta, S., Pal, P., Bam, N., Biswas, B., Barek, V., Ondrisik, P., Skalicky, M., & Hossain, A. (2022). Prospects of hydrogels in agriculture for enhancing crop and water productivity under water deficit condition. *International Journal of Polymer Science*, 2022, 1–15.

Perera, K. Y., Hopkins, M., Jaiswal, A. K., & Jaiswal, S. (2024). Nanoclays-containing bio-based packaging materials: Properties, applications, safety, and regulatory issues. *Journal of Nanostructure in Chemistry*, 14(1), 71–93.

Perera, K. Y., Jaiswal, A. K., & Jaiswal, S. (2023). biopolymer-based sustainable food packaging materials: Challenges, solutions, and applications. *Foods*, *12*(12), 2422. https://doi.org/10.3390/foods12122422

Pooja, N., Chakraborty, I., Rahman, Md. H., & Mazumder, N. (2023). An insight on sources and biodegradation of bioplastics: A review. *3 Biotech*, 13(7), 220.

Poulose, A., Parameswaranpillai, J., George, J. J., Gopi, J. A., Krishnasamy, S., Dominic C. D., M., Hameed, N., Salim, N. v., Radoor, S., & Sienkiewicz, N. (2022). Nanocellulose: A fundamental material for science and technology applications. *Molecules*, 27(22), 8032.

Rahman, R. (2019). Bioplastics for food packaging: A review. *International Journal of Current Microbiology and Applied Sciences*, 8(03), 2311–2321.

Rahmati, F., Sethi, D., Shu, W., Asgari Lajayer, B., Mosaferi, M., Thomson, A., & Price, G. W. (2024). Advances in microbial exoenzymes bioengineering for improvement of bioplastics degradation. *Chemosphere*, 355, 141749.

Ramontja, J., Ray, S. S., Pillai, S. K., & Luyt, A. S. (2009). High-Performance carbon nanotube-reinforced bioplastic. *Macromolecular Materials and Engineering*, 294(12), 839–846.

Reichert, C. L., Bugnicourt, E., Coltelli, M.-B., Cinelli, P., Lazzeri, A., Canesi, I., Braca, F., Martínez, B. M., Alonso, R., Agostinis, L., Verstichel, S., Six, L., Mets, S. de, Gómez, E. C., Ißbrücker, C., Geerinck, R., Nettleton, D. F., Campos, I., Sauter, E., ..., Schmid, M. (2020). Bio-based packaging: Materials, modifications, industrial applications and sustainability. *Polymers*, 12(7), 1558.

Rismawati, F., Sriherfyna, F. H., & Syamani, F. A. (2023). Starch-based bioplastic reinforced by cellulose nanocrystal isolated from water hyacinth. *Jurnal Bahan Alam Terbarukan*, 12(1), 9–15.

Rosseto, M., Krein, D. D., Balbé, N. P., & Dettmer, A. (2019). Starch–gelatin film as an alternative to the use of plastics in agriculture: A review. *Journal of the Science of Food and Agriculture*, 99(15), 6671–6679.

Rosenboom, J.-G., Langer, R., & Traverso, G. (2022). Bioplastics for a circular economy. *Nature Reviews Materials*, 7(2), 117–137.

Safitri, A., Sinaga, P. S. D., Nasution, H., Harahap, H., Masyithah, Z., Iriany, & Hasibuan, R. (2022). The role of various plasticizers and fillers additions in improving tensile strength of starch-based bioplastics: A mini review. *IOP Conference Series: Earth and Environmental Science*, 1115(1), 012076.

Saharan, R., & Kharb, J. (2022). Exploration of bioplastics: A review. *Oriental Journal of Chemistry*, 38(4), 840–854.

Saleh, R. I., Kim, M., Baek, S. Y., & Cha, C. (2022). Chitosan-functionalized silica nanoparticles as a multifunctional coating material for improved water repellency, antimicrobial activity and mechanical strength of degradable bioplastics. *Cellulose*, 29(14), 7691–7701.

Samir, A., Ashour, F. H., Hakim, A. A. A., & Bassyouni, M. (2022). Recent advances in biodegradable polymers for sustainable applications. *Npj Materials Degradation*, 6(1), 68.

Serrano-Aguirre, Lara, and M. Auxiliadora Prieto. "Can Bioplastics Always Offer a Truly Sustainable Alternative to Fossil-based Plastics?" Microbial Biotechnology 17, no. 4 (April 1, 2024). https://doi.org/10.1111/1751-7915.14458.

Shaikh, S., Yaqoob, M., & Aggarwal, P. (2021). An overview of biodegradable packaging in food industry. *Current Research in Food Science*, 4, 503–520.

Siddiqui, S. A., Yang, X., Deshmukh, R. K., Gaikwad, K. K., Bahmid, N. A., & Castro-Muñoz, R. (2024). Recent advances in reinforced bioplastics for food packaging – A critical review. *International Journal of Biological Macromolecules*, 263, 130399.

Silva, R. R. A., Marques, C. S., Arruda, T. R., Teixeira, S. C., & de Oliveira, T. V. (2023). Biodegradation of polymers: Stages, measurement, standards and prospects. *Macromol*, 3(2), 371–399.

Suryanegara, L., Fatriasari, W., Zulfiana, D., Anita, S. H., Masruchin, N., Gutari, S., & Kemala, T. (2021). Novel antimicrobial bioplastic based on PLA-chitosan by addition of TiO_2 and ZnO. *Journal of Environmental Health Science and Engineering*, 19(1), 415–425.

Tennakoon, P., Chandika, P., Yi, M., & Jung, W.-K. (2023). Marine-derived biopolymers as potential bioplastics, an eco-friendly alternative. *iScience*, 26(4), 106404.

Thomas, A. P., Kasa, V. P., Dubey, B. K., Sen, R., & Sarmah, A. K. (2023). Synthesis and commercialization of bioplastics: Organic waste as a sustainable feedstock. *Science of the Total Environment*, 904, 167243.

Tsou, C.-H., Suen, M.-C., Yao, W.-H., Yeh, J.-T., Wu, C.-S., Tsou, C.-Y., Chiu, S.-H., Chen, J.-C., Wang, R. Y., Lin, S.-M., Hung, W.-S., De Guzman, M., Hu, C.-C., & Lee, K.-R. (2014). Preparation and characterization of bioplastic-based green renewable composites from Tapioca with acetyl tributyl citrate as a plasticizer. *Materials*, *7*(8), 5617–5632. https://doi.org/10.3390/ma7085617

Tsung, T.-H., Tsai, Y.-C., Lee, H.-P., Chen, Y.-H., & Lu, D.-W. (2023). Biodegradable polymer-based drug-delivery systems for ocular diseases. *International Journal of Molecular Sciences*, 24(16), 12976.

Vieira, M .G. A., da Silva, M. A., dos Santos, L. O., & Beppu, M. M. (2011). Natural-based plasticizers and biopolymer films: A review. *European Polymer Journal*, 47(3), 254–263. https://doi.org/10.1016/j.eurpolymj.2010.12.011

Weinrich, R., & Herbes, C. (2023). Consumer research on bioplastics: A systematic review. *Q Open*, 3(1). https://doi.org/10.1093/qopen/qoad013

Wu, Z., Tang, S., Deng, W., Luo, J., & Wang, X. (2021, November 30). Antibacterial chitosan composite films with food-inspired carbon spheres immobilized AgNPs. *Food Chemistry*, 363, 130342.

Yang, J., Ching, Y., & Chuah, C. (2019). Applications of lignocellulosic fibers and lignin in bioplastics: A review. *Polymers*, 11(5), 751.

Yu, H.-Y., Wang, C., & Abdalkarim, S. Y. H. (2017). Cellulose nanocrystals/polyethylene glycol as bifunctional reinforcing/compatibilizing agents in poly(lactic acid) nanofibers for controlling long-term in vitro drug release. *Cellulose*, 24(10), 4461–4477.

Yu, Y., & Flury, M. (2024). Unlocking the potentials of biodegradable plastics with proper management and evaluation at environmentally relevant concentrations. *Npj Materials Sustainability*, 2(1), 9.

Zaim, N. S. H. B. H., Tan, H. L., Rahman, S. M. A., Abu Bakar, N. F., Osman, M. S., Thakur, V. K., & Radacsi, N. (2023). Recent advances in seed coating treatment using nanoparticles and nanofibers for enhanced seed germination and protection. *Journal of Plant Growth Regulation*, 42(12), 7374–7402.

Zhang, W., Ahari, H., Zhang, Z., & Jafari, S. M. (2023). Role of silica (SiO_2) nano/micro-particles in the functionality of degradable packaging films/coatings and their application in food preservation. *Trends in Food Science & Technology*, 133, 75–86.

Zhao, X., Wang, Y., Chen, X., Yu, X., Li, W., Zhang, S., Meng, X., Zhao, Z.-M., Dong, T., Anderson, A., Aiyedun, A., Li, Y., Webb, E., Wu, Z., Kunc, V., Ragauskas, A., Ozcan, S., & Zhu, H. (2023). Sustainable bioplastics derived from renewable natural resources for food packaging. *Matter*, 6(1), 97–127.

Zhao, Y., Xue, L., Huang, Z., Lei, Z., Xie, S., Cai, Z., Rao, X., Zheng, Z., Xiao, N., Zhang, X., & Ma, F. (2024, October 28). Lignin valorization to bioplastics with an aromatic hub metabolite-based autoregulation system. *Nature Communications*, *15*(1), 9288.

4 Bioplastic Production and Food Safety

Mustapha Abdulsalam, Abdulhakeem Idris Abdulhakeem, Innocent Ojeba Musa, Miracle Uwa Livinus, Shehu-Alimi Elelu, Muhammed Abdulazeez, Maiyaki Taofiqat Damilola, Ibrahim Abdulrazaq, Umar Abdulrasheed, Abdulkareem Tajudeen Zakari, and Tijani Abiola Tajudeen

4.1 INTRODUCTION

Today's world is conscious of environmental pollution and the depletion of fossil resources, hence the call to bioplastics. These materials are synthesized from renewable resources inclusive of plant starches, vegetable oils, and algae, and can be regarded as renewable polymers that are most likely to replace conventional polymers (Rai et al., 2021). They are characterized as biodegradable and compostable, which makes this plastic derived from biomass, a good solution to the problem of plastic pollution in the world. In contrast to ordinary plastics derived from crude oil, bioplastics have a shorter life cycle in the environment and do not depend on non-renewable resources, thereby solving two problems at once (Rosenboom et al., 2022).

Nonetheless, given the fact that commodity bioplastics have a positive environmental profile, they have raised a significant concern about their potential impact as food-related packaging and container materials. Food packaging materials, especially, have specific laws that regulate them to reduce the effects that the materials have on the health of the consumers. This embraces the transfer of certain substances from the packaging into the food and depends on the chemical make-up of the bioplastic material besides the storage and the conditions under which the food is consumed (Shlush & Davidovich-Pinhas, 2023). Recently, several works have been related to the assessment of chemical transfers from bioplastics to food products with a particular interest in plasticizers, stabilizers, and other related additives used in the process (Riboni et al., 2023).

Food safety also means focusing not only on chemical migrations but also on the presence of microbes in food products. Interestingly, the microbial degradation rate seems to be higher for bioplastics, particularly those based on natural polymers such as starch, than for conventional plastics (Zoungranan et al., 2020). Therefore, it is important to guarantee that bioplastics are subject to tests on their capacity to preserve food safety for the designed life cycle. Another important characteristic of bioplastics for food applications is the legal requirement that impacts its production. There are disparities in the kinds of bioplastics and their set measures and laws in several nations particularly those applied in immediate contact with food. To tackle the issues more specifically, the European Union (EU) and the United States (US) have elaborated strict regulations over bioplastic consumption in food packaging which involves stringent migration testing, safety standards, and material certification (Thapliyal et al., 2024). These regulations make sure that bioplastics possess all the right safety measures necessary for them to qualify to be used with foods. Despite this, debates remain about the sufficiency of these standards and the requirement of updating them periodically

DOI: 10.1201/9781003602828-5

depending on new science and a diverse range of novel bioplastics appearing in the market (Parveen et al., 2024).

This study aims to provide an extended analysis of bioplastics with a specific focus on food safety. It covers the kinds of bioplastics, their manufacture, their use in food packaging, and the risks of direct food applications of such products. In accomplishing this goal of identifying key issues surrounding the current legal structures for bioplastic production and using this study, the recommendations for future enhancement of those structures provide an essential picture of bioplastics within the global food system that is safe, practical, and sustainable.

4.1.1 Background and Importance of Bioplastics

Bioplastics have been considered a highly favorable plastic substitute for conventional petroleum-based plastics, in the last few years especially due to the environmental issues regarding plastic waste. Conventional plastics respond to long-term pollution, and plastic objects have been noted to decompose in the environment within, at most, several centuries (MacLeod et al., 2021). On the other hand, bioplastics are produced based on renewable materials, which include plants, algae, and different types of agricultural residues where the articles made from bioplastic are biodegradable and compostable (Rai & Choure., 2023). It eliminates one of the most important features that hurt their environment, the increase in amounts of plastic waste and pollution.

Increasing environmental problems constantly linked to plastic waste have led to a focused interest in bioplastics, as they serve as a possible solution for making less use of petroleum-based products and reducing CO_2 emissions. For example, bioplastics derived from plant sources, polylactic acid (PLA), are said to have the capacity to remove carbon dioxide through plant growth, thus offering a form of carbon sink that could balance carbon emissions incurred in the production of bioplastics (Keith et al., 2024). These favorable environmental effects correlate with international sustainability goals whose aim includes a decrease in the effects of climate change and an increased reliance on renewable energy sources. For this reason, biodegradable plastics are steadily increasing their popularity in numerous sectors, such as food and agriculture packaging, and even in medicine and healthcare, as a material that is both environmental friendly and supports the principles of circular economy (Hussain et al., 2024).

In the field of packaging, and the food industry in particular, bioplastics have a great potential to replace the majority of traditional plastics used to pack foods such as a plastic foil, containers, and disposable cutlery. The use of bioplastics in foods is now growing due to consumer preference for green products and companies' obligations to contribute to sustainability targets (Notaro et al., 2022). Biopolymers like PLA and PHA are employed for packaging most food items since they are more biodegradable than conventional plastics and less likely to clog the environment. Besides, bioplastics could be utilized in packaging films and coatings for food to reduce the required amount of packaging and increase its efficiency through its utilization as an edible barrier in protecting against external impacts (Petkosk et al., 2021).

Although bioplastics have several benefits like environmental advantages and automotive industry, using bioplastics in food industries requires examination, regarding possible food safety issues. Bioplastics have to be safe to avoid the leaching out of hazardous chemicals, for instance, plasticizers and monomers into the foods that come into contact with the materials (Obuzor & Onyedikachi, 2023). This concern is important since bioplastics come into direct contact with food and chemical migration to the food chain may be lethal. Therefore, legislation which regulates the application of the bioplastics in food packaging is important to allow safe utilization of the bioplastics as well as consumers' health.

Bioplastics can thus be summarized to go beyond their role in the reduction of environmental impacts and as a measure toward the safety of food products. Thus, the use of innovative approaches to the high demand for supplies to reduce the environmentally active plastic particles that affect people's health becomes more critical. Consequently, the idea of using bioplastics as a solution to the adverse consequences of plastics and plastic products has potential to help reduce the effects of the problem most evidently seen in aquatic environments and on lands (Ogunola et al., 2018). In addition, the use of bioplastics establishes itself within food system realities attempting to combat food waste, and excessive packaging, and improve the circular economy.

Bioplastics are a relatively novel development in terms of seeking sustainable material sources in food chain contexts. Due to a lower environmental profile, promoting sustainable agendas, and improving food protection, bioplastics can prospectively transform approaches to food packaging and corresponding applications. This study seeks to establish global knowledge on the production, handling, and preparedness of these confined feeds so that they may effectively address the emerging worldwide food challenges.

4.2 TYPES OF BIOPLASTICS

Bioplastics sacked can be defined, with respect to their chemical basis, level of biodegradability, and the utilization they are designed for. In the topic of food safety and sustainability of food packaging and containers, it is important to look at the different categories of bioplastic materials to determine which kind may be most appropriate for food packaging and handling environments. Some of the most widely used bioplastic material types in the food packaging industry are biodegradable starch bioplastics, PLA bioplastics, PHA bioplastics, and cellulose bioplastics, with numerous properties.

4.2.1 Starch-Based Bioplastics

One of the oldest and most studied types of bioplastics is based on starch. Some of these bioplastics are sourced from corn, potatoes, or tapioca and are biodegradable, cheaper and have multiple uses. The biopolymers include starch that is derived from corn, potatoes, and rice through a process known as gelatinization, in which bio-based plastics contain starch in a plastic-like state when mixed with water and heat (Musa et al., 2024a). They are particularly applied as wrappers, trays, and containers for foods that are expected to have a relatively short shelf life on the market.

Nevertheless, there are some issues connected with starch-based bioplastics, for example, their low resistance to water and some other mechanical characteristics. It means that these materials are generally more vulnerable in terms of disintegration in the conditions of high humidity which does not make them appropriate for long-term storage of food products (Ziegler et al., 2021). However, starch-based bioplastics stand distinctive in single-use food packing since they can be degraded immediately after use. They are also incorporated into blends with other bioplastics or incorporated with biodegradable additives for modifying properties or fundamentally altering their characteristics (Yuvaraj et al., 2023).

4.2.2 Polylactic Acid (PLA)

A major type of bioplastics out in the market and prominently used in food packaging is PLA. PLA is a biopolymer extracted from fermented plant starch, primarily from corn or sugarcane, characterized by high transparency, rigidity, and heat resistance, and thus, used for packaging products such as bottles, containers, and films (Pandit et al., 2018). For PLA bioplastics to be biodegradable, they must be industrially compostable which makes them degrade in a few months if they are exposed to the appropriate conditions.

Another benefit of PLA is that it has a reduced environmental footprint primarily over petroleum-based plastics. PLA production results in the generation of comparatively lesser greenhouse gases, and they have a raw material that is naturally replenished (Mosomi et al., 2024). As much as PLA has these benefits when used in packaging food products, it has its flaws when used particularly in food packaging. While it is biodegradable, it needs to undergo some special composting conditions; therefore, it will stay in the landfills for more time when not well disposed of (Ayilara et al., 2020). However, PLA is sensitive to heat and this might be disadvantageous in some food packaging uses that require heat resistance such as microwave or hot food containers.

4.2.3 Polyhydroxyalkanoates (PHA)

PHA is defined as a set of bacterial biodegradable bioplastics synthesized via the fermentation of sugars or lipids. It is important for PHAs to be very versatile since it is possible to design the PHAs to have these variable properties, which may include flexibility, strength, and biocompatibility (Jurak et al., 2021). During the last few years, the use of PHAs has been linked to food packaging including single-use food packaging, trays, and cutlery because biodegradable features are important for such applications. While PLA undergoes biodegradation based on certain conditions, PHA can degrade itself in natural conditions such as in soil and marine environments (Meereboer et al., 2020).

The main drawback of PHA is its relatively higher production cost than other bioplastics because microbial fermentation, which is essential for its production, has yet to become cheaper (Meereboer et al., 2020). However, due to continuous innovation in the field of biotechnology and the cost-effectiveness of renewable feedstocks for PHA synthesis, the cost of the product is going to reduce and it has good market potentiality in the commodity market. One of the biggest advantages of the developed PHA is the fact that it can degrade in rather various conditions, which means that the material is one of the potential solutions to fight against such a problem as plastic pollution, particularly in ecosystems where plastic waste is a significant issue, such as oceans and rivers (Miha et al., 2021).

4.2.4 Cellulose-Based Bioplastics

Cellulose-based bioplastics are derived from plant fibers, typically from wood, cotton, or hemp. These bioplastics are valued for their biodegradability, strength, and ability to be processed into thin films and coatings, making them ideal for food packaging applications (Shlush & Davidovich-Pinhas, 2023). Cellulose-based films are already in use in applications such as edible films on fruits, vegetables, and confectioneries as natural substitutes for synthetic plastic films.

Since cellulose bioplastics are not usually toxic, they are safe for direct food contact, especially in the safety of food (Zimmerman et al., 2020). Cellulose-based bioplastics share the drawback that their mechanical properties are not always optimal: they are less flexible than other kinds of bioplastics. However, more studies are carried out on the development of the processing methods and on the property of cellulose-based products with the aims of increasing flexibility and strength (Zhao et al., 2021). Cellulose is also one of the most renewable materials, and using it encourages sustainable farming, which in turn would make it even more attractive as a bioplastic food-safe material.

4.3 APPLICATIONS OF BIOPLASTICS IN FOOD SYSTEMS

Increased use of bioplastics in food systems is perhaps most evident due to the need for environmentally friendly products that replace conventional black petrol plastics. There are the following benefits of bioplastics: biodegradability, compostability, and comparatively low level of influence on environmental conditions, which makes them suitable for the usage in food products industry. Such uses are in food packaging, edible films or coatings and food contacting articles such as containers or utensils. In the following sections, we elaborate on these key applications.

4.3.1 Packaging Materials

Food packaging is the largest and steadily growing sector that incorporates bioplastic products. Biodegradable and compostable packaging materials derived from bioplastics become a good solution to minimize quantities of petroleum-based plastics in the world. The frequent types of food packaging biopolymers include PLA, PHA, and biopolymers based on starch, normally packaged in the form of wraps, films, and containers (Agarwal et al., 2023a). PLA is more accepted for use in the packaging of beverages, snack food, and fresh produce due to its transparency, rigidness, and relatively lower environmental effect in its processing (Ncube et al., 2020).

The advantages associated with the application of packaging materials derived from bioplastics include decreased carbon footprint. Like other packaging materials, bioplastics have different end-of-life disposal solutions that encourage the practice of circularity. For instance, PLA and PHA have been witnessed to biodegrade under industrial composting conditions, hence lowering the impact of the food packaging waste on the environment. This is a major leap from the traditional plastics which may take over 300 years to degrade in landfill or aquatic ecosystem settings (Chamas et al., 2020). Similarly, biopolymers such as PHA can be used to replace the single-use plastics in food packaging with special respect for escalating the problem of plastic pollution, threatening wildlife and the environment globally (Mosomi et al., 2024).

4.3.2 Edible Films and Coatings

A novel concept in the use of bioplastic in food systems is the use of edible films and coatings. These films are made most often of bioplastics such as starch, cellulose, or protein and are intended to be disposed of in the same way as the food they encompass. Some advantages of edible films and coatings include the following: they extend the shelf life of the food they cover; they retain more moisture in foods than other conventional methods; and, most importantly, they restrict the entry of contaminants in the foods they cover. For example, plastic films made of cellulose and available for the protection of fruits and vegetables can be used to provide a barrier to moisture loss, slow down the process of ripening of the fruits and vegetables, and preserve the freshness without having to pack them in plastic (Agarwal et al., 2023).

Besides, the edible films and coatings are also functional in the purpose of the reduction of conventional plastics which will support the sustainability. Unlike traditional packaging that forms part of filth since it is mostly plastic based, edible bioplastics do not pollute since they are ingested with the product. It will also be most helpful in the food industry because the adoption of novel packaging solutions to combat plastic waste is rapidly gaining ground (Jurak et al., 2021). Furthermore, the application of edible coating from natural biopolymers including chitosan or alginate contains antimicrobial properties, which can further enhance food safety by reducing the growth of harmful bacteria on the surface of food products (Yuvaraj et al., 2023).

4.3.3 Food Containers and Utensils

The same applies for food packing containers, trays, and utensils, which are now being made from biodegradable plastics, hence replacing the normal plastics. Various items, including food containers, cups, plates, and cutlery manufactured from bio-based materials, including PLA and PHA, are slowly finding their way into the market as food service and disposable foodware (Tan et al., 2021). A variety of containers and utensils made from PLA are frequently employed in the take-out packaging and catering sectors as they are lightweight, sturdy, can affirm to the conditions generally experienced in the catering industry, and can in the end degrade, thus turning to compost.

The shift toward bioplastic for food containers and utensils would inevitably make a positive impact in eliminating disposable foodware as a major problem to plastic finds in oceans and other dump sites. Furthermore, bioplastics for food packaging and disposable cutleries generally entail lower energy consumption than conventional plastics and thus have a lighter carbon footprint (Ncube et al., 2020). PHA, particularly, is easily biodegradable and can decompose in various conditions such as soil and marine environments and thus can be useful in the food service industry where single-use products made from it can be easily disposed of (Idris et al., 2023).

4.4 PRODUCTION PROCESSES OF BIOPLASTICS

Bioplastic manufacturing includes several processes: obtaining the necessary raw materials and the manufacturing process that turns these materials into the needed bioplastic types (Figure 4.1). The process also entails what is called life cycle assessment (LCA) for the purpose of estimating the ecological footprint of bioplastic production. Knowledge of these processes is fundamental to deciding on the environmental impact, cost-effectiveness, and efficiency of bioplastics in food distribution.

4.4.1 Raw Material Sourcing

The first of the processes involved in producing bioplastics is to get materials for the production process. These materials are normally sourced from renewable sources such as crops, agricultural

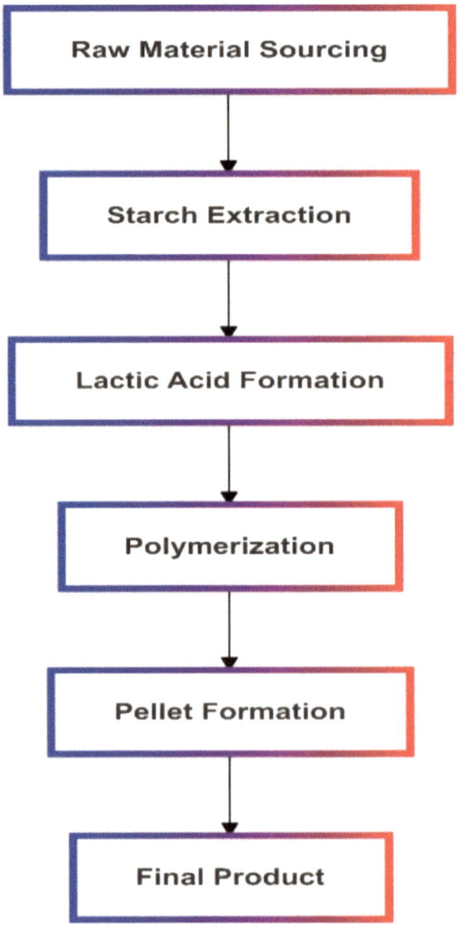

FIGURE 4.1 Schematic diagram of polylactic acid (PLA) production process.

residue, and plant-based oils. Such types of bioplastics include starch-based bioplastics that are derived from crops such as corn, potatoes, or wheat, while PLA is made from fermented sugars derived from corn or sugarcane (Abdulsalam et al., 2022). Other bioplastics, such as PHA, are produced through microbial fermentation, where microorganisms convert carbon sources into bioplastic polymers (Abd El-malek et al., 2020).

The sustainability of bioplastic production is highly, therefore, dependent on the type of raw material used. In the case of manufacturing, materials should be most fundamental, easily available, and renewable that is raw materials which should be transported less or not at all and at the same time processed in less time. Nonetheless, the adoption of agricultural food crops for the production of bioplastic has been a contentious issue due to its effects on food availability and the utility of land and water (Gontard et al., 2018). To address these issues, researchers are searching for other raw materials like agricultural residuals, algae, and non-food plants which eliminate competition between the food and bioplastic producers (Otoni et al., 2021).

4.4.2 Manufacturing Techniques

After acquiring raw materials, on the other hand, there are procedures applied to transform the raw materials into usable bioplastics. The dominant techniques are mainly extrusion, injection molding, and blow molding which are applicable for the various types of bioplastics and uses.

Extrusion is probably the most popular process in the fabrication of bioplastics where raw material is in the molten state and forced through a die to form co-extruded sheets or films. This technique is widely employed in the creation of PLA-based films which are used widely in the food packaging industry (Riboni et al., 2023). Injection molding is typically used to produce solid bioplastic items, such as food containers or cutlery, where hot melted bioplastic is injected into the mold and solidifies once it cools down (Negrete-Bolagay & Guerrero, 2024). Blow molding is a process of producing hollow bioplastic articles including bottles and containers through inflating a tube of the molten material into a casting mold (Idris et al., 2023).

These techniques are normally different and flexible to meet certain performance criteria for food compliance, toughness, and biodegradation. The selection of the manufacturing process also affects the energy requirement and eco-friendly feature of manufacturing which shows that extrusion requires less energy than injection molding (Ncube et al., 2020).

4.4.3 Life Cycle Assessment

LCA is another technique aimed at the assessment of bioplastic's environmental effects through its whole life cycle from the procurement of the raw material to its disposal (Figure 4.2). LCA assists in the evaluation of the environmental impacts of bioplastics relative to conventional ones and the potentiality of bioplastics to decrease carbon footprint, energy usage, and solid waste.

Research has confirmed that compared with the petroleum-based plastics, bioplastics, such as PLA and PHA, distinctly have the fewer carbon impact due to their renewable raw material sourcing and the ability for biodegradation at their ultimate life cycle (Keith et al., 2024). However, the use of bioplastics can have different impacts on the environment, the source of feedstock used, the method of manufacture, and/or the conditions that appertain to the disposal of bioplastics. For instance, PLA is typically biodegradable in industrial composting sites while when ends up in landfills it is unlikely to degrade quickly due to the absence of composting environment (Riboni et al., 2023). Like other kinds of bioplastics like the PHA mentioned above, it is still possible to use a significantly large amount of energy during the fermentation of the materials needed to make it, thereby negating some of the benefits of using the product in the long run (Agarwal et al., 2023).

LCA offers useful knowledge on ideas that have the potential to make bioplastics more sustainable and allow producers to build better and less resource-damaging bioplastic food packaging

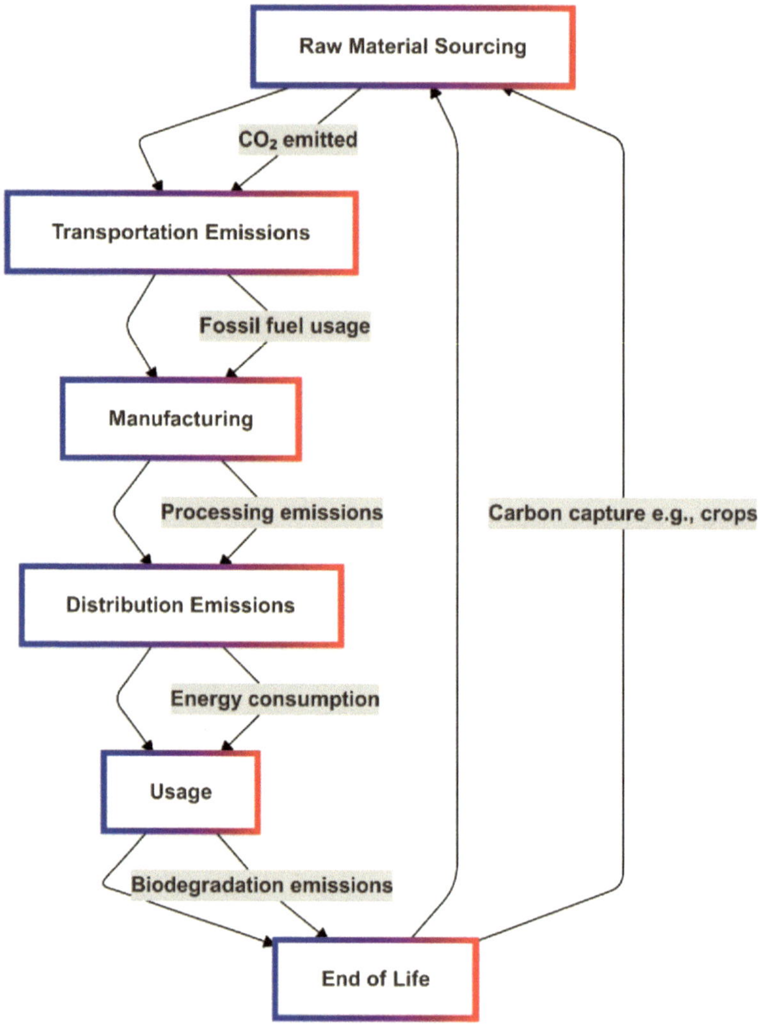

FIGURE 4.2 Circular life cycle analysis of carbon emissions in bioplastic production.

design. It also aids in forming policy decisions regarding the relative environmental impact of bioplastics to traditional plastics (Obuzor & Onyedikachi, 2023).

4.5 REGULATIONS AND GUIDELINES FOR BIOPLASTICS

The use and production of bioplastics in food industry has been guided by different laws and standards to meet food safety, environment, and health considerations. They include the types of materials that may be used in making bioplastics, methods of manufacturing, and circumstances under which bioplastics can be safely used in foods. Some of the most developed legislation regarding bioplastics is found in two major regions, i.e. the EU and the US.

4.5.1 EU Regulations on Food Contact Materials

To protect the consumer interest, the EU has put in place fully fledged regulation on the use of bioplastics in food contact materials. EU regulation of this area is anchored on Regulation (EC) No

1935/2004 which provides the overall requirements on the safety of the materials and articles in contact with food. According to this regulation, one cannot transfer any substances to food and this transfer cannot be in amounts that are dangerous to human health and affect the composition of the food or the organoleptic properties of the food (Paganizza, 2024).

Further, the EU adheres to certain directives like the Commission Regulation (EU) No 10/2011 that also introduced list of approved substances which are allowed to be used in the fabrication of plastics including bioplastics for the food contact application. This regulation also sets maximum levels for the transfer of chemical substances to food from bioplastics to protect the consumer (Cruz, 2022).

EU also supports the consumption of biodegradable and compostable bioplastics by undergoing standardized industrial compostability referring to EN 13432. The European norms of EN 13432 for compostability and ASTM D6400 for American bioplastics make compostable labeling possible, which improves the market acceptance of bioplastics and brings the bioplastics industry in line with the right strategies to deal with our waste (Nizamuddin et al., 2024).

4.5.2 US Regulations and Safety Standards

In the US, the bioplastic intended for food contact is mainly regulated by the Food and Drug Administration (FDA). The FDA has put down specific rules in CFR Title 21 that outline the standard that food contact substances such as bioplastics must meet. Bioplastics are assessed by the FDA concerning the chemical nature of the material, the application that the product is to be used in, and whether it is capable of migrating into food.

Any bioplastic used in food packaging has to be on the FDA list of approved food contact substances or be on the GRAS list. For instance, PLA is well known to be safe and has been authorized by the FDA for use in food packaging, containers, and cutlery (O'Loughlin et al., 2023).

The US also supports the growth of biodegradable bioplastic production through programs such as the United States Department of Agriculture (USDA) Certified Biobased Product label that promotes products based on renewable resources. But this certification is not obligatory for food safety and gives the consumer information about the bio-based content of bioplastics, thus making a sustainable choice easier (Koehler & Wiles, 2022).

4.5.3 Importance of Harmonized Regulations

There is problem in differences of regulations in the usage of bioplastics in the food systems across the world despite improvements in the regulations of bioplastics. A proper and complete acceptance of bioplastics might be possible when there are common regulations that have been established at an international level and creating the global standards for bioplastics might improve the acceptability and their trade flows, although common safety and environmental standards need to be met (Zhang et al., 2022).

There is a need for further study and cooperation between various authorities and various industries in addition to scientists to examine the shortcomings of the existing standards and to establish the recommendations for the proper application of bioplastic products in the food industry.

4.6 FOOD SAFETY CONSIDERATIONS

Bioplastics as food packaging materials pose significant safety questions in food systems and such concerns must be met to ensure consumer safety. Such considerations include sample trial of migration of substances, chemical safety, and the ecological and environmental effects of biopolymers.

4.6.1 Migration Testing and Limit Values

This study shows that migration testing is vital in assessing the migration of chemical substances from bioplastic materials into food. The substances, including monomers and additives or the products of their degradation, can penetrate food in some cases and become a threat to health (Čapla et al., 2023). In order to minimize these risks, some jurisdictions, including the European Food Safety Authority (EFSA) and the FDA, have set threshold values for substances in food contact materials (FCMs) (European Food Safety Authority (EFSA), 2023).

For example, PLA-based packaging materials are subjected to migration tests to prove that lactic acid and other additives do not migrate beyond the regulation limit under conditions of temperature, type of food, and storage time (Singh et al., 2018). Also, materials like plasticizers and stabilizers will undergo limits for migration into the food to cause negative effects on human health. Modern instrumental methods, including gas chromatography-mass spectrometry (GC-MS), are used to determine the presence and concentration of migrating substances as accurately as possible (Musa et al., 2022b).

4.6.2 Chemical Safety and Toxicological Assessments

Bioplastics need to be adapted to them in terms of chemical safety and compatibility of the proposed biopolymers, absent or present dangerous compounds, such as heavy metals, endocrine-disrupting substances, or monomers that have not undergone depolymerization (Jagaba et al., 2023). Toxicological evaluations are conducted to compare the abilities of these chemicals to have adverse effects when bioplastics are utilized in food contact products (Koehler & Wiles, 2022).

Literature review reveals that, for instance, starch-based materials to be used in the production of bioplastics hold certain impurities, which are natural but, if not washed properly, can be harmful to human health (Abdulsalam et al., 2024). However, the degradation products of some bioplastics like PHA have to be assessed for the generation of toxic by-products when the material is in use. Thus, chemical safety assessments are required to address biopolymeric products' whole life cycle from manufacturing processes to their disposal.

4.6.3 Environmental Impact and Ecotoxicity

It also has to do with environmental safety, specifically the biodegradable nature of these products or the possibility of their influence as ecotoxic materials. Though bioplastics are known eco-friendly substitutes to traditional plastics, the breakdown of bioplastics in a natural setting can generate production of microplastics or emit chemical compounds that are destructive for ecosystems (Zhang et al., 2022).

For instance, the littered bioplastics harm the water, ground, and food chain of animals and wildlife. Biodegradable bioplastics such as PLA and PHA decompose only efficiently under industrial composting conditions; otherwise, they stay around environmentally to counterbalance their sustainability proclamation (Geyer et al., 2023).

In addition, the production of the raw materials used to make bioplastics including corn or sugarcane has reported effects on the environment, including the forests, the soil, and the use of pesticides. These impacts must be carefully managed to ensure that the environmental benefits of bioplastics outweigh their potential drawbacks (Koehler & Wiles, 2022).

4.7 CHALLENGES IN BIOPLASTIC IMPLEMENTATION

Several setbacks prevent bioplastics from becoming a widespread material to produce sustainable alternatives to conventional plastics. These are not resolved issues but issues that revolve on several aspects, including the consumer view, performance constraints, and economic consequences.

4.7.1 Consumer Perception and Market Acceptance

However, consumer perception and acceptance is one of the major challenges that bioplastics face and this variability complicates things. Although bioplastics today are said to be environmentally friendly, it is still hard for people to trust their performance and sustainability claims. Bioplastics themselves are not bad; in fact, consciousness regarding the advantages and disadvantages of bioplastics has not been particularly widespread among the public (Chen et al., 2023).

The public perceives biopolymers as being more costly or otherwise of inferior quality to those based on crude oil. To undo this misperception, labeling needs to be clear, consumers must be informed and marketing strategies must be strong to help demonstrate the environmental advantages and performance equivalence of bioplastics in food supply chains (Musa et al., 2024a). Furthermore, certifications like "biodegradable" or "compostable" also require classification and legal standards so that consumers will not be misled by false advertisements (Zhang et al., 2022).

4.7.2 Performance vs Sustainability Trade-Offs

Another problem is to consider aspects of bioplastics' functional performance and environmental credentials simultaneously. Biodegradable and compostable bioplastics may be a different material from conventional plastics, mechanically and thermally. For example, PLA has low heat stability which poses it to be ideal for use in heat-sensitive applications such as packaging hot meals (EFSA, 2023).

To overcome these trade-offs, scientists are discovering a novel group of bioplastics that integrate the use of biodegradable polymers blended with additional components of synthetic structures and characteristics. However, this type of thinking leads to a violation of biodegradability and complicates the topic of recycling even more (Geyer et al., 2023). For these gaps to be dealt with, advances in materials science are important to guarantee the potential of bioplastics to transform the kinetics of food systems.

4.7.3 Cost Implications for Producers

Bioplastic manufacturing costs exceed costs per unit of conventional plastics because the main biopolymer materials and manufacturing techniques are costlier. For instance, the fermentation process of PHA needs enhanced technological development which is expensive for most producers (Koehler & Wiles, 2022).

These economic factors are magnified by volatility in supply and cost of agricultural feedstocks required to make bioplastic. While competition between food and non-food applications of crops such as corn or sugarcane may pose such difficulties, competition over limited resources for agriculture in developing nations complicates the subject (Ahmed et al., 2024).

Further, government grants and subsidies and increasing investment in research and development are the key drivers in controlling the cost of bioplastic and achieving the economic advantage over the conventional plastics. These economic challenges require policy, producer, and consumer cooperation to encourage greater use of bioplastics in the food supply chain (Chen et al., 2023).

4.8 RECOMMENDATIONS

To properly address the issues of using bioplastic materials and guarantee the successful application of bioplastics in the food chains, they have to follow some steps. These are stimulating innovations in materials science, improving safety systems, bettering policy backing, increasing consumer awareness, and international cooperation. In general, it may be argued that the development of solutions addressing both technical and societal challenges will require cooperation between researchers, policymakers, and industries.

4.8.1 Innovations in Materials Science

Material development is necessary in enhancing the performance constraints and drawbacks that have condemned bioplastics to stand below conventional plastics, including poor mechanical strength, thermal conductivity, and functional diversification. It is suggested that more attention should be given to the synthesis of hybrids and biopolymer composites. For instance, when PLA was strengthened by blistering PHA or incorporated with nanomaterials, such as nanocellulose or graphene, enhanced mechanical properties, thermal stability, and barrier characteristics that enlarge the application prospect of bioplastics in food chain systems are available (Livinus et al., 2024a).

In the same vein moderation of the feedstock dependence is called for, as is the shift toward the use of the various alternatives. Algae, agricultural residues, and food waste are renewable feedstock that has a less adverse effect on the environment as compared to corn or sugarcane. Main advantages of these sources include the fact that they can be used to avoid dependence on arable land, and secondly, they do not compete with food production, which raises concern over food security (Musa et al., 2024b). Enhancement of biotechnological developments, which were genetically engineered microbes to promote biopolymer production, may extend to advancing innovation in the area.

4.8.2 Enhancements in Food Safety Protocols

This brought a major need for ensuring food safety is a key factor to the acceptance of bioplastics in food systems. The guidelines must also consider such properties as an attribute of bioplastics. It is recommended that migration testing should cover the possibility of leaching undesirable migratory compounds under different food contact conditions. For example, elevated storage temperatures and specific foods, with low pH levels, might increase the migration velocity of certain additives or degradation products, which means that an appropriate testing procedure should be developed (EFSA, 2023).

Furthermore, the assessment of the toxicity of new biopolymers, with reference to bioplastic produced from unexpected sources, deserves further emphasis. Sophisticated technological tools like the high-performance liquid chromatography (HPLC) and nuclear magnetic resonance (NMR) spectroscopy offer the exact chemical safety information. Information transparency on what goes into bioplastic formulation including additives or stabilizers also increases consumers' and regulatory authorities' confidence (Koehler & Wiles, 2022). Attempts at establishing international reference databases containing the safety characteristics of different bioplastics may help to harmonize evaluation efforts and promote the use of more secure products more effectively.

4.8.3 Policy Support and Consumer Education

Policy actions play a revolutionary role in promoting the use of bioplastics. Governments must provide incentives to manufacturers in the form of tax exemptions, funds, and subvention since costs are relatively high. These incentives can decrease the cost burden on the producers and therefore increase the bioplastic's competitiveness in the market (Lawal et al., 2023).

Hopefully, requiring the utilization of bioplastics within selected industries, including food containment and disposable utensils, may enhance utilization of bioplastics and lower overall pollution levels. Moreover, the measures changing the legislation to allow for and encourage the availability of industrial composting facilities and the recycling infrastructure constitute important measures to facilitate the suitable end-of-life management of bioplastics.

Awareness crusades are known to contribute to the same extent. Such campaigns should also aim at helping consumers understand the advantages of bioplastics to the environment and human health, clarify the myth surrounding bioplastics, and explain how to dispose of them correctly. The necessary information regarding biodegradability, compostability, and recyclability should be incorporated into the containers of bioplastics to enhance purchaser choices (Livinus et al., 2024b). In a school setting, in workplace settings, and within community centers, education could help enhance awareness of bioplastics and their function in eco-friendly lifestyles.

4.8.4 International Collaboration and Harmonization

Bioplastics should engage in global corporate social responsibility for them to exert the maximum benefits. The implementation and adoption of the best and most efficient standards and regulations across the world will likewise warrant standards, safety, and sustainability to be worldwide consistent, thus helping the different companies (pages 1 & 2). International research partnerships can enhance the pace at which countries develop value-added bioplastics and deal with issues unique to their country including supply of raw materials and waste disposal (Zhang et al., 2022).

There is need to have forums that will include government, academic institutions and industries to enable them to share new knowledge, practices, technologies, and policies. The local collaborations can also strengthen the capacity in areas that require logistic or economic challenges in the use of bioplastic. For instance, progressing nations may benefit from collective funds and asset establishment, like industrial composting centers, which can be financed by international funding organizations.

Finally, supporting international campaigns linked to the purpose of sustainable campaigns status of bioplastic material support the actions of United Nations sustainable development goals (SDGs).

4.9 CONCLUSION

The use of bioplastics as an avenue toward the amelioration of negative impacts of traditional plastics in the current food complexes is quite vivid. As this study has postulated, bioplastics' mere value is in their biodegradability, lower carbon footprint, and the capability to harness renewables. Its uses are extended to health-related products such as food packaging, cutlery, and edible coatings, concluding that they are relevant in meeting the SDGs.

However, there are some problems that bioplastics encounter which cannot be solved individually. These include customer awareness, imposed efficiencies, legal uncertainties, and cost concerns which make the adoption of these factors challenging. Solving these challenges requires a combination of solutions from new materials, safe practices, and policies. For example, increased performance of biopolymer formulation can improve functionality as well as act sustainably to provide benefits, whereas strong laws protect the consumer as well as the environment. In addition, it is critical to provide an importance of the international cooperation. Correct standardization, global databases, and forums along with international consortium research are important for the acceptance and practice across borders. Likewise, for public awareness the education awareness forms the part of the intervention strategy and would also enhance consumer trust as well as the proper usage and disposal.

Nonetheless, bioplastics cannot provide solution to all the existing environmental problems but they have reliable and beneficial solutions that meet the circular economy change currently sweeping across the world. Further commitments to funding research, policy, and outreach will help bioplastics deliver their maximum benefit toward creating more sustainable food systems. Thus, the way toward establishing and organizing the bioplastic product as an organic part of consumers' life and everyday use might be long and arduous, but such task is crucial for the health of our planet.

REFERENCES

Abd El-malek, F., Khairy, H., Farag, A., & Omar, S. (2020). The sustainability of microbial bioplastics, production and applications. *International Journal of Biological Macromolecules*, *157*, 319–328.

Abdulsalam, M., Fari, H. I., Tiamiyu, B. B., & Salam, O. L. (2022). Optimizing α-amylase production from locally isolated *Aspergillus* species using selected agro waste as substrate. *Bioscience Biotechnology Research Communications*, *15*(3), 424–430.

Abdulsalam, M., Musa, I. O., Livinus, M. U., Elelu, S. A., Ibrahim, G. O., Salami, O. L., & Pal, S. K. (2024). Blue Bioeconomy and Biomedical Innovation. In *Marine Bioprospecting for Sustainable Blue-Bioeconomy* (pp. 143–157). Cham: Springer Nature Switzerland.

Agarwal, A., Shaida, B., Rastogi, M., & Singh, N. B. (2023a). Food packaging materials with special reference to biopolymers-properties and applications. *Chemistry Africa*, *6*(1), 117–144.

Agarwal, S., Singhal, S., Godiya, C. B., & Kumar, S. (2023b). Prospects and applications of starch-based biopolymers. *International Journal of Environmental Analytical Chemistry*, *103*(18), 6907–6926.

Ahmed, A. A., Yusuf, U. A., Fadlallah, Z. H., Innocent, M. O., & Abdulsalam, M. (2024). Developing an omics-driven computational framework for next-generation microbiome therapeutics. *Biological Sciences*, *4*(4), 1–19.

Ayilara, M. S., Olanrewaju, O. S., Babalola, O. O., & Odeyemi, O. (2020). Waste management through composting: Challenges and potentials. *Sustainability*, *12*(11), 4456.

Čapla, J., Zajác, P., Ševcová, K., Čurlej, J., & Fikselová, M. (2022). Milk and dairy products – Summary of European legislation, hygiene manuals, ISO standards and Codex Alimentarius standards. *Slovak Journal of Food Sciences*, *16*, 431–462.

Chamas, A., Moon, H., Zheng, J., Qiu, Y., Tabassum, T., Jang, J. H., & Suh, S. (2020). Degradation rates of plastics in the environment. *ACS Sustainable Chemistry & Engineering*, *8*(9), 3494–3511.

Chen, L., Zhang, X., & Wang, H. (2023). Bioplastics from agricultural residues: Opportunities, challenges, and future perspectives. *Journal of Sustainable Materials and Systems*, *15*(3), 45–62.

Cruz, R. M., Krauter, V., Krauter, S., Agriopoulou, S., Weinrich, R., Herbes, C., & Varzakas, T. (2022). Bioplastics for food packaging: Environmental impact, trends and regulatory aspects. *Foods*, *11*(19), 3087.

European Food Safety Authority (EFSA). (2023). Guidelines on safety assessment for food contact materials: Bioplastics focus. *EFSA Journal*, *21*(5), 2134–2156.

Geyer, R., Jambeck, J. R., & Law, K. L. (2023). The environmental toll of plastics: A decade-long analysis of trends and solutions. *Science Advances*, *9*(4), eabc1234.

Gontard, N., Sonesson, U., Birkved, M., Majone, M., Bolzonella, D., Celli, A., & Sebok, A. (2018). A research challenge vision regarding agricultural waste management in a circular bio-based economy. *Critical Reviews in Environmental Science and Technology*, *48*(6), 614–654.

Hussain, S., Akhter, R., & Maktedar, S. S. (2024). Advancements in sustainable food packaging: From eco-friendly materials to innovative technologies. *Sustainable Food Technology*, *2*(5), 1297–1364.

Idris, S. N., Amelia, T. S. M., Bhubalan, K., Lazim, A. M. M., Zakwan, N. A. M. A., Jamaluddin, M. I., ..., & Ramakrishna, S. (2023). The degradation of single-use plastics and commercially viable bioplastics in the environment: A review. *Environmental Research*, *231*, 115988.

Jagaba, A. H., Lawal, I. M., Birniwa, A. H., Affam, A. C., Usman, A. K., Soja, U. B., ..., & Yaro, N. S. A. (2024). Sources of Water Contamination by Heavy Metals. In *Membrane Technologies for Heavy Metal Removal from Water* (pp. 3–27). Boca Raton, FL: CRC Press.

Jurak, M., Wiącek, A. E., Ładniak, A., Przykaza, K., & Szafran, K. (2021). What affects the biocompatibility of polymers? *Advances in Colloid and Interface Science*, *294*, 102451.

Keith, M., Koller, M., & Lackner, M. (2024). Carbon recycling of high value bioplastics: A route to a zero-waste future. *Polymers*, *16*(12), 1621.

Koehler, A., & Wiles, C. (2022). Toxicological assessment and risk management of biodegradable plastics. *Environmental Science & Technology*, *56*(18), 11234–11245.

Lawal, T. O., Abdulsalam, M., Aliyu, M., & Sundararajan, S. (2023). Economic and environmental implications of sustainable agricultural practices in Arid regions: A cross-disciplinary analysis of plant science, management, and economics. *International Journal of Membrane Science*, *10*(3), 3100–3114.

Livinus, M. U., Bala, S. Z., Abdulsalam, M., Innocent, M. O., Hassan, M., Elelu, S. A., & Kini, P. (2024a). Role of Microbes in Soil Food Webs and Vegetation Development. In *Soil Microbiome in Green Technology Sustainability* (pp. 107–132). Cham: Springer Nature Switzerland.

Livinus, M. U., Bala, S. Z., Abdulsalam, M., Musa, I. O., Hassan, M., & Kini, P. (2024b). Marine Genetic Resources of Blue Economy. In *Marine Bioprospecting for Sustainable Blue-Bioeconomy* (pp. 159–180). Cham: Springer Nature Switzerland.

MacLeod, M., Arp, H. P. H., Tekman, M. B., & Jahnke, A. (2021). The global threat from plastic pollution. *Science*, *373*(6550), 61–65.

Meereboer, K. W., Misra, M., & Mohanty, A. K. (2020). Review of recent advances in the biodegradability of polyhydroxyalkanoate (PHA) bioplastics and their composites. *Green Chemistry*, *22*(17), 5519–5558.

Mihai, F. C., Gündoğdu, S., Markley, L. A., Olivelli, A., Khan, F. R., Gwinnett, C., …, & Molinos-Senante, M. (2021). Plastic pollution, waste management issues, and circular economy opportunities in rural communities. *Sustainability*, *14*(1), 20.

Mosomi, E. K., Olanrewaju, O. A., & Adeosun, S. O. (2024). Pivotal role of polylactide in carbon emission reduction: A comprehensive review. *Engineering Reports*, 6(8), e12909.

Musa, I. O., Patrick, O. I., Job, O. S., Abdulsalam, M., Adams, M., Sikirula, A., Oluwafemi, A. O., Ifeoluwa, I. K., & Hephzibah, A. (2024a). Introduction to Nanotoxicology. In *Environmental Nanotoxicology: Combatting the Minute Contaminants* (pp. 1–22). Cham: Springer Nature Switzerland.

Musa, I. O., Samuel, J. O., Adams, M., Abdulsalam, M., Abdulazeez, M., Muhammad, M. S., & Akande, S. A. (2024b) Cases of Micro-and Nanoplastics in Food and Food Product. In *Emerging Contaminants in Food and Food Products* (pp. 90–105). United Kingdom: CRC Press.

Ncube, L. K., Ude, A. U., Ogunmuyiwa, E. N., Zulkifli, R., & Beas, I. N. (2020). Environmental impact of food packaging materials: A review of contemporary development from conventional plastics to polylactic acid based materials. *Materials*, *13*(21), 4994.

Negrete-Bolagay, D., & Guerrero, V. H. (2024). Opportunities and challenges in the application of bioplastics: Perspectives from formulation, processing, and performance. *Polymers*, *16*(18), 2561.

Nizamuddin, S., Baloch, A. J., Chen, C., Arif, M., & Mubarak, N. M. (2024). Bio-based plastics, biodegradable plastics, and compostable plastics: Biodegradation mechanism, biodegradability standards and environmental stratagem. *International Biodeterioration & Biodegradation*, *195*, 105887.

Notaro, S., Lovera, E., & Paletto, A. (2022). Consumers' preferences for bioplastic products: A discrete choice experiment with a focus on purchase drivers. *Journal of Cleaner Production*, *330*, 129870.

O'Loughlin, J., Doherty, D., Herward, B., McGleenan, C., Mahmud, M., Bhagabati, P., & Gaughran, J. (2023). The potential of bio-based polylactic acid (PLA) as an alternative in reusable food containers: A review. *Sustainability*, *15*(21), 15312.

Obuzor, G. U., & Onyedikachi, U. B. (2023). Chemical Leaching into Food and the Environment Poses Health Hazards. In *Modernity in Health and Disease Diagnosis: The Account from STEM Women* (pp. 129–148). Cham: Springer Nature Switzerland.

Ogunola, O. S., Onada, O. A., & Falaye, A. E. (2018). Mitigation measures to avert the impacts of plastics and microplastics in the marine environment (a review). *Environmental Science and Pollution Research*, *25*, 9293–9310.

Otoni, C. G., Azeredo, H. M., Mattos, B. D., Beaumont, M., Correa, D. S., & Rojas, O. J. (2021). The food–materials nexus: Next-generation bioplastics and advanced materials from agri-food residues. *Advanced Materials*, *33*(43), 2102520.

Paganizza, V. (2024). Regulatory Aspects of Food Analogues in Food Sciences. Food Analogues in the European Union: A Legal Perspective. In *Food Analogues: Emerging Methods and Challenges* (pp. 465–489). Cham: Springer Nature Switzerland.

Pandit, P., Nadathur, G. T., Maiti, S., & Regubalan, B. (2018). Functionality and Properties of Bio-Based Materials. In *Bio-Based Materials for Food Packaging: Green and Sustainable Advanced Packaging Materials* (pp. 81–103). Singapore: Springer.

Parveen, N., Naik, S. C. S., Vanapalli, K. R., & Sharma, H. B. (2024). Bioplastic packaging in circular economy: A systems-based policy approach for multi-sectoral challenges. *Science of the Total Environment*, *945*, 173893.

Petkoska, A. T., Daniloski, D., D'Cunha, N. M., Naumovski, N., & Broach, A. T. (2021). Edible packaging: Sustainable solutions and novel trends in food packaging. *Food Research International*, *140*, 109981.

Rai, P. K., & Choure, K. (2023). Agriculture Waste to Bioplastics: A Perfect Substitution of Plastics. In *Value-Addition in Agri-Food Industry Waste through Enzyme Technology* (pp. 299–314). United Kingdom: Academic Press.

Rai, P., Mehrotra, S., Priya, S., Gnansounou, E., & Sharma, S. K. (2021). Recent advances in the sustainable design and applications of biodegradable polymers. *Bioresource Technology*, *325*, 124739.

Riboni, N., Bianchi, F., Cavazza, A., Piergiovanni, M., Mattarozzi, M., & Careri, M. (2023). Mass spectrometry-based techniques for the detection of non-intentionally added substances in bioplastics. *Separations*, *10*(4), 222.

Rosenboom, J. G., Langer, R., & Traverso, G. (2022). Bioplastics for a circular economy. *Nature Reviews Materials*, *7*(2), 117–137.

Shlush, E., & Davidovich-Pinhas, M. (2023). Bioplastics for food packaging. *Trends in Food Science & Technology*, *125*, 66–80.

Singh, S., Gaikwad, K. K., Lee, M., & Lee, Y. S. (2018). Temperature-regulating materials for advanced food packaging applications: A review. *Journal of Food Measurement and Characterization*, *12*, 588–601.

Tan, J., Tiwari, S. K., & Ramakrishna, S. (2021). Single-use plastics in the food services industry: Can it be sustainable? *Materials Circular Economy*, *3*(1), 7.

Thapliyal, D., Karale, M., Diwan, V., Kumra, S., Arya, R. K., & Verros, G. D. (2024). Current status of sustainable food packaging regulations: Global perspective. *Sustainability*, *16*(13), 5554.

Yuvaraj, G., Ramesh, M., & Rajeshkumar, L. (2023). Carbon and cellulose-based nanoparticle-reinforced polymer nanocomposites: A critical review. *Nanomaterials*, *13*(11), 1803.

Zhang, Y., Li, Q., & Liu, H. (2022). Nanotechnology in bioplastics: A new frontier for food packaging and safety. *Advanced Materials Interfaces*, *9*(12), 2200423.

Zhao, D., Zhu, Y., Cheng, W., Chen, W., Wu, Y., & Yu, H. (2021). Cellulose-based flexible functional materials for emerging intelligent electronics. *Advanced Materials*, *33*(28), 2000619.

Ziegler, V., Paraginski, R. T., & Ferreira, C. D. (2021). Grain storage systems and effects of moisture, temperature and time on grain quality – A review. *Journal of Stored Products Research*, *91*, 101770.

Zimmermann, L., Dombrowski, A., Völker, C., & Wagner, M. (2020). Are bioplastics and plant-based materials safer than conventional plastics? In vitro toxicity and chemical composition. *Environment International*, *145*, 106066.

Zoungranan, Y., Lynda, E., Dobi-Brice, K. K., Tchirioua, E., Bakary, C., & Yannick, D. D. (2020). Influence of natural factors on the biodegradation of simple and composite bioplastics based on cassava starch and corn starch. *Journal of Environmental Chemical Engineering*, *8*(5), 104396.

5 Microbial Production of Bio-Based Plastics

A.O. Abdulkareem, Salamatu S. Machunga-Mambula, Sesan Abiodun Aransiola, O.L. Okwute, and B.O. Akanbi

5.1 INTRODUCTION

Since their inception in the 19th century, plastics have pervaded numerous facets of human life due to their lightweight, durable, and corrosion-resistant properties. The per capita annual usage of plastics has been reported to have increased beyond 140 kilograms per year (Filiciotto and Rothenberg, 2021). Nonetheless, while the widespread use of plastics enhances convenience, it also poses significant environmental degradation (Musa et al., 2024; Babaniyi et al., 2023). The indiscriminate utilization of plastics yields copious plastic waste, with less than 10% amenable to recycling, and the bulk thereof is indiscriminately disposed of in the natural environment (Rhodes et al., 2018; Daza et al., 2024; Aransiola et al., 2024a). Traditional plastics exhibit enduring stability in the environment, with instances of the earliest plastic samples produced not yet to be degraded completely (Filiciotto and Rothenberg, 2021; Aransiola et al., 2023b). Consequently, the deleterious environmental repercussions of plastics endure over protracted periods. However, certain plastics undergo degradation in natural surroundings via physical or chemical processes, yielding microplastic byproducts that infiltrate food chains, potentially permeating animal and human biological systems (Law et al., 2014). Recent studies confirm the detection of microplastic elements in some human blood samples (Musa et al., 2024). Consequently, there is a pressing need for new eco-friendly materials to replace the conventional plastics and solve myriads of problems from the environment.

A promising alternative to regular plastics is biodegradable plastics, sharing similar properties and breaking down quickly by microorganisms into CO_2, water, and usable compost, which are harmless substances (Moshood et al., 2021). Most biodegradable plastics which are petroleum-based possess resource consumption issues during production. Additionally, modern or recent raw materials from which bio-based plastics often come are food crops, thereby competing for valuable agricultural resources (Aransiola et al., 2023a). Therefore, there's a pressing need for affordable and sustainable alternatives in the biodegradable plastics market.

Waste-derived lignocellulosic biomass serves as a sustainable alternative to traditional petroleum-based materials. Globally, approximately 150 billion tonnes of lignocellulosic biomass are sourced annually from waste streams like straw and wood chips (Vigneswari et al., 2021; Aransiola et al., 2024b). Direct disposal or incineration of these biomass wastes poses environmental risks, making recycling imperative. Cost-wise, lignocellulosic biomass obtained from waste proves more economical than prevalent bio-based feedstocks, thereby offering the potential of cost-control benefits. Moreover, this is a renewable substrate and does not require a separate allocation of resource, although production of bioplastics from lignocellulosic biomass which is obtained from waste could be demanding. Lignocellulose has a tough structure, making it hard to break down efficiently through environmentally friendly methods (Lorenci et al., 2020). However, recent advancements

DOI: 10.1201/9781003602828-6

in metabolic engineering and synthetic biology have led to significant progress. New bio-based synthesis techniques have emerged for key monomers used in petroleum-based biodegradable plastics, like adipic acid (Zhao et al., 2018) and 1,4-butanediol (Burgard et al., 2016). Additionally, microorganisms play a significant role in making the degradation of lignocellulose both effective and eco-friendly (Shahab et al., 2018). Therefore, the use of microbial cell factories to produce biodegradable plastics from lignocellulosic biomass which is obtained from waste describes a likely green plastic technique.

5.2 DEFINITION AND SIGNIFICANCE OF BIO-BASED PLASTICS

Microbial production of bio-based plastics refers to the use of microorganisms, such as bacteria or fungi, to produce plastics from renewable biomass sources. This method provides a sustainable alternative to the conventional plastics, which are obtained from finite fossil fuel resources and contribute to environmental pollution.

In microbial production, microorganisms are engineered or selected to metabolize specific substrates, such as sugars or plant-derived biomass, and convert them into precursor molecules for plastic synthesis (Auta et al., 2022). These key precursor molecules known as monomers, such as hydroxyalkanoic acid, lactic acid, adipic acid, and 1,4-butanediol, are then polymerized to form biodegradable plastics.

Polymers that can be degraded by microorganisms are biodegradable plastics. When conditions are right, they can degrade into carbon dioxide and water relatively quickly, helping to reduce environmental pollution (Moshood et al., 2021). This makes them eco-friendly alternatives to regular plastics. There are various types of plastics that are biodegradable, such as poly(hydroxyalkanoate) (PHA) (Brandl et al., 1988), poly(lactic acid) (PLA) (Kulkarni et al., 1971), and poly(butylene adipate-co-terephthalate) (PBAT) (Herrera et al., 2002). These plastics made from either natural or petroleum sources display significant differences in biodegradation rates, heat resistance, and mechanical properties and have the possibilities to replace conventional plastics in various uses.

5.3 SOME MICROORGANISMS USED FOR BIOPLASTIC PRODUCTION

PHAs have gained much attention as a potential replacement for conventional thermoplastics. They have been discovered in over 240 bacteria strains and exist in over 100 distinct structural forms (Carlson and Srienc, 2006).

5.3.1 Alcaligenes eutrophus

Above 250 different bacteria strains have been reported to accumulate different PHAs. Several bacteria produce and accumulate PHA as energy and carbon storage materials (Steinbüchel, 1991; Byrom, D., 1994; Yu, 2001; Du and Yu, 2002b; El-Kady, 2008). An example of bacterium that has the ability to achieve this is called *Alcaligenes eutrophus*. The "Biopol", which is a commercial polyester product, is a copolyester containing randomly arranged units of [R]-3-hydroxybutyrate (HB) and [R]-3-hydroxyvalerate (HV) (Lenz and Maechessault, 2005). Large amount of poly-3-hydroxybutyrate (PHB) from simple carbon sources can be accumulated by *A. eutrophus* (Khanna and Srivastsva, 2005). This microorganism grows well in a minimal medium at 30°C on many carbon sources but does not thrive well on glucose medium (Lee et al., 2005).

5.3.2 Pseudomonas

Pseudomonas such as *Pseudomonas oleovorans*, *Pseudomonas putida*, *Pseudomonas aeruginosa*, *Pseudomonas stutzeri*, *Pseudomonas nitroreducens*, and *Pseudomonas mendocina* (Qiang et al.,

2001; Ribera et al., 2001; El-Kady, 2008; El-Sawah et al., 2008) can synthesize PHAs with various monomer structures. A significant amount of PHB can be synthesized by *Pseudomonas fluorescens*, reaching up to 70% of the weight of its dry cell, using inexpensive sugarcane liquor instead of costly yeast extract (Jiang et al., 2008). This substitution of substrates could potentially reduce the production cost of PHB. Numerous researchers have investigated the production of PHB by *Pseudomonas* sp. (Roo et al., 2002), (Ewering et al., 2002), and (Sheu and Lee, 2004).

5.3.3 Azospirillum brasilense and Azospirillum lipoferum

The above organisms can pile up high level of PHB under deficient nutrition. *Azospirillum* sp. can accumulate PHB, reaching up to 40% of its dry cell. In most PHB-producing organism such as *Azospirillum*, minimal PHB is produced during the active stage of cell growth. This means it takes a considerable amount of time for the bacteria to grow into a high-density biomass without PHB before significant PHB accumulation takes place. However, during the exponential stage of growth, *Alcaligenes latus*, *P. putida*, and *Azotobacter vinelandii* are able to accumulate a substantial amount of PHB (Sun et al., 2002).

5.3.4 Archaea

Archaea are ancient microorganisms that thrive in an extreme environmental condition such as high temperature, high salinity, and pH levels, earning them the tag of "extreme halophile". Certain groups of Halobacteriaceae family have been found to amass substantial quantities of the intracellular reserve polymer PHB (Hezayen et al., 2002). Bacteria which produce PHA naturally grow slowly in nature and prefer lower temperature for satisfactory growth. These bacteria are often hard to break open and have pathways that degrade PHA. Unlike these natural PHA producers, bacteria like *Escherichia coli* cannot produce or break down PHA. However, *E. coli* has a high growth rate, even at high temperatures, and breaks open quite easily. This rapid growth rate of *E. coli* allows the accumulation of a large amount of polymer, and its easy disintegration, thereby saving the cost of purifying PHA granules. Scientists are actively researching metabolic engineering to introduce new pathways that expand the range of substrates usable for PHA production and enhance PHA synthesis to create novel forms of the polymer. Recombinant *E. coli* strains have been engineered to carry the PHA-cloned genes from *A. eutrophus* in a stable, high-copy-number plasmid, resulting in high production of PHA. Since *E. coli* can thrive on different sources of carbon such as glucose, lactose, sucrose, and xylose, there is potential for further cost reduction by using cheap substrates like whey, molasses, and hemicellulose hydrolysate (Reddy et al., 2003; Ganduri et al., 2005).

5.3.5 Actinomycetes

The production of PHB was observed in *Nocardia* sp. and *Streptomyces* sp. (Ramachander and Rawal, 2005). Several species of *Streptomyces* (*S. albus*, *S. venezuelae*, *S. olivaceus*, *S. fradiae*, *S. lividans*, *S. kanamyceticus*, *S. coelicolor*, *S. rosa*, and *S. parvus*) produce PHB, which was carried out and observed that all the tested isolates produced PHB with wide difference in the PHB produced. Only a few bacteria which produce PHA were used in the production of bioplastics due to the different growth condition requirements.

5.4 BIOPLASTIC PRODUCTION PROCESS

PHA is produced and stored within the cells of many bacteria under adverse growth condition by limiting the availability of macronutrient, such as nitrogen, oxygen, phosphorus, or magnesium, under an abundant carbon supply (Du et al., 2001; Du and Yu, 2002a; and Lee et al., 2005). Strategies are underway in order to stimulate conditions for effective production of PHAs (Yu, 2001; Du et al., 2001; and Du and Yu, 2002b).

Some bacteria that can accumulate PHA during growth in the absence of nutrient limitation are *A. eutrophus*, *A. latus*, and a mutant strain of *A. vinelandii.*

Several factors must be considered when selecting microorganisms for industrial PHA production, including their capability to exploit inexpensive sources of carbon, its producing rate of polymer, growth rate, and the maximum amount of polymer accumulation achievable based on the substrate. Some researchers have developed equations to predict PHA yield from various carbon sources, which can be helpful to calculate the preliminary probable yields of PHA (Yu and Wang, 2001).

In order to minimize overall costs, it is crucial to achieve high productivity and yield in PHA production. Various methods, including continuous cultivations and fed-batch culture, have been employed to enhance production (Du and Yu, 2002a, 2002b; Du et al., 2001 and Yu and Wang, 2001). While only a few prominent PHAs, such as PHB, poly(3-hydroxybutyrate-co-3-hydroxyvalerate), and poly(3-hydroxyhexanoate-co-3-hydroxyoctanoate), have been produced in high concentrations with high productivity, recent efforts have focused on investigating strategies of cultivation using cheap, renewable carbon substrates to reduce production costs while maintaining high yields (Yu and Wang, 2001).

5.4.1 Production of PHA on a Commercial Scale

Several companies worldwide are now industrially producing PHA for its use as natural, biocompatible, and biodegradable thermoplastics, which have various potential applications. Among the PHAs, PHB is the most well-known. It has been demonstrated that bacteria can completely biodegrade PHB into carbon dioxide and water (and also, the release of methane under anaerobic conditions) in natural environments such as soil, compost, and water (Lee et al., 2005). Many research have been directed toward making PHA production more efficient by using low-cost carbon sources. This is because the production cost of PHA is still relatively high compared to chemically produced plastics. By making use of sugars as cheap sources of carbon, researchers have successfully achieved high productivity in PHB production through optimized fermentation processes. Despite numerous efforts, PHB production from sugars has a significant drawback: an insufficient yield of PHA from the carbon source. The yield of PHB from glucose typically ranges from 0.30 to 0.40 grams of PHB per gram of glucose used. To further reduce production costs, it is essential to replace the sugar with a more affordable carbon source that allows bacteria to achieve a greater yield of PHA. Plant oils serve as promising feedstock for PHA production since they offer cost-effective sources of carbon. Plant oils, with their higher carbon atom content per unit weight, offer a theoretical yield coefficient for PHA production that can surpass 1.0 gram of PHA per gram of oil, unlike other sources of carbon (Kahar et al., 2004).

5.4.2 Decomposition of PHB by Fungi

Fungi, such as *Aspergillus fumigatus*, *Fusarium solani*, *Phanerochaete chrysosporium*, and *Penicillium* sp. isolated from various samples obtained from the environment, demonstrated the ability to degrade PHB. They are consistent with their inherent role as degraders of organic polymer. Samples with diverse fungal isolates, such as hay, compost, and garden soil, likely contained greater bacterial diversity and thus supported the existence of PHB-degrading fungi. Surprisingly, lichen was also found to harbor PHB-degrading fungi, which makes sense given lichen's nutritional conditions favor bacterial synthesis of PHB. Identifying the specific bacteria associated with lichen involved in PHB synthesis could help understand carbon flow perturbations in lichens, organisms negatively impacted by air pollution (Lee et al., 2005).

PHA degradation in aerobic environments result in the release of carbon dioxide and water, while under anaerobic conditions methane is also released. Various studies have investigated the rate at which PHB and poly(HB-HV) degrade in diverse environments. Degradation is fastest in anaerobic sewage while it is slowest in seawater. For instance, poly(HB-HV) degrades completely after 6 weeks in anaerobic sewage, 75 weeks in soil, and 350 weeks in seawater (Ojumu et al., 2004).

5.5 APPLICATION AND BENEFITS OF BIO-BASED PLASTICS

PHAs are natural thermoplastic polyesters mainly used as substitutes for petroleum-based polymers, especially in packaging and coating. PHAs are typically biodegradable and exhibit excellent biocompatibility, which makes them appealing for use as biomaterials in tissue engineering.

Biodegradable polymers find wide applications in medical, agriculture, and consumer goods packaging, as shown in Figure 5.1. Medical devices, such as sutures, bone plates, and tissue repair devices, have seen rapid development due to their specialized nature and higher value. PHA and its composites are utilized in a range of medical devices, providing appropriate mechanical properties, biocompatibility, and degradation rates that align with specific physiological conditions (Chen and Wu, 2005).

Besides their various material applications and properties, PHAs have the potential to yield small molecules. Through chemical hydrolysis, PHA can be degraded into monomers, which can then be converted into commercially valuable compounds like β-hydroxy acids, 2-alkenoic acids, β-amino acids, β-hydroxyalkanoates β-acyl lactones, and β-hydroxyacid esters (Williams and Peoples, 1996).

5.6 BIODEGRADABLE PLASTICS

Biodegradable polymers, a type of environmentally friendly plastic, are sourced from renewable materials like biomass, starch, fats, and oils (Babu et al., 2013). They are derived from waste materials such as food, agriculture, and vegetable wastes, as well as renewable feedstock like biomass. These bioplastics offer the advantage of not causing environmental pollution, depleting fossil fuel resources, or posing any danger to humans (Paul et al., 2021). Microbial populations can be harnessed to yield biodegradable plastics from bio-wastes, offering a sustainable solution (Kalia et al., 2000). Biodegradable polymers naturally decompose in the environment, facilitated by microorganisms metabolizing and breaking down their molecular structure into harmless molecules. There are two types: those that degrade through aerobic processes like composting and those that degrade in anaerobic environments like landfills. This degradation process results in the conversion of the substance into carbon dioxide, methane, water, and biomass by microorganisms within specific conditions and timeframes (Leja and Lewandowicz, 2010). Examples of biodegradable plastics include PHB (Babaniyi and Adelaja, 2020), PLA, and thermoplastic starch (Adelaja and Babaniyi, 2020; Kharb and Saharan, 2022).

Biodegradable plastics offer a viable alternative to traditional plastics as they can be used in similar ways (Onen et al., 2020). These materials provide a practical solution for achieving environmental sustainability because they easily degrade and are compatible with living organisms. Moreover, they are sourced from renewable and natural resources like wood pulp, starch, and vegetable oils, thereby making them highly versatile for various industries, including biomedical, agriculture, packaging, textiles, and more (Prasanth et al., 2021). The key benefits of using biodegradable plastics lie in their utilization of sustainable feedstock and their ability to break down naturally (Mohammadi et al., 2013) (Figure 5.2).

FIGURE 5.1 Some applications of bioplastics.

5.6.1 Environmental Impacts

Using materials like bio-based fibers and biopolymers from natural and sustainable sources can help solve environmental problems caused by regular plastics. Biodegradable plastics are a big step forward that could boost the bio-economy and reduce our need for fossil fuels, switching to more eco-friendly alternatives (Nanda et al., 2022). The mildness of biodegradable polymers make them environmentally friendly since their production lead to the emission of less carbon dioxide, which contribute to global warming. The use of renewable feedstock in plastic manufacture can help decrease the emission of greenhouse gases, including CH_4 and CO_2, into the atmosphere compared to traditional plastic syntheses from coal or petroleum (Barker, 2009). Table 5.1 highlights some comparisons between bioplastics and conventional plastics.

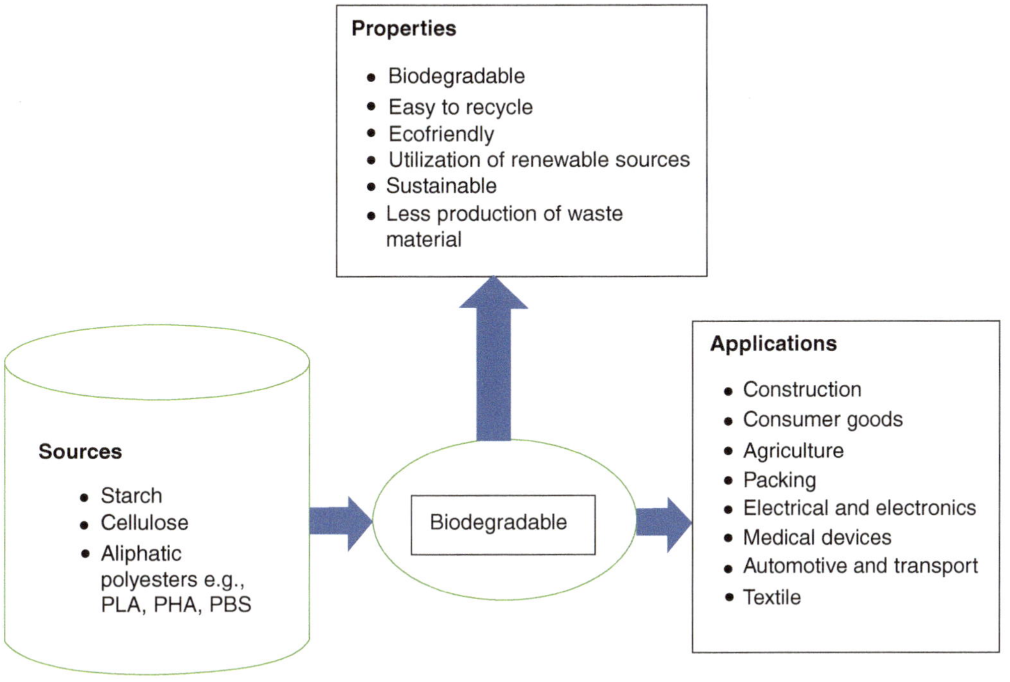

FIGURE 5.2 Sources, properties, and uses of biodegradable plastics.

TABLE 5.1
Biodegradable versus Conventional Plastics

Biodegradable Plastics	Conventional Plastics (Petroleum-Based)
• Obtained from renewable sources e.g., sugarcane, corn, and microbes.	• Obtained from non-renewable fossil fuels e.g., crude oil.
• Designed to breakdown into natural substances like water, carbon dioxide, and biomass.	• Breaks down into microplastics, which persist in the environment and can be harmful to wildlife.
• Biodegradable in specific conditions such as industrial composting; some can degrade in marine environments.	• Non-biodegradable; persists in the environment and contribute to landfill accumulation.
• It can degrade within months to years under specific conditions (e.g., composting).	• It takes several hundreds to thousands of years to degrade, leading to prolonged environmental pollution.
• Produced through microbial fermentation or chemical processes using renewable feedstocks.	• Produced through the refining and polymerization of petrochemical.
• Examples are as follows: polyhydroxyalkanoates (PHA), polylactic acid (PLA), and starch-based plastic.	• Examples include the following: polyethelene (PE), polypropelene (PP), polyvinyl chloride (PVC), and polysterene (PS).

5.7 POTENTIAL INDUSTRIAL USES

There are various applications of biodegradable plastics, the dominant being packaging business. PLA, PHAs, polybutylene succinate, starch, and cellulose polymers have inherent biodegradability and sustainability, making them highly suitable for replacing synthetic plastics in numerous applications (Babaniyi et al., 2020; Kharb and Saharan, 2022). Other sectors where biodegradable plastics have wide application include but are not limited to textiles, consumer products, transportation and automotive, horticulture and agriculture, building, electrical and electronics, adhesives and coatings.

5.8 GENETIC ENGINEERING'S ROLE IN BIOPLASTIC PRODUCTION

In 1988, some scientists at James Madison University successfully cloned the whole set of genes responsible for synthesis of PHB from acetyl CoA in *Ralstonia eutropha*. These three genes which are organized into one operon were then transferred into *E. coli*. The genetically modified *E. coli* strain, containing this operon, can produce all the three enzymes necessary for the synthesis of PHB from numerous organic compounds. Some engineered strains of *E. coli* can even produce the HB/HV copolymer. Alternatively, as reported by researchers at the Massachusetts Institute of Technology in 1994, strains with only the synthase gene can produce this protein in sufficient quantities for isolation and purification (Jung et al., 2004 and Lenz and Marchessault, 2005).

5.9 CHALLENGES OF BIOPLASTICS

Bioplastics are often described as eco-friendly alternatives to traditional plastics. However, their production poses challenges, especially concerning potential competition with food sources. To mitigate this, second-generation bioplastics are made from non-food resources. Yet, manufacturing these requires processes like extrusion, compression, and injection molding. The environmental implications of bioplastics remain incompletely understood, necessitating further research to address limited resources, improve efficiency, and minimize environmental impact.

Moreover, certain bioplastics, derived from bacterial polymer PLA, only biodegrade under specific temperature and humidity conditions, limiting their degradation potential (Zulkafli et al., 2014). Overcoming this constraint is essential to ensure bioplastics degrade effectively in various landfill environments. Utilizing agricultural fibers for bioplastic production presents lucrative market opportunities. However, economic factors alone are insufficient to drive widespread adoption of this technology. Enhancements in the performance of natural fiber composites and green composites are necessary to expand their applications across industries (Misra et al., 2009).

The advancement of bioplastics hinges largely on the credibility attained through new technologies and the integrity of companies involved in manufacturing, marketing, and promoting sustainable practices (Thakur et al., 2018). Achieving sustainability entails engaging in dialogue with communities about the role of bioplastics in future service provision. It also involves enhancing biodegradability, exploring agricultural applications, and establishing practical recycling facilities (Iles et al., 2013).

5.10 CONCLUSION

The rising concern over environmental impact from non-degradable waste is driving research into innovative biodegradable materials made from natural sources like biomass, plants, and bacteria. These advancements in bioplastics promise increased production efficiency, expanded applications, and new opportunities in the future. Moreover, the growing emphasis on sustainability is expected to boost the market for bioplastics. Additionally, advancements in biotechnology, particularly in microbial processes, offer promising prospects for bioplastic manufacturing across various industries, including agriculture, medical, pharmaceutical, veterinary, and more.

Therefore, it is essential to establish new guidelines and standards for the production, usage, and waste management of bioplastics worldwide. This entails strengthening labeling legislation based on factors such as substrate usage, energy consumption, and emissions from manufacturing and usage. Bio-based and biodegradable materials hold significant potential for composting purposes. Continuous technological advancements, ongoing innovation, and global support are crucial for the commercialization and demonstration of bioplastics. However, it is imperative that bioplastic development aligns with integrated environmentally friendly practices to enhance sustainability throughout their lifecycle. Additionally, bioplastic materials should not compete with traditional sources and should aim to reduce reliance on non-renewable resources in the long term.

REFERENCES

Adelaja, O.A., and Babaniyi, B.R. (2020). Mechanical and morphological characterization of plasticized low-density polyethylene (LDPE) granules blend with polyhydroxybutyrate (PHB). *International Journal of Advanced Materials Research*, 6(2): 29–36.

Aransiola, S.A., Atta, H.I., and Maddela, N.R. (Eds.). (2024a). *Soil Microbiome in Green Technology Sustainability*. Springer. https://link.springer.com/book/9783031718434

Aransiola S.A., Babaniyi, B.R., Aransiola, A.B., and Maddela, N.R. (Eds.). (2024b). *Prospects for Soil Regeneration and Its Impact on Environmental Protection*. (1st ed.). Springer. https://link.springer.com/book/9783031532696

Aransiola, S.A., Ikhumetse, A.A., Babaniyi, B.R., Abioye O.P., Oyedele, O.J., and Falade, N.O. (2023a). Phytoaccumulation of Micro and Nanoplastics: Root Uptake. In Maddela, N.R., Reddy, K.V., Ranjit, P. (eds), *Micro and Nanoplastics in Soil*. Springer. https://doi.org/10.10007/978-3-031-21195-9_8

Aransiola, S.A., Victor-Ekwebelem, M.O., Ajiboye, A.E., Leh-Togi Zobeashia, S. S., Ijah, U.J.J., and Oyedele, O.J. (2023b). Ecological Impacts and Toxicity of Micro and Nanoplastics in Agroecosystem. In Maddela, N.R., Reddy, K.V., Ranjit, P. (eds), *Micro and Nanoplastics in Soil*. Springer. https://doi.org/10.10007/978-3-031-21195-9_10

Auta, H.S., Aboyeji, D.O., Bala, J.D., Abioye, O.P., Adabara, N.U., Aransiola, S.A., Hassana, A., and Azize, A. (2022). Marine Microbial Enzymes – An Overview. In Maddela, N.R., Aransiola S.A., Prasad, R. (eds), *Ecological Interplays in Microbial Enzymology* (1st ed.). Springer Nature Singapore Pte Ltd. ISSN: 2662-1681. www.springer.com/series/16324

Babaniyi, B.R., and Adelaja O.A. (2020). Production of Poly-β-hydroxybutyric acid (PHB) by *Bacillus cereus* on pineapple peels. *GSC Advanced Research and Reviews*, 4(1): 24–30.

Babaniyi, B.R., Afolabi, F., and Obagunwa, M.P. (2020). Recycling of used polyethylene through solvent blending of plasticized polyhydroxybutyrate and its degradation potential. *Composites Part C: Open Access*, 2(2020): 100021.

Babaniyi, B.R., Ogundele, O.D., Bisi-omotosho, A., Babaniyi, E.E., and Aransiola, S.A. (2023). Remediation Approaches in Environmental Sustainability. In Maddela, N.R., Kretli Winkelstroter Eller, L., Prasad, R. (eds), *Microbiology for Cleaner Production and Environmental Sustainability*. CRC Press. ISBN: 9781032496061. www.routledge.com/Microbiology-for-Cleaner-Production-and-Environmental-Sustainability/Maddela-Eller-Prasad/p/book/9781032496061

Babu, R.P., O'connor, K., and Seeram, R. (2013). Current progress on bio-based polymers and their future trends. *Progress in Biomaterials*, 2: 1–16.

Barker, M. (2009). *Industrial Uses for Crops: Bioplastics*. HGCA.

Brandl, H., Gross, R.A., Lenz, R.W., and Fuller, R.C. (1988). *Pseudomonas oleovorans* as a source of poly(beta-hydroxyalkanoates) for potential applications as biodegradable polyesters. *Application of Environmental Microbiology*, 54: 1977–1982.

Burgard, A., Burk, M.J., Osterhout, R., Van Dien, S., and Yim, H. (2016). Development of a commercial scale process for production of 1,4-butanediol from sugar. *Current Opinion in Biotechnology*, 42: 118–125.

Byrom, D. (1994). Polyhydroxyalkanoates. In D. P. Mobley (ed), *Plastics from Microbes: Microbial Synthesis of Polymers and Polymer Precursors* (pp. 5–33).

Carlson, R., and Srienc, F. (2006). Effects of recombinant precursor pathway variations on poly-R-3-hydroxybutyrate synthesis in Saccharomyces cerevisiae. *Journal of Biotechnology*, 124: 561–573.

Chen, G.Q., and Wu, Q. (2005). The application of polyhydroxyalkanoates as tissue engineering materials. *Biomaterials*, 26: 6565–6578.

Daza, B.X.D., Mendoza, A.J.D., Zambrano, J.J.Z., Aransiola, S.A., and Maddela, N.R. (2024). Effects of Soil Contaminants on Soil Microbiome. In Aransiola, S.A., Atta, H.I., Maddela, N.R. (eds), *Soil Microbiome in Green Technology Sustainability*. Springer. https://doi.org/10.1007/978-3-031-71844-1_7

Du, G., Si, Y., and Yu, J. (2001). Inhibitory effect of medium-chain-length fatty acid on synthesis of polyhydroxyalkanoates from volatile fatty acid by *Ralstonia eutrophus*. *Biotechnology Letters*, 23: 1613–1617.

Du, G., and Yu, J.(2002a). Production of Polyhydroxyalkanoates by *Ralstonia eutropha* from various carbon sources. *Bioresource Technology*, 84(3): 207–213.

Du, G., and Yu, J. (2002b). Metabolic analysis on fatty acid utilization *by Pseudomonas oleovorans*: mcl-poly(3-hydroxyalkanoates) synthesis versus β-oxidation. *Process Biochemistry*, 38(3): 325–332.

El-Kady, S. (2010). Bioplastic production from inexpensive sources: Bacterial biosynthesis, cultivation system, production, and biodegradability.

El-Sawah, A.A., El-Sakhawy, M.A., and El-Sakhawy, M.A.M. (2008). Production of polyhydroxyalkanoates (PHAs) by Pseudomonas species. *Egyptian Journal of Microbiology*, 43(1): 1–14.

Ewering, C., Eversloh, T.L., Luftmann, H., and Steinbüchel, A. (2002). Identification of novel sulfur containing bacterial polyesters: Biosynthesis of poly(3-hydroxy-S-propyl-x-thioalkanoates) containing thioether linkages in the side chains. *Microbiology*, 148: 1397–1406.

Filiciotto, L., and Rothenberg, G. (2021). Biodegradable plastics: Standards, policies, and impacts. *Chemistry and Sustainability, Energy and Materials*, 14: 56–72.

Ganduri, V.S.R.K., Ghosh, S., and Patnaik, P.R. (2005). Mixing control as a device to increase PHB production in batch fermentations with co-cultures of *Lactobacillus delbrueckii* and *Ralstonia eutropha. Process Biochemistry*, 40: 257–264.

Herrera, R., Franco, L., Rodrı´guez-Gala´ n, A., and Puiggalı´, J. (2002). Characterization and degradation behavior of poly(butylene adipate-co-terephthalate)s. *Journal of Polymer Science Part A. Polymer Chemistry*, 40, 4141–4157.

Hezayen, F.F., Steinbuchel, A., and Rehm, B.H.A. (2002). Biochemical and enzymological properties of the polyhydroxybutyrate synthase from the extremely halophilic archaeon strain 56. *Archives of Biochemistry and Biophysics*, 403: 284–291.

Iles, A., and Martin, A.N. 2013. Expanding bioplastics production: Sustainable business innovation in the chemical industry. *Journal of Cleaner Production*, 45: 38–49.

Jiang, Y., Song, X., Gong, L., Li, P., Dai, C., and Shao, W. (2008). High poly-β-hydroxybutyrate production by *Pseudomonas fluorescens* A2a5 from inexpensive substrates. *Enzyme and Microbial Technology*, 42: 167–172.

Jung, Y.M., Lee, J.N., Shin, H.D., and Lee, Y.H. (2004). Role of *tktA* gene in pentose phosphate pathway on odd-ball biosynthesis of poly-β-hydroxybutyrate in transformant *Escherichia coli* Harboring *phbCAB* operon. *Journal of Bioscience and Bioengineering*, 98(3): 224–227.

Kahar, P., Tsuge, T., Taguchi, K., and Doi, Y. (2004). High yield production of polyhydroxyalkanoates from soybean oil by *Ralstonia eutropha* and its recombinant strain. *Polymer Degradation and Stability*, 83: 79–86.

Kalia, V.C., Raizada, N., and Sonakya, V. (2000). Bioplastics. *Journal of Scientific and Industrial Research*, 59(6): 433–445.

Khanna, S., and Srivastava, A.K. (2005). Statistical media optimization studies for growth and PHB production by *Ralstonia eutropha. Process of Biochemistry*, 40: 2173–2182.

Kharb, J., and Saharan, R. (2022). Sustainable biodegradable plastics and their applications: A mini review. IOP Conference Series. *Materials Science and Engineering*, 1248(1), 12008.

Kulkarni, R.K., Moore, E.G., Hegyeli, A.F., and Leonard, F. (1971). Biodegradable poly (lactic acid) polymers. *Journal of Biomedical Materials Research*, 5: 169–181.

Law, K.L., and Thompson, R.C. (2014). Oceans. Microplastics in the seas. *Science*, 345: 144–145.

Lee, K.M., Gimore, D.F., and Huss, M.J. (2005). Fungal degradation of the bioplastic PHB (Poly-3-hydroxybutyric acid). *Journal of Polymers and Environment*, 13(3): 213–219.

Leja, K., and Lewandowicz, G. (2010). Polymer biodegradation and biodegradable polymers-a review. *Polish Journal of Environmental Studies*, 19(2): 255–266.

Lenz, R.W., and Marchessault, R.H. (2005). Bacterial polyesters: Biosynthesis, biodegradable plastics and biotechnology. *American Chemical Society*, 6(1): 1–8.

Lorenci Woiciechowski, A., Dalmas Neto, C.J., Porto de Souza Vandenberghe, L., de Carvalho Neto, D.P., Novak Sydney, A.C., Letti, L.A.J., Karp, S.G., Zevallos Torres, L.A., and Soccol, C.R. (2020). Lignocellulosic biomass: Acid and alkaline pretreatments and their effects on biomass recalcitrance-conventional processing and recent advances. *Bioresource Technology*, 304: 122848.

Misra, M., Nagarajan V., Reddy, J., and Mohanty, A. K. (2009). Bioplastics and green composites from renewable resources: Where we are and future directions. *International Conference on Composite Materials*, 18: 1–5.

Mohammadi Nafchi, A., Moradpour, M., Saeidi, M., and Alias, A.K. (2013). Thermoplastic starches: Properties, challenges, and prospects. *Starch-Stärke*, 65(1–2): 61–72.

Moshood, T.D., Nawanir, G., and Mahmud, F. (2021). Sustainability of biodegradable plastics: A review on social, economic, and environmental factors. *Critical Reviews in Biotechnology*, 16: 1–21.

Musa, I.O., Auta, H.S., Ilyasu, U.S., Aransiola, S.A., Makun, H.A., Adabara, N.U., Abioye, O.P., Aziz, A., Jayanthi, B., Maddela, N.R., and Prasad, R. (2023). Micro- and nanoplastics in environment: Degradation, detection, and ecological impact. *International Journal of Environmental Research*, 17(1), Article 551.

Nanda, S., Patra, B.R., Patel, R., Bakos, J., and Dalai, A.K. (2022). Innovations in applications and prospects of bioplastics and biopolymers: A review. *Environmental Chemistry Letters*, 20(1): 379–395.

Ojumu, T.V., Yu, J., and Solomon, B.O. (2004). Production of Polyhydroxyalkanoates, a bacterial biodegradable polymer. *African Journal of Biotechnology*, 3(1): 18–24.

Onen Cinar, S., Chong, Z.K., Kucuker, M.A., Wieczorek, N., Cengiz, U., and Kuchta, K. (2020). Bioplastic production from microalgae: A review. *International Journal of Environmental Research and Public Health*, 17(11): 3842.

Paul, S., Sen, B., Das, S., Abbas, S.J., Pradhan, S.N., Sen, K., and Ali, S.I. (2021). Incarnation of bioplastics: Recuperation of plastic pollution. *International Journal of Environmental Analytical Chemistry*, Article 1–24.

Prasanth, S.M., Kumar, P.S., Harish, S., Rishikesh, M., Nanda, S., and Vo, D.-V.N. (2021). Application of biomass derived products in mid-size automotive industries: A review. *Chemosphere*, 280, 130723.

Qiang, C.G., Jun, X., Qiong, W., Zengming, Z., and Ping, H.K. (2001). Synthesis of copolyesters consisting of medium-chain-length β-hydroxyalkanoates by *Pseudomonas stutzeri* 1317. *Reactive Functional Polymer*, 48: 107–112.

Ramachander, T.V.N., and Rawal, S.K. (2005). PHB synthase from *Streptomyces aureofaciens* NRRL 2209. *FEMS Microbiol Letters*, 242: 13–18.

Reddy, C.S.K., Ghai, R., and Kalia, R.V.C. (2003). Polyhydroxyalkanoates: An overview. *Bioresource Technology*, 87: 137–146.

Rhodes, C.J. (2018). Plastic pollution and potential solutions. *Science Progress*, 101: 207–260.

Ribera, R.G., Monteoliva-Sánchez, M., and Ramos-Cormenzana, A. (2001). Production of polyhydroxyalkanoates by Pseudomonas putida KT2442 harboring pSK2665 in wastewater from olive oil mills (alpechin). *Electronic Journal of Biotechnology*, 4(2): 116–119.

Roo, G., Kellerhals, M.B., Ren, Q., Witholt, B., and Kessler, B. (2002). Production of chiral R-3-hydroxyalkanoic acid and R-3-hydroxyalkanoic acid methylesters via hydrolytic degradation of polyhydroxyalkanoate synthesized by Pseudomonads. *Biotechnology and Bioengineering*, 77(6): 717–722.

Shahab, R.L., Luterbacher, J.S., Brethauer, S., and Studer, M.H. (2018). Consolidated bioprocessing of lignocellulosic biomass to lactic acid by a synthetic fungal-bacterial consortium. *Biotechnology and Bioengineering*, 115: 1207–1215.

Sheu, D.S., and Lee, C.Y. (2004). Altering the substrate specificity of polyhydroxyalkanoate synthase derived from *Pseudomonas putida* GPo by localized semirandom mutagenesis. *Journal of Microbiology*, 186 (13): 4177–4184.

Steinbüchel, A. (1991). Polyhydroxyalkanoic Acids. In Byrom, D. (ed), *Biomaterials: Novel Materials from Biological Sources*. Stockton 124–213.

Sun, J., Dommelen, A.V., Impe, J.V., and Vanderleyden, J. (2002). Involvement of glnB, glnZ, and glnD genes in the regulation of poly-3-hydroxybutyrate biosynthesis by ammonia in *Azospirillum brasilense* Sp7. *Applied and Environmental Microbiology*, 68(2): 985–988.

Thakur, S., Chaudhary, J., and Sharma, B. 2018. Sustainability of bioplastics: Opportunities and challenges. *Current Opinion in Green and Sustainable Chemistry*, 13: 68–75.

Vigneswari, S., Noor, M.S.M., Amelia, T.S.M., Balakrishnan, K., Adnan, A., Bhubalan, K., Amirul, A.A.A., and Ramakrishna, S. (2021). Recent advances in the biosynthesis of polyhydroxyalkanoates from lignocellulosic feedstocks. *Life*, *11*, 807.

Williams, S.F., and Peoples, O.P. (1996). Biodegradable plastics from plants. *CHEMTECH*, 26(7): 3–44.

Yu, J. (2001). Production of PHA from starch wastewater via organic acids. *Journal of Biotechnology*, 86: 105–112.

Yu, J., and Wang, J. (2001). Metabolic flux modelling of detoxification of acetic acid by *Ralstonia eutrophus* at slightly alkaline pH levels. *Biotechnology and Bioengineering*, 73: 458–464.

Zhao, M., Huang, D., Zhang, X., Koffas, M.A.G., Zhou, J., and Deng, Y. (2018). Metabolic engineering of *Escherichia coli* for producing adipic acid through the reverse adipate-degradation pathway. *Metabolic Engineering*, 47: 254–262.

Zulkafli, N.N. (2014). Production of bioplastic from agricultural waste (Master's thesis). Universiti Malaysia Pahang, Faculty of Chemical and Natural Resources Engineering. .

Section II

Interplay of Additives with Agriculture

6 Soil Microbial Diversity in Bioplastics Remediation

Aisha Eniola Adegbite[1], Amina Yahaya, Victoria Oyinlola Olayiwola, Sesan Abiodun Aransiola, and Naga Raju Maddela

6.1 INTRODUCTIONS

The prevalence of plastic pollution has become one of the most pressing environmental challenges of the 21st century (Musa *et al.*, 2024). Conventional petroleum-based plastics, characterized by their durability and resistance to degradation, have been widely used across industries, leading to significant ecological consequences. According to Geyer *et al.* (2017), an estimated 8 million metric tons of plastic wastes enter the oceans annually, while large quantities accumulate in terrestrial environments, and agricultural soil. This persistent plastic waste disrupts ecosystems, threatens diversity, and contributes to microplastic pollution. As global awareness of these issues rises, bioplastics have emerged as a promising alternative.

Bioplastics, derived from renewable resources such as starch, polylactic acid (PLA), and polyhydroxyalkanoates (PHAs), are designed to be biodegradable or compostable under certain conditions (Rillig *et al.*, 2017; Ahsan *et al.*, 2023). These materials offer a potential solution to reduce reliance on fossil fuels and mitigate the accumulation of non-degradable plastics in the environment. Nevertheless, their degradation dynamics in natural settings, particularly in soil ecosystems, are influenced by various factors, including the composition of the bioplastic, environmental conditions, and the diversity of soil microbial communities (Lackner *et al.*, 2023; Tokiwa *et al.*, 2009). Understanding the microbial processes involved in bioplastics remediation is crucial to evaluate their true ecological impact and optimize their use as a sustainable alternative (Rajvanshi *et al.*, 2023).

Soil is one of the most biologically diverse habitats on earth, harbouring an estimated 10^9 microbial cells per gram and thousands of distinct microbial species (Aransiola *et al.*, 2023b). These microorganisms play a crucial role in nutrient cycling, organic matter decomposition, and the maintenance of soil health (Schimel and Schaeffer, 2012; Aransiola *et al.*, 2023a). In bioplastic remediation, microbes possess enzymatic capabilities to break down complex polymers into simpler compounds that can be assimilated into microbial metabolic pathways (Wani *et al.*, 2023). For instance, microbial enzymes such as depolymerases, oxireductases, and esterases have been instrumental in the degradation of PLA and PHA (Pathak and Navneet, 2017). The efficiency of these processes is often linked to the diversity and functional potential of microbial communities present in the soil.

Microbial diversity in soil can be classified into taxonomic, functional, and phylogenetic diversity (Di Lenola *et al.*, 2017). Taxonomic diversity refers to the variety of microbial species or operational taxonomic units (OTUs), while functional diversity is the range of metabolic function and processes that these microorganism can perform (Mishra, 2015). Phylogenetic diversity encompasses the evolutionary relationships among microbial taxa. Each of these dimensions of diversity influences

DOI: 10.1201/9781003602828-8

the capacity of soil microbial communities to degrade bioplastics (Brown and Jumpponen, 2015). Highly diverse microbial communities are more likely to harbour organisms with specific enzymes required for bioplastic degradation, thereby enhancing the overall efficiency of the remediation process (Othman *et al.*, 2021).

Despite the potential of bioplastics to alleviate plastic pollution, their degradation in soil environments remains inconsistent. Research indicates that factors such as soil temperature, moisture content, pH, and nutrient availability significantly affect the activity of microbial communities and, consequently, the degradation rate of bioplastics (Adhikari *et al.*, 2016). Moreover, the chemical compositions of bioplastics, including the presence of additives or copolymers, can either enhance or inhibit microbial degradation. For example, studies have shown that while PLA degrades efficiently in industrial composting conditions, its breakdown in natural soil environment is much slower due to insufficient microbial activity or suboptimal environmental conditions (Castro-Aguirre *et al.*, 2018).

Recent advances in molecular biology have provided new insights into the microbial ecology of bioplastics degradation. Metagenomics, transcriptomics, and proteomics are powerful tools to unravel the genetic and functional potential of soil microbial communities involved in the process (Abiraami *et al.*, 2020). These approaches enable the identification of key microbial taxa, enzymes, and metabolic pathways associated with bioplastics degradation. For instance, members of the genera Pseudomonas, Bacillus, and Actinobacteria have been frequently reported to exhibit biodegradation capabilities for various bioplastics (Malik *et al.*, 2023). However, there remains a gap in understanding how these microorganisms interact within complex soil ecosystems and the extent to which environmental factors mediate these interactions (Chandra and Enespa, 2019).

Another critical aspect to consider is the potential ecological impact of bioplastics' degradation products on soil health and microbial communities. The breakdown of bioplastics generates by-products, including monomers, oligomers, and organic acids, which may influence microbial community dynamics and soil chemistry (Sun *et al.*, 2022). While some of these by-products may be readily assimilated into microbial pathways, other could potentially disrupt soil ecosystems if they accumulate to toxic levels. Thus, it is essential to assess the long-term implications of bioplastics remediation on soil ecosystems to ensure their environmental sustainability (Fan *et al.*, 2022).

6.2 OVERVIEW OF BIOPLASTICS

Bioplastics are a wide category of biomaterials that can be derived from degradable biomass sources like sugarcane and cellulose. These biomaterials are suitable alternatives to conventional plastics derived from fossil fuels and these biomaterials are said to be sustainable. Bioplastics are often considered an environmentally friendly alternative to conventional plastics.

According to Dilshad *et al.* (2021), it is said that plastic materials derive their properties from the combination of polymer(s) and additives, which also applies to fossils and bio-based plastics. Bioplastic(s) is/are a derivation of biopolymers like the combination of PLA and additives for processing and colouration, starch and plasticizer or fibres (natural) reinforced with poly-3-hydroxybutyrate (P3HB).

According to Green Compostables (2019), bioplastics are categorized into three main groups, each is further subdivided based on its composition and properties as shown in Figure 6.1:

i. Bio-Based Plastics: These bioplastics are made out of bio polyesters. Examples include polyethylene (PE), polyethylene terephthalate (PET) and certain polyamides (PA), and polyurethanes (PUR).
ii. Bio-Based and Biodegradable Plastics: This group consists of plastics like PLA, polybutylene succinate (PBS), and PHAs.

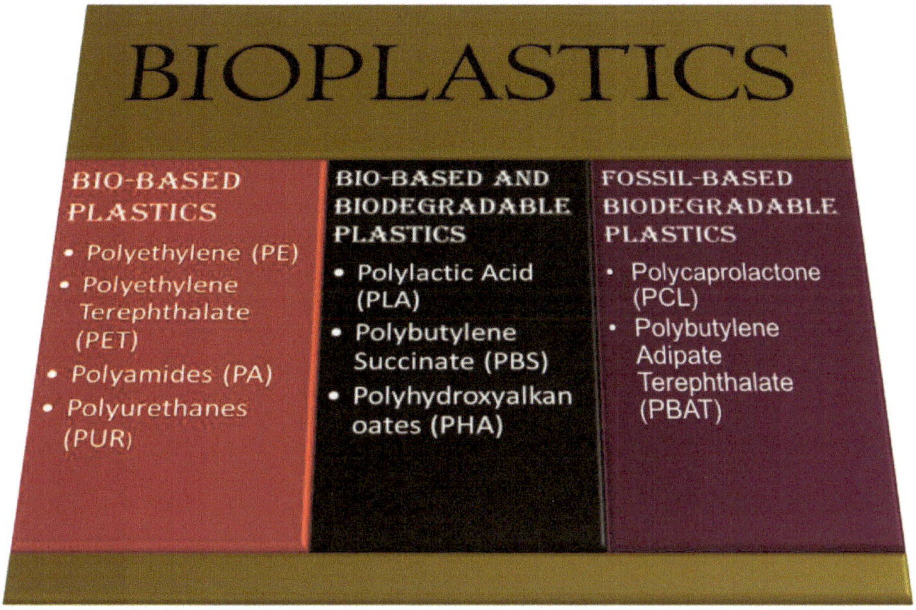

FIGURE 6.1 Categories of bioplastics.

iii. Fossil-Based Biodegradable Plastics: These are plastics made from fossil fuels but are designed to be biodegradable. Examples include polycaprolactone (PCL) and polybutylene adipate terephthalate (PBAT).

Among the most commonly used bioplastics are PHAs, PLA, polybutylene succinate (PBS), PBAT, and PCL, each with unique properties and degradation behaviours.

6.2.1 Polyhydroxyalkanoates (PHAs)

PHAs are biodegradable polyesters naturally synthesized by a variety of microorganisms, such as *Cupriavidus necator* and *Halomonas elongata*, as carbon and energy storage materials under nutrient-limiting conditions such as nitrogen or phosphorus while carbon sources like glucose, glycerol, or fatty acid are in excess (Choi *et al.*, 2020). This process leads to the accumulation of PHA molecules/granules within the cells of the microbes, which are later extracted and purified for utilization (Sedlacek *et al.*, 2019). The PHAs stems are classified based on their monomer composition, which determines their properties. Examples are polyhydroxybutyrate (PHB), which is highly crystalline yet brittle polymer, making it ideal for rigid applications (Yeo *et al.*, 2019). Meanwhile, Poly (3-hydroxybutyrate-co-3-hydroxyvalerate) (PHBV) is copolymer with improved flexibility and toughness, suited for packaging and agricultural uses (Ibrahim *et al.*, 2021b). Despite the advantages of PHAs, there are challenges to the production of PHAs such as high cost of production due to cultivation of required microbes, their specificity and intensive energy required for the extraction of PHAs, and their brittleness and thermal instability (Colombo *et al.*, 2020).

6.2.2 Polylactic Acid (PLA)

According to Ghasemolu *et al.* (2024), PLA is an aliphatic polyester derived from lactic acid that has undergone condensation polymerization. PLA has a transparent, non-toxic, and good mechanical strength properties which can be extracted from starch, sugarcane, and other natural substances

(Younis *et al.*, 2024). PLA has been used in packaging of food (food-grade disposables) such as cups, trays, and films (Mulla *et al.*, 2021). PLA is also a favoured material in additive manufacturing for creating prototypes, models, and end-use products due to its low melting point and ease of processing (Fabra Rivera *et al.*, 2022).

The adoption of bioplastics like PLA is driven by increasing awareness of the ecological consequences of plastic pollution (Raab and Bonger, 2020). Traditional plastics contribute significantly to environmental degradation, as they are often non-biodegradable and accumulate in landfills and oceans (Ali *et al.*, 2021). In contrast, bioplastics are designed to decompose under specific conditions, thereby mitigating long-term waste. The biodegradability of PLA, for instance, is contingent upon the presence of microorganisms and appropriate environmental conditions, which can facilitate its breakdown into natural substances (Xia *et al.*, 2022). According to (Garg *et al.*, 2019a), PLA requires specific conditions such as high temperature and humidity, for its degradation. Although in natural environments like soil and marine, its degradation is really slow leading to accumulation concerns (Muralidharan *et al.*, 2023).

6.2.3 Polybutylene Succinate (PBS)

PBS is an aliphatic polyester that is biodegradable and produced through the polycondesation reaction of succinic acid and 1,4-Butanediol (Qi *et al.*, 2019). Historically sourced from fossil fuel derivatives, recent innovations have focused on bio-based manufacturing utilizing renewable feedstocks such as sugarcane and corn (Pallathadka *et al.*, 2015). PBS demonstrates remarkable biodegradability in both terrestrial and composting settings, positioning it as a promising solution for mitigating plastic waste in agricultural and packaging applications (Bher *et al.*, 2023). PBS has shown high tensile strength and elongation, with thermal stability, which allows it to be used in various applications. PBS is used in the production of compostable bags, agricultural films, and disposable tableware (Darwesh *et al.*, 2023).

6.2.4 Polybutylene Adipate Terephthalate (PBAT)

PBAT is a biodegradable aromatic–aliphatic copolymer produced from adipic acid, 1, 4-butanediol, and terephthalic acid (Dong *et al.*, 2022). The combination of aromatic and aliphatic segments in its structure provides a balance between flexibility, durability, and biodegradability (Wang *et al.*, 2023a). PBAT is said to degrade effectively under industrial composting conditions, although its breakdown in natural environments is slower compared to fully aliphatic polyesters like PBS (Azevedo *et al.*, 2023). Due to PBAT flexibility, elongation at break, and toughness, it is used in flexible packaging and agricultural films (Alpaslan Guler and Deniz, 2023). Its mechanical properties closely resemble those of low-density polyethylene (LDPE), enabling it to replace traditional plastics in similar applications. PBAT is often blended with other biodegradable polymers like PLA to enhance specific properties, such as toughness (Kumar *et al.*, 2022).

6.2.5 Polycaprolactone (PCL)

PCL is a biodegradable polyester commonly used in medical devices. While it breaks down more readily in natural environments compared to PLA, its complete degradation still depends on specific enzymes and optimal environmental conditions (Joseph *et al.*, 2023).

6.3 CHALLENGES IN BIOPLASTICS DEGRADATION

1. Incomplete Degradation in Natural Environments: Many bioplastics, including PLA, require industrial composting facilities to fully degrade, which limits their sustainability

in uncontrolled environments. In marine ecosystems, even biodegradable plastics degrade slowly, contributing to microplastic formation (Kirstein *et al.*, 2018).

2. Contamination and Separation Issues: Non-biodegradable and biodegradable plastics often mix in recycling streams, complicating waste management. The inability to properly separate these materials can reduce the quality of recycled products (Odobel *et al.*, 2021).
3. High Production Costs: The cost of raw materials and specialized manufacturing processes for bioplastics, such as PHAs, remains a barrier to widespread adoption. Improvements in fermentation technology and enzymatic treatments are under exploration to address this issue (Chen *et al.*, 2020).
4. Environmental Trade-Offs: While bioplastics reduce dependence on fossil fuels, the agricultural inputs required for feedstock production (e.g., water and land) can contribute to environmental degradation and compete with food production (Levytskyi *et al.*, 2021).
5. Microbial Diversity Challenges: Non-biodegradable bioplastics like Bio-PET often host diverse microbial biofilms that can complicate degradation processes in aquatic and soil environments. Conversely, biodegradable plastics may harbour fewer microbial species, slowing their breakdown (Mercier *et al.*, 2017).

6.4 SOIL MICROBIAL DIVERSITY

Soil microbial diversity can be said to be rich with varieties of bacteria, fungi, archea, and other microorganism living in the soil. Thus, microbial diversity is crucial for nutrient cycling, soil structure, and mitigating climate change impacts (Garg *et al.*, 2019b; Khursheed, 2016).

6.4.1 Importance of Soil Microbial Diversity

6.4.1.1 Nutrient Cycling and Plant Growth

Organic matters are broken down by microbial communities to release nutrient like nitrogen, phosphorus, and potassium for plant uptake (Aguilar-Paredes *et al.*, 2020). According to example given by Carrara *et al.* (2018), nitrogen-fixing bacteria convert atmospheric nitrogen to forms that can be utilized by plants, while fungi help to decompose complex organic matters.

6.4.1.2 Carbon Sequestration and Climate Regulation

Microbial communities help in carbon cycling by aiding sequestration of organic carbon within soil layers, which in return reduces greenhouse gas emissions, contributing to climate change mitigation (Ramin and Allison, 2019).

6.4.1.3 Ecosystem Resilience

Diverse microbial populations confer ecological redundancy, thereby ensuring the continuity of essential soil functions amidst environmental stressors or perturbations (Jiao *et al.*, 2019).

6.4.2 Factors Influencing Soil Microbial Diversity

6.4.2.1 Land Use and Management Practices

Reduction in microbial diversity can be caused by agricultural practices such as monocropping or intensive use of chemical. By contrast, practices like crop rotation, organic farming, and reduced tillage improve diversification of microbial communities (Chen *et al.*, 2018).

6.4.2.2 Soil Physiochemical Properties

Microbial population is directly affected by physiochemical characteristics of soil such as pH, temperature, moisture content, and availability of nutrients. This can be seen in soils that are abundant in organic materials, which is conducive for nurturing more diverse microbial populations (Bonner *et al.*, 2017).

6.4.2.3 Environmental Conditions

Microbial community dynamics can be affected by climatic factors, including precipitation and temperature. Extreme conditions, like drought, can significantly change the population of microbes, frequently resulting in a decrease in their functional diversity (Ibrahim *et al.*, 2021a).

6.4.3 Approaches to Studying Soil Microbial Diversity

6.4.3.1 Culture-Based Methods

Traditional methodologies that have been employed in the field of microbiology typically consist of the process of cultivating a diverse array of microbial life forms on an assortment of specialized media designed for growth (Rouf *et al.*, 2017). While these conventional methodologies may indeed prove to be beneficial in specific context, it is also essential to acknowledge that they only succeed in capturing and isolating a miniscule fraction of the extensive and intricate diversity of soil microbial life, as a considerable proportion of these microbes regrettably are not amendable to cultivation under specific conditions of laboratory environment (Ouyang *et al.*, 2021).

6.4.3.2 Culture-Independent Techniques

Molecular techniques, which encompass a spectrum of sophisticated approaches such as DNA sequencing and phospholipid fatty acid profiling, provide an exceptionally comprehensive and nuanced understanding of the vast and intricate landscape of microbial diversity that exists in various soil environment (Kumar *et al.*, 2021). These advanced methodologies possess the capability to identify not only the active microbial populations present in a given habitat but also to elucidate their specific functional roles and contributions to the overarching ecological dynamics, all without necessitating reliance on traditional cultivation techniques that may often impose limitations and fail to accurately represent the genuine diversity of microbial life (Johnson and Pomati, 2020).

6.4.3.3 Biochemical Assays

Community-level physiological profiling (CLPP) represents a sophisticated technique employed to evaluate the functional diversity of microbial communities by systematically assessing their capacity to utilize a variety of specific carbon sources (Su *et al.*, 2022). This method is pivotal in elucidating the metabolic capabilities of microbial populations, providing a comprehensive understanding of their ecological roles and interactions within various environments (Zhou *et al.*, 2022). By examining the differential utilization patterns of carbon substrates, CLPP offers valuable insights into the physiological responses of microbial communities to a range of environmental changes, including variations in nutrient availability, pH, temperature, and other ecological stressors (Jones *et al.*, 2018). Moreover, the utilization of CLPP goes beyond simple characterization, functioning as an essential instrument for observing alterations in microbial community composition and functionality as a response to human activities and natural disturbances; given that microbial communities are vital to biogeochemical cycles, their functional diversity is inherently connected to the health and resilience of ecosystems, making it imperative to comprehend how these communities adjust to changes through their metabolic adaptability in order to anticipate the repercussions of environmental shifts on ecosystem dynamics (Moscatelli *et al.*, 2018; Perujo *et al.*, 2020).

6.4.4 Threats to Soil Microbial Diversity

6.4.4.1 Soil Degradation

Erosion, soil compaction, and the depletion of organic matter significantly diminish the habitats available for microorganisms, which in turn results in a marked decline in both the diversity and functional capabilities of these vital organisms (Herath *et al.*, 2017). The loss of microbial diversity can have profound implications for ecosystem health, as microorganisms play crucial roles in nutrient cycling, soil fertility, and overall ecosystem resilience (Yang *et al.*, 2023).To mitigate these detrimental impacts and restore the ecological balance, the implementation of restorative practices is of paramount importance. Such practices are designed to rehabilitate degraded environments, enhance soil structure, and promote the regeneration of organic matter, thereby fostering a more conducive habitat for microbial populations and reinstating their essential functions within the ecosystem (Issoufou *et al.*, 2022).

6.4.4.2 Pollution and Chemical Inputs

The over-application of fertilizers, pesticides, and various industrial pollutants has the potential to significantly disrupt the delicate balance of microbial communities within ecosystems (Shen *et al.*, 2022). This disruption occurs as these substances introduce toxic compounds into the environment or alter the inherent chemistry of the soil (Oboh, 2014). Such alterations can lead to detrimental effects on the diversity and functionality of microbial populations, which play a crucial role in maintaining soil health and ecosystem integrity (Dave *et al.*, 2022). To address and mitigate these adverse impacts, the adoption of sustainable agricultural practices is essential. These practices not only aim to reduce the reliance on harmful chemicals but also promote the restoration and preservation of microbial diversity, ultimately fostering a more resilient and productive agricultural system (Saikanth *et al.*, 2023).

6.4.4.3 Climate Change

The increase in global temperatures, coupled with changes in precipitation patterns, has the potential to significantly alter the composition of microbial communities within various ecosystems (Wang *et al.*, 2023b). Such shifts in microbial diversity and functionality may have profound implications for essential ecological processes, including nutrient cycling and carbon storage (Shang *et al.*, 2023). Specifically, the efficiency of these critical processes could be compromised, leading to disruptions in the overall health and stability of the environment (Shang *et al.*, 2023). As microbial communities play a pivotal role in mediating these processes, understanding the consequences of climate change on their dynamics is vital for predicting future ecological outcomes and developing strategies for environmental management and conservation (Qayoom *et al.*, 2022).

6.5 ROLES OF SOIL MICROBES IN BIOPLASTIC REMEDIATION: ENZYMES AND THEIR OCCURRENCE

The growing concern over the environmental impact of traditional plastic waste has shifted global attention towards bioplastics as a more sustainable alternative. Despite being promising, challenges of bioplastic degradation remain a dire issue. Bioplastics, such as PLA and PHAs, are engineered to degrade more readily than conventional plastics, which are derived from petrochemicals. However, their decomposition process is quite complex and heavily relies on the activity of specific soil microorganisms (García-Depraect *et al.*, 2021), which often leads to the accumulation of bioplastics in the soil.

Bioplastic remediation refers to the natural process where soil microorganisms aid the breakdown of bioplastics into simpler, non-toxic molecules. This microbial activity is essential for bioplastics to fulfil their environmental purpose, which includes reducing plastic waste and lessening ecological harm (Ahmed *et al.*, 2018). Effective degradation depends on the ability of the microorganism to produce specific enzymes that target bioplastic polymers.

6.6 SOIL MICROBES IN BIOPLASTIC REMEDIATION

Soil microbes, including bacteria, fungi, and actinomycetes, play an essential role in bioplastic remediation by breaking down large polymer chains into smaller, usable molecules through various

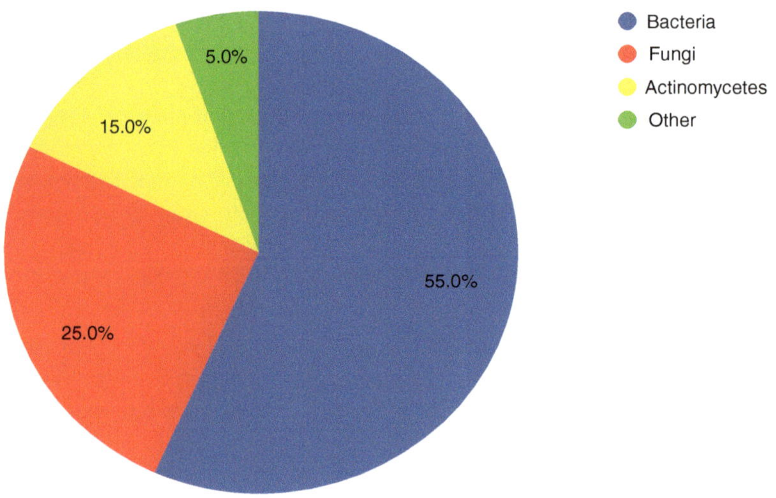

FIGURE 6.2 The percentage of soil microbes in bioplastic remediation (pie chart).

enzymatic processes. The main microbial groups involved in this process are bacteria, fungi, and actinomycetes due to their metabolic diversity and efficiency in degrading different types of bioplastics (Figure 6.2).

6.6.1 Bacteria

Bacteria are the most commonly studied microbes in bioplastic degradation due to their metabolic diversity. Certain species, like Pseudomonas and Bacillus, have demonstrated their capability to degrade PHAs and PLA, which are commonly used in producing biodegradable plastics by using their enzymes, such as lipases and esterases (Dilkes-Hoffman *et al.*, 2019). For instance, *Pseudomonas putida* has shown significant potential in degrading polymers like PCL and PLA (Safiyanu *et al.*, 2022). The degradation efficiency of these bacteria depends on environmental factors, such as temperature and pH, which influence the activity of their enzymes.

Some examples of bacteria involved in bioplastic remediation include the following:

i. Pseudomonas and Bacillus Species: Both genera are known for their versatility in degrading bioplastics. *P. putida* strains, for example, produce enzymes that can break down polyurethane (PU), which is a widely used polymer in various products (Ellis *et al.*, 2021). *Bacillus megaterium*, on the other hand, has been studied for its ability to degrade bioplastics in both soil and aquatic environments, making it suitable for large-scale remediation purposes (Schneier *et al.*, 2024).
ii. Engineered *Escherichia coli*: Although not capable of degrading plastics, genetically modified *E. coli* strains have been engineered to express enzymes like polyethylene terephthalate hydrolase (PETase). According to Ru *et al.* 2020, these engineered strains exhibit higher degradation rates and could potentially be applied in industrial settings for controlled bioplastic recycling processes.
iii. *Ideonella sakaiensis*: This bacterium has gained significant attention for its ability to degrade PET plastics. It produces PETase, an enzyme that hydrolyses the ester bonds in PET, breaking it down into its monomers, which can further be broken down (Klauer *et al.*, 2024).

6.6.2 Fungi

Fungi are highly effective in degrading bioplastics due to their ability to produce enzymes, particularly lignocellulolytic enzymes like laccases, manganese peroxidases, and cellulases. These enzymes enable fungi to break down complex biopolymer structures found in many bioplastics. Fungi including *Aspergillus niger*, *Penicillium* spp., and *Trichoderma reesei* have shown significant potential in degrading various bioplastic types, especially PHAs and PLA (Tomer *et al.*, 2021; Okal *et al.*, 2023).

Certain fungi can degrade plastics in low-nutrient environments, making them suitable for remediation in soil and aquatic environments. An example is *Aspergillus tubingensis*, which is known for breaking down polyester polyurethane and has enzymes that have the ability to attack bioplastic polymers, reducing them to smaller, less harmful monomers that can easily be absorbed into the environment (Okal *et al.*, 2023).

Fungi also play a role in recycling bioplastic waste through a process called fungal bio-upcycling. This involves converting bioplastic components into valuable bioproducts such as organic acids and biofuels. Recent research suggests that enzymes, particularly laccases, can be incorporated into certain fungi to optimize performance so as to increase degradation and upcycling rates (Ru *et al.*, 2020; Temporiti *et al.*, 2022).

6.6.3 Actinomycetes

Actinomycetes, a group of gram-positive, filamentous bacteria, are also involved in bioplastic degradation. Species such as Streptomyces secrete various hydrolytic enzymes, such as lipases, proteases, and polyesters that aid in the breakdown of complex polymers under aerobic conditions (Singh *et al.*, 2020; Oliveira *et al.*, 2022). Genera such as Streptomyces, Rhodococcus, and Nocardia have shown impressive ability to degrade polyesters like PLA and PHA in different environments (Devanshi *et al.*, 2022). Actinomycetes, which are known to survive in nutrient-rich soils, make them important contributors to bioplastic remediation in agricultural environments.

Actinomycetes are invaluable in soil environment, where they contribute to the natural biodegradation of bioplastics by releasing extracellular enzymes. For example, Streptomyces species produce extracellular enzymes that hydrolyse plastic polymers into smaller compounds. These smaller molecules can then be further metabolized or mineralized by other soil microorganisms. Additionally, according to recent studies, *Rhodococcus ruber* has been observed to form biofilms on plastic surfaces, enhancing contact and degradation efficiency (Li *et al.*, 2024).

Actinomycetes are also known for their role in synergistic degradation, where they can work alongside fungi or other bacteria to hasten the breakdown of plastic. This synergy combines the diverse enzyme systems of mixed microbial communities, leading to faster degradation, specifically in composting and soil remediation (Oliveira *et al.*, 2022; Devanshi *et al.*, 2022).

6.6.4 Other

Although not as common in plastic degradation as bacteria, fungi, or actinomycetes, algae – particularly microalgae – are an emerging potential source for bioplastics production because they do not require nutrient sources to grow (Arora *et al.*, 2023). Being photosynthetic organisms, they can capture carbon dioxide from the environment and use it as a carbon source. Some species of algae used in bioplastic remediation due to their crack resistance and high biomass production include Chlorella (green algae) and Spirulina (bluish-green algae) (Arora *et al.*, 2023).

The following pie chart shows the percentage of each microorganism when it comes to bioplastic remediation. Each microbial group has unique properties which allows for its contribution to remediation processes.

6.7 BIOCHEMICAL PATHWAYS IN MICROBIAL BIOREMEDIATION

The breakdown of bioplastics by soil microbes involves various enzymatic pathways, each contributing to the conversion of complex bioplastics into simpler, non-toxic forms.

i. Bacterial Enzymes: Bacteria produce several enzymes, including lipases that break down ester bonds in PHAs and esterases that break down the ester bonds in PLA to degrade bioplastic polymers. For instance, the bacterium *Pseudomonas aeruginosa* has been shown to produce an esterase that hydrolyses PLA, by breaking it down into lactic acid monomers which can be absorbed by the environment (Vaishnav *et al.*, 2024).
ii. Fungal Enzymes: Fungi play a significant role in bioplastic degradation by producing laccases and manganese peroxidases, which are important in breaking down bioplastics, especially under anaerobic or anoxic conditions typical of soil environments. These enzymes have the ability to oxidize complex bipolymers, which allows the bioremediation of PLA and other polymers (Temporiti *et al.*, 2022).
iii. Actinomycetes Enzymes: Actinomycetes, particularly species such as Streptomyces, secrete enzymes like polyesterases and cutinases which are important for degrading aliphatic polyesters. These enzymes cleave the polyester linkage in bioplastics, breaking them down into smaller compounds that can be further degraded by other microorganisms in the soil (Devanshi *et al.*, 2022).

The efficiency of these enzymes depends on various environmental factors such as soil pH, temperature, and moisture content. For instance, bacterial lipases are more active at neutral to slightly alkaline pH, while fungal esterases function best in slightly acidic soils (Vaishnav *et al.*, 2024).

6.8 STAGES OF BIOPLASTIC REMEDIATION

The bioremediation of bioplastics involves a process of various steps in which microbes convert complex polymers into simpler, eco-friendly compounds.

6.8.1 Initiation of Biodegradation by Polymer Hydrolysis

The first stage of bioplastic remediation entails the breakdown of complex polymer structures. Microorganisms such as Pseudomonas and Bacillus species produce extracellular enzymes, including lipases, esterases, and cutinases, that cleave the ester bonds within bioplastics, initiating degradation (Atiwesh *et al.*, 2021; Wei and Zimmermann, 2017). This action of enzymes converts long-chain polymers into smaller molecules that can be easily metabolized by microbes (Joseph *et al.*, 2023).

6.8.2 Polymer Fragmentation

After initiation, microorganisms begin polymer fragmentation, breaking down the complex structures of bioplastics, such as PLA and PHAs, into oligomers and monomers. Actinomycetes, especially Streptomyces, are effective in this process due to their ability to release various extracellular enzymes under different environmental conditions (Narancic *et al.*, 2020). The presence of optimal temperature, pH, and oxygen levels enhances fragmentation and boosts microbial activity, as demonstrated by Pathak and Navneet (2017). However, the intermediate by-products from the breakdown of bioplastics can be toxic. To address this, microbes such as Pseudomonas and Aspergillus produce detoxifying enzymes that neutralize these compounds, preventing the accumulation of harmful residues in the soil. Detoxification stops by-products from bioplastic degradation from accumulating in the soil, which could otherwise hinder plant and microbial growth (Wei and Zimmermann, 2017).

6.8.3 Monomer Metabolization

Once the polymers are fragmented, microorganisms metabolize the resultant monomers, such as lactic acid from PLA, using metabolic pathways such as glycolysis and the citric acid cycle. This process effectively converts monomers into energy and biomass, fuelling microbial growth. Fungi such as Aspergillus and Penicillium species are particularly effective in this stage, metabolizing intermediate products that bacteria alone may not fully degrade (Brdlik *et al.*, 2021). These microbes utilize monomers as carbon sources, fuelling their growth and helping in the elimination of bioplastic residues from the environment.

6.8.3 Assimilation and Mineralization

The final stage of bioplastic remediation is assimilation and mineralization which involves the conversion of bioplastic monomers into inorganic compounds such as carbon dioxide, water, methane, and mineral salts, depending on the environmental conditions (aerobic or anaerobic). In aerobic conditions, bacterial genera such as Bacillus and Pseudomonas facilitate the production of carbon dioxide and water as end products (Hasnat *et al.*, 2024). Conversely, in anaerobic conditions, methanogens contribute to methane production and other by-products by converting monomers (Pooja *et al.*, 2023). This mineralization process effectively removes the bioplastics impact on the soil environment, restoring the environment to a near-natural state without harmful residues (Prendiz *et al.*, 2019).

6.9 FACTORS INFLUENCING MICROBIAL BIOREMEDIATION EFFICIENCY

Many environmental and microbial factors can have an impact on the effectiveness of bioremediation in soil environments.

i. Soil pH and Moisture: Soil pH can significantly impact enzyme activity and microbial growth. Many bioplastic-degrading enzymes, such as laccases, are most effective in slightly acidic to neutral conditions. In acidic soils, microbial activity may slow down, thereby reducing the rate of bioremediation (Joseph *et al.*, 2023). The microbial growth of fungi mostly occurs within a slightly acidic to neutral pH, while bacteria thrive in neutral to alkaline soils. The second influencing factor is the moisture content – soil with more water increases the rate of degradation of bioplastic.
ii. Temperature: Microbial bioremediation of bioplastics is completely dependent on temperature. Higher temperatures can accelerate microbial metabolism and enzyme activity, thus speeding up the breakdown of bioplastics. However, extreme temperatures can inhibit microbial growth, especially for fungi that prefer moderate climates (Srikanth *et al.*, 2022). The optimum temperature at which most microbial degradation occurs is between 20°C and 30°C. When the temperature is lower, enzymatic activities slow down, and there is a reduction in the rates of degradation.

iii. Nutrient Availability: The presence of nitrogen, phosphorus, and other nutrients supports microbial growth and enzyme production. Nutrient-rich soils promote higher microbial activity and efficient bioremediation of bioplastics, as microbes have the resources needed to produce bioplastic-degrading enzymes (Vaishnav *et al.*, 2024).
iv. Biofilm Formation: Certain microbes, such as *R. ruber*, form biofilms on plastic surfaces, enhancing bioremediation. Biofilms facilitate better adherence to plastic surfaces, allowing microbes to secrete enzymes directly onto bioplastics and enhancing degradation (Hasnat *et al.*, 2024).

6.10 FUTURE GUIDELINES PERSPECTIVES IN AI BIOPLASTICS REMEDIATION

6.10.1 The Role of Artificial Intelligence in Environmental Science

Artificial intelligence (AI) is becoming an essential tool in tackling some of our most pressing environmental challenges, especially when it comes to biodegradation and waste management (Aniza *et al.*, 2023). Researchers are leveraging AI technologies to develop innovative solutions aimed at reducing the negative impacts of plastic and bioplastic waste on our planet (Chen *et al.*, 2021). This exploration looks at how AI is transforming biodegradation processes, focusing on areas like predictive analytics, the study of microbial ecosystems, the remediation of bioplastics, and the creation of sustainable waste management strategies. The goal is to not only address the current issues but also to pave the way for a healthier, more sustainable future (Ye *et al.*, 2020).

6.10.2 Applications of AI in Biodegradation and Waste Management

AI is becoming an essential partner in modern waste management, helping us tackle the growing challenges of waste disposal and environmental sustainability. Think of it as a smart assistant that uses data to predict how quickly waste materials can break down in different conditions (Strollo *et al.*, 2020). By analysing factors like material properties and the environment, AI can suggest the best times and methods for intervention, making the waste degradation process more efficient and sustainable (Hardian *et al.*, 2020).

When it comes to bioplastics, AI is working hand-in-hand with the natural world. It helps scientists identify specific microbes that can break down these materials effectively (Massarelli *et al.*, 2021). By sifting through vast amounts of microbial data, AI can highlight which microorganisms have the right enzymes to tackle bioplastics. This not only speeds up the remediation process but also supports the development of better strategies for managing plastic waste (Anitha *et al.*, 2022).

As the use of bioplastics continues to rise around the globe, the collaboration between AI and microbial studies is becoming increasingly vital (Maraveas *et al*, 2023). It's a promising approach to addressing the environmental issues linked to plastic waste, ensuring that we can enjoy the benefits of these materials while also protecting our planet (Ghasemlou *et al.*, 2024).

6.10.3 AI-Driven Approaches to Bioplastic Remediation

In Figure 6.3, it can be shown that there are different approaches to bioplastic remediation by the use of AI.

6.10.3.1 AI-Driven Predictive Models in Bioplastic Degradation

Recent advancements in AI have significantly enhanced our understanding of the interactions between microorganisms and bioplastics. By employing predictive models, researchers can simulate these interactions, allowing for the optimization of environmental parameters – such as pH levels and temperature – that are crucial for effective degradation (Parsaeimehr *et al.*, 2023). Furthermore,

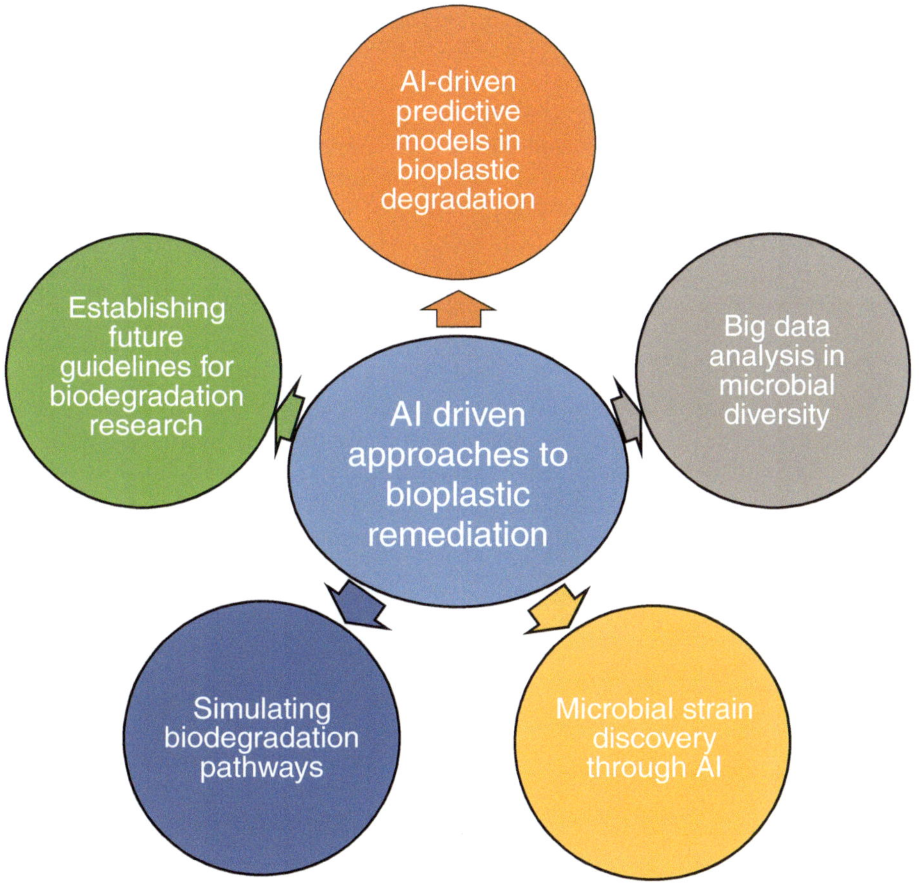

FIGURE 6.3 The various AI-driven approaches to bioplastic remediation.

these models provide valuable insights into the ecological consequences of the by-products produced during the degradation process (Singh *et al.*, 2023). This foresight empowers researchers to make informed decisions regarding the use of bioplastics, ultimately promoting long-term sustainability while minimizing ecological disruption (Coppola *et al.*, 2021).

6.10.3.2 Big Data Analysis in Microbial Diversity

The fields of metagenomics and metatranscriptomic have yielded vast datasets pertaining to microbial communities. The integration of AI technologies facilitates the processing and analysis of these extensive datasets, revealing important correlations between microbial diversity and biodegradation rates (Zampieri *et al.*, 2023). This innovative approach enables the identification of specific microbial consortia that are particularly efficient at degrading various types of bioplastics, thereby advancing targeted bioremediation strategies (Salinas *et al.*, 2023).

6.10.3.3 Microbial Strain Discovery through AI

AI, especially neural networks, plays a pivotal role in the discovery and design of enzymes essential for the degradation of bioplastics. By mapping out the metabolic pathways involved in the breakdown of polymers, AI assists researchers in engineering microorganisms with enhanced biodegradation capabilities (Maraveas *et al.*, 2023). This progress not only deepens our understanding of the underlying biochemical processes but also fosters the development of novel microbial solutions aimed at mitigating the environmental impact of plastic waste (Orlando *et al.*, 2023).

6.10.3.4 Simulating Biodegradation Pathways

AI simulations offer critical insights into the mechanisms underlying bioplastic degradation. These simulations can predict the by-products generated during the degradation process and assess their potential ecological impacts (Jiang *et al.*, 2023). By optimizing these degradation pathways, researchers can minimize environmental harm while enhancing process efficiency. Such analyses are vital for comprehending how bioplastics interact with ecosystems, thereby guiding the development of materials that align with conservation goals (Lukas *et al.*, 2023).

6.10.3.5 Establishing Future Guidelines for Biodegradation Research

The absence of standardized biodegradation protocols has posed challenges to the comparability of data across various research efforts. To address this issue, global initiatives are focusing on the creation of shared databases that integrate microbial, enzymatic, and environmental data (Cernava *et al.*, 2022). The success of AI-driven innovations is significantly bolstered by these harmonized datasets, which enhance the quality of research and promote sustainable remediation practices (Raviteja and Reddy, 2023).

6.10.4 Advancements in AI-Enhanced Biodegradation Technologies

The scalability of AI solutions in field applications represents a significant area of focus in current research efforts. A notable example of this advancement is the development of AI-powered bioreactor systems tailored for the industrial-scale remediation of bioplastics (Aniza *et al.*, 2023). These innovative systems are equipped with capabilities for real-time monitoring and operational adjustments, which collectively enhance the efficiency of degradation processes (Singh *et al.*, 2022).

Moreover, AI plays a critical role in evaluating soil health throughout the biodegradation process. This ensures that the breakdown of bioplastics contributes positively to ecological sustainability (Deorankar and Rohankar, 2020). By implementing such proactive measures, we not only address immediate waste management challenges but also protect soil ecosystems for future generations (Elbeltagi *et al.*, 2022).

To further encourage the integration of AI in this domain, it is essential to establish supportive policy frameworks that promote research and development. Governments can play a pivotal role in fostering innovation by providing incentives that align with technological advancements, ultimately leading to more sustainable solutions for bioplastics remediation (Cassar and Vassallo, 2022).

6.10.5 Challenges and Ethical Considerations

The integration of AI into the field of biodegradation presents a promising opportunity for advancing environmental sustainability. However, several significant challenges must be addressed to fully realize its potential (Singh *et al.*, 2021).

Firstly, there are technical issues that hinder progress. AI systems rely heavily on high-quality and diverse datasets, which are often expensive and difficult to obtain. This limitation is particularly pronounced in resource-constrained environments, where computational demands can further complicate the implementation of AI solutions (Vaidyanathan *et al.*, 2021).

In addition to technical barriers, there are environmental and societal risks associated with the use of engineered microbes for biodegradation. The ecological consequences of these interventions remain largely uncertain, highlighting the need for thorough evaluation and monitoring (Yaashikaa *et al.*, 2022). Furthermore, it is essential to ensure equitable access to AI-driven technologies. Without careful consideration, there is a risk that these advancements could exacerbate existing global inequalities (Liu, 2020).

Lastly, ethical concerns surrounding AI in biodegradation cannot be overlooked. Ensuring transparency and accountability within AI systems is crucial, particularly in sensitive applications such as the degradation of bioplastics (Ukhurebor *et al.*, 2021). By addressing these challenges, we can foster a more responsible and effective application of AI in biodegradation, ultimately contributing to a more sustainable future (Maraveas et al., 2023).

6.11 FUTURE PERSPECTIVES

The role of AI in the field of biodegradation is on the brink of significant expansion, largely due to recent advancements that allow for more personalized and scalable solutions. By customizing AI models to align with specific local environmental conditions and the diversity of microbial life, we can greatly enhance the effectiveness of regional efforts aimed at remediating bioplastics (Aniza *et al.*, 2023). Furthermore, AI-driven formulations for bioplastics could be designed with a focus on facilitating easier degradation, thereby ensuring that material design is closely aligned with overarching sustainability objectives (Ukhurebor *et al.*, 2021). The long-term success of these initiatives will hinge on the successful integration of AI within circular economy frameworks, which advocate for zero-waste strategies through improved biodegradation processes. Achieving these ambitious goals will require a collaborative approach that draws on expertise from multiple disciplines, including microbiology, AI, materials science, and environmental engineering (Onyeaka *et al.*, 2023). By employing hybrid methodologies that merge biological insights with computational tools, we stand to make significant strides in the realm of sustainable waste management practice (Nanez Alonso *et al.*, 2021).

6.12 CONCLUSIONS

The significance of soil microbial diversity in the bioremediation of bioplastics is fundamental in tackling the environmental issues associated with plastic pollution. Bioplastics, which are produced from renewable resources and engineered for biodegradability, present a sustainable alternative to traditional plastics. However, their degradation is largely dependent on the functional abilities of soil microorganisms, including bacteria, fungi, and actinomycetes. These microbes synthesize enzymes such as esterases and depolymerases that are essential for decomposing complex polymers into environmentally benign compounds. Environmental conditions, including pH, temperature, and nutrient availability, play a critical role in determining the efficiency of microbial activity. Recent advancements in molecular techniques, such as metagenomics and proteomics, have improved our comprehension of the roles that microbes play in the degradation of bioplastics. Nevertheless, challenges remain, particularly concerning the variable rates of bioplastic degradation in natural soil environments and the ecological consequences of the by-products generated during this process.

Looking ahead, there is potential for integrating AI to optimize biodegradation pathways, forecast microbial interactions, and facilitate the discovery of new microbial strains. AI-driven methodologies could transform bioplastic remediation by addressing existing knowledge gaps and developing scalable solutions. Overcoming these challenges necessitates interdisciplinary collaboration that merges microbiology, environmental engineering, and computational sciences. In summary, while the diversity of soil microbes holds significant promise for bioplastic remediation, a holistic approach that integrates advanced technologies and sustainable practices is crucial for maximizing environmental advantages and ensuring long-term ecological stability.

NOTE

1 Corresponding author: aishaadegbite@gmail.com

REFERENCES

Abiraami, T.V., Singh, S. and Nain, L. (2020). Soil metaproteomics as a tool for monitoring functional microbial communities: Promises and challenges. *Review in Environmental Science and Biotechnology*, 19:73–102. https://doi.org/10.1007/s11157-019-09519-8

Adhikari, D., Mukai, M., Kubota, K., Kai, T., Kaneko, N., Araki, K.S. and Kubo, M. (2016). Degradation of bioplastics in soil and their degradation effects on environmental microorganisms. *Journal of Agricultural Chemistry and Environment*, 5:23–34. https://doi.org/10.4236/jacen.2016.51003

Aguilar-Paredes, A., Valdes, G. and Nuti, M. (2020). Ecosystem functions of microbial consortia in sustainable agriculture. *Agronomy*, 10:1902. https://doi.org/10.3390/agronomy10121902

Ahmed, T., Shahid, M., Azeem, F., Rasul, I., Shah, A.A., Noman, M., Hameed, A., Manzoor, N., Manzoor, I. and Muhammad, S. (2018). Biodegradation of plastics: Current scenario and future prospects for environmental safety. *Environmental Science and Pollution Research International*, 25(8):7287–7298. https://doi.org/10.1007/s11356-018-1234-9

Ahsan, W.A., Hussain, A., Lin, C. and Nguyen, M.K. (2023). Biodegradation of different types of bioplastics through composting—A recent trend in green recycling. *Catalysts*, 13(2):294. https://doi.org/10.3390/catal13020294

Ali, S.S., Elsamahy, T., Koutra, E., Kornaros, M., El-Sheekh, M., Abdelkarim, E.A., Zhu, D. and Sun, J. (2021). Degradation of conventional plastic wastes in the environment: A review on current status of knowledge and future perspectives of disposal. *Science of the Total Environment*, 771:144719. https://doi.org/10.1016/j.scitotenv.2020.144719

Alpaslan Guler, C.A. and Deniz, S. (2023). The effect of different clay types on the barrier properties of PLA/PBAT/clay bio-nanocomposite packaging films. *Journal of Dispersion Science and Technology*, 44(13):2367–2378. https://doi.org/10.1080/01932691.2023.2225609

Anitha, R., Maruthi, R. and Sudha, S. (2022). Automated segregation and microbial degradation of plastic wastes: A greener solution to waste management problems. *Global Transitions Proceedings*, 3(1):100–103. https://doi.org/10.1016/j.gltp.2022.04.021

Aniza, R., Chen, W.H., Petrissans, A., Hoang, A.T., Ashokkumar, V. and Petrissans, M. (2023). A review of biowaste remediation and valorization for environmental sustainability: Artificial intelligence approach. *Environmental Pollution*, 324:121363. https://doi.org/10.1016/j.envpol.2023.121363

Aransiola, S.A., Ikhumetse, A.A., Babaniyi, B.R., Abioye, O.P., Oyedele, O.J. and Falade, N.O. (2023a). Phytoaccumulation of Micro and Nanoplastics: Root Uptake. In: Maddela, N.R., Reddy, K.V. and Ranjit, P. (Eds) *Micro and Nanoplastics in Soil*, Springer, Cham, pp. 165–188. https://doi.org/10.1007/978-3-031-21195-9_8

Aransiola, S.A., Victor-Ekwebelem, M.O., Ajiboye, A.E., Leh-Togi Zobeashia, S.S., Ijah, U.J.J. and Oyedele, O.J. (2023b). Ecological Impacts and Toxicity of Micro and Nanoplastics in Agroecosystem. In: Maddela, N.R., Reddy, K.V. and Ranjit, P. (Eds) *Micro and Nanoplastics in Soil*, Springer, Cham, pp. 221–236. https://doi.org/10.1007/978-3-031-21195-9_10

Arora, Y., Sharma, S. and Sharma, V. (2023). Microalgae in bioplastic production: A comprehensive review. *Microalgae in Bioplastic Production: A Comprehensive Review*, 48(6):7225–7241. https://doi.org/10.1007/s13369-023-07871-0

Atiwesh, G., Mikhael, A., Parrish, C.C., Banoub, J. and Le, T.A.T. (2021). Environmental impact of bioplastic use: A review. *Heliyon*, 7(9). https://doi.org/10.1016/j.heliyon.2021.e07918

Azevedo, J.V.C., Hausnerova, B., Moginger, B. and Sopik, T. (2023). Effect of chain extending cross-linkers on the disintegration behaviour of composted PBAT/PLA blown films. *International Journal of Molecular Sciences*, 24(5):4525. https://doi.org/10.3390/ijms24054525

Bher, A.R., Cho, Y. and Auras, R. (2023). Boosting degradation of biodegradable polymers. *Macromolecular Rapid Communications*, 44(5):e2200769. https://doi.org/10.1002/marc.202200769

Bonner, M.T.L, Shoo, L.P., Brackin, R. and Schmidt, S. (2017). Relationship between microbial composition and substrate use efficiency in tropical soil. *Geoderma*, 314:96–103. https://doi.org/10.1016/j.geoderma.2017.11.026

Brdlík, P., Borůvka, M., Běhálek, L. and Lenfeld, P. (2021). Biodegradation of poly(lactic acid) biocomposites under controlled composting conditions and freshwater biotope. *Polymers*, 13(4):594. https://doi.org/10.3390/polym13040594

Brown, S.P. and Jumpponen, A. (2015). Phylogenetic diversity analyses reveal disparity between fungal and bacterial communities during microbial primary succession. *Soil Biology and Biochemistry*, 89:52–60. https://doi.org/10.1016/j.soilbio.2015.06.025

Carrara, J.E., Walter, C.A., Hawkins, J.S., Peterjohn, W.T., Averill, C. and Brzostek, E.R. (2018). Interactions among plants, bacteria, and fungi reduce extracellular enzyme activities under long-term N fertilization. *Global Change Biology*, 24(6):2721–2734. http://dx.doi.org/10.1111/gcb.14081

Cassar, C. and Vassallo, M.T. (2022). Mapping Out Cross-Sectoral Collaboration and Multi-level Governance within Single Use Plastic Policy: An Analysis. In: *Insurance and Risk Management for Disruptions in Social, Economic and Environmental System: Decision and Control Allocations within New Domains of Risk*, Emerald Publishing Limited, pp. 1–28. http://dx.doi.org/10.1108/978-1-80117-139-720211001

Castro-Aguirre, E., Auras, R., Selke, S., Rubino, M. and Marsh, T. (2018). Enhancing the biodegradation rate of poly(lactic acid) films and PLA bio-nanocomposites in simulated composting through bioaugmentation. *Polymer Degradation and Stability*, 154:46–54. https://doi.org/10.1016/j.polymdegradstab.2018.05.017

Cernava, T., Rybakova, D., Buscot, F., Clavel, T., McHardy, C.A., Meyer, F., Meyer, F., Overmann, J., Stecher, B., Sessitsch, A., Schloter, M., Berg, G. and The Microbiome Support Team. (2022). Metadata harmonization – Standards are the key for a better usage of omics data for integrative microbiome analysis. *Environmental Microbiome*, 17(1):33. https://doi.org/10.1186/s40793-022-00425-1

Chandra, P. and Enespa. (2019). Soil–Microbes–Plants: Interactions and Ecological Diversity. In: Varma, A., Tripathi, S., and Prasad, R. (Eds.) *Plant Microbe Interface*, Springer, Cham, pp. 145–176. https://doi.org/10.1007/978-3-030-19831-2_6

Chen, G.Q., Chen, X.Y., Wu, F.Q. and Chen, J.C. (2020). Polyhydroxyalkanoates (PHA) toward cost competitiveness and functionality. *Advanced Industrial and Engineering Polymer Research*, 3(1):1–7. https://doi.org/10.1016/j.aiepr.2019.11.001

Chen, H., Xia, Q., Yang, T. and Shi, W. (2018). Eighteen-year farming management moderately shapes the soil microbial community structure but promotes habitat-specific taxa. *Frontier Microbiology*, 9:1776. http://dx.doi.org/10.3389/fmicb.2018.01776

Chen, J., Huang, S., Murugan, S.B. and Tamizharasi, G.S. (2021). Artificial intelligence based e-waste management for environmental planning. *Environmental Impact Assessment Review*, 87:106498. https://doi.org/10.1016/j.eiar.2020.106498

Choi, S.Y., Cho, I.J., Lee, Y., Kim, Y.J., Kim, K.J. and Lee, S.Y. (2020). Microbial polyhydroxyalkanoates and non-natural polyesters. *Advanced Materials*, 32(35):1907138. https://doi.org/10.1002/adma.201907138

Colombo, B., Pereira, J., Martins, M., Torres-Acosta, M.A., Dias, A.C.R.V., Lemos, P.C., Ventura, S.P.M., Eisele, G., Alekseeva, A., Adani, F. and Serafim, L.S. (2020). Recovering PHA from mixed microbial biomass: Using non-ionic surfactants as a pre-treatment step. *Separation and Purification Technology*, 253:117521. https://doi.org/10.1016/j.seppur.2020.117521

Coppola, G., Gaudio, M.T., Lopresto, C.G., Calabro, V., Curcio, S. and Chakraborty, S. (2021). Bioplastic from renewable biomass: A facile solution for a greener environment. *Earth Systems and Environment*, 5:231–251. https://doi.org/10.1007/s41748-021-00208-7

Darwesh, A.H.A., Mohammed, P.A., Mamand, S.M., Hussen, S.A., Aziz, S., Brza, M.A., Abdullah, R.M. and Karim, W.O. (2023). Investigation of structural and optical characteristics of biopolymer composites based on polyvinyl alcohol inserted with PBS nanoparticles. *Coatings*, 13(3):578. https://doi.org/10.3390/coatings13030578

Dave, M., Panwar, K., Dadhich, I., Sharma, Y., Malodia, A. and Shaik, N. (2022). Microbial diversity reduction in soil by long-term climate warming. *Advances in Materials Science and Engineering*, 2022(1):7670890. https://doi.org/10.1155/2022/7670890

Deorankar, A.V. and Rohankar, A.A. (2020). Soil health monitoring system using AI. *Journal of Emerging Technologies and Innovative Research*, 7(1). http://dx.doi.org/10.6084/m9.jetir.JETIR1908649

Devanshi, S., Shah, K.R., Arora, S. and Saxena, S. (2022). *Actinomycetes as an Environmental Scrubber*, IntechOpen. http://dx.doi.org/10.5772/intechopen.99187

Di Lenola, M., Grenni, P., Proenca, D.N., Morais, P.V. and Carracciolo, A.B. (2017). Comparison of Two Molecular Methods to Assess Soil Microbial Diversity. In: Lukac, M., Grenni, P., and Gamboni, M.

(Eds.) *Soil Biological Communities and Ecosystem Resilience*. Sustainability in Plant and Crop Protection, Springer, Cham, pp. 25–42. https://doi.org/10.1007/978-3-319-63336-7_3

Dilkes-Hoffman, L., Ashworth, P., Laycock, B., Pratt, S. and Lant, P. (2019). Public attitudes towards bioplastics – Knowledge, perception and end-of-life management. *Resources, Conservation and Recycling*, 151:104479.

Dilshad, E., Waheed, H., Ali, U., Amin, A. and Ahmed, I. (2021). General Structure and Classification of Bioplastics and Biodegradable Plastics. In: Kuddus, M., and Roohi (Eds.) *Bioplastics for Sustainable Development,* Springer, Singapore, pp. 61–82. https://doi.org/10.1007/978-981-16-1823-9_2

Dong, Y., Wang, J., Yang, Y., Wang, Q., Zhang, X., Hu, H. and Zhu, J. (2022). Bio-based poly(butylene diglycolate-co-furandicarboxylate) copolyesters with balanced mechanical, barrier and biodegradable properties: A prospective substitute for PBAT. *Polymer Degradation and Stability*, 202:110010. https://doi.org/10.1016/j.polymdegradstab.2022.110010

Elbeltagi, A., Kushwaha, N.L., Srivastava, A. and Zoof, A.T. (2022). Artificial Intelligent-Based Water and Soil Management. In: Poonia, R.C., Singh, V., and Nayak, S.R. (Eds.) *Deep Learning for Sustainable Agriculture*, Academic Press, pp. 129–142. https://doi.org/10.1016/B978-0-323-85214-2.00008-2

Ellis, L.D., Rorrer, N.A., Sullivan, K.P., Otto, M., McGeehan, J.E., Román-Leshkov, Y., Wierckx, N. and Beckham, G.T. (2021). Chemical and biological catalysis for plastics recycling and upcycling. *Nature Catalysis*, 4(7):539–556. https://doi.org/10.1038/s41929-021-00648-4

Fabra Rivera, A.D.P., De Castro Magalhaes, F. and Rubio, J.C.C. (2022). Experimental characterization of PLA composites printed by fused deposition modelling. *Journal of Composite Materials*, 57(5):941–954. http://dx.doi.org/10.1177/00219983221146619

Fan, P., Yu, H., Xi, B. and Tan, W. (2022). A review on the occurrence and influence of biodegradable microplastics in soil ecosystems: Are biodegradable plastics substitute or threat? *Environment International*, 163:107244. https://doi.org/10.1016/j.envint.2022.107244

García-Depraect, O., Bordel, S., Lebrero, R., Santos-Beneit, F., Börner, R. A., Börner, T. and Muñoz, R. (2021). Inspired by nature: Microbial production, degradation and valorization of biodegradable bioplastics for life-cycle-engineered products. *Biotechnology Advances*, 53:107772. https://doi.org/10.1016/j.biotechadv.2021.107772

Garg, M., White, S.R. and Sottos, N.R. (2019a). Rapid degradation of polylactic acid with organometallic catalysts. *ACS Applied Materials and Interfaces*. 11(49):46226–46232. https://doi.org/10.1021/ACSAMI.9B17599

Garg, N., Saroy, K., Cheema, A. and Bisht, A. (2019b). Microbial Diversity in Soil: Biological Tools for Abiotic Stress Management in Plants. In: Varma, A., Tripathi, S., and Prasad, R. (Eds.) *Plant Biotic Interactions*, Springer, Cham. https://doi.org/10.1007/978-3-030-26657-8_17

Geyer, R., Jambeck, J.R., and Law, K.L. (2017). Production, use, and fate of all plastics ever made. *Science Advances*, 3(7):e1700782. http://dx.doi.org/10.1126/sciadv.1700782

Ghasemlou, M., Barrow, C.J. and Adhikari, B. (2024). The future of bioplastics in food packaging: An industrial perspective. *Food Packaging and Shelf Life*, 43:101279. https://doi.org/10.1016/j.fpsl.2024.101279

Hardian, R., Liang, Z., Zhang, X. and Szekely, G. (2020). Artificial intelligence: The silver bullet for sustainable materials development. *Green Chemistry*, 22:7521. http://dx.doi.org/10.1039/d0gc02956d

Hasnat, A., Zahura, F.T., Hasan, M., Hasan, M.R. and Rashid, T.U. (2024). Bioremediation of Microplastics by Microorganisms: Trends, Challenges, and Perspectives. In: *Royal Society of Chemistry eBooks*, pp. 49–95. https://doi.org/10.1039/bk9781837673131-00049

Herath, L., Seneviratne, G., Jayasinghe, J.A.W.W. and Senanayake, D.M.N. (2017). Microbial biofilms and mitigation of loss of agro-biodiversity in degraded soils. *Journal of the National Science Foundation of Sri Lanka*, 45(4):329–335. http://dx.doi.org/10.4038/jnsfsr.v45i4.8226

Ibrahim, A.D., Uhuami, A.O., Abdulkadir, N. and Uzoh, I.M. (2021a). Climate Change Alters Microbial Communities. In: Choudhary, D.K., Mishra, A., and Varma, A. (Eds.) *Climate Change and the Microbiome*. Soil Biology, Springer, Cham, 63, pp. 13–29. https://doi.org/10.1007/978-3-030-76863-8_2

Ibrahim, M.I., Alsafadi, D., Alamry, K.A. and Hussein, M.A. (2021b). Properties and applications of poly(3-hydroxybutyrate-co-3-hydroxyvalerate) biocomposites. *Journal of Polymers and the Environment*, 29(4):1010–1030. https://doi.org/10.1007/s10924-020-01946-x

Issoufou, A.A., Younoussa, B.H., Soumana, I. and Mahamane, A. (2022). Soil restoration practices on priming effect intensity and carbon fluxes. *Applied and Environmental Soil Science*, 2022(1):1038514. https://doi.org/10.1155/2022/1038514

Jiang, R., Shang, L., Wang, R., Wang, D. and Wei, N. (2023). Machine learning based prediction of enzymatic degradation of plastics using encoded protein sequence and effective feature representation. *Environmental Science and Technology Letters*, 10(7):557–564. https://pubs.acs.org/doi/10.1021/acs.estlett.3c00293

Jiao, S., Chen, W. and Wei, G. (2019). Resilience and assemblage of soil microbiome in response to chemical contamination combined with plant growth. *Applied Environmental Microbiology*, 85(6):e02523–18. https://doi.org/10.1128/AEM.02523-18

Johnson, D.R. and Pomati, F. (2020). A brief guide for the measurement and interpretation of microbial functional diversity. *Environmental Microbiology*, 22(8):3039–3048. https://doi.org/10.1111/1462-2920.15147

Jones, D.L., Hill, P., Smith, A., Farrell, M., Ge, T. and Murphy, D.V. (2018). Role of substrate supply on microbial carbon use efficiency and its role in interpreting soil microbial community level physiological profiles (CLPP). *Soil Biology and Biochemistry*, 123:1–6. https://doi.org/10.1016/j.soilbio.2018.04.014

Joseph, T.M., Unni, A.B., Joshy, K.S., Kar Mahapatra, D., Haponiuk, J. and Thomas, S. (2023). Emerging bio-based polymers from lab to market: Current strategies, market dynamics and research trends. *Journal of Carbon Research,* 9(1):30.

Khursheed, S. (2016). Soil biodiversity and climate change. *Advances in Plants and Agriculture Research*, 3(5):166–168. http://dx.doi.org/10.15406/apar.2016.03.00113

Kirstein, I.V., Wichels, A., Krohne, G. and Gerdts, G. (2018). Mature biofilm communities on synthetic polymers in seawater – Specific or general? *Marine Environmental Research*, 142:147–154. https://doi.org/10.1016/j.marenvres.2018.09.028

Klauer, R.R., Alex Hansen, D., Wu, D., Oliveira, M., Solomon, K.V., and Blenner, M.A. (2024). Biological upcycling of plastics waste. *Annual Review of Chemical and Biomolecular Engineering*, 15(1):315–342.

Kumar, P., Singh, A., Sachan, A. and Sachan, S.G. (2021). Molecular Tools for Microbial Diversity Analysis. In: Shah, M.P., Sarkar, A., and Mandal, S. (Eds.) *Wastewater Treatment*, Elsevier, pp. 55–65. https://doi.org/10.1016/B978-0-12-821881-5.00003-9

Kumar, S., Krishnan, S., Mohanty, S. and Nayak, S.K. (2022). PBAT-Based Blends and Composites. In: Rangappa, S.M., Parameswaranpillai, J., Siengchin, S., and Ramesh, M. (Eds.) *Woodhead Publishing Series in Composites Science and Engineering, Biodegradable Polymers, Blends and Composites*, Woodhead, pp. 327–354. https://doi.org/10.1016/B978-0-12-823791-5.00018-1

Lackner, M., Mukherjee, A. and Koller, M. (2023). What are "Bioplastics"? Defining renewability, biosynthesis, biodegradability, and biocompatibility. *Polymers*, 15:4695. https://doi.org/10.3390/polym15244695

Levytskyi, V., Katruk, D., Masyuk, A., Kysil, K., Bratychak Jr, M. and Chopyk, N. (2021). Resistance of polylactide materials to water mediums of the various natures. *Chemistry and Chemistry Technology*, 15(2):191–197. https://doi.org/10.23939/chcht15.02.191

Li, S., Li, F., Bao, Y., Peng, A. and Lyu, B. (2024). Polyethylene and sulfa antibiotic remediation in soil using a multifunctional degrading bacterium. *The Science of the Total Environment*, 930:172619–172619. https://doi.org/10.1016/j.scitotenv.2024.172619

Liu, X. (2020). Microbial technology for the sustainable development of energy and environment. *Biotechnology Reports*, 27:e00486. https://doi.org/10.1016/j.btre.2020.e00486

Lukas, M., Kittner, M., Isernhinke, L., Altmann, K. and Braun, U. (2023). A new concept for the ecotoxicological assessment of plastics under consideration of aging processes. *Applied Research*, 3(2):e202200124. https://doi.org/10.1002/appl.202200124

Malik, S., Maurya, A., Khare, S.K. and Srivastava, K.R. (2023). Computational exploration of bio-degradation patterns of various plastic types. *Polymers*, 15(6):1540. https://doi.org/10.3390/polym15061540

Maraveas, C., Kotzabasaki, I.M. and Bartzanas, T. (2023). Intelligent technologies, enzyme-embedded and Microbial Degradation of Agricultural Plastics. *AgriEngineering*, 5(1):85–111. https://doi.org/10.3390/agriengineering5010006

Massarelli, C., Campanale, C. and Uricchio, V.F. (2021). A handy open-source application based on computer vision and machine learning algorithms to count and classify microplastics. *Water*, 13(15):2104; https://doi.org/10.3390/w13152104

Mercier A., Gravouil K., Aucher W., Brosset-Vincent S., Kadri L., Colas J., Bouchon, D. and Ferreira, T. (2017). Fate of eight different polymers under uncontrolled composting conditions: Relationships between deterioration, biofilm formation, and the material surface properties. *Environmental Science and Technology*, 51:1988–1997. https://doi.org/10.1021/acs.est.6b03530

Mishra, M. (2015). Microbial Diversity: Its Exploration and Need of Conservation. In: Kaushik, G. (Ed.) *Applied Environmental Biotechnology: Present Scenario and Future Trends*, Springer, New Delhi, pp. 43–58. https://doi.org/10.1007/978-81-322-2123-4_4

Moscatelli, M.C., Secondi, L., Marabottini, R., Papp, R., Stazi, S.R., Mania, E. and Marinari, S. (2018). Assessment of soil microbial functional diversity: Land use and soil properties affect CLPP-MicroResp and enzymes responses. *Pedobiologia*, 66:36–42. https://doi.org/10.1016/j.pedobi.2018.01.001

Mulla, M.Z., Rahman, R.T., Marcos, B., Tiwari, B. and Pathania, S. (2021). Poly lactic acid (PLA) nanocomposites: Effect of inorganic nanoparticles reinforcement on its performance and food packaging applications. *Molecules*, 26(7):1967. https://doi.org/10.3390/molecules26071967

Muralidharan, M., Gayathri, K.V., Kumar, P.S., Preethi, D.S., Kavitha, R., Rajagopal, R. and Rangasamy, G. (2023). Mixed polyaromatic hydrocarbon degradation by halotolerant bacterial strains from marine environment and its metabolic pathway. *Environmental Research*, 216(2):114464. https://doi.org/10.1016/j.envres.2022.114464

Musa, I.O., Samuel, J.O., Adams, M., Abdulsalam, M., Abdulazeez, M., Muhammad, M.S., Pal, S.K., Mahmud, M.E., Adebayo, O,A., Akande, S.B., and Tiamiyu, A.G.T. (2024). Cases of Micro- and Nanoplastics in Food and Food Product. In: Aransiola, S.A., Akinsola, R.O., Abioye, O.P., and Maddela, N.R. (Eds.) Emerging Contaminants in Food and Food Products, CRC Press, Boca Raton, 90–105. https:// doi.org/10.1201/978100 3438 236

Nanez Alonso, S.L., Forradellas, R.F.R., Pi Morell, O. and Jorge-Vasquez, J. (2021). Digitalization, circular economy and environmental sustainability: The application of artificial intelligence in the efficient self-management of waste. *Sustainability*, 13(4):2092. https://doi.org/10.3390/su13042092

Narancic, T., Cerrone, F., Beagan, N., and O'Connor, K.E. (2020). Recent advances in bioplastics: Application and biodegradation. *Polymers*, 12(4):920. https://doi.org/10.3390/polym12040920

Odobel, C., Dussud, C., Philip, L., Derippe, G., Lauters, M., Eyheraguibel, B., Burgaud, G., Ter Halle, A., Meistertzheim, A.L., Bruzaud, S., Barbe, V. and Ghiglione, J.F. (2021). Bacterial abundance, diversity and activity during long-term colonization of non-biodegradable and biodegradable plastics in seawater. *Frontier in Microbiology*, 12:734782. http://dx.doi.org/10.3389/fmicb.2021.734782

Oboh, I.O. (2014). The Chemistry of Removal of Inorganic Compounds from the Soil. In: Parmar, N., and Singh, A. (Eds.) *Geomicrobiology and Biogeochemistry*. Soil Biology, Springer, Berlin, Heidelberg, 39, pp. 247–259. https://doi.org/10.1007/978-3-642-41837-2_13

Okal, E.J., Heng, G., Magige, E.A., Khan, S., Wu, S., Ge, Z., Zhang, T., Mortimer, P.E. and Xu, J. (2023). Insights into the mechanisms involved in the fungal degradation of plastics. *Ecotoxicology and Environmental Safety*, 262:115202.

Oliveira, J., Almeida, P.L., Sobral, R.G., Lourenço, N.D. and Gaudêncio, S.P. (2022). Marine-derived actinomycetes: Biodegradation of plastics and formation of PHA bioplastics: A circular bioeconomy approach. *Marine Drugs* [online], 20(12):760. https://doi.org/10.3390/md20120760

Onyeaka, H., Tamasiga, P., Nwauzoma, U.M., Miri, T., Uche, C.J., Nwaiwu, O. and Akinsemolu, A.A. (2023). Using artificial intelligence to tackle food waste and enhance the circular economy: Maximizing resource efficiency and minimizing environmental impact: A review. *Sustainability*, 15(13):10482. https://doi.org/10.3390/su151310482

Orlando, M., Molla, G., Castellani, P., Pirillo, V., Torretta, V. and Ferronato, N. (2023). Microbial enzyme biotechnology to reach plastic waste circularity: Current Status, Problems and Perspectives. *International Journal of Molecular Sciences*, 24(4):3877. https://doi.org/10.3390/ijms24043877

Othman, A.R., Hasan, H.A., Muhamad, M.H., Ismail, N.I. and Abdullah, S.R.S. (2021). Microbial degradation of microplastics by enzymatic processes: A review. *Environmental Chemistry Letter*, 19:3057–3073. https://doi.org/10.1007/s10311-021-01197-9

Ouyang, Y., Chen, D., Fu, Y., Shi, W., Provin, T., Han, A., Shaik, E.V., Samuel, J.E., Figueiredo, P., Zhou, A. and Zhou, J. (2021). Direct cell extraction from fresh and stored soil samples: impact on microbial viability and community compositions. *Soil Biology and Biochemistry*. 155:108178. https://doi.org/10.1016/j.soilbio.2021.108178

Pallathadka, P.K., Koh, X.Q., Khatta, A., Luckachan, G.E. and Mittal, V. (2015). Characteristics of biodegradable poly(butylene succinate) nanocomposites with thermally reduced graphene nanosheets. *Polymer Composites*, 38(S1):E42–E48. https://doi.org/10.1002/pc.23824

Parsaeimehr, A., Miller, C.M. and Ozbay, G. (2023). Microplastics and their interactions with microbiota. *Heliyon*, 9(4):e15104. http://dx.doi.org/10.1016/j.heliyon.2023.e15104

Pathak, V.M. and Bithel, N. (2017). Review on the current status of polymer degradation: A microbial approach. *Bioresources and Bioprocessing*, 4(1). https://doi.org/10.1186/s40643-017-0145-9

Perujo, N., Romani, A.M. and Martin-Fernandez, J.A. (2020). Microbial community-level physiological profiles: Considering whole data set and integrating dynamics of colour development. *Ecological Indicators*, 117:106628. https://doi.org/10.1016/j.ecolind.2020.106628

Pooja, N., Chakraborty, I., Rahman, M.H., and Mazumder, N. (2023). An insight on sources and biodegradation of bioplastics: A review. *3 Biotech*, 13(7):220. https://doi.org/10.1007/s13205-023-03638-4

Prendiz, J., Vega-Baudrit, J. and Mena, M. (2019). Polylactic acid (PLA) as a bioplastic and its possible applications in the food industry. *Food Science and Nutrition*, 5(2):1–6. https://doi.org/10.24966/fsn-1076/100048

Qayoom, I., Mohi-Ud-Din, H., Khursheed, A., Altaf, A. and Bandh, S.A. (2022). Climate Change and Its Influence on Microbial Diversity, Communities, and Processes. In: Parray. J.A., Bandh, S.A., and Shameem, N. (Eds.) *Climate Change and Microbes*, Apple Academic Press, New York, p. 44. http://dx.doi.org/10.1201/9781003189725-4

Qi, J., Wu, J., Chen, J. and Wang, H. (2019). An investigation of the thermal and (bio) degradability of PBS copolyesters based on isosorbide. *Polymer Degradation and Stability*, 160:229–241. https://doi.org/10.1016/j.polymdegradstab.2018.12.031

Raab, P. and Bonger, F.X. (2020). Microplastics in the environment: Raising awareness in primary education. *The American Biology Teacher*, 82(7):478–487. https://doi.org/10.1525/abt.2020.82.7.478

Rajvanshi, J., Sogani, M., Kumar, A., Arora, S., Syed, Z., Sonu, K., Gupta, N.S. and Kalra, A. (2023). Perceiving biobased plastics as an alternative and innovative solution to combat plastic pollution for a circular economy. *Science of the Total Environment*, 874:162441. https://doi.org/10.1016/j.scitotenv.2023.162441

Ramin, K.I. and Allison, S.D. (2019). Carbon Cycle Implications of Soil Microbial Interactions. In: Hurst, C. (Ed.) *Understanding Terrestrial Microbial Communities*. Advances in Environmental Microbiology, Springer, Cham, 6, pp. 1–29. https://doi.org/10.1007/978-3-030-10777-2_1

Raviteja, K. and Reddy, K.R. (2023). Application of Artificial Intelligence, Machine Learning, and Deep Learning in Contaminated Site Remediation. In: Shukla, S.K., Raman, S.N., Bhattacharjee, B., and Singh, P. (Eds.) *Recent Developments in Geotechnics and Structural Engineering*. TRACE 2022. Lecture Notes in Civil Engineering, Springer, Singapore, 338:393–407. https://doi.org/10.1007/978-981-99-1886-7_33

Rillig, M.C., Ziersch, L., and Hempel, S. (2017). Microplastic transport in soil by earthworms. *Scientific Reports*, 7(1):1362. https://doi.org/10.1038/s41598-017-01594-7

Rouf, A., Kanojia, V., Naik, H.R., Naseer, B. and Qadri, T. (2017). An overview of microbial cell culture. *Journal of Pharmacognosy and Phytochemistry*. 6(6):1923–1928.

Ru, J., Huo, Y., and Yang, Y. (2020). Microbial degradation and valorization of plastic wastes. *Frontiers in Microbiology*, 11(442). https://doi.org/10.3389/fmicb.2020.00442

Safiyanu, I., Abdulwahid, I.A, Mudi, Z.R., Ya'u, S.A. and Rita, S. M. (2022) Bioremediation on oil spills using bacterium (*Pseudomonas putida*). *Iconic Research and Engineering Journal*, 5(8):272–276.

Saikanth, D.R.K., Gupta, K., Srivastava, P., Saryam, M., Rani, K.S., Jena, P. and Rout, S. (2023). Environmental sustainability and food security of traditional agricultural practices in India: A review. *International Journal of Environment and Climate Change*. 13(8):1847–1856. http://dx.doi.org/10.9734/ijecc/2023/v13i82138

Salinas, J., Carpena, V., Martinez-Gallardo, M.R., Segado, M., Estrella-Gonzalez, M., Toribio, A.J., Jurado, M.M., Lopez-Gonzalez, J.A., Suarez-Estrella, F. and Lopez, M.J. (2023). Development of plastic-degrading microbial consortia by induced selection in microcosms. *Frontier in Microbiology*, 14:1143769. https://doi.org/10.3389/fmicb.2023.1143769

Schimel, J.P. and Shaeffer, S.M. (2012). Microbial control over carbon cycling in soil. *Frontier in Microbiology*, 3:348. https://doi.org/10.3389/fmicb.2012.00348

Schneier, A., Melaugh, G. and Sadler, J. C. (2024). Engineered plastic-associated bacteria for biodegradation and bioremediation. *Biotechnology for the Environment*, 1(7). https://doi.org/10.1186/s44314-024-00007-0

Sedlacek, P., Slaninova, E., Koller, M., Nebesarova, J., Marova, I., Krzyzanek, V. and Obruca, S. (2019). PHA granules help bacterial cells to preserve cell integrity when exposed to sudden osmotic imbalances. *New Biotechnology*, 49:129–136. https://doi.org/10.1016/j.nbt.2018.10.005

Shang, J., Zhang, W., Li, Y., Zheng, J., Ma, X., Wang, L. and Niu, L. (2023). How nutrient loading leads to alternative stable states in microbially mediated N-cycle pathways: A new insight into bioavailable nitrogen removal in urban rivers. *Water Research*, 236:119938. https://doi.org/10.1016/j.watres.2023.119938

Shen, M.C., Zhang, Y.Z., Bo, G.D., Yang, B., Wang, P., Ding, Z.Y., Wang, Z.B., Yang, J.M., Zhang, P. and Yuan, X.L. (2022). Microbial responses to the reduction of chemical fertilizers in the rhizosphere soil of flue-cured tobacco. *Frontier in Bioengineering and Biotechnology*, 9:812316. http://dx.doi.org/10.3389/fbioe.2021.812316

Singh, A.K., Bilal, M., Iqbal, H.M.N. and Raj, A. (2021). Trends in predictive biodegradation for sustainable mitigation of environmental pollutants: Recent progress and future outlook. *Science of the Total Environment*, 770:144561. https://doi.org/10.1016/j.scitotenv.2020.144561

Singh, A.K., Bilal, M., Iqbal, H.M.N. and Raj, A. (2022). In silico analytical toolset for predictive degradation and toxicity of hazardous pollutants in water sources. *Chemosphere*, 292:133250. https://doi.org/10.1016/j.chemosphere.2021.133250

Singh, G., Yadav, S., Chowdhary, K. and Sharma, S. (2020). Role of Microbes in Plastic Degradation. In: Prasad, R., Kumar, V., Singh, J., and Upadhyaya, C.P. (Eds.) *Recent Developments in Microbial Technologies*. Environmental and Microbial Biotechnology, Springer, Singapore, pp. 281–299. https://doi.org/10.1007/978-981-15-4439-2_13

Singh, P., Pani, A., Mujumdar, A.S. and Shirkole, S.S. (2023). New strategies on the application of artificial intelligence in the field of phytoremediation. *International Journal of Phytoremediation*, 25(4):505–523. https://doi.org/10.1080/15226514.2022.2090500

Srikanth, M., Sandeep, T.S.R.S., Sucharitha, K. and Godi, S. (2022). Biodegradation of plastic polymers by fungi: A brief review. *Bioresources and Bioprocessing*, 9(1). https://doi.org/10.1186/s40643-022-00532-4

Strollo, E., Sansonetti, G., Mayer, M.C., Limongelli, C. and Micarelli, A. (2020). An AI-Based Approach to Automatic Waste Sorting. In: Stephanidis, C., and Antona, M. (Eds.) *HCI International 2020 – Posters. HCII 2020*. Communications in Computer and Information Science, Springer, Cham, 1124, pp. 662–669. https://doi.org/10.1007/978-3-030-50726-8_86

Su, D., Li. X.Y., Liu S.K. and Bu Y.Y. (2022). Research progress on the effects of salt stress on photosynthesis and lipids of microalgae, *Molecular Soil Biology*, 13(1):1–6. https://doi.org/10.5376/msb.2022.13.0001

Sun, Y., Duan, C., Cao, N., Ding, C., Huang, Y. and Wang, J. (2022). Biodegradable and conventional microplastics exhibit distinct microbiome, functionality, and metabolome changes in soil. *Journal of Hazardous Materials*, 424(A):127282. https://doi.org/10.1016/j.jhazmat.2021.127282

Temporiti, M.E.E., Nicola, L., Nielsen, E. and Tosi, S. (2022). Fungal enzymes involved in plastics biodegradation. *Microorganisms*, 10(6):1180. https://doi.org/10.3390/microorganisms10061180

Tokiwa, Y., Calabia, B.P., Ugwu, C.U. and Aiba, S. (2009). Biodegradability of plastics. *International Journal of Molecular Sciences*, 10(9):3722–3742. https://doi.org/10.3390/ijms10093722

Tomer, A., Singh, R., Singh, S.K., Dwivedi, S.A., Reddy, C.U., Keloth, M.R.A., and Rachel, R. (2021). Role of Fungi in Bioremediation and Environmental Sustainability. Fungal Biology. *Mycoremediation and Environmental Sustainability*, 187–200. https://doi.org/10.1007/978-3-030-54422-5_8

Ukhurebor, K.E., Nwankwo, W., Adetunji, C.O. and Makinde, A.S. (2021). Artificial Intelligence and Internet of Things in Instrumentation and Control in Waste Biodegradation Plants: Recent Developments. In: Adetunji, C.O., Panpatte, D.G., and Jhala, Y.K. (Eds.) *Microbial Rejuvenation of Polluted Environment*. Microorganisms for Sustainability, Springer, Singapore, 27:265–279. https://doi.org/10.1007/978-981-15-7459-7_12

Vaidyanathan, S., Sivakumar, M. and Kaliamourthy, B. (2021). Challenges of Developing AI Applications in the Evolving Digital World and Recommendations to Mitigate Such Challenges: A Conceptual View. In: Misra, S., Arumugam, C., Jaganathan, S., and Saraswathi, S. (Eds.) *Confluence of AI, Machine, and Deep Learning in Cyber Forensics*, IGI Global Scientific Publishing, pp. 177–198. https://doi.org/10.4018/978-1-7998-4900-1.ch011

Vaishnav, A., Lal, J., Singh, N.S., Pati, B.K., Mehta, N.K., and Priyadarshini, M.B. (2024). Role of microbial enzymes and their modification for plastic biodegradation. *Advanced Strategies for Biodegradation of Plastic Polymers*, 373–403. https://doi.org/10.1007/978-3-031-55661-6_16

Wang, H., Cheng, Z., Djouonkep, L.D.W., Wang, L., Cal, S. and Gauthier, M. (2023a). Synthesis and properties of biodegradable aliphatic–aromatic polyesters derived from 4-hydroxybenzaldehyde. *Journal of Applied Polymer Science*, 140(29):e54063. https://doi.org/10.1002/app.54063

Wang, S., Jiang, X., Li, J., Zhao, X., Han, E., Qu, H., Ma, X. and Lian, J. (2023b). Increasing precipitation weakened the negative effects of simulated warming on soil microbial community composition in a semi-arid sandy grassland. *Frontier in Microbiology*, 13:1074841. http://dx.doi.org/10.3389/fmicb.2022.1074841

Wani, A.K., Akhtar, N., Naqash, N., Rahayu, F., Djajadi, D., Chopra, C., Singh, R., Mulla, S.I., Sher, F. and Pine Americo-Pinheiro, J.H. (2023). Discovering untapped microbial communities through metagenomics for microplastic remediation: recent advances, challenges, and way forward. *Environmental Science and Pollution Research*. 30:81450–81473. https://doi.org/10.1007/s11356-023-25192-5

Wei, R., and Zimmermann, W. (2017). Microbial enzymes for the recycling of recalcitrant petroleum-based plastics: How far are we? *Microbial Biotechnology*, 10(6):1308–1322.

Xia, C., Lam, S.S., Zhong, H., Fabbri, E. and Sonne, C. (2022). Assess and reduce toxic chemicals in bioplastics. *Science*. 378(6622):842. http://dx.doi.org/10.1126/science.ade9069

Yaashikaa, P.R., Devi, M.K. and Kumar, P.S. (2022). Engineering microbes for enhancing the degradation of environmental pollutants: A detailed review on synthetic biology. *Environmental Research*, 214(1):113868. https://doi.org/10.1016/j.envres.2022.113868

Yang, X., Cheng, J., Franks, A.E., Huang, X., Yang, Q., Cheng, Z., Liu, Y., Ma, B., Xu, J. and He, Y. (2023). Loss of microbial diversity weakens specific soil functions, but increases soil ecosystem stability. *Soil Biology and Biochemistry*, 117:108916. https://doi.org/10.1016/j.soilbio.2022.108916

Ye, Z., Yang, J., Zhong, N., Tu, X., Jia, J. and Wang, J. (2020). Tackling environmental challenges in pollution controls using artificial intelligence: A review. *Science of the Total Environment*, 699:134279. https://doi.org/10.1016/j.scitotenv.2019.134279

Yeo, J.C.C., Muiruri, J.K., Thitsartarn, W., Li, Z. and He, C. (2019). Recent advances in the development of biodegradable PHB-based toughening materials: Approaches, advantages and applications. *Materials Science and Engineering: C*, 92:1092–1116. https://doi.org/10.1016/j.msec.2017.11.006

Younis, H., Abdelrahman, F., Samer, M. and Adellatif, H. (2024). An Overview of Biodegradable Polymer and Types of Bioplastics: Properties and Applications. In: *Bioplastics within the Circular Bioeconomy*, IntechOpen. http://dx.doi.org/10.5772/intechopen.1007621

Zampieri, G., Campanaro, S., Angione, C. and Treu, L. (2023). Metatranscriptomics-guided genome-scale metabolic modelling of microbial communities. *Cells Reports Methods*, 3(1):100383. http://dx.doi.org/10.1016/j.crmeth.2022.100383

Zhou, Z., Tran, P.Q., Breister, A.M. Liu, Y., Kieft, K., Cowley, E.S., Karaoz, U. and Anantharaman, K. (2022). METABOLIC: High-throughput profiling of microbial genomes for functional traits, metabolism, biogeochemistry, and community-scale functional networks. *Microbiome*, 10:33. https://doi.org/10.1186/s40168-021-01213-8

7 Microbes and Microbial Products in Bioplastic Degradation

Mustapha Abdulsalam,
Abdulhakeem Idris Abdulhakeem, Innocent Ojeba Musa,
Miracle Uwa Livinus, Shehu-Alimi Elelu,
Muhammed Abdulazeez, Maiyaki Taofiqat Damilola,
Ibrahim Abdulrazaq, Umar Abdulrasheed,
Abdulkareem Tajudeen Zakari, and Tijani Abiola Tajudeen

7.1 INTRODUCTION

The use of plastic material has been identified as one of the causes of environmental issues in dominated societies. Synthetic plastics, primarily produced from non-renewable fossil resources, are voracious and robust in natural degradation processes, leading to their eventual disking in terrestrial and aquatic habitats (Trinh *et al.*, 2023). Such plastics lead to 'big ocean' pollution, and invasion of natural ecosystems, and are potentially hazardous to the lives of abundant animals and the human population through the entry of micro-plastics and effluence diffusion (Lakshmayya *et al.*, 2024). The world uses over 400 million tons of plastics annually, and its use has become an international concern to find improved forms of recycling or better still, environment-friendly substitutes (Musa *et al.*, 2024a).

Bioplastics are a relatively new development as a material that can replace regular plastics with desired properties such as being made from renewable resources and biodegradable. Bioplastic sources can be raw materials derived from plants including starch and cellulose and those produced by microbial fermentation including polyhydroxyalkanoates (PHAs) and PLA. All these materials are increasingly used in packaging, medical devices, and agriculture because of their lower environmental impact and ability to integrate into the circular economy (Hahladakis *et al.*, 2024). Nevertheless, the environmental advantages of bioplastics mainly provide packaging that can decompose naturally with the help of microbial action.

Most of the biodegradation of bioplastics depends on microorganisms such as bacteria and fungi. This is achieved through enzymatic degradation of complex polymer frameworks, in order to produce simple, non-hazardous divisions including CO_2, water, CH_4, and biomass. These microbes either act alone or in combinations and are known to degrade all forms of bioplastics through distinct mechanisms that depend on the structure of the polymer (Müller *et al.*, 2021). Knowledge about the processes occurring between bioplastics and microbiota will allow enhancing the biodegradation rate and biocompatibility.

This study aims to introduce the readers to the concept of the bioplastic degradation process, primarily utilizing microbial community products, the enzymes. The relationship between the categories of bioplastic, the physicochemical characteristics of these polymers, and microorganisms capable of their degradation is also investigated. Moreover, it investigates all the conditions that

 DOI: 10.1201/9781003602828-9

empower or hinder biodegradation, including environment and polymer, and provides examples of microbial biodegradation of individual types of bioplastics. The discussion goes further concerning utilization of microbial products in increasing degradation and current issues like technical constrains and other hurdles that are likely to affect this field.

7.1.1 Background on Bioplastics

Bioplastics can be defined as any polymer that is formed from renewable biological sources or which may be formulated for biodegradation. Bioplastics on the other hand are a relatively new family of plastics that are intended to accommodate the problems observed with traditional plastics (Goel *et al.*, 2021). Combined with the fact that the plastic production and disposal process is a highly prevalent environmental problem also imposes a necessity for bioplastic materials and product development to consider the functionality and versatility of plastics alongside their negative effects on the environment. There are two main categories of bioplastics: biopolymers, which are plastics produced from bio-based materials like starch, and cellulose, and products of microbial fermentation and biodegradable plastics that break down by the actions of microorganisms. For example, PLA and PHAs come from renewable resources and are biodegradable or susceptible to biodegradation, and thus suitable for bioapplications (Narancic *et al.*, 2020).

The worldwide use of bioplastics is informed by the fact that they offer an opportunity to decrease the levels of emissions of greenhouse gases, decrease dependency on fossil resources, and find sustainable solutions to the problem of persistent plastic waste. For instance, the life cycle assessments done on the bioplastics reveal that emissions of carbon related to the production of bioplastics are lower than the emissions related to the production of conventional plastics (Karan *et al.*, 2019). In addition, the incorporation of agricultural by-products and waste streams as source materials was also a win–win situation for the environment and the economy as well as for the bioplastics. However, the use of bioplastics as a material is not without its problems: it is more costly to produce, there are few facilities where it can be adequately biodegraded, and the rates of degradation differ in various locations. Bioplastic degradation can be relatively slower depending on the molecular structure and the conditions that prevail where the product is disposed of or whether suitable microbial enzymes to break the product are available or not. These limitations highlight the need for further development of the microbial degradation process for the effective utilization of fully biodegradable bioplastics as a sustainable replacement for standard plastic products (Bhatia *et al.*, 2021).

7.2 IMPORTANCE OF BIODEGRADATION

Biodegradation is among the most effective methods for handling the impact of plastics on the environment. It is the process by which complex substances are decomposed into simpler and harmless forms by the action of enzymes of microorganisms. In the case of bioplastics, biodegradation makes those materials return to the natural cycle and minimize pollution (Musa *et al.*, 2024b).

The significance of biodegradation thus leads from the fact that there has been presented an environment-friendly mode of waste disposal by contrasting it with landfilling and incineration. The traditional method increases greenhouse gas emissions and soil, and air pollution, while biodegradation is the friendly, efficient, and non-polluting solution using microbial action for the disposal of plastics at its end use (Samir *et al.*, 2022). This is especially essential in the prevention of the buildup of the plastic particles that when they find their way into the aquatic and the terrestrial environment, inflict lots of harm to the biochemical succession of life. In addition, biodegradation aligns with circular economy concept by offering a process through which bioplastics are transformed into useful products like carbon dioxide, methane where the process is anaerobic and organic matter. They can be utilized for generating energy or in farming or even as feedstock to synthesize other materials which makes it a completely cyclic system (Wang *et al.*, 2021). The biodegradation process depends on several factors, many of which are discussed here; the environmental conditions in the disposal site, temperature, pH values, availability of moisture and the chemical content of the bioplastics, and the existence of adequate microbial populations to help in the degradation of the material. Studies conducted in these areas have allowed bioplastic type and degradation conditions to be tailored to maximize degradation rates. Moreover, the possibilities for introducing new approaches of biotechnological level such as genetically engineered microbes and the enzymes as a cocktail provide important potential for enhancing the biodegradation processes (Kun *et al.*, 2019).

7.3 ROLE OF MICROBES IN BIODEGRADATION

Since biopolymers and bioplastics are polymers and long-chain structures, they are decomposed by the action of microorganisms because microorganisms have all the enzymes capable of degrading polymers. This process involves the use of bioplastics as carbon and energy by bacteria and fungi, the common microbial degraders. Their function is crucial in the process of converting polymers to non-hazardous products such as water, carbon dioxide, methane under anaerobic environments, and biomass (Kumar *et al.*, 2023).

Microbes also start bioplastic degradation through enzyme action where esterases, lipases, and proteases act on the polymer chains to split and depolymerize them into oligomers and monomers. These smaller units are then transported into microbial cells where intracellular enzymes oxidize them via multiple biochemistry pathways (Walsh *et al.*, 2018). The biodegradation rate and how it occurs also depends on some of such factors; the nature of the bioplastic used, chemical structures of the bioplastic, and the nature of the microorganisms that are in operation. Specific microbial forms are unique in their ability to degrade specific types of bioplastics. For example, *Pseudomonas* spp. and *Bacillus* spp. are capable of degrading PHAs, while *Aspergillus niger* and *Trichoderma reesei* are for starch-based plastics (Lim *et al.*, 2022). These microbes can thrive in different environments and capture polymers from various systems, which gives them broad applicability as biodegraders in both natural and engineered systems.

In natural ecosystems, microbes uniformly synergize to increase rate constants, where the action is concerted rather than consecutive. For instance, the bipartite relations between bacteria and fungi in the process of bioplastic degradation can be explained by the synergistic action of their enzymes. This sample illustrates the interactions and interdependence of interactions of microbial ecosystems to create synergy in the biodegradation process (Müller *et al.*, 2021). Newer developments in microbial biotechnology have further underlined the use of microbes in the biodegradation process. Some of the strategies for underutilization are genetic engineering to produce recombinant microbes which can degrade environmental pollutants better than the natural microbial strains in terms of enzyme synthesis and substrate selectivity. These engineered microbes hold promise for large-scale applications in waste management and environmental remediation (Sharma *et al.*, 2021).

7.4 TYPES OF BIOPLASTICS

Bioplastics include many materials that can be distinguished by their source, chemical composition, and ability to biodegrade. Their classification generally falls into two broad categories: first, the bio-based bioplastics which are sourced from biomass resources, and second, the biodegradable bioplastics which will degrade under standard natural conditions. Some bioplastics possess both qualities, which are convenient regarding both sustainability and biocompatibility (Abd El-malek *et al.*, 2020).

7.4.1 Starch-Based Bioplastics

The most common types include bioplastics derived from starch since it is easily and cheaply available and biodegradable. Category one bioplastics are made from raw materials generated from crops that include maize, potato as well as cassava where starch is a major input. Amylose and amylopectin simplicity of the starch structure allows the degradation by various microbes like *Amylomyces rouxii* and *Bacillus subtilis* (Anal, 2019). However, starch bioplastics are brittle and not waterproof; they are improved by the use of plasticizers like glycerin and blending with other polymers like PLA. Such changes improve their relevance in use for packaging purposes, in agriculture, as well as for medical purposes (Han *et al.*, 2018).

7.4.2 Polyhydroxyalkanoates (PHAs)

PHAs are microbial plastics produced exclusively inside the bacterial cells as carbon and energy reserves in situations of starvation. Some of the examples include *Cupriavidus necator* and *Pseudomonas putida*. PHAs are rapidly degradable polymers with thermoplastic characteristics; therefore, their favorable uses include packaging, biomedical, and agricultural films (Meereboer *et al.*, 2020). Some members of the PHA family are PHAs, and polyhydroxybutyrate (PHB), which is the most broadly described type of PHA and is highly biodegradable in soil and water. However, it remains highly brittle, and even its manufacture is expensive, which hampers its applicability on a large scale. Microbial engineering and fermentation improvement have been targeted to overcome these challenges (Livinus *et al.*, 2024a).

7.4.3 Polylactic Acid (PLA)

PLA is a biodegradable thermoplastic aliphatic polyester polymer produced through fermenting sugars, primarily from plants such as corn and sugarcane. Due to its transparency, mechanical strength, and general biodegradability under industrial conditions of composting, it is used extensively in food packaging, disposable cutlery, and medical applications such as implants (La Fuente *et al.*, 2023). PLA like most polyesters undergoes hydrolysis and further biodegradation by the action of microorganisms on the lactic acid monomers formed. There is specific knowledge that few microbes like *Amycolatopsis* sp. are efficient in degrading PLA under particular conditions of their natural environment (Xu *et al.*, 2022).

7.4.4 Other Emerging Bioplastics

New inventions in the course of time have given birth to new bioplastics of the known normal bioplastics including polybutylene succinate (PBS), polycaprolactone (PCL), and bio-based polyethylene (bio-PE). The used material has adaptability for functionality, biodegradability, and the effects on the environment. For instance, PBS is produced from succinic acid and is eco-friendly in different ecosystems for sale in packaging and agriculture (Abdulsalam *et al.*, 2023). Another emerging aspect includes the use of plant biomass mainly the lignocellulosic feed for the production of bio-based bioplastics. The development of the bioplastic industry is the main factor increasing opportunities for renewable resource use in bioplastic production from agricultural residues, and non-food plants (Aliotta *et al.*, 2022).

7.5 MICROBIAL COMMUNITIES INVOLVED IN BIOPLASTIC DEGRADATION

Microorganisms are essential components of the degradation of bioplastics due to metabolic versatility and enzymatic adroitness in the decomposition of both synthetic and natural polymers.

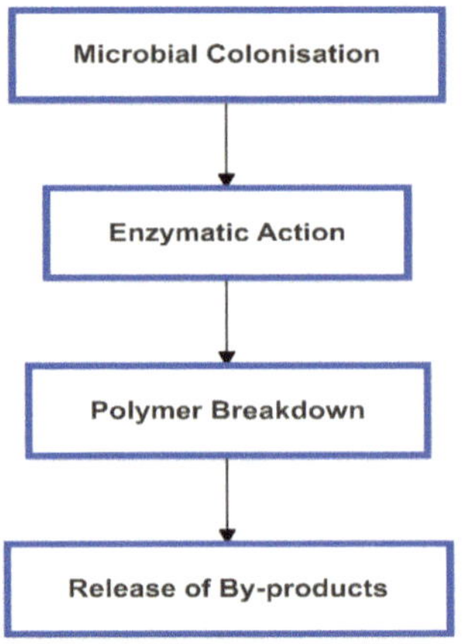

FIGURE 7.1 Microbial degradation process of bioplastics.

Bacterial communities within these biofilms comprise bacteria, fungi, and sometimes archaea, and work cooperatively to accomplish polymer degradation (Figure 7.1). Flemming & Wuertz (2019) found that their activity depends on factors like the structure of the polymer, physical conditions, and nutrient concentration. The specific focus of this section is the principal biodegradation agents, bacteria and fungi, with emphasis on the species and their actions.

7.5.1 Bacteria

Bacteria are among the microorganisms that can be considered the most efficient in the degradation of bioplastics because they are hard-adapted to the changes in the environment in which they exist: soil, sea, or compost.

7.5.1.1 Key Bacterial Species

Many bacterial genera are acknowledged to possess an ability for bioplastic degradation. Major species for PHAs are *Pseudomonas aeruginosa*, *Alcaligenes faecalis*, and *Bacillus megaterium.* These species secrete extracellular depolymerases which catabolize PHA polymers into hydroxyalkanoic acid monomers that are then metabolized within their cellular context (Ciebiada *et al.*, 2020). The degradation of polylactic acid (PLA) has been well-researched in microbes such as *Amycolatopsis orientalis* and *Thermobifida fusca.* Some of these organisms are very efficient in depolymerizing PLA into lactic acid monomers, particularly under industrial composting conditions such as temperatures of 50–70°C, as pointed out by Taib *et al.* (2023). Similarly, *Clostridium botulinum* has been reported for its potential in hydrolyzing PLA under the anaerobic condition that exists in landfill sites, and in the process producing methane as a by-product (O'Connor *et al.*, 2021). Starch-derived bioplastics tend to accumulate species like *Escherichia coli* and *Lysinibacillus fusiformis* that catabolize the starch derivatives to glucose units for their nutrition. These bacteria grow well in environments containing elements of nutrient value, for instance, agricultural tracts and wastewater treatment systems, where the tendency to dispose of starch-based plastics prevails (Okeke *et al.*, 2022).

7.5.1.2 Mechanisms of Action

External and internal processes are used by bacteria to break down bioplastics. Firstly, the extracellular enzymes, including hydrolyzing enzymes like hydrolases and enzymes that depolymerize biomass like depolymerase, break the long polymer chains. For the same, PHA depolymerases break ester bonds in the PHA backbone and produce hydroxy-alkanoic acids. These monomers are then internalized into bacterial cells where they undergo pathways such as the tricarboxylic acid (TCA) cycle, where they gain energy and biomass (Yañez *et al.*, 2020).

The depolymerization of PLA includes hydrolysis of the ester linkages, thereby yielding oligomers and monomers of lactic acid. *T. fusca* optimizes this process under high-temperature conditions, whereby thermophilic enzymes play an important role in the breakdown of PLA (Sharma *et al.*, 2019).

7.5.2 Fungi

Because fungi can manage a large variety of enzymatic activities and colonize numerous types of substrata, bioplastics are vulnerable to their action. Their ability to survive in low-nutrient environments, such as landfills and forest floors, enhances their role in natural bioplastic decomposition (Desrochers *et al.*, 2020).

7.5.2.1 Key Fungal Species

The most effective fungal species reported in the bioplastic degradation include *A. niger*, *T. reesei*, and *Fusarium oxysporum*. It is these species that are best suited for the disintegration of starch-based plastics by the formulation of amylases that catalyze the breakdown of amylose and amylopectin into glucose. The ability of *A. niger* to occupy polymer surfaces and produce extracellular enzymes thus makes it efficient in the degradation of starch bioplastic (Akromah *et al.*, 2024). PLA belongs to aliphatic polymers for which fungi such as *Penicillium roqueforti* and *Fusarium moniliforme* have been identified to be capable of degrading the polymer into lactic acid monomers under acidic environments. Other wood-decay fungi for example *Phanerochaete chrysosporium* are also capable of degrading complex polymers; it secretes oxidative enzymes such as laccases and manganese peroxidases (Pournou, 2020).

7.5.2.2 Mechanisms of Action

By evaluating the biodrainage of bagasse bioplastic in the presence of different culture fungi, we found that microbial degradation most significantly depends on the hydrolytic and oxidative enzyme. In positivist, the case of starch-based plastics, fungi release amylase and glucosidase enzymes that hydrolyze the glycosidic bond present within amylose and amylopectin (Compart *et al.*, 2023). Out of this, glucose is formed which is used by fungi for their growth and development or reproduction.

Fungal PLA degradation includes the depolymerization of the PLA material by hydrolytic enzymes that cleave ester bonds on the polymer backbone while producing lactic acid. Fungi have shown the advantage of performing PLA degradation in acidic and thermophilic conditions, making them highly efficient in industrial composting (Sankhla *et al.*, 2020). In more complex polymers like bio-PE, fungi can use oxidative enzymes like laccases and peroxidases to oxidize the polymer, introducing functional groups that make it available to be attacked again by enzymatic hydrolysis (Polman *et al.*, 2021).

7.5.2.2.1 Synergistic Role of Microbial Communities

In microbial consortia, several bacterial and fungal species may form synergistic relationships in improving bioplastic degradation efficiency. For instance, bacteria may cause polymer fragmentation so that fungi can metabolize the derived substrates. The several metabolic activities of these communities promote the mineralization of bioplastics to CO_2, water, and biomass especially under compost and soil environments (Brown *et al.*, 2023).

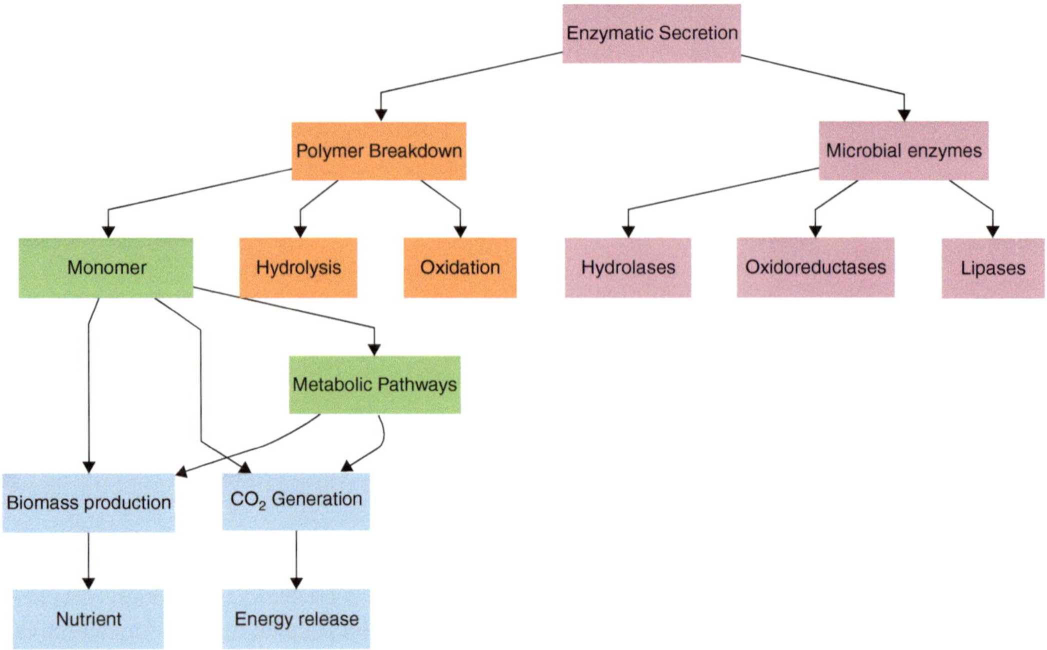

FIGURE 7.2 Mechanisms of microbial action.

7.6 MICROBIAL ENZYMES IN BIOPLASTIC DEGRADATION

Microbial enzymes present a critical intermediate step in bioplastic degradation since they are enzyme catalysts of the process by which polymers are cleaved into simpler structures. These enzymes are generated by bacteria, fungi, and other microorganisms which are inherent in ecological systems to which bioplastics are disposed, for example, compost piles, soil, and water bodies (Nandakumar *et al.*, 2021). The enzymatic process depends on the specialized functions of the enzymes to hydrolyze, oxidize, and cleave the polymer profile of bioplastics into simpler forms more easily digestible by the bacteria (Meereboer *et al.*, 2020). This section aims at explaining different forms of enzymes used in bioplastic degradation, how they work, and different factors determining production and optimization of these enzymes.

7.6.1 Types of Enzymes

Several enzyme classes are implicated in the biodegradation of bioplastics by microbes through the known biodegradation pathways. These enzymes can be summed up into hydrolases, lipases, and oxidative enzymes, all of which are involved in the degradation process (Figure 7.2) but in different ways.

7.6.1.1 Esterases

Esterases are enzymes belonging to the category of hydrolases that split ester bonds, characteristic of several bioplastics, including PHAs and PLA. These enzymes break the ester bonds within monomers polymerized in the bioplastics and reduce the long-chained bioplastics into smaller units of oligomers or monomers that can be metabolized by microorganisms. Pathogenic bacteria which include *P. aeruginosa* and *B. subtilis* secrete depolymerases which are a kind of esterase and will break down the PHA plastics into their basic structural unit termed hydroxy-alkanoic acids (Megha *et al.*, 2024). In the same way, the PLA degradation performed by esterases triggers the ester bond of the polymer to release lactic acid which can be further metabolized by the microorganisms.

Esterases primarily contribute significantly to the biodegradation of abundant biopolymers with big molecular mass: they cut polymer chains into pieces that could be processed further by enzymes from other classes, which is critical for the total biodegradation process (Bher *et al.*, 2022).

7.6.1.2 Lipases

Lipases are enzymes that modify lipid molecules and are involved in the biodegradation of fatty ester-based bioplastics and other types of bioplastics such as PHAs. Lipases act on the ester linkages in the polymer backbone and cause hydrolysis of long-chain fatty acids, which can be utilized as a carbon source of microbial growth (Lai *et al.*, 2023). Specific bacterial species such as *Pseudomonas* sp. have been identified to synthesize lipases which contribute to the degradation of the synthetic and biodegradable polyesters. These bacteria secrete lipase enzymes that facilitate the breakdown of ester bonds in bioplastics such as PCL and PHB, which are normally used in bioplastic materials (Al Hosni, 2019).

7.6.1.3 Laccases and Peroxidases

Laccases and peroxidases are oxidative enzymes acting in the lignin and other complex biopolymer's biodegradation process. They are especially relevant to biodegradation of aromatic-containing bioplastics or PLA where ester bond cleavage under oxidative conditions promotes plastics' decomposition (Larrañag & Lizundia, 2019). Sing and Yogalakshm (2020) have reported that laccases from fungi like *Trametes versicolor* and *P. chrysosporium* oxidize aromatic structures using oxygen, which leads to the beginning of bioplastic polymer degradation into simpler forms. Likewise, peroxidases are enzymes that trigger reactions containing hydrogen peroxide, to enhance the oxidation of polymer chains in bioplastics. These enzymes are used by the lignin-degrading fungi to break down lignocellulosic materials and can also be involved in the biodegradation of specific categories of bioplastics like PLA through the disruption of the ester linkage in the polymer chain (Ahmad *et al.*, 2023).

7.6.2 Enzyme Production and Optimization

The synthesis of microbial enzymes for the degradation of biopolymers depends on the microbial strains, availability of substrate, and conditions of the environment. If the biodegradation of bioplastics is to be efficient for commercial purposes, it is evident that the rate of enzyme production needs to be optimized (Figure 7.3).

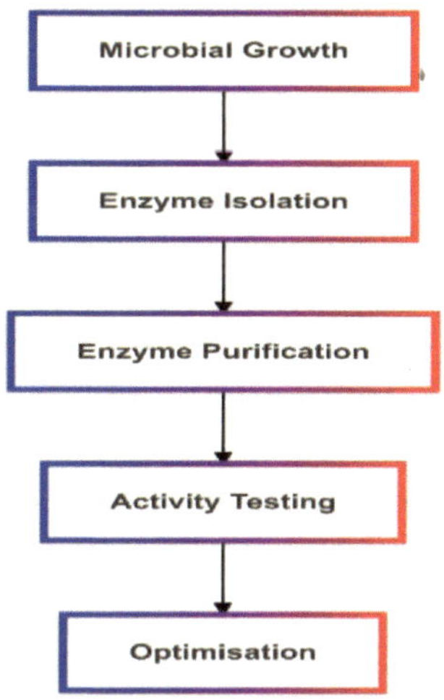

FIGURE 7.3 Enzyme production and optimization processes.

7.6.2.1 Factors Influencing Enzyme Production

The production of degradation enzymes is regulated by factors such as temperature, pH, and availability of nutrients. For example, PLA degradation by the thermophilic bacterium *T. fusca* at composting temperature involves the production of thermostable enzymes that help degrade PLA under the required conditions (Finore *et al.*, 2023). Likewise, *T. reesei*, a common saprotroph of autumn, secretes its enzymes on condition of proper temperature and moisture fondness for the biodegradation of starch-based plastics and other biopolymers in the field as well as in industrial compost sites. It was also found that certain specific carbon sources can induce enzyme synthesis in bacterial cells. Generally, bacteria and fungi secrete higher degradation enzyme levels when grown on bioplastic polymers or their monomers. Further, the inoculation of different microbial species supports the cooperative engagement of various species on the enzymes involved in bioplastic degradation (Schaerer *et al.*, 2023).

7.6.2.2 Biotechnological Approaches to Enzyme Optimization

To increase the efficiency of the degradation of bioplastics, techniques of genetic engineering, immobilized enzymes, and metabolic engineering are becoming common and popular. They work through recombinant technology to improve the traits of microorganisms to produce and secrete the large number of enzymes that degrade bioplastics. For instance, the expression of esterases or lipases in genetically modified strains of *E. coli* or *Pseudomonas* has demonstrated the potential to improve the degradation rate of polyesters and PHA by-products (El-Malek & Steinbüchel, 2022). For example, the addition of genes encoding for esterases or lipases into the genetically modified sink of *E. coli* or *Pseudomonas* has certain potential effects on enhancing the degradation. Enzyme immobilization can also be useful since the enzymes, after use, can be reused and thus increase the efficiency of bioreactor systems in breaking down bioplastics. Moreover, for large-scale production of enzymes, certain methods like solid-state fermentation or submerged fermentation provide ways for the optimization in the production of enzymes (Londoño-Hernandez *et al.*, 2020).

7.7 FACTORS INFLUENCING MICROBIAL DEGRADATION OF BIOPLASTICS

The biodegradation of bioplastics is a complicated process, which depends upon many factors such as environmental, chemical, and biological factors. Such factors are important for the enhancement of bioplastic degradation to increase its rate and efficiency with a focus on management of waste issues. These include effects of temperature, pH, moisture, and oxygen levels, and the chemical nature of the bioplastics. These factors are described in more detail in the subsections, in relation to their impact on the level of microbial activity as well as the rates of bioplastic degradation.

7.7.1 Environmental Conditions

Microbial degradation of bioplastics depends on the environmental characteristics of the location in which the bioplastics are placed. Bioplastics are broken down by microbes that function optimally at certain environmental conditions, determining regulation of enzymes and thus efficiency of biodegradation.

7.7.1.1 Temperature and pH

Temperature is an important aspect of the microbial degradative process because it influences the microorganisms' metabolic efficiency, as well as the activity of enzymes. The majority of bioplastics are biodegraded more effectively at moderate to high temperatures about 30–60°C where thermophilic bacteria and fungi are commonly present (Urbanek *et al.*, 2021). At these temperatures, thermophilic strains possess heat-stable enzymes that are capable of degrading the bioplastics at a faster rate than at lower temperatures. However, high or low temperature increases can slow down microbial activity, and thus, the decomposition processes (Joo *et al.*, 2021). The microbial enzymatic activity is also affected to a large extent by pH. Degradation of bioplastics is known to be generally mesophilic and mostly operates under neutral to slightly basic conditions for pH. These microorganisms, most of which involve species of *Pseudomonas* and *Aspergillus*, usually perform optimally at an outlined pH of between 6 and 8 (Rani *et al.*, 2020). It is important to maintain this optimal pH range because anything outside this range slows down enzyme synthesis and the rate at which bioplastic is being degraded since extreme temperatures are capable of denaturing enzymes together with having a negative effect on microbial growth (Atanasova *et al.*, 2021).

7.7.1.2 Moisture Content and Oxygen Availability

Moisture serves a vital role in different metabolic activities of microorganisms existing in gaseous, liquid as well as solid phases. The microbial cell needs an appropriate measure of moisture since it allows the transportation of nutrients and the enzymes within the cell. As for other bioplastics like PHAs and PLA, for example, water is required for the degradation of ester bonds, which are a part of the materials' structure. However, in dry weather, the germination and growth of microorganisms is slow and the rate of degradation of bioplastic is also slow (Chah *et al.*, 2022). Another important factor is the relative availability of oxygen to the fuels which are 69 percent carbon and 9 percent hydrogen. Aerobic conditions are generally favorable to rapid degradation since many of the degradation enzymes including the esterases and lipases require oxygen for their activity from the aerobic bacteria and fungi. However, some bioplastic-degrading microorganisms, especially belong to genus that degrade PHA, are capable to perform under anaerobic conditions but are a bit slower compared to aerobic one (García-Depraect *et al.*, 2021). Whether oxygen is present or absent determines the whole microbial degradation cascade of bioplastics and therefore the general efficiency of bioplastic degradation.

7.7.2 Chemical Composition of Bioplastics

The type and configuration of the basic polymer or polymers used in the creation of bioplastics significantly define their rate of disintegration by microbial action. Monomers and polymers of which bioplastics are made include different chemical bonds including ester, amide, and ether bonds which determine area of interaction of different microbial enzymes.

7.7.2.1 Polymer Structure and Molecular Weight

The kind, molecular weight, and polymer structure of the bioplastics affect their degradation rates. The high molecular weight incorporating the PHAs and PLA makes bioplastic highly resistant to microbial degradation because the long polymer chains cannot easily be degraded by microbes (Fairley, 2016). But when the long polymer chains are broken down into monomers or oligomers by microbial enzymes, then they become available for assimilation by microbes. On the other hand, it has been found out that bioplastics with lower molecular weight are readily biodegradable by microbial action, as they are already in a form that is accessible to microbial enzymes (Solanki *et al.*, 2021).

7.7.2.2 Presence of Additives and Stabilizers

Another factor with bioplastics is that they contain additives and stabilizers which also influence their biodegradation. Most biopolymer-based commercial products include plasticizers, stabilizers, and pigments, which exert negative impact on the activity of microorganisms and therefore hinder its decomposition process (Gupta *et al.*, 2022). For instance, plasticizers can improve the flexibility of the polymer; however, they can also improve the chemical resistance of the material to a certain degree, thus reducing susceptibility of the material to microbial degradation. Likewise, UV stabilizers, and other chemicals incorporated to enhance the biodegradation of bioplastics, compromise the rate at which microorganisms can degrade the polymer backbone (Tyagi *et al.*, 2022).

7.7.2.3 Biopolymer Blends

In cases when the bioplastics are made from the blends of different biopolymers, e.g., starch-PLA blends or utilizing the PHA-based composites, the degradation rate might have completely different characteristics compared to the pure bioplastics. Some properties of composites may affect the biodegradation rate depending on the structure, surface features, and the molecular weight of the polymer composition. At times, some biopolymers improve the microbial degradation of bioplastics because the microorganisms attack the weaker biopolymer first, thus breaking down the entire blend (Aaliya *et al.*, 2021). But when the two biopolymers are used, then their biodegradation could be limited by the level of difference in chemical structure and biodegradation rates between them.

7.8 APPLICATIONS OF MICROBIAL PRODUCTS IN ENHANCING DEGRADATION PROCESSES

Microbial enzymes and metabolites are strategically important in the breakdown and degradation of bioplastics. The use of these products seems to present a viable method of enhancing the biodegradation of bioplastic materials in several ecological niches, biodegradation effectiveness and minimizing the detrimental effects of plastic pollution. The subsections review bioaugmentation techniques and microbial metabolites that can be used in boosting microbial degradation of bioplastics.

7.8.1 Bioaugmentation Strategies

In this method, microorganisms are introduced to an environment to enhance the degradation of pollutants. This strategy is particularly helpful in a condition where a native microbial community cannot sufficiently decompose an item. Therefore, bioplastics bioaugmentation can be effective in biodegradation by supplementing the environment with microbes that are capable of synthesizing enzymes that can break down polymeric structures to boost the speed of biodegradation.

Bioaugmentation can therefore be defined as the exercise of adding specific microorganisms or microbial assemblage to a particular environment to increase the rate at which pollutants such as bioplastics are degraded. This approach is often applied to enhance biodegradation activity, especially when the indigenous microbial population fails or is ineffective to degrade a given material. Musa *et al.* (2024b) investigated the application of bioaugmentation with bacteria in PLA degradation in a case study. The finding of the study showed that the rate of degradation of PLA was significantly enhanced when a microbial consortium of *Pseudomonas* spp., *Bacillus* spp., and *Rhodococcus* spp. was incorporated when compared to natural degradation. Thus, the augmented community was able to effectively degrade the polymer as well as synthesize PLA depolymerase as well as other enzymes that acted upon the ester bonds of PLA structure (Ellis *et al.*, 2021). According to this study, the concept of bioaugmentation may reduce the degradation time of PLA even when it is applied in low microbial diversity environments.

Likewise, bioaugmentation was used to increase the ability to degrade PHA bioplastics in the environment. Shah *et al.* (2022) came up with a study that showed that use of *Bacillus* and *Pseudomonas* species in the PHA-contaminated soil prompted natural biodegradation of the polymer. The microbial consortium employed the depolymerases to hydrolyze the polymer into monomer forms which were utilized by the microorganisms as direct carbon source. This research also pointed out that the process of bioaugmentation could assist in increasing the decline rate of PHAs and eliminate them in the environment where the microbes degrading these polymers are scarce.

7.8.2 Use of Microbial Metabolites

Some of the microbial metabolites that have been found key to boosting the degradation of bioplastics include enzymes as well as organic acids. These metabolites may either hydrolyze the polymer directly or else modify its structure to make it vulnerable for the fragmentation by microbes.

SHA-1 depolymerase is another metabolite that is synthesized by several bacterial species, for the purpose of breaking down PHAs. Liu *et al.* (2021) have focused their study on utilizing the strains of *Pseudomonas* and *Bacillus* for the depolymerization enhancement of marine PHA. The work demonstrated thereby that the utilization of PHA depolymerase enhanced the rate of PHA degradation in seawater. Some of the microorganisms also secrete surfactants, which weaken the hydrophobic characteristic of bioplastics and make them more vulnerable to attack by depolymerases. It has also been established that surfactants synthesized by *Rhodococcus* spp. and *Sphingomonas* spp. enhance PLA degradation through reduction of its crystallinity and subsequent surface area to increase microbial attachment (Popelka *et al.*, 2020).

Another potential method is the utilization of the microorganisms which derived organic acids. Singh *et al.* (2020) observed that fungal organic acids including acetic acid, citric acid, and lactic acid greatly improved the mechanical and chemical depolymerization of starch-based bioplastics. The presence of these acids can decrease the molecular weight of the polymer, hence making it easier to be broken down by enzymes. The study also pointed out that when employing organic acids and microbial consortiums, biodegradation process of starch-based plastics can significantly be improved since the companion organic acids mollify the beneficiary surroundings of the microbial group in an improved way.

7.8.3 Enhancing Degradation through Microbial Community Engineering

Microbial community manipulation is still a novel procedure with the aim to optimize the breakdown of plastics by altering the microbial community. This can be done by either isolating bacteria or changing existing ones, which are beneficial and can efficiently break down bioplastics due to their specific enzymes or metabolites. Another investigation by Shalem et al. (2024) sought to investigate the possibility of the biodegradation of PLA using genetically modified *Pseudomonas*. The researchers genetically engineered derivatives of *Pseudomonas* to secrete relatively higher levels of PLA depolymerase and therefore better PLA degradation capabilities. This study showed that increasing enzyme production through genetic modification helped to enhance the correct degradation rate of bioplastics under different conditions.

Microbial community engineering may also include synthetic ecosystems where specific microbial species are combined with organisms possessing complementary degradative capacities. A recent case study by Wei *et al.* (2023) dealt with the use of a synthetic microbial system of *Bacillus*, *Pseudomonas*, *Rhodococcus* species for biodegradation of mixed bioplastic waste of PLA and starch-based plastics. The synthetic consortia outcompete the individual strains and, thus, reduce a broad range of bioplastics, suggesting that community engineering possesses significant potential for enhancing the biodegradation process.

7.8.3.1 Bacterial Bioaugmentation

Bacterial bioaugmentation entails the addition of certain bacterial strains or a combination of bacterium that has the potential to break down bioplastics such as PLA, PHA, and starch-blended plastics. Some research has made it possible to show that bacterial species can enhance the degradation of bioplastics. For example, Eronen-Rasimus *et al.* (2022) have reviewed the role of *Bacillus* spp. and *Pseudomonas* spp. in boosting the degradation of PLA in soil. Thus, the study established that these bacteria could produce PLA-degrading enzymes, which enhance the degradation of PLA and minimize its accumulation in the environment. Moreover, the present study demonstrated that bioaugmentation with *P. putida* and *Rhodococcus* spp. improved the degradation of PHA in soil because the bacteria used the polymer as a carbon source and produced substances that benefited the polymer's biodegradation (Livinus *et al.*, 2024). Furthermore, the ability of bacterial consortia to break down bioplastic has been found to be enhanced by the wider range of enzymes available for this purpose. For instance, the bioaugmentation with the mixed culture comprising *Bacillus* spp., *Rhodococcus* spp., and *Pseudomonas* spp. promoted the degradation of PLA and PCL since the species acted in combination to degrade the polymers using different enzymes such as esterases and proteases (Zaaba *et al.*, 2020). Perhaps the greatest strength of this study was that it focused on bioaugmentation using a mixed population of bacteria because it observed that such a composite addend would improve the overall rate of biodegradation in contrast to a single bacterial species.

7.8.3.2 Fungal Bioaugmentation

Fungal bioaugmentation has also been considered as the method to improve bioplastic degradation prospects. Among the numerous fungi that can produce a variety of enzymes, *Aspergillus* spp., *Penicillium* spp., and *Trichoderma* spp. are notable for their capability to release ligninases, cellulases to address the degradation of complex bioplastics. Plastics are susceptible to biodegradation by enzymatic mechanism which exist intrinsically in fungi. Singh *et al.* (2023) conducted a study of degradation of starch-based bioplastics using *A. niger* and *Trichoderma harzianum*. The study proved that both genera of the fungi had the capacity to hydrolyze the starch polymers by producing enzymes such as amylase and others. These fungi were inoculated into bioplastic-contaminated nurseries to increase the rate of degradation, thereby demonstrating the possibilities of fungal bioaugmentation for biodegradation, especially of bioplastics with carbohydrate-based matrices (Hussain *et al.*, 2022). Bioaugmentation involving fungi has also been used in the degradation of synthetic bioplastics including PLA. A study by Li *et al.* (2022) established that *P. chrysosporium* was capable of degrading PLA through the secretion of PLA-degrading enzymes inclusive of esterases and hydrolases. The study also suggested that in habitats where bacteria are unsuccessful in the breakdown of selected bioplastics, fungal bioaugmentation is helpful.

7.8.3.3 Combined Bacterial and Fungal Bioaugmentation

Bioaugmentation with bacterial and fungal system in converting bioplastic has a complementary advantage where both bacteria and fungi can work together. Research done reveal that synergy between bacteria and fungi enhances the efficiency and effectiveness of bioplastic degradation. For instance, simultaneous use of *Bacillus* spp. and *T. harzianum* has been found to enhance the further degradation of starch-based plastic to a very high level. A part of the polymer was degraded by bacteria through released enzymes; the fungal activity complemented the process through more enzymes as well as metabolic products that supported the destruction of the polymer (Wei *et al.*, 2023). Another work revealed that a community of *Pseudomonas* spp. and wood-decay fungus *P. chrysosporium* was more effective for PLA biodegradation than the pure culture. The bacterial species formed the PLA-degrading enzymes and the fungus carried out the biodegradation of lignocellulosic materials, thereby enhancing the breakdown process (Shalem et al., 2024). The bioaugmentation aging strategy shown with this combination approach proves that this process has the potential to enhance the degradation of bioplastics in different environments such as soil, water, and composting.

7.9 CHALLENGES AND LIMITATIONS IN THE FIELD

There are still some restrictions and complications which prevent the use of microbial degradation of bioplastics on a large scale. These disadvantages include the various technical, regulatory, and environmental hitches that have the potential to slow down the process of biodegradation or posing a threat to the large-scale application of microbial bioremediation. The following are some of the major challenges that need to be covered to increase the efficacy of microbial biodegradation of plastics.

7.9.1 Technical Challenges

One of the primary technical challenges is the fact that different types of bioplastics have different degradation rates. Therefore, PLA is more resistant to microbial degradation than PHAs because it's a polymer and it is crystalline in nature. This structural variability puts significant challenges to the development of general microbial consortia, adequate for selective and rapid degradation of all bioplastics. A study has also revealed that the degradation of one kind of bioplastic may effectively be carried out under the action of certain microbes, while the other kind of bioplastic may not be degraded due to the variation in the polymer contents (Ruggero *et al.*, 2021). Besides, the degradation process usually occupies a more time, especially where the microorganisms are likely to be active in worst environmental conditions. Microbes may need specific conditions and do not always find them in large scale or in nature degradation environment such as temperature, pH, and moisture. These variations in degradation conditions somewhat restrain the usage of microbial bioremediation (Zhao *et al.*, 2021). One technical issue they identified is the ability to control the conditions and times that microorganisms can work on bioplastics most efficiently. Other conditions include nutrient availability, oxygen level, moisture, all of which greatly affects the microbial activities necessary for bioplastic degradation. In many cases, these optimal conditions are difficult to achieve and thus additional complex and costly bioreactor systems are used to enhance the biodegradation effects (Karahaliloglu *et al.*, 2021).

7.9.2 Regulatory and Environmental Considerations

The current environment on regulation of bioplastic biodegradation and microbial-applications is still dynamic. However, the employing of genetically modified organisms (GMOs) for biodegradation-purposes is quite restricted in many regions due to regulation that may prohibit or restrict the use of some microbial strains or consortia. Indeed, bioplastics' biodegradation efficiency is high when using genetically engineered microorganisms as the biodegraders, but the application of those microorganisms in the field is limited due to safety-concerns and risk-associated with GMOs release into natural ecosystems (Gaharwar *et al.*, 2021). Also missing is a standardized protocol for evaluating the success of microbial bioremediation approaches for bioplastics, leading to variations in how different microbes or bioplastic degradation processes are compared and assessed. Modern microbial bioremediation standards require universally acceptable methods for assessing the biodegradability of bioplastics and for studying the effectiveness and safety of microbial consortia. This lack of standardization creates the problem of inadequate information for industries and regulatory agencies concerning the applications of microbial degradation technologies (Chandra *et al.*, 2021). Microbial degradation may be one of the most effective solutions for controlling plastic waste, but it is not the ultimate solution to the worldwide issue. Some bioplastics have long degradation times; moreover, microbial remediation is available only in a few regions of the globe coupled with the fact that there is a lot of plastic production annually. Also, some bioplastics disintegrate into smaller bioplastics in the process, which means that the problem of plastic waste remains topical (Dolci *et al.*, 2023).

7.9.3 Economic and Scale-Up Issues

Another major problem is the challenge to sue microbial bioremediation from pilot scale experiments to large-scale experiments. Applying microbial consortia in industrial or environmental bioprocesses involves massive production of microorganisms which is costly and time-consuming. It can be costly to cultivate and sustain these microbial cultures as well as to provide suitable conditions that will enhance their functionality; this factor definitively establishes a disadvantage relative to common waste treatment methods such as burning or recycling (Xiao *et al.*, 2022).

In addition, microbial bioreactors used for the degradation of bioplastics have their drawbacks such as higher initial and operational costs of bioreactors. The economic viability of large-scale application of bioremediation would require significant improvement in the cost of construction of the bioreactor, media, and microbial cultures. Moreover, obtaining large-scale permits for the application of GMOs and, to a greater extent, non-GMO microbial strains may also hinder the economic rationale for microbial bioplastics degradation at scale (Degli Esposti *et al.*, 2021).

7.9.4 Public Awareness and Acceptance

Another challenge can be referred to as the overall societal acceptance of extracted microbial degradation technologies. Although the majority of bioplastics manufactured today can be biodegradable, compostable, or both, the public remains largely unaware of the complications and environmental consequences that may be attached to their degradation (Karahaliloglu *et al.*, 2021). There is therefore need to create awareness of the viability of microbial bioremediation technologies either as a single process or in combination with other technologies, depending on feasibility and applicability of any particular technology. Therefore, views of the public can introduce changes or impact policy approaches and large-scale adoption and utilization of bioplastics and biodegradation plans (Zhao *et al.*, 2021). Hence, to make bioplastic waste accepted and incorporate them in global waste management strategies, there is the need to increase understanding of microbial degradation processes (Abraham *et al.*, 2021).

7.10 RECOMMENDATION

With the global emphasis on proper waste management, especially the practice of sustainable waste management practices improving, microbial degradation of bioplastics is now looking more and more like one of the most effective ways that can be used to deal with the problem of plastic pollution. Nevertheless, many residual areas remain for the developmental investigation that can focus on how to optimize and advance microbial bioremediation technology. Here are some of the general research prospects/future developments and prospects in this field.

7.10.1 Genetic Engineering Approaches

Exploring other options for the creation of organisms might be a topic of interest for researchers in the future, for example, the use of genetic engineering to produce microorganisms capable of more active bioplastic degradation. The involvement of microbes in the degradation of bioplastic may be enhanced through genetic engineering to turn up the enzymatic activity and the pathways involved (Abdulsalam *et al.*, 2024). For instance, the strain enhancement of microorganisms for the synthesis of enzymes with the ability to degrade diverse bioplastics would enhance biodegradation efficiency (Amobonye *et al.*, 2021). Furthermore, improving the tolerance of such engineered microbes to a number of environmental conditions could contribute to the up-scaling of the microbial bioremediation process to industrial scale (Filote *et al.*, 2021). Technological improvements also encompass future opportunities detailing microorganisms created to specifically biodegrade certain kinds of bioplastics, the final wastes resulting from the bioplastic biodegradation processes outlined by Van Roijen *et al.* (2022). For instance, the engineered microorganisms that can metabolize structures of complex biopolymers like PLA or PCL could be considered a more effective way of managing unresponsive bioplastics (Cucina *et al.*, 2021). Although this approach produces positive results in the test tube method, it is necessary to assess the effect of the use of genetically developed microorganisms in the natural environment.

7.10.2 Development of Novel Biodegradable Materials

Other research directions deemed worthy in the future include coming up with more biodegradable bioplastics by microbial communities. While bioplastics are mostly seen as being more environmentally friendly than other plastics, their biodegradation property depends on the polymer type. Existing research is focused on creating bioplastics which are even more biodegradable by designing those that will decompose faster in a natural environment including marine environment or soil (Atiwesh *et al.*, 2021). For example, bioplastics derived from starch, cellulose, and other polysaccharides are biodegradable as opposed to other synthetics like PLA and HPA and require particular conditions to be broken down (Raj *et al.*, 2022). Profound changes in the molecular and engineering properties of biopolymers and in biodegradation processing technology should produce enhanced biodegradable plastics that are more readily degradable by native microbial populations. Furthermore, the kind of bioplastics which have been developed to degrade the wider condition including the effect of temperature, humidity, and acidity of the environment will enhance the efficiency of the microbial degradation process (Yuan *et al.*, 2020). These could go a long way in alleviating the problem of plastic wastes and expand the field of microbial degradation strategies in waste disposal.

7.10.3 Integration of Microbial Degradation with Recycling Systems

The use of microbial degradation processes in blend with conventional recycling methods is another prospective research area that has a lot of promise for bioplastic waste management. The mechanical or chemical recycling process currently in place faces some challenges in managing bioplastics appropriately. To illustrate, PLA and other bioplastics pollute standard plastics recycling channels, so recycling them by conventional techniques is a problem (Moshood *et al.*, 2022). Perhaps we can develop a symbiotic system where, through microbial degradation, bioplastics that cannot be recycled mechanically or chemically are classified as biodegradable materials that can be decomposed by a particular microbial consortium. With such a platform, it could be easier to minimize wastage and at the same time enhance the effectiveness of the bioplastic recycling circles. Future work in this field can be aimed at defining better microorganisms for the biodegradation of particular biopolymers and implementing such microorganisms into waste recycling plants to improve bioplastic waste disposal (Awasthi *et al.*, 2022).

7.10.4 Optimization of Bioreactor Systems for Large-Scale Biodegradation

Factors limiting microbial degradation have shifted from the laboratory system to large-scale bioreactors meant for industrial application. Another future aspect of research and development includes designing improved bioreactor systems for large-scale microbial bioremediation more efficiently at a reasonable cost. It will therefore be vital for the specific parameters of the bioreactors to be optimized as regards the microbial culture conditions, the nutrients available for the microbes as well as the environmental factors such as temperature and moisture to increase the efficiency of bioplastic degradation as proposed by Chah *et al.* (2022). Further studies should be devoted to enhancing the construction of bioreactors and microbial community management in a bioreactor environment. The usage of microbial consortia instead of a single bacterial or fungal strain has efficiency in breaking down complex bioplastics and can be attributed to the particular species or strains that are designed to break down different phases or parts of bioplastic polymers (Dilshad *et al.*, 2021). Improvements in the future for the bioreactor processes could make the continuous, industrial-scale degradation of bioplastics both more efficient and cost-effective.

7.11 CONCLUSION

Microbial degradation of the bioplastics is a plausible method for tackling the increasing problem of plastic pollution more so when addressing the bioplastic pollution. Bacteria and fungi have been proven to possess the capability of degrading one and/or more types of bioplastic – PHAs, PLA, and starch plastics. These natural processes are environmentally friendly and also are less expensive when compared to other situations of waste disposal. However, there are still some important factors that have to be addressed to optimize microbial degradation. The irregular biodegradation of the bioplastics, the condition of use, and the periodicity of these plastics still present several problems of use and implementation at large scale. Further, lack of public awareness, availability of regulatory policies, and technical constraints act as imposing factors to the diffusion of microbial bioremediation technologies. However, it is encouraging to note that literature in this field is growing at a very fast rate. Techniques of genetic engineering have increased the performance of microorganisms and new biodegradable materials are being synthesized which have better degradability features. Combining microbial degradation with the conventional recycling process and improving bioreactor technology are other areas that are ripe for scale-up in bioplastic waste management systems. This study shows enormous potential for microbial degradation of bioplastics in the future. Further studies accompanied by progress in technology, as well as by increased community awareness of the benefits of such processes, will probably result in better application of these technologies. Finally, there lies a great potential for microbial biodegradation in the fight against plastics and their pollution and offers possibilities to create lighter and more sustainable types of economy.

REFERENCES

Aaliya, B., Sunooj, K. V., & Lackner, M. (2021). Biopolymer composites: A review. *International Journal of Biobased Plastics*, *3*(1), 40–84.

Abd El-malek, F., Khairy, H., Farag, A., & Omar, S. (2020). The sustainability of microbial bioplastics, production, and applications. *International Journal of Biological Macromolecules*, *157*, 319–328.

Abdulsalam, M., KalandarAmeer, S., Balakrishnan, S., & Pal, S. K. (2023). Maximizing sustainability: Leveraging indigenous fungal cellulases for sugarcane bagasse bioethanol production and agricultural waste management in Kano. *Journal of Advanced Zoology*, *44*(S2), 2379–2393.

Abdulsalam, M., Salihu, A. T., Usman, H. Y., & Usman, M. Y. (2024). Protein biosynthesis in microorganisms: Mechanisms, regulation, and biotechnological applications. *World Journal of Advanced Research and Reviews*, *21*(1), 869–881.

Abraham, A., Park, H., Choi, O., & Sang, B. I. (2021). Anaerobic co-digestion of bioplastics as a sustainable mode of waste management with improved energy production – A review. *Bioresource Technology*, *322*, 124537.

Ahmad, N., Aslam, S., Hussain, N., Bilal, M., & Iqbal, H. M. (2023). Transforming lignin biomass to value: Interplay between ligninolytic enzymes and lignocellulose depolymerization. *BioEnergy Research*, *16*(3), 1246–1263.

Akromah, S., Chandarana, N., & Eichhorn, S. J. (2024). Mycelium composites for sustainable development in developing countries: The case for Africa. *Advanced Sustainable Systems*, *8*(1), 2300305.

Al Hosni, A. S. K. (2019). *Biodegradation of Polycaprolactone Bioplastic in Comparison with Other Bioplastics and Its Impact on Biota*. The University of Manchester (United Kingdom).

Aliotta, L., Seggiani, M., Lazzeri, A., Gigante, V., & Cinelli, P. (2022). A brief review of poly (butylene succinate) (PBS) and its main copolymers: Synthesis, blends, composites, biodegradability, and applications. *Polymers*, *14*(4), 844.

Amobonye, A., Bhagwat, P., Singh, S., & Pillai, S. (2021). Plastic biodegradation: Frontline microbes and their enzymes. *Science of the Total Environment*, *759*, 143536.

Anal, A. K. (2019). Quality ingredients and safety concerns for traditional fermented foods and beverages from Asia: A review. *Fermentation*, *5*(1), 8.

Atanasova, N., Stoitsova, S., Paunova-Krasteva, T., & Kambourova, M. (2021). Plastic degradation by extremophilic bacteria. *International Journal of Molecular Sciences*, *22*(11), 5610.

Atiwesh, G., Mikhael, A., Parrish, C. C., Banoub, J., & Le, T. A. T. (2021). Environmental impact of bioplastic use: A review. *Heliyon*, *7*(9), e07918.

Awasthi, S. K., Kumar, M., Kumar, V., Sarsaiya, S., Anerao, P., Ghosh, P., & Awasthi, M. K. (2022). A comprehensive review of recent advancements in biodegradation and sustainable management of biopolymers. *Environmental Pollution*, *307*, 119600.

Bhatia, R., Sharma, R., & Malhotra, M. (2021). A review on biodegradability of bioplastics: Recent updates and challenges. *Environmental Science and Pollution Research*, *28*(18), 22711–22725.

Bher, A., Mayekar, P. C., Auras, R. A., & Schvezov, C. E. (2022). Biodegradation of biodegradable polymers in mesophilic aerobic environments. *International Journal of Molecular Sciences*, *23*(20), 12165.

Brown, R. W., Chadwick, D. R., Zang, H., Graf, M., Liu, X., Wang, K., & Jones, D. L. (2023). Bioplastic (PHBV) addition to soil alters microbial community structure and negatively affects plant-microbial metabolic functioning in maize. *Journal of Hazardous Materials*, *441*, 129959.

Chah, C. N., Banerjee, A., Gadi, V. K., Sekharan, S., & Katiyar, V. (2022). A systematic review on bioplastic-soil interaction: Exploring the effects of residual bioplastics on the soil geoenvironment. *Science of the Total Environment*, *851*, 158311.

Chandra, R., Soni, S. K., & Tiwari, P. (2021). Biodegradation of bioplastics and its impact on the environment. *Environmental Science and Pollution Research*, *28*(17), 21418–21430.

Ciebiada, M., Kubiak, K., & Daroch, M. (2020). Modifying the cyanobacterial metabolism as a key to efficient biopolymer production in photosynthetic microorganisms. *International Journal of Molecular Sciences*, *21*(19), 7204.

Compart, J., Singh, A., Fettke, J., & Apriyanto, A. (2023). Customizing starch properties: A review of starch modifications and their applications. *Polymers*, *15*(16), 3491.

Cucina, M., De Nisi, P., Trombino, L., Tambone, F., & Adani, F. (2021). Degradation of bioplastics in organic waste by mesophilic anaerobic digestion, composting, and soil incubation. *Waste Management*, *134*, 67–77.

Degli Esposti, M., Morselli, D., Fava, F., Bertin, L., Cavani, F., Viaggi, D., & Fabbri, P. (2021). The role of biotechnology in the transition from plastics to bioplastics: An opportunity to reconnect global growth with sustainability. *FEBS Open Bio*, *11*(4), 967–983.

Desrochers, A., Cox, S. A., Oxenford, H. A., & van Tussenbroek, B. (2020). *Sargassum Uses Guide: A Resource for Caribbean Researchers, Entrepreneurs, and Policymakers*. Bridgetown, Barbados: Centre for Resource Management and Environmental Studies (CERMES), University of the West Indies, Cave Hill Campus.

Dilshad, E., Waheed, H., Ali, U., Amin, A., & Ahmed, I. (2021). General structure and classification of bioplastics and biodegradable plastics. In *Bioplastics for Sustainable Development* (pp. 61–82). Springer, Singapore.

Dolci, G., Rigamonti, L., & Grosso, M. (2023). The challenges of bioplastics in waste management. *Waste Management & Research*, *41*(8), 1281–1282.

Ellis, L. D., Rorrer, N. A., Sullivan, K. P., Otto, M., McGeehan, J. E., Román-Leshkov, Y., & Beckham, G. T. (2021). Chemical and biological catalysis for plastics recycling and upcycling. *Nature Catalysis*, *4*(7), 539–556.

El-malek, F. A., & Steinbüchel, A. (2022). Post-synthetic enzymatic and chemical modifications for novel sustainable polyesters. *Frontiers in Bioengineering and Biotechnology*, *9*, 817023.

Eronen-Rasimus, E. L., Nakki, P. P., & Kaartokallio, H. P. (2022). Degradation rates and bacterial community compositions vary among commonly used bioplastic materials in a brackish marine environment. *Environmental Science & Technology*, *56*(22), 15760–15769.

Fairley, P. (2016). Biodegradable plastics: A growing solution to the plastic pollution problem. *Nature*, *531*(7593), 440–441.

Filote, C., Roşca, M., Hlihor, R. M., Cozma, P., Simion, I. M., Apostol, M., & Gavrilescu, M. (2021). Sustainable application of biosorption and bioaccumulation of persistent pollutants in wastewater treatment: Current practice. *Processes*, *9*(10), 1696.

Finore, I., Feola, A., Russo, L., Cattaneo, A., Di Donato, P., Nicolaus, B., …, & Romano, I. (2023). Thermophilic bacteria and their thermozymes in composting processes: A review. *Chemical and Biological Technologies in Agriculture*, *10*(1), 7.

Flemming, H. C., & Wuertz, S. (2019). Bacteria and archaea on Earth and their abundance in biofilms. *Nature Reviews Microbiology*, *17*(4), 247–260.

Gaharwar, U. S., Verma, A., Singh, R., Prasad, T., & Rajamani, P. (2021). Health risks and environmental concerns of GM crop adoption. In *Policy Issues in Genetically Modified Crops* (pp. 371–400). Academic Press.

García-Depraect, O., Bordel, S., Lebrero, R., Santos-Beneit, F., Börner, R. A., Börner, T., & Muñoz, R. (2021). Inspired by nature: Microbial production, degradation, and valorization of biodegradable bioplastics for life-cycle-engineered products. *Biotechnology Advances*, *53*, 107772.

Goel, V., Luthra, P., Kapur, G. S., & Ramakumar, S. S. V. (2021). Biodegradable/bio-plastics: Myths and realities. *Journal of Polymers and the Environment*, *29*, 3079–3104.

Gupta, I., Cherwoo, L., Bhatia, R., & Setia, H. (2022). Biopolymers: Implications and application in the food industry. *Biocatalysis and Agricultural Biotechnology*, *46*, 102534.

Hahladakis, J. N., Iacovidou, E., & Gerassimidou, S. (2024). Plastic waste in a circular economy. In *Environmental Materials and Waste* (pp. 99–134). Elsevier.

Han, J. W., Ruiz-Garcia, L., Qian, J. P., & Yang, X. T. (2018). Food packaging: A comprehensive review and future trends. *Comprehensive Reviews in Food Science and Food Safety*, *17*(4), 860–877.

Hussain, C. M., & Kadeppagari, R. K. (Eds.). (2022). *Biotechnology for Zero Waste: Emerging Waste Management Techniques*. John Wiley & Sons.

Joo, S., Cho, I. J., Seo, H., Son, H. F., Sagong, H. Y., Shin, T. J., & Kim, K. J. (2021). Structural insight into molecular mechanisms of polyethylene terephthalate degradation by a bacterial hydrolase. *Nature Communications*, *12*(1), 1–12. https://doi.org/10.1038/s41467-020-20489-2

Karahaliloglu, Z., Duman, T., & Yilmaz, L. (2021). Biodegradation of bioplastics in environmental conditions: Challenges and prospects. *Journal of Environmental Management*, *280*, 111672.

Karan, H., Funk, C., Grabert, M., Oey, M., & Hankamer, B. (2019). Green bioplastics as part of a circular bioeconomy. *Trends in Plant Science*, *24*(3), 237–249. https://doi.org/10.1016/j.tplants.2018.11.010

Kumari, A., Yadav, M., & Gupta, S. (2023). Microbial biopolymers and their role in bioplastic degradation: A review. *Biodegradation*, *34*(2), 101–118.

Kun, R. S., Gomes, A. C. S., Hildén, K. S., Cerezo, S. S., Mäkelä, M. R., & de Vries, R. P. (2019). Developments and opportunities in fungal strain engineering for the production of novel enzymes and enzyme cocktails for plant biomass degradation. *Biotechnology Advances*, *37*(6), 107361.

La Fuente, C. I. A., Maniglia, B. C., & Tadini, C. C. (2023). Biodegradable polymers: A review about biodegradation and its implications and applications. *Packaging Technology and Science*, *36*(2), 81–95.

Lai, J., Huang, H., Lin, M., Xu, Y., Li, X., & Sun, B. (2023). The enzyme catalyzes ester bond synthesis and hydrolysis: The key step for sustainable usage of plastics. *Frontiers in Microbiology*, *13*, 1113705.

Lakshmayya, N. S., Panday, A., Yadavalli, R., Reddy, C. N., Mandal, S. K., Agrawal, D. C., & Mishra, B. (2024). Food contamination with micro-plastics: Occurrences, bioavailability, human vulnerability, and prevention. *Current Nutrition & Food Science*, *20*(7), 797–810.

Larrañaga, A., & Lizundia, E. (2019). A review of the thermomechanical properties and biodegradation behavior of polyesters. *European Polymer Journal*, *121*, 109296.

Li, X., Wang, W., & Zhang, X. (2022). Fungal degradation of polylactic acid (PLA): Mechanisms and applications. *Journal of Industrial Microbiology and Biotechnology*, *49*(4), 559–572.

Lim, B. K. H., & San Thian, E. (2022). Biodegradation of polymers in managing plastic waste—A review. *Science of the Total Environment*, *813*, 151880.

Liu, H., Wei, L., & Zhang, M. (2021). Biodegradation of bioplastics: Current research and future perspectives. *Journal of Hazardous Materials*, *419*, 126505.

Livinus, M. U., Bala, S. Z., Abdulsalam, M., Innocent, M. O., Hassan, M., Elelu, S. A., & Kini, P. (2024a). Role of microbes in soil food webs and vegetation development. In *Soil Microbiome in Green Technology Sustainability* (pp. 107–132). Springer Nature Switzerland.

Livinus, M. U., Bala, S. Z., Abdulsalam, M., Innocent, M. O., Hassan, M., & Kini, P. (2024b). Natural Occurrences of Soil Dilapidation. In *Prospects for Soil Regeneration and Its Impact on Environmental Protection* (pp. 205–223). Cham: Springer Nature Switzerland.

Londoño-Hernandez, L., Ruiz, H. A., Toro, C. R., Ascacio-Valdes, A., Rodriguez-Herrera, R., Aguilera-Carbo, A., & Aguilar, C. N. (2020). Advantages and progress innovations of solid-state fermentation to produce industrial enzymes. In *Microbial Enzymes: Roles and Applications in Industries* (pp. 87–113). Springer, Singapore.

Meereboer, K. W., Misra, M., & Mohanty, A. K. (2020). Review of recent advances in the biodegradability of polyhydroxyalkanoate (PHA) bioplastics and their composites. *Green Chemistry*, *22*(17), 5519–5558.

Megha, M., Kamaraj, M., Nithya, T. G., GokilaLakshmi, S., Santhosh, P., & Balavaishnavi, B. (2024). Biodegradable polymers – Research and applications. *Physical Sciences Reviews*, *9*(2), 949–972.

Moshood, T. D., Nawanir, G., Mahmud, F., Mohamad, F., Ahmad, M. H., & AbdulGhani, A. (2022). Sustainability of biodegradable plastics: New problem or solution to solve the global plastic pollution? *Current Research in Green and Sustainable Chemistry*, *5*, 100273.

Müller, R. J., Kleeberg, I., & Deckwer, W. D. (2021). Biodegradation of polyesters containing aromatic constituents. *Journal of Biotechnology*, *86*(2), 87–95.

Musa, I. O., Auwal, A. I., Abdulsalam, M., Livinus, M. U., Abdulhakeem, A. I., Muhammed, A., & Muhammad, A. S. (2024a). Microbial bioprospecting products of marine economy. In *Marine Bioprospecting for Sustainable Blue-Bioeconomy* (pp. 181–204). Springer Nature Switzerland.

Musa, I. O., Samuel, J. O., Adams, M., Abdulsalam, M., Abdulazeez, M., Muhammad, M. S., & Akande, S. A. (2024b). Cases of micro-and nanoplastics in food and food product. In *Emerging Contaminants in Food and Food Products* (pp. 90–105). CRC Press.

Nandakumar, A., Chuah, J. A., & Sudesh, K. (2021). Bioplastics: A boon or bane? *Renewable and Sustainable Energy Reviews*, *147*, 111237.

Narancic, T., Verstichel, S., Chaganti, S. R., Morales-Gamez, L., Kenny, S. T., De Wilde, B., …, & O'Connor, K. E. (2020). Biodegradable plastics: Standards, policies, and impacts. *Current Opinion in Green and Sustainable Chemistry*, *21*, 1–8.

O'Connor, J., Hoang, S. A., Bradney, L., Dutta, S., Xiong, X., Tsang, D. C., …, & Bolan, N. S. (2021). A review on the valorisation of food waste as a nutrient source and soil amendment. *Environmental Pollution*, *272*, 115985.

Okeke, E. S., Olagbaju, O. A., Okoye, C. O., Addey, C. I., Chukwudozie, K. I., Okoro, J. O., …, & Abesa, S. (2022). Microplastic burden in Africa: A review of occurrence, impacts, and sustainability potential of bioplastics. *Chemical Engineering Journal Advances*, *12*, 100402.

Polman, E. M., Gruter, G. J. M., Parsons, J. R., & Tietema, A. (2021). Comparison of the aerobic biodegradation of biopolymers and the corresponding bioplastics: A review. *Science of the Total Environment*, *753*, 141953.

Popelka, A., Abdulkareem, A., Mahmoud, A. A., Nassr, M. G., Al-Ruweidi, M. K. A., Mohamoud, K. J., …, & Kasak, P. (2020). Antimicrobial modification of PLA scaffolds with ascorbic and fumaric acids via plasma treatment. *Surface and Coatings Technology*, *400*, 126216.

Pournou, A., & Pournou, A. (2020). Wood deterioration by terrestrial microorganisms. In *Biodeterioration of Wooden Cultural Heritage: Organisms and Decay Mechanisms in Aquatic and Terrestrial Ecosystems* (pp. 345–424). Cham: Springer Nature Switzerland.

Raj, T., Chandrasekhar, K., Kumar, A. N., & Kim, S. H. (2022). Lignocellulosic biomass as renewable feedstock for biodegradable and recyclable plastics production: A sustainable approach. *Renewable and Sustainable Energy Reviews*, *158*, 112130.

Rani, S., Sharma, M., & Gupta, R. (2020). Recent advances in microbial degradation of bioplastics: Challenges and prospects. *Biotechnology Advances*, *44*, 107608.

Ruggero, F., Onderwater, R. C., Carretti, E., Roosa, S., Benali, S., Raquez, J. M., …, & Wattiez, R. (2021). Degradation of film and rigid bioplastics during the thermophilic phase and the maturation phase of simulated composting. *Journal of Polymers and the Environment*, *29*, 3015–3028.

Samir, A., Ashour, F. H., Hakim, A. A., & Bassyouni, M. (2022). Recent advances in biodegradable polymers for sustainable applications. *npj Materials Degradation*, *6*(1), 68.

Sankhla, I. S., Sharma, G., & Tak, A. (2020). Fungal degradation of bioplastics: An overview. In *New and Future Developments in Microbial Biotechnology and Bioengineering* (pp. 35–47). Elsevier.

Schaerer, L., Putman, L., Bigcraft, I., Byrne, E., Kulas, D., Zolghadr, A., …, & Techtmann, S. (2023). Coexistence of specialist and generalist species within mixed plastic derivative-utilizing microbial communities. *Microbiome*, *11*(1), 224.

Shah, A. A., Hasan, F., Hameed, A., & Ahmed, S. (2022). Biological degradation of plastics: A comprehensive review. *Biotechnology Advances*, *40*, 12762.

Shalem, A., Yehezkeli, O., & Fishman, A. (2024). Enzymatic degradation of polylactic acid (PLA). *Applied Microbiology and Biotechnology*, *108*(1), 413.

Sharma, A., Sharma, A., Singh, S., Kuhad, R. C., & Nain, L. (2019). Thermophilic fungi and their enzymes for biorefineries. In *Fungi in Extreme Environments: Ecological Role and Biotechnological Significance* (pp. 479–502).

Sharma, P., Sirohi, R., Tong, Y. W., Kim, S. H., & Pandey, A. (2021). Metal and metal(loids) removal efficiency using genetically engineered microbes: Applications and challenges. *Journal of Hazardous Materials*, *416*, 125855.

Singh, J., Das, A., & Yogalakshmi, K. N. (2020). Enhanced laccase expression and azo dye decolourization during co-interaction of *Trametes versicolor* and *Phanerochaete chrysosporium*. *SN Applied Sciences*, *2*, 1–8.

Singh, P., Sharma, R., & Singh, N. (2023). Fungal bioaugmentation for the degradation of starch-based bioplastics. *Biodegradation*, *34*(3), 245–258.

Solanki, P., Putatunda, C., Kumar, A., Bhatia, R., & Walia, A. (2021). Microbial proteases: Ubiquitous enzymes with innumerable uses. *3 Biotech*, *11*(10), 428.

Taib, N. A. A. B., Rahman, M. R., Huda, D., Kuok, K. K., Hamdan, S., Bakri, M. K. B., ..., & Khan, A. (2023). A review on poly lactic acid (PLA) as a biodegradable polymer. *Polymer Bulletin*, *80*(2), 1179–1213.

Trinh, B. M., Chang, B. P., & Mekonnen, T. H. (2023). The barrier properties of sustainable multiphase and multicomponent packaging materials: A review. *Progress in Materials Science*, *133*, 101071.

Tyagi, P., Agate, S., Velev, O. D., Lucia, L., & Pal, L. (2022). A critical review of the performance and soil biodegradability profiles of biobased natural and chemically synthesized polymers in industrial applications. *Environmental Science & Technology*, *56*(4), 2071–2095.

Urbanek, A. K., Strzelecki, M. C., & Mirończuk, A. M. (2021). The potential of cold-adapted microorganisms for biodegradation of bioplastics. *Waste Management*, *119*, 72–81.

Van Roijen, E. C., & Miller, S. A. (2022). A review of bioplastics at end-of-life: Linking experimental biodegradation studies and life cycle impact assessments. *Resources, Conservation and Recycling*, *181*, 106236.

Walsh, C. T., Tu, B. P., & Tang, Y. (2018). Eight kinetically stable but thermodynamically activated molecules that power cell metabolism. *Chemical Reviews*, *118*(4), 1460–1494.

Wang, F., Ouyang, D., Zhou, Z., Page, S. J., Liu, D., & Zhao, X. (2021). Lignocellulosic biomass as sustainable feedstock and materials for power generation and energy storage. *Journal of Energy Chemistry*, *57*, 247–280.

Wei, L., Zhang, W., & Chen, J. (2023). Synergistic effects of bacterial and fungal bioaugmentation for accelerated degradation of starch-based bioplastics. *Environmental Pollution*, *288*, 117728. https://doi.org/10.1016/j.envpol.2021.117728

Xiao, Z., Pramanik, A., Basak, A. K., Prakash, C., & Shankar, S. (2022). Material recovery and recycling of waste tyres – A review. *Cleaner Materials*, *5*, 100115.

Xu, B., Chen, Y., He, J., Cao, S., Liu, J., Xue, R., ..., & Jiang, M. (2022). New insights into the biodegradation of polylactic acid: From degradation to upcycling. *Environmental Reviews*, *30*(1), 30–38.

Yañez, L., Conejeros, R., Vergara-Fernández, A., & Scott, F. (2020). Beyond intracellular accumulation of polyhydroxyalkanoates: Chiral hydroxyalkanoic acids and polymer secretion. *Frontiers in Bioengineering and Biotechnology*, *8*, 248.

Yuan, J., Ma, J., Sun, Y., Zhou, T., Zhao, Y., & Yu, F. (2020). Microbial degradation and other environmental aspects of microplastics/plastics. *Science of the Total Environment*, *715*, 136968.

Zaaba, N. F., & Jaafar, M. (2020). A review on degradation mechanisms of polylactic acid: Hydrolytic, photodegradative, microbial, and enzymatic degradation. *Polymer Engineering & Science*, *60*(9), 2061–2075.

Zhao, X., Liu, X., & Liu, C. (2021). Challenges and opportunities in microbial degradation of bioplastics: A review. *Environmental Science and Pollution Research*, *28*(17), 21298–21312.

8 Agricultural Implications of Bioplastics and Their Additives

Gabriel Gbenga Babaniyi, Femi Ibrahim, and Ulelu Jessica Akor

8.1 INTRODUCTION TO BIOPLASTICS IN AGRICULTURE

8.1.1 Definition and Overview

Bioplastics, derived from renewable biomass sources, represent an evolving alternative to traditional petroleum-based plastics. These biopolymers are not only biodegradable but also designed to reduce environmental footprints. In other words, bioplastics are a class of materials derived from renewable biomass sources, such as plant starch, cellulose, or microorganisms. They can be either biodegradable or bio-based, offering a sustainable alternative to conventional petroleum-based plastics, aiming to reduce environmental impact while maintaining similar functional properties (Biswas et al., 2024). More so, bioplastics represent a complex combination of scientific advancement and environmental necessity born from the pursuit of reducing the ecological footprint of traditional plastics (Magni et al., 2024). These materials are distinct because they can either be bio-based, biodegradable, or sometimes both, depending on their chemical composition and origin.

8.1.1.1 Bio-Based Plastics

Bio-based plastics are derived from renewable biological resources such as corn starch, sugarcane, or even algae rather than fossil fuels. However, not all bio-based plastics are biodegradable (Biswas et al., 2024). They may have a similar chemical structure to traditional plastics, meaning their degradation rates could be similar to non-renewable plastics.

- **Sources and Composition:** Bio-based plastics can be synthesized from carbohydrates like cellulose or fermentation byproducts like lactic acid. For instance, polylactic acid (PLA) is created by fermenting sugars (Biswas et al., 2024).
- **Non-Biodegradable Bio-Based Plastics:** Certain bio-based plastics, like bio-polyethylene (bio-PE), share structural similarities with fossil fuel-based counterparts, making them non-biodegradable, but renewable in terms of sourcing (Magni et al., 2024).

8.1.1.2 Biodegradable Plastics

Biodegradable plastics are designed to decompose under specific environmental conditions such as heat, moisture, and microbial action within a relatively short timeframe, unlike conventional plastics, which persist for centuries. However, biodegradability does not necessarily imply these plastics, which are made from biological sources; some are synthesized from petroleum yet still degrade over time (Magni et al., 2024).

DOI: 10.1201/9781003602828-10

Biodegradation Mechanism: These plastics are broken down by microorganisms into carbon dioxide, water, and biomass under industrial composting or specific natural conditions (Ferreira-Filipe et al., 2024).

Popular Biodegradable Plastics: Examples include polyhydroxyalkanoates (PHAs), which are microbial polyesters, and starch blends, which mix traditional polymers with plant-based starch to enhance decomposition (Ferreira-Filipe et al., 2024; Magni et al., 2024).

8.1.1.3 The Intersection: Bio-Based and Biodegradable

Some bioplastics exist at the intersection of being both bio-based and biodegradable, making them particularly attractive from an environmental standpoint. These plastics degrade completely into non-toxic residues while being sourced from renewable materials (Quilliam et al., 2023; Biswas et al., 2024). For example, PLA is both bio-based and biodegradable but requires specific industrial conditions for effective decomposition, complicating its sustainability profile.

8.1.1.4 Complexity of Bioplastic Classification

Bioplastic classification defies linear logic, weaving through layers of contradiction and complexity. The term "bio" conjures images of nature and sustainability, yet not all bio-based plastics align with this vision. Surprisingly, some bio-based variants, though derived from renewable sources, persist in the environment like traditional plastics, still adding to the mounting waste crisis (Magni et al., 2024). In a paradoxical twist, certain biodegradable plastics designed to decompose swiftly are derived from fossil fuels, blurring the line between ecological benefit and environmental harm. This creates a sustainability conundrum where being bio-based doesn't guarantee eco-friendliness, and biodegradability doesn't always mean green (Biswas et al., 2024; Ferreira-Filipe et al., 2024). The relationship between source, function, and disposal is fraught with confusion, challenging assumptions of what's truly sustainable (Table 8.1).

8.1.2 Evolution and Technological Advancements

Historically, agriculture heavily relied on synthetic plastics. However, in recent decades, innovations have propelled the rise of bioplastics, responding to sustainability demands. How did the shift happen? The gradual integration of eco-friendly materials redefined agricultural tools and packaging systems. Therefore, evolution and technological advancements in bioplastics are a saga of different innovations, where scientific progress and environmental necessity collide, sparking a relentless pursuit of materials that balance performance and sustainability.

8.1.2.1 Early Stages: The Emergence of Bioplastics

The journey began in the early 20th century, with the development of cellophane, one of the first cellulose-based plastics, and later Bakelite, a synthetic polymer from renewable sources. These early attempts were rudimentary, more focused on replacing scarce fossil resources than addressing environmental degradation (Shah et al., 2024). Bioplastics existed in the shadows, a niche innovation overshadowed by the rise of cheap, durable, petroleum-based plastics during the post-war economic boom.

8.1.2.2 The Rise of Environmental Awareness

The environmental crises of the 1960s and 1970s highlighted by the stark realities of plastic pollution triggered a reevaluation of plastics. Scientific communities began exploring biodegradable alternatives, realizing that the plastics which dominate the world were not sustainable. It was in this context that starch-based polymers and PHAs, derived from microbial fermentation, gained traction, offering a glimmer of hope in a petrochemical-dominated world (Shah et al., 2024).

TABLE 8.1
Difference between Bio-Based and Biodegradable Plastics

Characteristic	Bio-Based Plastics	Biodegradable Plastics
Definition	Derived from renewable biological resources such as plants (e.g., starch, sugarcane) or microorganisms.	Plastics designed to decompose into natural elements through biological processes under specific conditions.
Source	Made from biomass (e.g., corn, potatoes, cellulose) but can also be synthesized to mimic conventional plastics.	Can be derived from either fossil fuels or biomass; focus is on degradation rather than source material.
Biodegradability	Not necessarily biodegradable. Some bio-based plastics (e.g., bio-PE) are chemically identical to petroplastics.	Always biodegradable, but biodegradation depends on environmental conditions (e.g., temperature, moisture, microbes).
Environmental impact	Reduces reliance on fossil fuels but may persist in the environment like traditional plastics if not biodegradable.	Mitigates plastic pollution by breaking down into non-toxic residues, yet the source may still be petrochemical-based.
Common types	Polylactic acid (PLA), bio-polyethylene (bio-PE), and bio-polypropylene (bio-PP).	Polyhydroxyalkanoates (PHAs), starch-based plastics, polybutylene succinate (PBS), and some types of PLA.
Industrial processing requirements	Similar to traditional plastics; it often requires no specialized facilities for production but might need them for disposal.	Some require specific composting facilities (e.g., PLA), while others degrade naturally under certain conditions.
Examples	Bio-PE (from sugarcane, non-biodegradable), PLA (bio-based and biodegradable, but industrial conditions required).	Starch blends (biodegradable under natural conditions), PHA (biodegradable but from microbial fermentation).
Life cycle assessment (LCA)	Lower carbon footprint during production due to renewable source material, but disposal remains problematic.	LCA depends on degradation conditions, can reduce environmental load but may produce methane in anaerobic settings.
End-of-life disposal	Can be recycled like conventional plastics, but some may contribute to microplastic pollution if not biodegradable.	Decomposes into CO_2, water, and biomass, though some additives or incomplete degradation can cause residual waste.
Applications	Packaging, bottles, agricultural films (focus on renewable sourcing rather than degradation).	Mulch films, compostable bags, disposable cutlery, and packaging (focus on end-of-life environmental integration).
Complexity in perception	Misconception: bio-based equals eco-friendly, though they may persist like conventional plastics if non-biodegradable.	Misconception: biodegradable equals safe for the environment, though conditions for degradation are often specific.
Regulatory framework	More focus on encouraging renewable sourcing in plastics, less on degradation pathways.	Growing regulations around compostability and biodegradation standards to ensure safe environmental breakdown.

Increasing Environmental Concerns over Traditional Plastics: The global ubiquity of traditional plastics has escalated into an environmental emergency of colossal proportions. These petroleum-derived polymers, though cheap and durable, have created a staggering waste crisis due to their resistance to degradation, persisting for centuries in ecosystems. As plastic waste infiltrates oceans, rivers, and soil, it disrupts fragile ecosystems, harms wildlife, and introduces microplastics into food chains, ultimately affecting human health (Babaniyi et al., 2024a; Rahmati et al., 2024). The accumulation of plastic in landfills releases toxic chemicals and greenhouse gases, exacerbating climate change. Moreover, the energy-intensive production process of plastics further compounds the environmental toll. Public awareness has skyrocketed in recent years, with viral images of polluted beaches and entangled marine life serving as grim reminders of the ongoing disaster (Ogunlade et al., 2021; Ibrahim et al., 2024). This heightened concern is fueling demands for alternatives bioplastics, reusable materials, and zero-waste strategies that promise to mitigate the damage caused by traditional plastics, yet the solution remains multifaceted, with technological, economic, and societal challenges ahead.

Global Trends Pushing for Sustainable Agricultural Practices: Amid escalating climate change, soil degradation, and resource depletion, global agricultural practices are undergoing a transformative shift toward sustainability. The unsustainable use of synthetic fertilizers, pesticides, and intensive farming methods has contributed to biodiversity loss, soil erosion, and freshwater depletion (Babaniyi et al., 2024b). To counter these effects, a wave of sustainable agricultural innovations is sweeping across the globe. Precision farming, driven by data analytics and Internet of Thing (IoT) technologies, enables farmers to optimize inputs, reducing waste and environmental impact. Organic farming is regaining prominence, rejecting chemical inputs in favor of natural methods that regenerate soil health. Furthermore, agroforestry integrates trees into croplands, fostering biodiversity while sequestering carbon. Simultaneously, regenerative agriculture emphasizes soil restoration and ecosystem health (Francis & Parayil, 2023; Ibrahim et al., 2024). International policy frameworks such as the Paris Agreement and UN Sustainable Development Goals are pushing governments and industries to adopt practices that reduce the environmental footprint of agriculture, ensuring food security for future generations while safeguarding the planet's ecosystems.

8.1.2.3 Technological Breakthroughs and Diversification

The modern era witnessed a profound leap in the sophistication of bioplastic technology. With innovations in biopolymer chemistry, scientists engineered new materials such as PLA, derived from plant sugars. PLA became a game-changer, offering both bio-based origins and the potential for biodegradation under industrial composting conditions (Francis & Parayil, 2023). Yet the complexity deepened: bio-PE, synthesized from renewable sources like sugarcane, mimicked traditional plastics in durability but was non-biodegradable (Rahmati et al., 2024). This duality between renewable origin and non-degradable end-of-life performance threw the bioplastic world into a paradox, leaving questions about true sustainability.

8.1.2.4 Nanotechnology and Additives: Enhanced Performance

The incorporation of nanotechnology into bioplastic manufacturing has unlocked new frontiers. Nanomaterials like cellulose nanocrystals (CNCs) and clay nanoparticles dramatically improved the mechanical, thermal, and barrier properties of bioplastics, making them competitive with conventional plastics (Babaniyi et al., 2024a; Ibrahim et al., 2024). Alongside this, advancements in additives such as biodegradable plasticizers and natural fillers like lignin ushered in a new era of performance enhancement, where bioplastics could now withstand the rigorous demands of packaging, agriculture, and even automotive industries.

8.1.2.5 Overcoming Challenges: Controlled Degradation and Compostability

One of the critical hurdles in bioplastic technology has been controlling biodegradability. Initial innovations struggled to ensure consistent degradation across various environments. However, recent advancements in the design of compostable bioplastics tailored to specific degradation environments, such as industrial composting, have provided new pathways for large-scale implementation (Dailin et al., 2022; Singhal et al., 2024). Smart polymers that degrade on demand, triggered by environmental cues like UV exposure or microbial activity, are pushing the boundaries of what's possible.

8.1.2.6 Future Prospects: Biofabrication and Synthetic Biology

The frontier of bioplastic evolution now lies in the realms of synthetic biology and biofabrication. Scientists are exploring genetically modified microorganisms to synthesize plastics from waste feedstocks such as agricultural residues or even CO_2 transforming waste into valuable resources. The development of self-repairing bioplastics, materials that can mend minor damages through bioactive components, could revolutionize industries dependent on durable plastics (Singhal et al., 2024). The evolution of bioplastics reflects a path fraught with contradictions and breakthroughs, where each technological advancement seems to unravel a new complexity (Dailin et al., 2022). As innovation pushes bioplastics toward higher performance and sustainability, their journey remains a balancing act between nature and technology, with the next leap always around the corner but never without its paradoxes.

8.1.3 Purpose and Scope in Agricultural Applications

As bioplastics permeate diverse agricultural domains, mulching, greenhouse coverings, irrigation systems, their purpose extends far beyond simple substitution for conventional plastics. These innovative materials unlock a myriad of functionalities that reshape the very fabric of agricultural practices (Da Silva et al., 2019). In mulching, bioplastics not only suppress weeds and regulate soil moisture but also enhance nutrient cycling as they decompose, feeding the soil with organic matter. In greenhouse coverings, advanced biopolymers provide improved light diffusion, optimizing photosynthesis while maintaining controlled climates, fostering better crop yields (Zhang et al., 2017; Fuller et al., 2022). Meanwhile, bioplastic-based irrigation systems reduce water loss through precision delivery, simultaneously minimizing environmental impact (Zhang et al., 2017). Unlike traditional plastics, which introduce long-lasting pollutants, these biodegradable materials actively contribute to soil health and ecosystem resilience. The intersection of technology and sustainability transforms bioplastics into critical agents of change, encouraging a holistic approach to farming where productivity and ecological stewardship are no longer in conflict. This evolution not only addresses environmental concerns but also drives the future of farming, making bioplastics indispensable in achieving both short-term agricultural efficiency and long-term ecosystem balance.

Applications: Bioplastics are revolutionizing agriculture with their multifaceted applications, particularly in mulching films, seed coatings, and packaging. Biodegradable mulching films, when applied to soil, offer weed suppression, moisture retention, and temperature regulation, all while naturally decomposing, eliminating the need for post-season removal (Dailin et al., 2022). Seed coatings, another innovative application, utilize bioplastics to encapsulate seeds with protective layers that enhance germination rates by controlling moisture absorption and nutrient release. As the coating breaks down, it integrates seamlessly into the soil, promoting healthy plant growth. Furthermore, bioplastic packaging for agricultural products provides a sustainable alternative to petroleum-based plastics (Da Silva et al., 2019; Fuller et al., 2022). These materials not only protect goods during transportation but also reduce plastic waste through compostable or biodegradable options (Zhang et al., 2017). This comprehensive approach helps minimize environmental harm while enhancing farming efficiency, demonstrating how bioplastics are not merely replacing conventional plastics but redefining agricultural practices to better align with eco-friendly principles, creating a synergy between productivity and environmental stewardship.

Bioplastics with Circular Economy and Sustainability Goals Alignment: Bioplastics epitomize the principles of the circular economy, wherein materials are designed to be reused, recycled, or returned to the environment with minimal ecological disruption. By sourcing plastics from renewable resources such as corn, sugarcane, or even agricultural waste, bioplastics reduce dependence on finite fossil fuels, closing the loop between production and resource regeneration (Zhang et al., 2017). More critically, many bioplastics are designed to be compostable, transforming into organic matter that enriches soil rather than polluting it. This aligns seamlessly with sustainability goals, which prioritize reducing waste and enhancing resource efficiency. In agriculture, bioplastics degrade without toxic residues, contributing to ecosystem health while addressing the plastic pollution crisis (Fuller et al., 2022). By fostering materials that work in harmony with natural cycles decomposing into harmless substances, bioplastics help mitigate the environmental burden of plastic waste. Thus, they support global sustainability frameworks like the UN Sustainable Development Goals, particularly in promoting responsible production and consumption patterns.

8.2 COMPOSITION AND ADDITIVES OF BIOPLASTICS

8.2.1 Raw Materials for Bioplastics

Bioplastics can be derived from a range of materials, including corn starch, sugarcane, and even microbial fermentation products. Each source lends distinct properties to the final product. The raw materials for bioplastics are as diverse and complex as the technologies driving their development. Derived from renewable biological sources, these materials typically fall into two categories: natural polymers and bio-based monomers. Key raw materials include starch obtained from corn, potatoes, or wheat, which forms the backbone of many biodegradable plastics like thermoplastic starch (TPS) (Savva et al., 2023). Another prominent resource is cellulose, the most abundant organic polymer on Earth, extracted from plants to create films, fibers, and composite materials. Beyond plant-based sources, microbial fermentation has unlocked new possibilities, producing polymers like PHAs directly from bacteria (Schnabl et al., 2023). Lactic acid, synthesized through the fermentation of sugars from crops like sugarcane or cassava, is used to create PLA, a highly versatile bioplastic. Additionally, researchers are exploring algal biomass, capturing the potential of fast-growing algae to serve as a bio-feedstock (Zarubina et al., 2023). The choice of raw material directly influences the biodegradability, mechanical properties, and environmental impact of the final product, reflecting the intricate balance between renewable sourcing, sustainability, and functionality in the evolving world of bioplastics.

Starch to TPS: Starch $\left(C_6H_{10}O_5\right)_n$ undergoes gelatinization and plasticization in the presence of water and plasticizers like glycerol. The process involves breaking intermolecular hydrogen bonds within the starch, altering its structure into a thermoplastic material.

Equation:

$$\text{Starch} + H_2O + (\text{Plasticizer}) \rightarrow \text{Thermoplastic starch (TPS)}$$

Cellulose Extraction for Bioplastics: Cellulose $\left(C_6H_{10}O_5\right)_n$ is dissolved and reformed into films or fibers. Though this process doesn't involve a classical chemical reaction, it involves the dissolution of cellulose in solvents like N-methylmorpholine N-oxide (NMMO) or ionic liquids, followed by regeneration.

Equation:

$$\text{Cellulose}\left(C_6H_{10}O_5\right)_n \rightarrow \text{Cellulose-base film/fiber}$$

PHAs Production via Microbial Fermentation: Microorganisms synthesize PHAs by converting substrates like glucose ($C_6H_{10}O_6$) into biopolymer chains. One example is the production of poly(3-hydroxybutyrate) (PHB).

Equation:

$$\text{Glucose}\left(C_6H_{10}O_5\right) \rightarrow \text{Poly}\left(\text{3-hydroxybutyrate}\right)\left(\text{PHB}\right)\left(C_6H_{10}O_2\right)_n + CO_2 + H_2O$$

Lactic Acid to PLA: Lactic acid ($C_3H_6O_3$) is polymerized through either a direct condensation or ring-opening polymerization to form PLA.

Equation:
For direct condensation polymerization:

$$n\,C_3H_6O_3 \rightarrow \left(C_3H_4O_2\right)_n + nH_2O$$

For ring-opening polymerization of lactide (a cyclic dimer of lactic acid):

$$\text{Lactide}\left(C_6H_8O_4\right) \rightarrow \left(C_3H_4O_2\right)_n (\text{PLA})$$

Algal Biomass Conversion to Bio-Feedstock: Algal biomass is typically converted into bio-based monomers through hydrolysis or pyrolysis, producing sugars or oils. The sugars can be fermented into lactic acid, similar to other crop-based bioplastics, though this step is more complex.
General process for algal hydrolysis:

$$\text{Algal biomass} + H_2O \rightarrow \text{Sugars } (C_6H_{12}O_6) + \text{other compounds}$$

8.2.2 Types of Bioplastic Additives

Additives, often necessary for functionality, improve mechanical strength, flexibility, or resistance to environmental stressors. These may include plasticizers, stabilizers, and cross-linkers. Therefore, additives in bioplastics are critical for tailoring material properties, spanning from mechanical strength to biodegradability. These additives fall into various categories: plasticizers, fillers, stabilizers, and even reinforcements. Plasticizers, such as glycerol, improve flexibility but can sometimes hinder biodegradation by forming rigid molecular interactions (Hou et al., 2023; Zhong et al., 2024). Stabilizers, like antioxidants, protect the polymer from degrading prematurely due to environmental exposure. Fillers, whether organic or inorganic, enhance strength or reduce cost, yet can complicate compostability (Kaur et al., 2023). The implications of these additives are profound while enhancing performance; they may introduce unintended environmental side effects. Certain chemical additives, for example, do not decompose easily, leaving behind microplastic residues that contaminate ecosystems. Balancing these enhancements with sustainability goals is a constant challenge in bioplastic design (Zhong et al., 2024). Furthermore, additives that leach into soil or water systems may affect microbial activity, potentially disrupting natural processes crucial for biodegradation. Thus, the choice of additives represents a delicate balancing act between functionality and ecological responsibility.

8.2.2.1 Biodegradable Additives

Biodegradable additives like PLA, PHA, and starch-based polymers are commonly used to enhance the breakdown of bioplastics. PLA degrades into lactic acid, while PHAs are produced by microbial fermentation and are known for their natural biodegradability. Starch-based polymers, derived from corn, wheat, or potatoes, decompose naturally, contributing to soil health (Kaur et al., 2023; Zhong et al., 2024). However, their interaction with environmental factors is the key: high temperatures accelerate degradation, while microbial activity plays a critical role in breaking down the bioplastic. Composting facilities often offer the ideal conditions such as high heat, moisture, and microbial diversity needed for these additives to degrade effectively (Herath et al., 2022; Savva et al., 2023). Yet, in colder or less microbially active environments, these same additives can persist longer than expected. This dynamic relationship between the additives and environmental conditions underscores the need for precise application strategies in agriculture, where material longevity must be balanced against the requirement for swift biodegradation under specific ecological circumstances.

8.2.2.2 Chemical versus Natural Additives

The distinction between chemical and natural additives in bioplastics presents contrasting environmental implications. Chemical plasticizers, such as phthalates, and stabilizers, like UV blockers, improve flexibility and durability but often carry toxicological risks. Their resistance to degradation can lead to the accumulation of harmful byproducts in ecosystems, contributing to long-term environmental pollution (Schnabl et al., 2023). Conversely, natural additives like cellulose and lignin are more compatible with sustainable agriculture. Derived from plant materials, these additives reinforce bioplastics while ensuring the biodegrade in soil or compost environments without releasing harmful substances (Herath et al., 2022). Natural fibers, such as hemp or flax, can also be used to increase mechanical strength, offering a greener alternative to synthetic fillers (Zhong et al., 2024). The trade-offs between these types of additives revolve around performance versus environmental impact: while chemical additives may offer superior material properties, natural alternatives prioritize biodegradability and ecological harmony, aligning better with sustainability objectives (Table 8.2).

TABLE 8.2
Differences between Chemical and Natural Additives in Bioplastics

Criteria	Chemical Additives	Natural Additives
Origin	Synthetically produced, often from petrochemicals.	Derived from renewable, organic sources like plants, animals, or microbes.
Types	Plasticizers (e.g., phthalates), stabilizers (UV blockers, antioxidants), synthetic fillers (silica).	Cellulose, lignin, natural fibers (hemp, flax), chitosan, starch.
Degradability	Limited or slow biodegradability, may leave behind microplastics or toxic residues.	Fully biodegradable, decomposes into harmless organic compounds under appropriate conditions.
Impact on environment	Potential to leach harmful chemicals into ecosystems, contributing to soil and water pollution.	Minimal environmental impact; non-toxic byproducts that contribute to soil health.
Performance enhancement	Enhances flexibility, durability, thermal stability, and UV resistance.	Improves mechanical properties, adds strength, and can enhance biodegradability.
Processing energy requirements	Often requires energy-intensive processes for incorporation into bioplastics.	Typically requires less energy to process, making it a more eco-friendly option.
Interaction with biodegradability	Can hinder biodegradation by increasing material rigidity or resistance.	Promotes faster degradation, especially in compost or soil environments.
Cost efficiency	Generally cheaper due to mass production in the chemical industry.	May be more expensive depending on source and extraction process.
Compatibility with sustainable agriculture	Less aligned with sustainability goals due to environmental risks.	Highly compatible, supports sustainable agricultural practices and soil regeneration.

8.2.2.3 Performance Enhancement Additives

Additives in bioplastics significantly improve their mechanical, thermal, and barrier properties, making them viable for agricultural applications. For instance, nanofillers like CNCs enhance tensile strength and durability, while clay nanoparticles improve gas barrier properties, making the material more effective for mulching films or greenhouse covers. Biodegradable plasticizers increase flexibility, allowing the material to endure temperature fluctuations without cracking (Herath et al., 2022; Kaur et al., 2023). However, these enhancements come with trade-offs. Increasing strength and durability may reduce the plastic's ability to biodegrade efficiently, as the structural rigidity makes microbial degradation more difficult. Additionally, some performance-enhancing additives may require energy-intensive processing methods, raising the overall environmental impact (Herath et al., 2022; Zhong et al., 2024). Balancing material performance with environmental sustainability becomes a nuanced challenge, especially in agricultural contexts where the bioplastic must withstand harsh environmental conditions but also integrate back into the ecosystem without leaving harmful residues.

8.2.3 Influence of Additives on Biodegradability

Though bioplastics are inherently biodegradable, certain additives can either hasten or inhibit this degradation process. The delicate interplay between base polymers and additives requires a careful balancing act, impacting both product lifespan and ecological decomposition. The influence of additives on the biodegradability of bioplastics is both intricate and paradoxical, creating a dynamic interplay between functionality and environmental degradation (Herath et al., 2022; Hou et al., 2023). On one hand, natural additives like cellulose, lignin, and starch promote biodegradability, enabling bioplastics to break down into harmless organic components. These additives integrate seamlessly into composting processes, enhancing microbial activity and accelerating the decomposition of the plastic matrix. For example, starch-based additives increase water absorption, fostering microbial access and faster breakdown in soil environments. However, the story shifts when it comes to chemical additives (Schnabl et al., 2023). Chemical plasticizers, stabilizers, and synthetic fillers though improving flexibility, durability, and UV resistance often inhibit biodegradability by forming stronger, more resistant molecular structures. These additives can reduce microbial interaction with the polymer, slowing down the natural decomposition process (Cao et al., 2023). In some cases, these chemicals leave behind microplastics or toxic residues, creating long-term environmental hazards despite the original intent of bioplastic. Moreover, external environmental factors like temperature, moisture, and microbial diversity further complicate the degradation process (Schnabl et al., 2024). Additives that improve performance in harsh agricultural conditions may inadvertently delay biodegradation in less controlled environments. Thus, the balance between additive selection and environmental interaction is critical to ensuring bioplastics contribute positively to sustainability goals without lingering ecological costs.

8.2.4 Agricultural Benefits of Bioplastics

Bioplastics offer transformative benefits to agriculture by integrating sustainability into critical farming practices. Mulching films made from biodegradable plastics help conserve soil moisture, regulate temperature, and suppress weeds without the need for post-harvest removal, reducing labor costs. Furthermore, bioplastic seed coatings enhance germination by providing controlled-release nutrients while breaking down naturally, leaving no toxic residues. Bioplastics also enhance greenhouse coverings, allowing for better light diffusion and temperature control, promoting optimal plant growth (Cao et al., 2023; Schnabl et al., 2024). Compared to traditional plastics, which persist in the environment, bioplastics decompose, contributing organic matter to the soil, boosting microbial activity, and fostering healthier ecosystems. Moreover, the renewable nature of bioplastics, derived from sources like starch or sugarcane, aligns with global shifts toward sustainable farming (Herath et al., 2022; Hou et al., 2023). As agricultural practices face pressure to reduce their carbon footprint, bioplastics provide an eco-friendly alternative that not only supports efficient farming but also mitigates the environmental burden, making agriculture more sustainable and regenerative in the long run.

Soil Health and Biodegradability: Biodegradable plastics break down in soil through microbial action, where bacteria, fungi, and other microorganisms consume the material as a carbon source, breaking it into water, carbon dioxide, and biomass. This process is accelerated in environments with high moisture and warm temperatures. When bioplastics decompose, they can improve soil structure, aeration, and water retention by leaving behind organic matter, potentially enhancing soil fertility (Herath et al., 2022). The presence of biodegradable plastics can also stimulate microbial diversity, as microorganisms thrive on the decomposing materials, thus enriching the soil's biological activity. Unlike traditional plastics that linger and disrupt ecosystems, biodegradable variants integrate into natural cycles, promoting ecosystem health. However, factors like temperature, pH, and soil composition can influence the rate of biodegradability. If conditions aren't ideal, breakdown can slow, leaving residual materials (Hou et al., 2023). Nevertheless, when designed appropriately, biodegradable plastics not only safeguard soil health but also align with broader agricultural sustainability goals by contributing positively to the biogeochemical cycles of ecosystems.

Reduction of Plastic Pollution: Bioplastics offer a promising solution to combat long-term plastic pollution in farming environments, an issue exacerbated by the widespread use of traditional, petroleum-based plastics. These conventional plastics, often employed in mulching films, irrigation systems, and packaging, persist in the environment for decades, fragmenting into microplastics that contaminate soil, water, and even food chains (Cao et al., 2023). In contrast, bioplastics degrade into organic matter through natural microbial processes, significantly reducing residual waste and pollution. By decomposing into non-toxic byproducts, bioplastics eliminate the need for costly plastic disposal and clean-up, which plague modern agriculture (Cao et al., 2023; Schnabl et al., 2024). Furthermore, unlike their petroleum-based counterparts, which rely on finite fossil fuels and release greenhouse gases, bioplastics are sourced from renewable resources such as starch or cellulose, aligning with the principles of sustainability (Cao et al., 2023). Although challenges remain in ensuring complete degradation under diverse environmental conditions, the adoption of bioplastics in agriculture represents a pivotal step toward mitigating the pervasive problem of plastic pollution.

Efficiency in Agricultural Practices: Bioplastics significantly enhance efficiency in agricultural practices through their use in mulching films, greenhouse covers, and irrigation systems. These materials offer critical benefits such as retaining soil moisture, regulating soil temperature, and suppressing weed growth, all of which reduce the need for herbicides and excess watering. Bioplastic mulching films act as a protective layer, creating a controlled microclimate that encourages faster seed germination and better root development (Cao et al., 2023; Schnabl et al., 2023). The reduction in water evaporation ensures more efficient water use, a vital aspect in drought-prone regions. By regulating temperature and moisture levels, these bioplastics lead to higher crop yields while cutting down on resource consumption. The fact that they decompose naturally after the growing season eliminates the need for manual removal, saving both time and labor (Kaur et al., 2023; Zhong et al., 2024). While traditional plastics present disposal challenges and environmental hazards, bioplastics provide a seamless, eco-friendly alternative that maximizes farming productivity while minimizing ecological impact.

8.3 ENVIRONMENTAL IMPACTS AND SUSTAINABILITY

8.3.1 Reduced Carbon Footprint

The reduced carbon footprint of bioplastics emerges from a delicate interplay between renewable sourcing, production efficiency, and end-of-life biodegradability. Unlike traditional plastics, derived from fossil fuels, bioplastics harness carbon-neutral resources like starch, cellulose, or plant oils, which absorb CO_2 during their growth phase. This creates a closed carbon loop where emissions released during production or degradation is partially offset by the carbon sequestered during plant growth (Yaakub et al., 2015). Moreover, many bioplastics such as PLA and PHA require lower energy input during polymerization, further reducing greenhouse gas emissions compared to petrochemical-based processes. A major selling point of bioplastics lies in their ability to lower greenhouse gas emissions compared to conventional plastics (Bare, 2014; Oke et al., 2021). The renewable origin of bioplastics means less fossil fuel reliance, reducing carbon output during production. End-of-life scenarios amplify these reductions. Bioplastics, when disposed of in compostable environments, break down into CO_2, water, and biomass without emitting methane, a potent greenhouse gas often released in traditional plastic landfills. Furthermore, eliminating the need for intensive plastic recycling and disposal processes lessens overall energy consumption (Sharma & Rao, 2019; Oke et al., 2021). However, while promising, the full carbon reduction potential depends on factors like feedstock cultivation practices and the energy efficiency of bioplastic manufacturing, making bioplastics a significant yet complex contributor to carbon footprint reduction in the global transition toward sustainable materials.

8.3.2 Biodegradability and Microplastic Formation

The relationship between biodegradability and microplastic formation in bioplastics is complex and often misunderstood. Biodegradable plastics are designed to break down into organic matter through natural microbial processes, ideally leaving no harmful residues. However, under less-than-optimal conditions such as low microbial activity, insufficient moisture, or cooler temperatures, these plastics may not fully degrade (Bare, 2014; Sharma & Rao, 2019; Yadav et al., 2021). Instead, they can fragment into microplastics, small plastic particles that persist in ecosystems much like traditional plastics. While bioplastics such as PLA and PHA are marketed as eco-friendly, incomplete degradation in natural environments raises concerns about microplastic formation. These tiny fragments can accumulate in soil and water, impacting wildlife and potentially entering the food chain (Anuradha et al., 2013; Oke et al., 2021). While the goal of bioplastics is biodegradability, improper disposal or incomplete degradation can lead to microplastic formation, causing unexpected long-term environmental harm.

The key difference lies in environmental conditions and the plastic's composition. True biodegradability depends on specific conditions such as heat, moisture, and microbial diversity, whereas microplastic formation occurs when degradation stalls, leading to fragmentation instead of full decomposition (Anuradha et al., 2013; Yaakub et al., 2015; Said et al., 2024). Thus, ensuring bioplastics meet composting standards under diverse conditions is crucial to prevent microplastic pollution (Table 8.3).

8.3.3 Soil Health and Ecosystem Effects

The impact of bioplastics on soil health and ecosystems is both nuanced and far-reaching, offering a spectrum of benefits and potential challenges. Bioplastics, particularly those designed to biodegrade, contribute positively to soil by breaking down into organic matter, enhancing soil structure, aeration, and water retention. Materials like starch-based polymers and PLA decompose through microbial activity, enriching the soil with nutrients that foster biodiversity and plant growth. This decomposition process promotes microbial diversity, stimulating the soil ecosystem, which is

TABLE 8.3
Differences between Biodegradability and Microplastic Formation

Aspect	Biodegradability	Microplastic Formation
Definition	Breaks down into natural, organic compounds via microbial action.	Fragmentation into tiny, persistent plastic particles under suboptimal conditions.
End product	CO_2, water, and biomass.	Micro-sized plastic fragments that remain in the environment.
Environmental impact	Positive, contributes to soil health and reduces pollution.	Negative, contributes to long-term pollution and can affect ecosystems.
Timeframe	Decomposes fully over weeks to months in the right conditions.	May take years or indefinitely persist in soil or water.
Conditions required	High microbial activity, moisture, and temperature.	Poor environmental conditions or mechanical breakdown.
Potential risks	Minimal if fully degraded.	Potential accumulation in ecosystems, affecting organisms and food chains.

vital for healthy agricultural cycles (Corwin & Bradford, 2008; Anuradha et al., 2013). Bioplastic residues, especially in the form of agricultural films or mulch, can alter soil health by influencing microbial activity, nutrient cycling, and water retention. The residual effects of additives must also be examined, particularly in terms of long-term soil integrity. However, the interaction of bioplastics with soil ecosystems is highly dependent on external factors such as temperature, moisture, and microbial presence (Bare, 2014; Said et al., 2024). In optimal composting conditions, bioplastics degrade rapidly, leaving minimal ecological footprint. But under less favorable conditions, partial degradation can occur, leading to the formation of microplastics, which may disrupt soil ecosystems (Yaakub et al., 2015). These fragments can hinder water infiltration and nutrient exchange, negatively affecting plant roots and microbial habitats. Furthermore, bioplastics containing chemical additives or non-biodegradable components might leach harmful substances into the soil, impacting long-term fertility. Thus, while bioplastics present a promising alternative to traditional plastics, their influence on soil health is deeply contextual, requiring careful design and monitoring to ensure they contribute positively to ecosystems.

8.3.4 Potential Risks and Challenges

The potential risks and challenges associated with bioplastics present a complex web of uncertainties, particularly in their real-world agricultural applications. One key issue is degradation uncertainty while bioplastics are designed to break down under natural conditions, the actual rate and completeness of this process vary dramatically depending on factors such as temperature, moisture, and microbial activity (Zhong et al., 2024). In controlled, industrial composting environments, biodegradation may be efficient, but in outdoor fields or harsher climates, the process can slow down or even halt, leading to the formation of microplastics rather than complete decomposition (Da Silva et al., 2019; Hou et al., 2023). This inconsistency poses a serious environmental paradox, as materials intended to reduce waste may instead persist in the ecosystem. Another pressing concern is the additive residue left in soils. Chemical additives used to enhance the performance of bioplastics like plasticizers or stabilizers may not degrade fully, leaving harmful residues in the soil that could impact microbial health, soil fertility, and crop yields (Dailin et al., 2022). These chemicals can leach into water systems, creating long-term environmental risks. Therefore, the economic feasibility of bioplastics remains a challenge, particularly for small-scale farmers. The higher production costs and limited availability of bioplastics make widespread adoption difficult, hindering the transition to sustainable farming practices.

8.3.4.1 Degradation Uncertainty

The degradation of bioplastics in agricultural settings presents a critical challenge due to the unpredictable nature of breakdown processes. While bioplastics are designed to decompose into organic matter, their performance varies widely depending on environmental conditions such as temperature, moisture, and microbial activity (Zhang et al., 2017; Francis & Parayil, 2023). In ideal composting environments high heat and rich microbial diversity bioplastics may fully biodegrade within a few months. However, in less controlled agricultural fields or cooler climates, the degradation process can stall, leaving behind plastic fragments that persist in the soil. This inconsistency raises significant concerns about their long-term ecological impact, as incomplete biodegradation may lead to microplastic formation (Rahmati et al., 2024; Singhal et al., 2024). Ensuring full biodegradation is particularly challenging in outdoor agricultural applications, where conditions are far less predictable than in industrial composting facilities. Moreover, different bioplastic types such as PLA, PHA, and starch-based polymers have varying degradation rates, adding complexity to the process (Singhal et al., 2024). This uncertainty creates an environmental paradox, where the promise of eco-friendliness collides with the reality of non-ideal conditions, potentially undermining the sustainability goals they aim to achieve.

8.3.4.2 Additive Residue in Soil

A critical concern surrounding the use of bioplastics in agriculture is the potential for additive residues to linger in the soil, contaminating ecosystems. Chemical additives such as plasticizers, stabilizers, or fillers, though designed to enhance performance, may not fully degrade and could leach into the soil, affecting its fertility and microbial balance (Zhong et al., 2024; Fuller et al., 2022). Long-term exposure to these residues poses a risk to crop health, as chemicals can disrupt nutrient absorption, alter soil pH, and impact the delicate balance of the soil's microbiome. This contamination risk extends beyond the immediate environment, potentially infiltrating water systems through runoff, further spreading the pollutants (Dailin et al., 2022; Hou et al., 2023). Over time, the accumulation of these additives could diminish soil fertility, leading to reduced crop yields and compromising the integrity of sustainable farming practices (Hou et al., 2023). While natural additives like cellulose and lignin pose fewer risks, chemical-based counterparts represent a significant challenge. The long-term environmental impacts of these residues remain an open question, demanding rigorous monitoring and regulation.

8.3.4.3 Cost and Economic Feasibility

The economic feasibility of adopting bioplastics in agriculture remains a formidable hurdle, particularly for small-scale farmers. Bioplastics production involves higher costs due to the complex processes needed to extract raw materials from renewable sources like starch or cellulose, and the integration of performance-enhancing additives further raises the price. In contrast to traditional petroleum-based plastics, bioplastics often require more energy-intensive processing and specialized manufacturing, driving up production expenses (Zhang et al., 2017; Da Silva et al., 2019). For farmers, especially in developing regions, these higher costs create a significant barrier to adoption, making bioplastics economically unattainable without subsidies or governmental support. Small-scale farmers, who already operate on thin margins, may find it difficult to justify the investment in bioplastic-based solutions despite their long-term environmental benefits (Savva et al., 2023; Rahmati et al., 2024; Singhal et al., 2024). Additionally, the infrastructure required for biodegradable waste management, such as composting facilities, may not be readily available in all regions, compounding the cost burden (Fuller et al., 2022; Dailin et al., 2022). Thus, while bioplastics offer environmental promise, their economic challenges must be addressed to ensure widespread adoption in the agricultural sector.

8.4 AGRICULTURAL PRODUCTIVITY AND BIOPLASTIC UTILIZATION

8.4.1 Role in Crop Protection and Yield Enhancement

The role of agricultural productivity in conjunction with bioplastic utilization for crop protection and yield enhancement represents an evolving nexus of sustainability and technological innovation. As global demand for food intensifies, the agricultural sector is tasked with not only increasing yields but also doing so in an environmentally conscious manner (Chen et al., 2023). Here, bioplastics are rapidly emerging as pivotal tools in transforming how crops are protected and cultivated, offering a sustainable alternative to conventional agricultural inputs while driving yield optimization. Bioplastics, when used as mulch films or for irrigation systems, can significantly influence crop growth, minimizing weed proliferation and optimizing water usage (Huang et al., 2022). Their ability to create favorable microclimates further bolsters crop productivity. Mulching films, for instance, made from biodegradable plastics, have revolutionized soil management. These films help regulate soil temperature, retain moisture, and suppress weed growth, creating an optimized microenvironment for crops (Sintim & Flury, 2017). By reducing evaporation rates, bioplastic mulches enhance water-use efficiency, which is especially crucial in arid regions or during periods of drought. The breakdown of these mulching films directly into the soil post-harvest adds organic matter, further improving soil fertility without the labor or environmental costs of removal, as seen with traditional polyethylene films. In the realm of crop protection, bioplastic films act as barriers against pests, reducing the reliance on chemical pesticides (Koskei et al., 2021). They also shield crops from extreme weather conditions, safeguarding young plants during vulnerable growth stages. Additionally, seed coatings made from bioplastics offer controlled-release systems for fertilizers and nutrients, ensuring that essential inputs are delivered precisely when needed, minimizing waste and environmental runoff while boosting germination rates and overall crop health (Zulkafli, 2014). Further bolstering yields, biodegradable irrigation systems crafted from bioplastics provide an eco-friendly solution to water distribution, ensuring that water is efficiently delivered to plant roots without the need for costly removal or replacement of non-biodegradable piping. These systems degrade over time, leaving behind no trace of plastic waste, thus contributing to sustainable soil and water management practices (Zulkafli, 2014; Koskei et al., 2021).

However, the use of bioplastics goes beyond immediate crop protection and extends to the broader goal of yield enhancement through ecosystem balance. By reducing soil disruption and encouraging microbial activity, bioplastics promote a healthy, thriving agricultural ecosystem. This not only boosts short-term productivity but also builds long-term soil resilience, crucial for sustained agricultural success (Chen et al., 2023). Bioplastics' ability to decompose into organic matter means they actively improve soil structure, nutrient availability, and water retention, all of which contribute to higher yields over multiple growing seasons (Maraveas et al., 2020). Hence, the integration of bioplastics into modern agricultural practices is more than just a sustainable alternative; it is a transformative force, enhancing productivity while reducing environmental impact. Their multifunctional role, from crop protection to yield enhancement, signifies a forward-thinking approach to farming, balancing the immediate needs of food production with the long-term health of the environment. As these technologies continue to evolve, the synergy between agricultural productivity and bioplastic innovation will undoubtedly shape the future of farming.

8.4.2 Greenhouse Films and Controlled Environments

The impact of agricultural productivity paired with the utilization of bioplastics, particularly in the form of greenhouse films and controlled environments, represents a groundbreaking intersection of sustainability and efficiency in modern farming practices. As the global food supply struggles to keep pace with rising populations and unpredictable climate conditions, farmers increasingly turn to controlled environments, like greenhouses to create the optimal conditions for plant growth (Anugrahwidya et al., 2021). Here, the incorporation of bioplastics offers an unprecedented opportunity to elevate productivity while reducing environmental degradation. Bioplastics also find application in greenhouse films, where they help regulate temperature, light diffusion, and moisture levels, creating ideal conditions for plant growth (Merino et al., 2019). However, bioplastic greenhouse films crafted from materials such as PLA or PHAs have begun to replace traditional polyethylene films, bringing with them several advantages. These films provide an ideal blend of durability, flexibility, and biodegradability, making them highly effective for controlled environments where crops require protection from external variables such as temperature fluctuations, UV radiation, and excessive moisture (Zulkafli, 2014; Chen et al., 2023). The ability of bioplastics to regulate the microclimate within greenhouses ensures that crops grow under stable conditions, leading to enhanced productivity and extended growing seasons critical factors for year-round agriculture (Huang et al., 2022). Furthermore, unlike traditional plastic films that degrade over time and require replacement, contributing to plastic waste accumulation, biodegradable bioplastic films can break down naturally (Anugrahwidya et al., 2021). This eliminates the costly and labor-intensive process of film disposal, while simultaneously reducing the plastic pollution associated with agricultural waste. As these films decompose, they contribute organic material to the soil, fostering improved soil fertility and enhancing the long-term sustainability of farming systems.

In addition, by merging crop protection with eco-conscious practices, bioplastics transform greenhouse agriculture into a circular process, where materials benefit the environment rather than burden it. In the context of controlled environments, bioplastic films offer superior light transmission and UV stabilization, which are key for optimizing photosynthetic activity and promoting plant growth (Landstrom & Wilkinson, 1986; Maraveas & Bartzanas, 2021). Many of these films are engineered with selective permeability, allowing them to control the passage of water vapor, gases, and heat. This level of control ensures that plants receive just the right amount of sunlight and moisture, reducing the risk of overheating or under-irrigation, and enabling farmers to fine-tune conditions for specific crops. The result is a significant boost in crop yield and quality, particularly for high-value crops such as tomatoes, peppers, and leafy greens, which thrive in such environments (Sprung, 1991; Ni et al., 2019). Moreover, the role of bioplastics extends into energy efficiency within these controlled settings. Greenhouses typically require significant energy inputs to maintain optimal growing conditions, especially in colder climates. However, advanced bioplastic films with thermal insulation properties help to retain heat within the greenhouse, reducing the energy required to maintain ideal temperatures. This not only lowers the carbon footprint of greenhouse operations but also cuts down on operational costs, making controlled-environment farming more economically feasible and sustainable (Moon et al., 2020; Maraveas & Bartzanas, 2021). An often-overlooked aspect of bioplastic utilization in greenhouses is its contribution to pest control. In traditional greenhouse settings, chemical pesticides are often used to protect crops from infestations. However, bioplastic films with integrated antimicrobial properties can create a more sterile environment, reducing the reliance on chemical inputs and fostering a more organic production system. This aligns with broader trends toward sustainable agriculture, which prioritize minimizing chemical use and improving food safety (Ni et al., 2019; Maraveas & Bartzanas, 2021). Nonetheless, challenges remain, particularly around the degradation rates of bioplastic films in specific environmental conditions. Factors such as humidity, UV exposure, and temperature can affect the longevity of these films, sometimes leading to earlier-than-expected breakdown or incomplete decomposition.

So, to address this, ongoing innovations in bioplastic formulations aim to strike the right balance between durability during use and biodegradability after use, ensuring that these materials fulfill their functional purpose while still contributing positively to the environment. In conclusion, the integration of bioplastic greenhouse films and other materials within controlled agricultural environments represents a transformative shift in how productivity and sustainability can be harmonized. These materials not only offer improved conditions for crop growth through precise climate regulation but also mitigate the environmental impact associated with traditional plastics. By enhancing energy efficiency, reducing chemical inputs, and contributing to a circular ecosystem where waste becomes a resource, bioplastics are poised to play a pivotal role in the future of agriculture. As these technologies advance, their potential to revolutionize greenhouse farming and controlled-environment agriculture will likely expand, pushing the boundaries of productivity while maintaining ecological integrity.

8.4.2.1 Environmental Impact of Bioplastics and Additives

The environmental impact of bioplastics and their additives is a double-edged sword, characterized by their potential to reduce pollution yet complicated by unintended consequences. Bioplastics, derived from renewable resources like starch, sugarcane, or cellulose, offer an eco-friendly alternative to conventional plastics, reducing dependency on fossil fuels. However, the environmental benefits hinge on additive selection (Moon et al., 2020; Anugrahwidya et al., 2021). Natural additives such as lignin or cellulose tend to enhance biodegradability and contribute to soil health, but synthetic additives like plasticizers and stabilizers can persist in the environment, leaving toxic residues. This incomplete breakdown could lead to microplastic formation, compromising soil and aquatic ecosystems (Maraveas & Bartzanas, 2021). Additionally, while bioplastics are often hailed as compostable, not all degrade uniformly across various environmental conditions. Inadequate composting infrastructure exacerbates the problem, as improperly discarded bioplastics could end up in landfills or waterways, where their eco-benefits diminish (Ni et al., 2019). Furthermore, the production process itself isn't always carbon-neutral, requiring significant energy inputs, sometimes from non-renewable sources, thus creating an environmental paradox where the very materials intended to reduce harm contribute indirectly to greenhouse gas emissions or pollution. Bioplastics present a promising yet complicated solution, and their environmental impact is far from straightforward.

8.4.2.2 Life-Cycle Assessment (LCA) of Bioplastics

A life-cycle assessment (LCA) evaluates the overall environmental footprint of bioplastics from production to disposal, providing a comprehensive picture of their true ecological impact. In terms of resource use, bioplastics have a clear advantage, sourced from renewable materials like corn, sugarcane, or cellulose, which absorb carbon dioxide during growth. However, the production phase can still be energy-intensive, particularly when it comes to polymerization and additive integration, potentially offsetting some environmental gains (Mahfuz et al., 2020). When compared to traditional plastics, bioplastics generally have lower carbon emissions and reduced fossil fuel dependency, but the picture becomes murkier when considering their end-of-life disposal (Ganzhur et al., 2020). Conventional plastics often last for centuries in the environment, but their recycling systems are more established, while bioplastics, though biodegradable, require specific industrial composting conditions to break down fully (Weng et al., 2020). LCA also reveals that the water usage in growing bio-feedstocks can be high, and land competition for food crops is another critical issue (Hemming & Baeza-Romero, 2018; Weng et al., 2020). Thus, while bioplastics appear to have a greener profile at face value, LCA highlights their complex trade-offs, showing that their environmental superiority depends heavily on production processes, disposal methods, and infrastructure.

8.4.2.3 Impact on Greenhouse Gas Emissions

The impact of bioplastics on greenhouse gas emissions is intricately tied to both their production and degradation phases. During the growth of bio-feedstocks like corn or sugarcane, carbon is absorbed from the atmosphere, contributing to a carbon-neutral lifecycle. However, the breakdown of bioplastics in the soil may release methane (CH_4) and carbon dioxide (CO_2), two potent greenhouse gases (Ganzhur et al., 2020). While bioplastics are designed to decompose naturally, the conditions under which this occurs can significantly affect the rate and type of emissions. In anaerobic conditions, such as landfills, the breakdown process may favor methane production, a greenhouse gas far more impactful than CO_2 (Anugrahwidya et al., 2021). This introduces a paradox where bioplastics, although reducing long-term pollution, could still contribute to climate change in the short term (Ni et al., 2019). Moreover, if bioplastics are improperly disposed of, they may not degrade as intended, leading to lingering microplastics or incomplete breakdown, further complicating their environmental footprint. Balancing the immediate biodegradability benefits with long-term climate impacts remains a key challenge in realizing the full potential of bioplastics as a solution to plastic waste.

8.4.2.4 Waste Management and Recycling

The waste management and recycling of bioplastics present a labyrinth of challenges that complicate their role as a sustainable alternative. Unlike conventional plastics, which have well-established recycling streams, bioplastics require specialized infrastructure for composting or biodegradation. Many bioplastics, like PLA, need industrial composting facilities with high temperatures and moisture levels to break down effectively, yet such facilities are not widely available, leading to confusion in disposal practices (Mahfuz et al., 2020; Moon et al., 2020). Consumers often mistakenly discard bioplastics into regular plastic recycling streams, contaminating the process and reducing the efficiency of both bioplastic and conventional plastic recycling (Maraveas & Bartzanas, 2021). Additionally, bioplastics like PHA may degrade in natural environments but still face collection and sorting challenges in waste management systems. The integration of compostable plastics into mixed waste streams creates a dilemma for waste managers, as not all bioplastics are truly compostable under household conditions (Hemming & Baeza-Romero, 2018). The current infrastructure for composting and biodegradation remains inadequate to handle the growing volume of bioplastic waste, revealing a critical gap between the promise of bioplastics and the reality of waste management.

8.4.3 Packaging and Storage Innovations

The intersection of agricultural productivity and bioplastic utilization in the domain of packaging and storage innovations represents a monumental shift in both sustainability and efficiency, profoundly affecting how food is preserved, transported, and ultimately consumed. As global agriculture grapples with the dual pressures of meeting growing food demands and reducing environmental impact, bioplastics are emerging as pivotal tools in revolutionizing agricultural packaging, ensuring that food reaches consumers in optimal condition while minimizing waste and resource use (Takakura, 2019). Beyond field applications, bioplastics offer improved agricultural packaging solutions. Their biodegradability reduces the burden of post-harvest waste while maintaining the quality of stored produce. One of the most significant impacts of bioplastic utilization is the reduction of petrochemical-based plastic in agricultural packaging, which has historically contributed to environmental degradation through its non-biodegradable nature and dependence on fossil fuels (El Ashmawy, 2019). By leveraging materials like PLA, PHAs, and starch-based bioplastics, agricultural packaging is evolving into a more eco-friendly, biodegradable solution. These materials break down under composting conditions, significantly reducing the volume of waste generated in food supply chains, which have long been dominated by non-recyclable plastics (Anugrahwidya et al., 2021). This shift is especially crucial in industries like fresh produce, where packaging is often discarded shortly after reaching consumers, contributing to substantial plastic pollution.

In terms of food preservation, bioplastic innovations are opening new doors by integrating functional additives that extend the shelf life of agricultural products. Bioplastics are not only passive containers; they can be designed with properties that actively interact with the food they store

(Al-Madani et al., 2024). For instance, some bioplastic packaging materials are engineered with antimicrobial agents to inhibit bacterial growth, thereby slowing down spoilage and reducing food waste. This is especially beneficial for perishable goods like fruits, vegetables, and dairy products, where spoilage is a major concern (Lu et al., 2023). By reducing the rate of spoilage, bioplastics indirectly boost agricultural productivity by ensuring that a higher percentage of harvested crops make it to market, thereby enhancing the efficiency of food distribution systems. Additionally, bioplastics are playing a key role in the development of intelligent packaging systems, which have the potential to further revolutionize agricultural storage. These systems often involve the incorporation of bio-sensors or indicator films within the packaging that monitor the freshness and condition of the stored products (Shi et al., 2022). Such technology allows for real-time tracking of food quality, alerting distributors and consumers when products are nearing spoilage or experiencing suboptimal conditions. This not only helps in minimizing food loss but also improves the overall management of supply chains, ensuring that products are stored and consumed at peak quality, thereby enhancing food safety and minimizing economic losses for farmers (Belhachemi et al., 2022).

Moisture control, another critical factor in food preservation, is being addressed by bioplastic packaging innovations. Certain bioplastic materials are designed to be breathable, allowing excess moisture to escape while maintaining the ideal humidity for the stored product. This is particularly important for products like grains, nuts, and cereals, which can deteriorate rapidly if exposed to improper moisture levels during storage (Zhu et al., 2024). By mitigating the risk of mold or degradation due to excess humidity, bioplastic packaging ensures that more of the harvested product remains viable for longer periods, directly contributing to improved food security. From an environmental perspective, bioplastics dramatically reduce the carbon footprint of agricultural packaging (Nemali, 2022). Traditional plastic production is resource-intensive, relying heavily on petroleum extraction, refining, and processing all activities associated with high carbon emissions. In contrast, bioplastics are derived from renewable resources, such as corn, sugarcane, or cellulose, which absorb carbon dioxide during their growth, offsetting emissions during production (Chimankare et al., 2023). Moreover, many bioplastics are biodegradable or compostable, meaning that their lifecycle concludes with a minimal environmental footprint, either by returning to the Earth as organic matter or through safe decomposition.

Bioplastic packaging also plays a role in reducing the weight of agricultural products during transport, as these materials are often lighter than conventional plastics. This reduction in weight translates to lower fuel consumption during transportation, further decreasing the environmental impact of food distribution (Shi et al., 2022; Lu et al., 2023). The lightweight nature of bioplastics, combined with their strength and flexibility, offers a dual advantage: packaging that is both sustainable and functional, capable of withstanding the rigors of transport while maintaining food integrity. However, the full-scale implementation of bioplastics in packaging and storage is not without its challenges. One key issue is the cost differential between traditional plastics and bioplastics (Zhu et al., 2024). Although bioplastics are becoming more competitively priced as technology advances, their production still tends to be more expensive than petroleum-based plastics, which can deter small-scale farmers and distributors from adopting these solutions (Lu et al., 2023). Additionally, the infrastructure for bioplastic waste management, particularly composting and biodegradation facilities, is not yet widespread, which limits the full potential of these materials in reducing environmental waste.

Nonetheless, the push toward circular economy models in agriculture, where resources are reused, recycled, and composted rather than discarded, has the potential to drive further innovations in bioplastic packaging and storage. By closing the loop between production and disposal, bioplastics could become integral to a system where agricultural productivity is enhanced not only through increased yields but also through resource efficiency, reduced waste, and improved environmental stewardship (Al-Madani et al., 2024). In conclusion, the integration of bioplastics in agricultural packaging and storage is shaping the future of food preservation and sustainability. From extending shelf life and improving moisture control to lowering the carbon footprint of food supply chains, these innovations are driving productivity while addressing critical environmental challenges. As bioplastic technologies continue to advance, they will likely play an increasingly essential role in reshaping the global food system, offering a path toward more sustainable, efficient, and resilient agricultural practices.

8.5 FUTURE PROSPECTS AND CHALLENGES

The future prospects of bioplastics in agriculture are filled with promise, yet marred by significant challenges. As innovations advance, bioplastics could revolutionize farming by improving soil health, reducing plastic waste, and enhancing crop yields. New materials are being developed to tailor biodegradability to different environments, enabling more efficient decomposition in soil while supporting sustainable agricultural practices. However, achieving large-scale adoption is fraught with obstacles (Di Bartolo et al., 2021; Xia et al., 2022). The inconsistency in biodegradation rates under varying conditions especially in non-ideal climates raises concerns about unintended microplastic formation. Moreover, the potential for chemical additives to leach into the soil could pose long-term risks to ecosystem health. Additionally, the high costs associated with bioplastic production and their incorporation into farming systems create barriers, particularly for small-scale farmers (Faris et al., 2014; Ibrahim et al., 2021; Skoczinski et al., 2021). The challenge lies in striking a delicate balance between performance enhancements, cost efficiency, and environmental sustainability, necessitating continuous innovation in both material science and policy frameworks.

8.5.1 Technological Innovations

Technological advancements in bioplastics are increasingly focused on refining polymer matrices to exhibit superior mechanical properties, such as enhanced tensile strength, elongation at break, and impact resistance, without compromising their biodegradability. Novel formulations are exploring copolymerization and the integration of nanofillers, such as CNCs or nanoclays, to significantly improve the thermal stability, flexibility, and barrier properties of bioplastics (Egon et al., 2024). This enables their application in sectors that require robust materials, such as agriculture, automotive, and packaging industries. In parallel, advancements in microbial degradation processes are accelerating the breakdown of bioplastics. A key area of research involves the optimization of extracellular hydrolytic enzymes like lipases, esterases, and cutinases, which catalyze the hydrolysis of complex polymer chains into monomers under controlled environmental conditions (Marturano et al., 2023). Genetic engineering of microbial strains such as *Pseudomonas putida* and *Cupriavidus necator* has also been pivotal in enhancing their metabolic pathways for faster degradation of materials like PLA and PHAs (Ruini et al., 2023). Furthermore, innovations in aliphatic–aromatic copolymers are being explored to balance the hydrophobicity and hydrophilicity within the polymer structure, thus enhancing the rate of microbial colonization and subsequent biodegradation. These breakthroughs, combined with improved enzymatic selectivity, promise to revolutionize the biodegradation kinetics of next-generation bioplastics.

8.5.2 Circular Economy and Life-Cycle Assessment

The integration of circular economy principles into bioplastic development is pivotal for achieving sustainability within agricultural and industrial ecosystems. Central to this alignment is the concept of LCA, a comprehensive methodology that evaluates the environmental impacts of bioplastics throughout their entire life cycle from feedstock cultivation, polymer synthesis, and manufacturing to usage, degradation, and disposal. LCA quantifies key metrics such as global warming potential (GWP), acidification, eutrophication, and non-renewable energy use, offering a holistic view of the bioplastic's carbon footprint and resource efficiency (Logan et al., 2024). The process also considers the biodegradability and compostability of these materials, assessing whether they meet end-of-life scenarios like industrial composting, anaerobic digestion, or organic recycling (Nhu et al., 2024). However, bio-based polymers like PLA or PHA must undergo comparative analysis against conventional plastics, which often outperform in durability but falter in post-consumer waste management. Challenges remain in scaling closed-loop recycling systems for bioplastics, as their chemical compositions often complicate mechanical or chemical recycling processes (Serrano-Aguirre & Prieto, 2024). Thus, advanced LCAs play a crucial role in refining bioplastic formulations to maximize circularity, minimize environmental trade-offs, and ensure alignment with sustainability metrics.

8.5.3 Unresolved Issues: Additives and Environmental Concerns

Despite the growing promise of bioplastics in agriculture, unresolved issues surrounding the toxicity and environmental impacts of certain additives present significant challenges. Additives such as plasticizers, stabilizers, and fillers are often integrated into bioplastic matrices to enhance mechanical properties or improve processability (Zellmer, 2005; Serrano-Aguirre & Prieto, 2024). However, some of these additives, particularly phthalates and bisphenol analogs, pose potential risks of leaching into soil or water systems, leading to toxicity that can disrupt microbial communities and adversely affect biodiversity. Furthermore, the incomplete degradation of bioplastics in suboptimal environmental conditions exacerbates concerns. Bioplastics like PLA or TPS require specific conditions such as elevated temperatures, moisture, and the presence of microbial consortia for complete degradation (Nhu et al., 2024). In environments lacking these factors, fragmentation rather than mineralization occurs, leading to the formation of microplastics, which persist in ecosystems and disrupt soil structure and fauna (Ruini et al., 2023). Addressing these issues demands advancements in the development of non-toxic, biodegradable additives derived from renewable sources like lignin or chitosan. Additionally, designing bioplastics with tailored degradation kinetics and optimizing biodegradation pathways in diverse soil environments are critical to aligning the benefits of bioplastics with sustainable agricultural practices without compromising ecosystem integrity.

8.6 CONCLUSION

The integration of bioplastics in agriculture marks a significant shift toward sustainable farming, promising to mitigate environmental degradation and improve resource efficiency. Bioplastics, particularly in forms like mulch films, seed coatings, and greenhouse covers, demonstrate clear benefits for crop productivity, soil health, and resource management. These materials not only provide protection for crops by regulating soil moisture and temperature but also contribute positively to ecosystems by decomposing into organic matter, reducing reliance on fossil-based plastics. However, challenges such as inconsistent degradation rates, the potential toxicity of chemical additives, and the high cost of production hinder their widespread adoption. The impact of additives remains a contentious issue, with chemical stabilizers and plasticizers posing risks of soil contamination and microplastic formation if not properly managed. To address these challenges, future research should focus on the development of non-toxic, biodegradable additives and innovations in controlled degradation. Additionally, policy interventions are critical. Governments should provide financial incentives and subsidies to offset the higher costs of bioplastics and enhance composting infrastructure. Effective LCAs must be integrated into both research and policy frameworks to ensure that bioplastics deliver on their promise of sustainability without compromising ecosystem integrity.

REFERENCES

Al-Madani, A. A., Al-Helal, I. M., & Alsadon, A. A. (2024). Assessing the effectiveness of reflective and diffusive polyethylene films as greenhouse covers in arid environments. *Agronomy*, *14*(5), 1082.

Anugrahwidya, R., Armynah, B., & Tahir, D. (2021). Bioplastics starch-based with additional fiber and nanoparticle: Characteristics and biodegradation performance: A review. *Journal of Polymers and the Environment*, *29*(11), 3459–3476.

Anuradha, B., Karthik, G., Padmapriya, V. S., & Shobana, P. (2013). Environmental impact assessment on reviving water storage structure for sustainable development. *Research Journal of Engineering and Technology*, *4*(4), 251–254.

Babaniyi, G. G., Akor, U. J., & Adebomi, J. I. (2024b). Legislation and programs supporting environmental protection. In S. A. Aransiola, B. R. Babaniyi, A. B. Aransiola, and N. R. Maddela (eds.), *Prospects for Soil Regeneration and Its Impact on Environmental Protection* (pp. 3–14). Springer.

Babaniyi, G. G., Ibrahim, F., Akor, U. J., & Daramola, O. E. (2024a). Regenerative agriculture for food security. In *Prospects for Soil Regeneration and Its Impact on Environmental Protection* (pp. 227–242). Springer Nature Switzerland.

Bare, J. C. (2014). Development of impact assessment methodologies for environmental sustainability. *Clean Technologies and Environmental Policy*, *16*, 681–690.

Belhachemi, A., Maatoug, M. H., & Canela-Garayoa, R. (2022). Comparative analysis by UV-vis and FT-IR spectroscopy of the chemical degradation of polyethylene used as greenhouse cover film. *Journal of Elastomers & Plastics*, *54*(6), 891–905.

Biswas, A., Pal, S., & Mitra, R. D. (2024). A review on—Preparation of bio plastics from agri-waste. *The Bioscan*, *19*(Supplement 1), 09–12.

Cao, X., Liang, Y., Jiang, J., Mo, A., & He, D. (2023). Organic additives in agricultural plastics and their impacts on soil ecosystems: Compared with conventional and biodegradable plastics. *TrAC Trends in Analytical Chemistry*, *166*, 117212.

Chen, B., Cui, J., Dong, W., & Yan, C. (2023). Effects of biodegradable plastic film on carbon footprint of crop production. *Agriculture*, *13*(4), 816.

Chimankare, R. V., Das, S., Kaur, K., & Magare, D. (2023). A review study on the design and control of optimised greenhouse environments. *Journal of Tropical Ecology*, *39*, e26.

Corwin, D. L., & Bradford, S. A. (2008). Environmental impacts and sustainability of degraded water reuse. *Journal of Environmental Quality*, *37*(S5), S1–S7.

Dailin, D. J., Low, L. Z. M. I., Nordin, N. Z., Azelee, N. I. W., Selvamani, S., Nayagam, V. M., …, & El Enshasy, H. A. (2022). Biotechnological advancements in the treatment of plastic wastes. In *Biotechnological Approaches in Waste Management* (pp. 91–113). CRC Press.

Da Silva, A. C., & Córdoba de Torresi, S. I. (2019). Advances in conducting, biodegradable and biocompatible copolymers for biomedical applications. *Frontiers in Materials*, *6*, 98.

Di Bartolo, A., Infurna, G., & Dintcheva, N. T. (2021). A review of bioplastics and their adoption in the circular economy. *Polymers*, *13*(8), 1229.

Egon, A., Bell, C., & Shad, R. (2024). Innovations in Polymer Additive Manufacturing: Research Advancements in Polymer Materials for Additive Manufacturing. Preprints.org. https://doi.org/10.20944/preprints202408.0061

El Ashmawy, N. M. (2019). Effect of glazing materials on warming up automated commercial greenhouses. *Journal of Soil Sciences and Agricultural Engineering*, *10*(10), 541–550.

Faris, N. A., Noriman, N. Z., Sam, S. T., Ruzaidi, C. M., Omar, M. F., & Kahar, A. W. M. (2014). Current research in biodegradable plastics. *Applied Mechanics and Materials*, *679*, 273–280.

Ferreira-Filipe, D. A., Paço, A., Pinho, B., Silva, R., Silva, S. A., Jesus, F., …, & Patrício-Silva, A. L. (2024). Microplastics from agricultural mulch films: Biodegradation and ecotoxicity in freshwater systems. *Science of the Total Environment*, *912*, 169287.

Francis, D., & Parayil, D. J. (2023). Microbial technology in bioplastic production and engineering. In *Handbook of Bioplastics and Biocomposites Engineering Applications* (pp. 121–148). Wiley Online Library.

Fuller, C. W., Padayatti, P. S., Abderrahim, H., Adamiak, L., Alagar, N., Ananthapadmanabhan, N., …, & Merriman, B. (2022). Molecular electronics sensors on a scalable semiconductor chip: A platform for single-molecule measurement of binding kinetics and enzyme activity. *Proceedings of the National Academy of Sciences*, *119*(5), e2112812119.

Ganzhur, M., Ganzhur, A., Kobylko, A., & Fathi, D. (2020). Automation of microclimate in greenhouses. In *E3S Web of Conferences* (Vol. 210, p. 05004). EDP Sciences.

Hemming, S. and Baeza-Romero, E.J. (2019). Effect of greenhouse films on climate, energy, light distribution and crop performance – measuring film properties and modelling results. *Acta Hortic, 1252*, 173–182. https://doi.org/10.17660/ActaHortic.2019.1252.23

Herath, H. M. N. A., Gamage, D. A. S., & Madhujith, T. (2022). Biopolymer-based food additives and their uses.

Hou, B., Wang, Y., Li, B., Li, N., Yang, D., Liu, H., & Li, J. (2023). Bio-based flame-retardant plasticizers derived from l-lactic acid: Effects of valence states of phosphorus structure on fire safety, flexibility and transparency of poly (lactic acid). *Chemical Engineering Journal*, *474*, 145563.

Huang, F., Liu, Z., Li, Z., Wang, B., Zhang, P., & Jia, Z. (2022). Is biodegradable film an alternative to polyethylene plastic film for improving maize productivity in rainfed agricultural areas?—Evidence from field experiments. *Agricultural Water Management*, *272*, 107868.

Ibrahim, F., Babaniyi, G. G., Ajao, B. H., Akor, U. J. & Daramola, O. E. (2024). Current trends and future research in emerging food contaminants. In S. A. Aransiola, R. O. Akinsola, O. P. Abioye, and N. R. Maddela (eds.), *Emerging Contaminants in Food and Food Products* (pp. 251–268). CRC Press. http://dx.doi.org/10.1201/9781003438236-18

Ibrahim, N. I., Shahar, F. S., Sultan, M. T. H., Shah, A. U. M., Safri, S. N. A., & Mat Yazik, M. H. (2021). Overview of bioplastic introduction and its applications in product packaging. *Coatings*, *11*(11), 1423.

Kaur, G., Thakur, B., Gill, R., Kaur, R., & Kaur, S. (2023). Biopolymeric conjugation with food additives. *Physical Sciences Reviews*, *12*(9), 2899–2923.

Koskei, K., Munyasya, A. N., Wang, Y. B., Zhao, Z. Y., Zhou, R., Indoshi, S. N., …, & Xiong, Y. C. (2021). Effects of increased plastic film residues on soil properties and crop productivity in agro-ecosystem. *Journal of Hazardous Materials*, *414*, 125521.

Landstrom, D. K., & Wilkinson, W. H. (1986). U.S. Patent No. 4,567,732. Washington, DC: U.S. Patent and Trademark Office.

Logan, H., Astrup, T. F., & Damgaard, A. (2024). Additive inclusion in plastic life cycle assessments part I: Review of mechanical recycling studies. *Journal of Industrial* Ecology, *28*(6), 1582–1597.

Lu, H., Liang, H., Liu, J., Zhou, H., Pi, M., Li, M., …, & Ran, R. (2023). Hydrogel-based intelligent greenhouse films for simultaneous fogging prevention and indoor environmental monitoring. *Chemical Engineering Journal*, *475*, 146204.

Magni, S., Fossati, M., Pedrazzani, R., Abba, A., Domini, M., Menghini, M., …, & Della Torre, C. (2024). Plastics in biogenic matrices intended for reuse in agriculture and the potential contribution to soil accumulation. *Environmental Pollution*, *349*, 123986.

Mahfuz, N., Jahan, R., Islam, M. M., Nigar, M., & Karmokar, S. (2020, December). Microcontroller based intelligent greenhouse environment monitoring and controlling system. In *2020 IEEE International Women in Engineering (WIE) Conference on Electrical and Computer Engineering (WIECON-ECE)* (pp. 418–421). IEEE.

Maraveas, C. (2020). Production of sustainable and biodegradable polymers from agricultural waste. *Polymers*, *12*(5), 1127.

Maraveas, C., & Bartzanas, T. (2021). Application of Internet of Things (IoT) for optimized greenhouse environments. *AgriEngineering*, *3*(4), 954–970.

Marturano, V., Marotta, A., Salazar, S. A., Ambrogi, V., & Cerruti, P. (2023). Recent advances in bio-based functional additives for polymers. *Progress in Materials Science*, *139,* 101186.

Merino, D., Mansilla, A. Y., Casalongué, C. A., & Alvarez, V. A. (2019). Performance of bio-based polymeric agricultural mulch films. *Polymers for Agri-Food Applications*, *1*, 215–240.

Moon, J. P., Park, S. H., Kim, J. G., Lee, J. H., Kang, Y. K., Lim, M. Y., & Kim, H. M. (2020). Yield increase and energy saving effect on plastic greenhouse covered with polyolefin film. *Journal of Bio-Environment Control*, *29*(4), 428–439.

Nemali, K. (2022). History of controlled environment horticulture: Greenhouses. *HortScience*, *57*(2), 239–246.

Nhu, T. T., Boone, L., Guillard, V., Chatellard, L., Reis, M., Matos, M., & Dewulf, J. (2024). Environmental sustainability assessment of biodegradable bio-based poly (3-hydroxybutyrate-co-3-hydroxyvalerate) from agro-residues: Production and end-of-life scenarios. *Journal of Environmental Management*, *356*, 120522.

Ni, M., Lan, D., Jahan, M. S., Wang, J., & Guo, S. (2019). A pilot study on the microclimate of a multi-span solar energy greenhouse. *Applied Engineering in Agriculture*, *35*(4), 601–616.

Ogunlade, T. M., Babaniyi, B. R., Afolabi, F. J., & Babaniyi, G. G. (2021). Physicochemical, heavy metals and microbiological assessment of wastewater in selected abattoirs in Ekiti State, Nigeria. *Journal of Environmental Treatment Techniques*, *9*(4), 788–795.

Oke, A., Aghimien, D., Aigbavboa, C., & Madonsela, Z. (2021). Environmental sustainability: Impact of construction activities. In *Collaboration and Integration in Construction, Engineering, Management and Technology: Proceedings of the 11th International Conference on Construction in the 21st Century, London 2019* (pp. 229–234). Springer International Publishing.

Quilliam, R. S., Pow, C. J., Shilla, D. J., Mwesiga, J. J., Shilla, D. A., & Woodford, L. (2023). Microplastics in agriculture – A potential novel mechanism for the delivery of human pathogens onto crops. *Frontiers in Plant Science*, *14*, 1152419.

Rahmati, F., Sethi, D., Shu, W., Lajayer, B. A., Mosaferi, M., Thomson, A., & Price, G. W. (2024). Advances in microbial exoenzymes bioengineering for improvement of bioplastics degradation. *Chemosphere*, *355*, 141749.

Ruini, C., Neri, P., Cavalaglio, G., Coccia, V., Cotana, F., Raspolli Galletti, A. M., …, & Rosa, R. (2023). Innovative bioplasticizers from residual *Cynara cardunculus* L. biomass-derived levulinic acid and their environmental impact assessment by LCA methodology. *ACS Sustainable Chemistry & Engineering*, *11*(32), 12014–12026.

Said, M. A., Hassan, F. U., Islam, M. S., Toriman, M. E. B., Rehim, A., Bashir, M. A., ... & Nawaz, R. (2024). Environmental sustainability. In *Global Challenges for the Environment and Climate Change* (pp. 1–15). IGI Global.

Savva, K., Borrell, X., Moreno, T., Pérez-Pomeda, I., Barata, C., Llorca, M., & Farré, M. (2023). Cytotoxicity assessment and suspected screening of plastic additives in bioplastics of single-use household items. *Chemosphere*, *313*, 137494.

Schnabl, K. B., Mandemaker, L. D., Ganjkhanlou, Y., Vollmer, I., & Weckhuysen, B. M. (2024). Green additives in chitosan-based bioplastic films: Long-term stability assessment and aging effects. *ChemSusChem*, *17*, e202301426.

Schnabl, K. B., Mandemaker, L. D., Nierop, K. G., Deen, O. V., Eefting, D. D., Vollmer, I., & Weckhuysen, B. M. (2023). Green additives in chitosan-based bioplastic films: Physical, mechanical, and chemical properties. *ChemSusChem*, *16*(20), e202300585.

Serrano-Aguirre, L., & Prieto, M. A. (2024). Can bioplastics always offer a truly sustainable alternative to fossil-based plastics?. *Microbial Biotechnology*, *17*(4), e14458.

Shah, Y. A., Bhatia, S., Al-Harrasi, A., & Khan, T. S. (2024). Advancements in the biopolymer films for food packaging applications: A short review. *Biotechnology for Sustainable Materials*, *1*(1), 2.

Sharma, R., & Rao, P. (2019). Relevance of impact studies on the environmental impacts of tourism and sustainability: A review and analysis. *Environmental Impacts of Tourism in Developing Nations*, 1–21.

Shi, J., Li, M., Tong, B., & Guo, Z. (2022). A new control strategy for greenhouse environment control system based on inverse model. *International Journal of Heat & Technology*, *40*(5), 1271.

Singhal, B., Kumar, S., Mehtab, S., Aggarwal, U., Kumar, V., Umesh, M., & Chakraborty, P. (2024). Biotechnological advancements in microplastics degradation in drinking water: Current insights and Future perspectives. *Case Studies in Chemical and Environmental Engineering*, *9*, 100640.

Sintim, H. Y., & Flury, M. (2017). Is biodegradable plastic mulch the solution to agriculture's plastic problem? *Environmental Science & Technology, 51,* 1068–1069.

Skoczinski, P., Krause, L., Raschka, A., Dammer, L., & Carus, M. (2021). Current status and future development of plastics: Solutions for a circular economy and limitations of environmental degradation. In *Methods in enzymology* (Vol. 648, pp. 1–26). Academic Press.

Sprung, P. D. (1991). U.S. Patent No. 5,001,859. Washington, DC: U.S. Patent and Trademark Office.

Takakura, T. (2019). Research exploring greenhouse environment control over the last 50 years. *International Journal of Agricultural and Biological Engineering*, *12*(5), 1–7.

Weng, Y., Wang, X., Hua, J., Wang, H., & Kang, M. (2020, July). Greenhouse environment control based on computational experiments. In *Proceedings of the 2020 4th High Performance Computing and Cluster Technologies Conference & 2020 3rd International Conference on Big Data and Artificial Intelligence* (pp. 238–245).

Xia, C., Lam, S. S., Zhong, H., Fabbri, E., & Sonne, C. (2022). Assess and reduce toxic chemicals in bioplastics. *Science*, *378*(6622), 842–842.

Yaakub, S., Subhan, M., Abdullah, N. A. H. N., & Gapar, N. (2015, August). Environmental sustainability strategies and impacts. In *2015 International Symposium on Technology Management and Emerging Technologies (ISTMET)* (pp. 401–406). IEEE.

Yadav, P., Singh, J., Srivastava, D. K., & Mishra, V. (2021). Environmental pollution and sustainability. In *Environmental Sustainability and Economy* (pp. 111–120). Elsevier.

Zarubina, A. N., Ivankin, A. N., Serdyukova, Y. V., & Oliferenko, G. L. (2023). Bioplastics based on copolymers of glycerin, phthalate anhydride, and lactic acid. *Polymer Science, Series D*, *16*(2), 410–414.

Zellmer, W. A. (2005). Unresolved issues in pharmacy. *American Journal of Health-System Pharmacy*, *62*(3), 259–265.

Zhang, B., Xie, F., Shamshina, J. L., Rogers, R. D., McNally, T., Wang, D. K., …, & Chen, L. (2017). Facile preparation of starch-based electroconductive films with ionic liquid. *ACS Sustainable Chemistry & Engineering*, *5*(6), 5457–5467.

Zhong, W., Xu, P., Niu, D., Wang, Q., Yang, W., Zhang, X., …, & Ma, P. (2024). Enhanced flame retardancy, ultraviolet shielding, and preserved mechanical properties of polylactic acid with fully biobased multifunctional additives by a green method. *ACS Sustainable Chemistry & Engineering*, *12*(10), 4017–4027.

Zhu, Z., Bi, C., & Tang, Y. (2024). Investigating precise decision-making in greenhouse environments based on intelligent optimization algorithms. *Processes*, *12*(5), 977.

Zulkafli, N. N. (2014). *Production of Bioplastic from Agricultural Waste* (Doctoral dissertation, UMP).

9 Role of Bioplastics in Soil Contamination and Soil Health

Stella Suanu Leh-Togi Zobeashia and
Gloria Imoh Bassey Obioh

9.1 INTRODUCTION

Pollution from plastics resulting from increased demand for plastic production and utilization is a major global challenge that poses severe ecological and health issues (Koelmans *et al.*, 2019). Plastic pollution has become a prevalent environmental problem, with an estimated 400 million metric tons of plastic produced annually worldwide (Lau *et al.*, 2020). Although focus historically was on marine pollution, terrestrial ecosystems, particularly soil, bear a significant and often underappreciated burden. Approximately 80% of global plastic waste ends up in landfills, agricultural soils, and natural landscapes, where it accumulates and contributes to long-term contamination and damage through disruption of the soil's biological and physicochemical properties (Chae & An, 2018, 2020; Rillig *et al.*, 2019). The growing concern about the environmental impact of conventional plastics has led to the need for an eco-friendly alternative. Bioplastics, having emerged as a promising alternative to conventional petrochemical-based plastics, are designed to address both the environmental and sustainability challenges posed by plastic pollution. Bioplastics are derived from renewable biomass sources, such as cornstarch, sugarcane, and vegetable oils, and are often marketed as biodegradable or compostable (Rujnic-Sokele & Pilipovic, 2017). As the global bioplastics market expands rapidly with production projected to reach 7.5 million tons by 2026 (Market Research Future, 2023), critical evaluation of their role in soil contamination, particularly in agricultural systems, becomes paramount. In agriculture, bioplastics are increasingly used in mulch films, controlled-release fertilizers, and packaging materials. Their biodegradability is a key advantage, as they are designed to decompose into carbon dioxide (CO_2), water, and biomass under appropriate environmental conditions (Haider *et al.*, 2019; Shen *et al.*, 2023). However, the soil environmental performance of bio-based plastics is not without challenges. Abiotic factors and microbial activity significantly influence their susceptibility and degradation rates. As estimated by European Bioplastics (2020), almost half of bioplastic polymers were not exposed for biological decomposition in the soil environment. Despite these concerns, bioplastics hold considerable potential to reduce plastic pollution compared to petroleum-based plastics when coupled with proper waste management practices. Bioplastics, synthesized from starch, cellulose, polylactic acid (PLA), or polyhydroxyalkanoates (PHAs), offer sustainable alternatives due to their biodegradability under certain conditions, which positions them as a potential solution to mitigate global plastic pollution (Emadian *et al.*, 2017; Yang *et al.*, 2022). However, the degradation of bioplastics in natural soil environments is highly variable and influenced by factors such as moisture, soil temperature, microbial activity, and the composition of the bioplastics (Huang *et al.*, 2023). For example, while some bioplastics like PHAs degrade efficiently in soil, others like PLA require industrial composting conditions for complete breakdown (Tokiwa *et al.*, 2009; Chen *et al.*, 2023).

DOI: 10.1201/9781003602828-11

Ironically, the improper use or disposal of bioplastics can result in soil contamination similar to that of fossil-derived plastics. Studies reported by Li *et al.* (2023) show that partially degraded bioplastics can result in microplastic (MP) (plastic particles <5 mm in size) formation and contribute to soil contamination, particularly in soils with suboptimal conditions for biodegradation (Qi *et al.*, 2022; Li *et al.*, 2023). These MPs can persist in the environment, posing risks to soil structure, water retention, and microbial diversity. Furthermore, some of the bioplastics in the global market contain modifiers/additives used to improve the physicochemical properties of the polymer. The additives reduce flammability, resist ultraviolet radiation, and increase pliability (Lambert *et al.*, 2014). When incorporated into the polymer during production, they can change the biological properties such as biodegradability and toxicity (Niaounakis, 2013; Rillig *et al.*, 2024). These additives (plasticizers, stabilizers, dyes, antioxidants, and pigments) can leach into the soil, environment, introducing potentially toxic substances that disrupt soil health and ecosystem services (Qi *et al.*, 2022; Yang *et al.*, 2022). While bioplastics offer an eco-friendly benefit to the environment, including reduction of greenhouse gas (GHG) emissions during production and reduce dependent on fossil fuels, their role in soil contamination remains complex issue that definitely reflects the challenge of managing plastic pollution in a sustainable manner. Bioplastics positively influence soil ecosystems through contribution to soil organic carbon, stimulation of microbial activity, release of essential nutrients, and improvement of soil structure. To optimize their benefits and mitigate adverse impacts (physical disruption, chemical alterations, and biological effects on soil organisms), further research is necessary, particularly on field-based degradation processes, additive safety, their cumulative effects on soil health, and proper management to fully realize their potential as an environmentally friendly alternative.

9.2 BIOPLASTIC AND SOIL CONTAMINATION

Bioplastics are promoted as eco-friendly alternatives to conventional plastics, with their biodegradability being a key advantage. However, their production, usage, and degradation can contribute to soil contamination under certain conditions. Soils have always been recognized as significant reservoir for bioplastics, with several studies exploring their effects on soil ecosystems (Yang *et al.*, 2018, 2022). An understanding of the sources of bioplastic contamination and toxicity requires assessing their role in soil. Bioplastics can enter soil environments through several pathways, including agricultural applications, improper disposal, and MP formation.

9.2.1 Agricultural Application

The major sources of contamination of bioplastics in agricultural soils include residual plastic fragments from mulching, sewage sludge, atmospheric deposition, municipal waste-derived compost, and plastic-coated fertilizers (Rodriguez-Seijo *et al.*, 2017; Blasing & Amelung, 2018; Liu *et al.*, 2018, 2024). Among these, agricultural plastic films used for mulching have been widely identified as the largest contributor (Hurley & Nizzetto, 2018; Qi *et al.*, 2022). Research by Yang *et al.* (2022) found that bioplastics used in agricultural mulch films contributed to soil contamination through residual fragments after harvest. These residues impacted crop productivity by altering soil texture and microbial activity. In China, He *et al.* (2018) reported a dramatic increase in the use of plastic film for mulching in agriculture, growing from 6,000 tons in 1982 to over 1.2 million tons in 2011. This surge is linked to China's agricultural intensification, where mulch films are applied to cover over 20 million hectares of farmland annually. The major setback is that recycling system is limited due to high labor costs, thereby leaving residual plastic in the soil to degrade into MPs through physical (UV radiation), chemical (oxidation), and biological (microbial action) processes. This creates long-term contamination risks.

Sewage sludge (biosolids) application as fertilizer can also increase bioplastic contamination in the soil environment. This treated sewage sludge contains significant amounts of fragments of plastics due to untreated wastewater input from urban area (Rodriguez-Seijo *et al.*, 2017), which shows an elevated MP concentrations, particularly polyethylene (PE) and polystyrene, which resist degradation and accumulate over time, potentially affecting soil fauna and microbial communities. In addition, atmospheric deposition in urban areas with airborne MPs concentration can also result in an upsurge of bio-based MP contamination in the soil, as highlighted by Liu *et al.* (2018). For instance, in Paris, atmospheric deposition was found to deliver over 300 particles per square meter per day, some of which inevitably settled on agricultural soils. In Germany, as documented by Blasing and Amelung (2018), plastic contamination on the soil environment originated from urban waste streams, including packaging materials in compost used as soil amendment, which spread over large tracts of farmland, posing risks to water infiltration and soil health. In the United States, Miles *et al.* (2017) studied slow-release fertilizers coated with plastics, commonly used in agriculture. These plastic coatings break down into microplastics, contributing to soil contamination over time. The use of such fertilizers, while improving crop productivity, inadvertently adds MP to soils, with limited solutions to recover or degrade the particles efficiently.

9.2.2 Improper Disposal

Bioplastic packaging materials and waste can end up in the soil environment, where their degradation rates depend on environmental factors (Rillig *et al.*, 2019; Zhang *et al.*, 2024). Improper disposal of bioplastics accounts for nearly 32% of all plastic waste globally, leading to environmental degradation (Geyer *et al.*, 2017; Chen *et al.*, 2023). Inadequate waste management systems, particularly in developing nations, exacerbate the issue as uncollected plastics are often dumped in open soil environments and water bodies. Annually, an estimated 8 million tons of plastic waste enter ecosystems (Jambeck *et al.*, 2015). Improper disposal of plastics not only harms ecosystems but also poses risks to human health. Some bioplastics leach hazardous chemicals like bisphenol A (BPA) and phthalates into the environment, which can contaminate food and water supplies and MPs are also found in the food chain especially in seafood, drinking water, salt, which raises concern about the health implications for humans (Smith *et al.*, 2018). This can be handle by developing multi-prolonged approach that includes enhancing recycling systems and implementing extended producer responsibility (EPR) programs to reduce plastic waste and create public awareness via educational campaigns to encourage responsible consumption and disposal practice. Governments can intervene by developing policy that must enforce bans on single-use plastics and incentivize the use of biodegradable alternatives.

9.2.3 Microplastic Formation

Bioplastics, despite being marketed as biodegradable, may not always fully degrade under natural soil conditions. Partial degradation can result in the formation of MPs, which persist in the soil and disrupt its structure, aeration, and water retention capacity (Rillig, 2018; Huang *et al.*, 2023). MPs are defined as plastic particles that are smaller than 5 mm and are increasingly recognized as significant environmental pollutants. While much attention has been given to their presence in aquatic environments, MP in soil ecosystems is a growing concern. The formation of MPs in soils primarily results from the degradation of larger plastic fragments, improper plastic waste disposal, and specific agricultural practices. This incomplete biodegradation of bioplastics contributes to MP contamination, similar to conventional plastics (Luo *et al.*, 2022), and has profound effects on soil health, nutrient cycles, and biodiversity. For instance, PLA, a commonly used bioplastic, requires industrial composting conditions for complete degradation, leaving residues when disposed of in soil (Qi *et al.*, 2022; Qi *et al.*, 2024). MP are formed in the soil when degradation of large plastic items such as mulch films from agriculture, packaging waste, and irrigation pipes degrade under physical processes such as UV radiation and temperature fluctuations, causing fragmentation of bioplastic. The chemical processes (oxidation and hydrolysis) promote its breakdown further decomposes into smaller particles size by microbial activity (He *et al.*, 2018). For instance, soil with a concentration of 0.1% MP is considered as polluted (Wang *et al.*, 2019). Cordova *et al.* (2019) reported that open dumping of plastic waste in nearby soils contributed to the formation of MPs, which leached into the soil and adjacent water systems, affecting agricultural productivity and leading to MP contamination in Indonesia.

The ecological impact of soil polluted with high concentrations of fragment of bio-based MPs cannot be overlooked. As documented by Li *et al.* (2020), incomplete degraded of some types of bioplastics can lead to bioturbation by migrating deeper into the soil profile, altering its characteristics, and disrupting its physical structure (Hurley & Nizzetto, 2018; Nizzetto *et al.*, 2024). This intrusion often results in blocked soil pores, which hampers essential processes such as water infiltration, nutrient cycling, and gas exchange and root penetration (Rillig, 2017; Liu *et al.*, 2018). Moreover, MPs can disrupt microbial communities by changing the composition and activity of soil microbes and interfere with microbial interactions, which are essential for organic matter decomposition and nutrient availability (Liu *et al.*, 2018; He *et al.*, 2018; Dong *et al.*, 2021; Rillig *et al.*, 2024).

MPs act as vectors for harmful pollutants such as heavy metals, pesticides, and persistent organic pollutants (POPs), intensifying their toxic effects on soil ecosystems (Rillig, 2017; Avio *et al.*, 2015; Dong *et al.*, 2021). Agricultural and urban soils are particularly susceptible due to activities like mulching, where residual plastic fragments degrade into MPs over time (Hurley & Nizzetto, 2018; Qi *et al.*, 2022). These biodegradable MPs can bind with other contaminants, forming complexes that exacerbate soil toxicity and disrupt microbial dynamics and plant health by hindering the growth of plants and reducing crop productivity by affecting the soil fertility, leading to poor crop yields (Hurley & Nizzetto, 2018; Zhang *et al.*, 2024). The long-lasting and persistent nature of MPs poses threat to the ecosystems of the soil (Yang *et al.*, 2018).

9.3 ROLES OF BIOPLASTICS IN SOIL CONTAMINATION

Bioplastics, particularly the biodegradable variants such as PLA and PHAs, offer a sustainable alternative to conventional plastics in mitigating soil pollution. Unlike petroleum-based plastics that persist in the soil environment for decades (Hurley & Nizzetto, 2018). Bioplastics degrade under suitable environmental conditions, breaking down into natural byproducts such as CO_2, water, and biomass (Rillig, 2018; Haider *et al.*, 2019; Shen *et al.*, 2023). This capability plays a crucial role in reducing the accumulation of non-degradable plastic residues in soil, thereby enhancing soil cleanliness and promoting long-term ecosystem health. Bioplastics decompose through microbial activity, where microorganisms in the soil produce enzymes that break down the polymer chains (Cui *et al.*, 2018; Zhang *et al.*, 2024; Nizzetto *et al.*, 2024). For instance, PLA degrades primarily in industrial composting conditions (high temperature and humidity), but it also shows partial degradation in soil environments with sufficient microbial activity. PHA is more naturally biodegradable, breaking down completely in soil, marine, and freshwater environments due to its microbial compatibility. Studies conducted by Haider *et al.* (2019) evaluated the degradation of starch-based bioplastics and conventional PE in agricultural soils over a 12-month period and indicated the starch-based bioplastics degraded by 60–80% within a year, leaving no detectable residues in the soil. While, the conventional PE film remained intact, showing no signs of degradation and contributed to soil contamination. The study therefore demonstrates that the bioplastics significantly reduced long-term plastic residues, promoting cleaner soil conditions by its ability to degrade and prevent the buildup of macro- and MPs in the soil, which are known to disrupt soil structure, water retention, and aeration. The research by Brodhagen *et al.* (2014) to evaluate the efficacy and environmental impact of PLA-based mulch films versus PE films in vegetable farming reported that PLA mulch films decomposed entirely within six

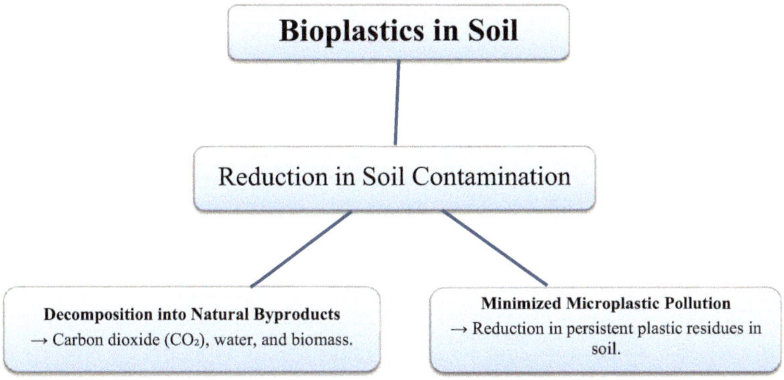

FIGURE 9.1 Role of bioplastics in soil contamination (Bandopadhyay *et al.*, 2018; Shen *et al.*, 2023).

months when left in the field post-harvest under compost-like conditions. No MP fragments were detected, and the soil's microbial activity improved due to the degradation byproducts but the PE mulch films required manual removal, with remnants often left in the soil, which therefore indicates that the PLA-based bioplastics reduced persistent plastic contamination while improving soil management practices. While the study documented by Qi *et al.* (2020) in the investigation of the degradation behavior of PHA in coastal soils compared to PE-reported PHA completely decomposed within ten months, leaving no residual fragments, while PE remained unchanged. The microbial diversity in the soil was unaffected or slightly enhanced due to PHA degradation. The study showed, PHA effectively reduces the burden of plastic waste in sensitive soil ecosystems.

Based on several studies, bioplastics offer a promising pathway to reduce soil contamination and improve sustainability through reduction of persistence plastic residue accumulation (Figure 9.1). Traditional plastics often break down into MPs (Rillig, 2018), which persist in the soil and disrupt its physical structure. Bioplastics, through their natural decomposition, eliminate this issue by leaving no long-term residues. Bioplastics also improved microbial activity in contrast to fossil-derived plastic, which hinders the activity of microbes (Liu *et al.*, 2018; He *et al.*, 2018; Dong *et al.*, 2021). The organic byproducts of bioplastic degradation stimulate soil microbial communities, which are essential for nutrient cycling and maintaining soil fertility. In addition, bioplastics enhance soil structure; the incorporation of biomass during bioplastic degradation improves soil aeration and water retention, fostering a healthier growing environment for plants (Luo *et al.*, 2022). By reducing persistent pollution, bioplastics contribute to more resilient and eco-friendly soil ecosystems (Lau, 2020; Luo *et al.*, 2022). However, further research and case-specific applications are necessary to optimize their benefits across different soil and environmental conditions.

9.4 ROLES OF BIOPLASTICS IN SOIL HEALTH

The degradation of bioplastics in soil environments can provide significant benefits to soil health and fertility. As bioplastics break down, they release organic byproducts such as CO_2, water, and biomass, which serve as substrates for soil microorganisms (Haider *et al.*, 2019; Yang *et al.*, 2022; Li *et al.*, 2024). This process not only enhances microbial activity but also improves soil structure, aeration, and water retention in contrast to conventional plastics (Luo *et al.*, 2022; Huang *et al.*, 2023). Certain bioplastic formulations may also release essential nutrients, directly promoting plant growth and boosting agricultural productivity. The significant roles of bioplastics in soil health and fertility include the addition of soil organic matter, the activation of microbial activity, the improvement of soil structure and water retention, the delivery of essential nutrients, and sustainable agriculture (Luo *et al.*, 2022; Yang *et al.*, 2022; Li *et al.*, 2024; Rillig *et al.*, 2024).

9.4.1 Addition to Soil Organic Matter

Organic matter is a critical component of fertile soils, influencing nutrient availability, soil structure, and water retention. Bioplastics made from natural polymers like starch, cellulose, and PLA decompose into organic matter that integrates seamlessly into the soil matrix. Study by Haider *et al.* (2019) reported a 70% degradation of starch-based bioplastics applied to agricultural plots within six months of the stipulated one year for monitoring the changes in the soil quality. The result shows that the bioplastic contributed significant organic matter that increased soil carbon content by 15%, which therefore suggest the bioplastic-enhanced soil aeration and microbial biomass and directly benefited the plant root systems (Figure 9.2).

9.4.2 Activation of Soil Microbial Activity

Microbial activity plays a central role in nutrient cycling and soil health. The degradation products of bioplastics, particularly PLA and PHAs, act as energy sources for inherent soil microorganisms, promoting their growth and activity (Bandopadhyay *et al.*, 2018; Huang et al., 2023; De *et al.*, 2019). Brodhagen *et al.* (2014) recorded a 25% increase of microbial biomass in soils with PLA mulch films compared to soil without mulch after one growing season during the study of the impact of PLA-based biodegradable mulch films on microbial activity in a vegetable farming system. The result therefore suggest that the degradation of PLA films released organic acids that stimulated microbial proliferation, improving soil respiration rates and nutrient availability, which is attributed to the mechanism that bioplastic-derived organic matter facilitates the growth of decomposer microbes, such as fungi and bacteria, which further break down soil organic residues, enriching soil fertility and health.

9.4.3 Delivers Essential Nutrients

Some bioplastics are formulated with renewable and biodegradable nutrient – rich materials such as starch- or protein-based addictive (soy protein, casein, gelatin, whey, zein), which helps to improve mechanical strength, flexibility, and biodegradability properties of the bioplastic making it versatile for various applications (Liu *et al.*, 2024). These additives release nutrients like nitrogen (N) and phosphorus (P) upon degradation (Mao *et al.*, 2024). Shirai *et al.* (2023) found that the study of bioplastics infused with urea applied to maize fields as mulch indicated the bioplastics degraded fully within one crop cycle, releasing nitrogen into the soil. Crop yields increased by 18% compared to fields with PE mulch. This implies nutrient-infused bioplastics that not only reduce plastic pollution but also function as slow-release fertilizers, combining environmental and agricultural benefits.

9.4.4 Improve Soil Structure and Water Retention

Bioplastics, when degraded or decomposed in the soil, can have several positive effects on soil structure and functionality especially in sandy or degraded soil. According to Qi *et al.* (2020), degradation of PHA-based bioplastics in sandy soil prone to water loss, increased soil water retention by 12% and reduced compaction, improving root penetration and crop resilience during drought conditions. The biodegradation process of bioplastics involves physical fragmentation, microbial assimilation, and chemical transformations, all of which contribute to soil aggregation, water retention, and erosion resistance in soil.

9.4.4.1 Contribution to Soil Aggregation

Decomposed bioplastics release organic carbon into the soil, which acts as a binding agent, promoting the formation of soil aggregates. Organic matter enhances the cohesiveness of soil particles, improving aggregation (Shen *et al.*, 2023). For instance, starch-based bioplastics provide readily available carbon for microbial processes, leading to the production of polysaccharides and other organic compounds that stabilize aggregates (Qi *et al.*, 2020, 2024).

The degradation process stimulates microbial populations, particularly those capable of metabolizing bioplastic byproducts. Microbes excrete extracellular polymeric substances (EPS), which act as "glues," binding soil particles into stable aggregates (Luo *et al.*, 2022; Huang *et al.*, 2023; Rillig *et al.*, 2024). The enhanced microbial activity around decomposed bioplastics supports the proliferation of fungi and bacteria that contribute to aggregate formation through physical entanglement and biochemical bonding (Luo *et al.*, 2022).

9.4.4.2 Contribution to Water Retention

Bioplastic fragments increase soil porosity by creating void spaces, which enhance water infiltration and retention. As the fragments degrade, they integrate with soil particles, forming micro- and macropores that trap water (Lozano *et al.*, 2021). A study on PHAs reported that soils with decomposed bioplastics retained 15% more water compared to control soils due to enhanced porosity and aggregation (Häkkinen *et al.*, 2023). In addition, the degraded byproducts, such as organic acids, enhance the soil's capacity to retain water by improving its cation exchange capacity (CEC). This helps soil particles hold onto water molecules, making them more available for plant uptake (Luo *et al.*, 2022; Huang *et al.*, 2023; Bandopadhyay et al., 2018).

9.4.4.3 Contribution to Erosion Resistance

Stable aggregates formed from bioplastic degradation resist disintegration under water or wind stress. A study showed that soils enriched with PLA degradation products exhibited a 25% increase in aggregate stability, reducing susceptibility to erosion (Shen *et al.*, 2023; Chen *et al.*, 2023). Bioplastics also support vegetation. Enhanced soil water retention from bioplastics improves plant growth, leading to more robust root systems; roots physically anchor soil particles, reducing erosion risks (Luo *et al.*, 2022). Microbial exudates resulting from bioplastic decomposition further bind soil particles, reinforcing the soil matrix against erosive forces.

9.5 SUSTAINABLE AGRICULTURE

Bioplastics, especially those designed for agricultural use, offer dual benefits of environmental protection and soil enhancement such as the biodegradable mulch films and the seed coatings (Qi *et al.*, 2022; Chaudhary *et al.*, 2024). These films suppress weeds, retain soil moisture, and decompose to contribute organic carbon and nutrients to the soil while the bioplastic-based seed coatings improve germination rates while degrading into beneficial organic matter in the Rhizosphere (Li *et al.*, 2024). In Italy, biodegradable mulch films were adopted across several agricultural regions, significantly decreasing plastic contamination and enhancing soil sustainability (European Bioplastics Report, 2023). The integration of bioplastics into agricultural systems helps enhance soil health, reduce pollution, and improve resource efficiency. Some of the ways in which bioplastic contribute to sustainable agriculture include the following.

9.5.1 Reduction of Pollution

Bioplastics replace non-biodegradable plastics in agricultural applications like mulch films, reducing long-term soil contamination by MPs. For instance, starch-based bioplastics degrade faster leaving minimal residues (Li *et al.*, 2023; Yang *et al.*, 2023; Chaudhary *et al.*, 2024). Bioplastics likewise lower ecotoxicity. Unlike petroleum-based plastics, bioplastics are designed to degrade into non-toxic byproducts such as water, CO_2, and biomass. This minimizes harm to soil organisms and plants while supporting a closed-loop agricultural system (Luo *et al.*, 2022).

9.5.2 Water Management

Bioplastics contribute to water retention by reducing water loss through evaporation and ensure a steady water supply to crops, particularly in arid regions (Lozano *et al.*, 2021). They control release systems applied in controlled-release fertilizers and pesticide coatings. These systems allow the slow and targeted delivery of nutrients and agrochemicals, reducing runoff and improving resource efficiency (Rillig, 2017; Qi *et al.*, 2022).

9.5.3 Circular Agriculture

Bioplastics are derived from biological-based renewable feedstocks such as corn, sugarcane, and cellulose. This reduces reliance on fossil fuels and integrates agricultural waste into the production cycle, promoting a circular economy (Bandopadhyay et al., 2018). The compostable properties of some of this imply; they can return nutrients to the soil after degradation. This eliminates waste and supports sustainable waste management practices on farms (Bandopadhyay *et al.*, 2018; European Bioplastics, 2023a,b).

9.5.4 Climate Resilience

Bioplastics are primarily produced from biological feedstocks that absorb atmospheric CO_2 during photosynthesis, offsetting the carbon footprint associated with bioplastic production. This "carbon sequestration" process contributes to a net reduction in GHG emissions compared to conventional petroleum-based plastics (Horton *et al.*, 2017; Shen *et al.*, 2023). European Bioplastics (2023a) reported that, PLA, a common bioplastic, has a carbon footprint that is approximately 60% lower than that of traditional PE when considering its life cycle.

The production of conventional plastics involves the extraction and processing of fossil fuels, leading to significant CO_2 and methane (CH_4) emissions. Bioplastics, by contrast, minimize reliance

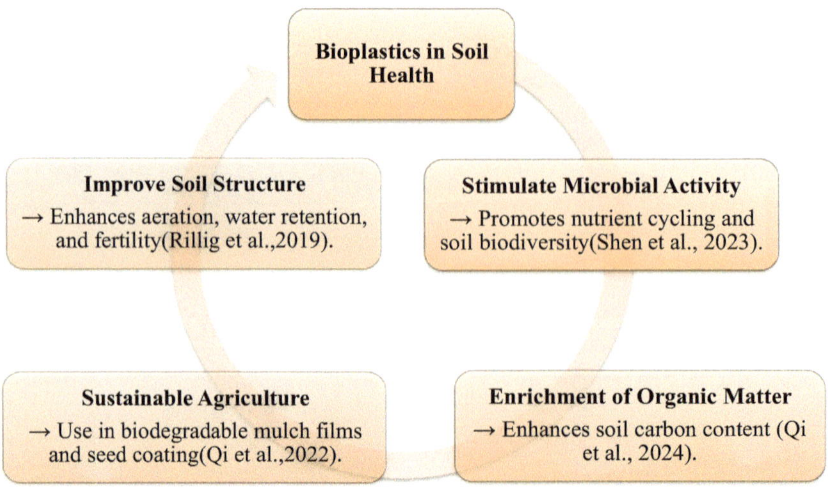

FIGURE 9.2 Role of bioplastic in soil health.

on fossil fuels, thereby lowering associated emissions (Rillig, 2018). It also saves energy, its production processes generally require less energy compared to the energy-intensive cracking and polymerization of fossil fuels (Rillig, 2018; Li *et al.*, 2023, 2024). The biodegradability potential of bioplastics, particularly compostable types like PLA and PHA, decompose into natural substances such as water, CO_2, and biomass under controlled conditions. This biodegradability reduces methane emissions typically generated by conventional plastics in anaerobic landfill conditions. Proper composting of bioplastics prevents the accumulation of plastics in landfills, reducing the formation of methane, a potent GHG (Qi *et al.*, 2020, 2022, 2024).

Due to the plant-based feedstock used for production, the lifecycle emissions are reduced compared to fossil-derived plastic. Bioplastics are also compatible with recycling systems, reducing the need for virgin material production and associated emissions (Luo *et al.*, 2022). Furthermore, some bioplastics, like those derived from algae, have the potential to be carbon-neutral or even carbon-negative. Algae-based feedstocks absorb large amounts of CO_2 during growth and require minimal land and water resources, making them an environmentally favorable option for reducing GHG emissions (Qi *et al.*, 2024).

9.5.5 Erosion Control and Innovative Agriculture

Bioplastic-based geotextiles and soil covers stabilize soil in vulnerable areas, preventing erosion and desertification. This enhances the resilience of agricultural land to extreme weather events (Rillig *et al.*, 2024). They are also used as seed coatings to protect seeds and enhance germination rates. These coatings degrade naturally, delivering micronutrients and reducing planting losses (Qi *et al.*, 2024). In addition, emerging innovative technologies such as integrating bioplastics into biodegradable sensors monitors soil moisture, pH, and nutrient levels. These sensors are decomposing after use, minimizing electronic waste, and supporting precision farming practices (Qi *et al.*, 2022; Shen *et al.*, 2023).

9.6 CONCLUSION

Bioplastics provide a viable solution to some of the most pressing environmental challenges associated with conventional plastics. By reducing persistent pollution, enhancing soil health through organic degradation byproducts, and supporting sustainable agricultural practices, bioplastics contribute to more resilient and eco-friendly soil ecosystems. However, further research and case-specific applications are necessary to optimize their benefits across different soil and environmental conditions.

REFERENCES

Avio, C. G., Gobri, S., Milan, M., Benedetti, M., Fattorini, D., D'errico, G., Pauletto, M., Bargelloni, L., & Regoli, F. (2015). Pollutants bioavailability and toxicological risk from micro plastics to marine mussels. *Environmental Pollution*, 198, 211–222.

Bandopadhyay, S., Martin-Closas, L., Pelacho, A. M., & DeBruyn, J. M. (2018). Biodegradable plastic mulch films: Impacts on soil microbial communities and ecosystem functions. *Frontiers in Microbiology*, 9, 819. https://doi.org/10.3389/FMICB.2018.00819

Bläsing, M., & Amelung, W. (2018). Plastics in soil: Analytical methods and possible sources. *Science of the Total Environment*, 612, 422–435.

Brodhagen, M., Moore-Kucera, J., Lee, J., Corbin, A., Miles, C., & Inglis, D. (2014). Effects of biodegradable mulch on soil quality. *Applied Soil Ecology*, 79, 59–69.

Chae, Y., & An, Y.-J. (2018). Current research trends on plastic pollution and ecological impacts on the soil ecosystem: a review. *Environmental Pollution*, 240, 387–395. https://doi. org/10.1016/j.envpol.2018.05.008

Chae, Y., & An, Y. J. (2020). Current research trends on plastic pollution and ecological impacts. *Environmental Science & Technology*, 54(8), 4709–4721.

Chaudhary, S., Kour, M., & Kumar, R. (2024). Bioplastic films from starch of *Colocasia esculenta* and its waste: A smart template for sensing applications. *International Journal of Biological Macromolecules*, 281, 136218.

Chen, G. Q., & Patel, M. K. (2023). Polyhydroxyalkanoates: Sustainable plastic for the future. *Nature Reviews Materials*, 8, 1–18.

Cordova, M. R., Purwiyanto, A. I. S., & Suteja, Y. (2019). Abundance and characteristics of microplastics in the northern coastal waters of Surabaya, Indonesia. *Marine Pollution Bulletin*, 142, 183–188.

Cui, Y., Fang, L., Guo, X., Wang, X., & Wang, Y., (2018). Responses of soil bacterial communities, enzymatic activities, and nutrient to agricultural-to-natural ecosystem conversion in the loess plateau, China. *Journal of Soils and Sediments*, 19, 1427–1440.

de Souza Machado, A. A., Lau, C. W., Kloas, W., Bergmann, J., Bachelier, J. B., Faltin, E., Becker, R., Görlich, A. S., & Rillig, M. C., (2019). Microplastics can change soil properties and affect plant performance. *Environmental Science & Technology*, 53, 6044–6052.

Dong, Y., Gao, M., Qiu, W., & Song, Z., (2021). Effect of microplastics and arsenic on nutrients and microorganisms in rice rhizosphere soil. *Ecotoxicology and Environmental Safety*, 211, 111899.

Emadian, S. M., Onay, T. T., & Demirel, B. (2017). Biodegradation of bioplastics in natural environments. *Waste Management*, 59, 526–536.

European Bioplastics & nova-Institute. (2020). Bioplastics market development update 2020. *Bioplastics market development update 2020: Bioplastics continue to become mainstream as the global bioplastics market is set to grow by 36 percent over the next 5 years*. Berlin, Germany: European Bioplastics. www.european-bioplastics.org/market-update-2020-bioplastics-continue-to-becomemainstream-as-the-global-bioplastics-market-is-set-to-grow-by-36-percent-overthe-next-5-years/

European Bioplastics & nova-Institute. (2023a). *Bioplastics market development update 2023: Global bioplastics market data* [PDF]. Berlin, Germany: European Bioplastics. Retrieved from https://docs.european-bioplastics.org/publications/market_data/2023/EUBP_Market_Data_Report_2023.pdf

European Bioplastics. (2023b). What are bioplastics? Retrieved from www.european-bioplastics.org/

Geyer, R., Jambeck, J. R., & Law, K. L. (2017). Production, use, and fate of all plastics ever made. *Science Advances*, 3(7), e1700782. https://doi.org/10.1126/sciadv.1700782

Haider, T. P., Völker, C., Kramm, J., Landfester, K., & Wurm, F. R. (2019). Plastics of the future? The impact of biodegradable polymers on the environment and on society. *Angewandte Chemie International Edition*, *58*(1), 50–62.

Häkkinen, T., et al. (2023). Impact of bioplastics on soil water retention. *Science of the Total Environment*, 857, 159883.

He, D., Luo, Y., Lu, S., Liu, M., Song, Y., & Lei, L., (2018). Microplastics in soils: Analytical methods, pollution characteristics and ecological risks. *Trends in Analytical Chemistry*, 109, 163–172.

Horton, A. A., Walton, A., Spurgeon, D. J., Lahive, E., & Svendsen, C. (2017). Microplastics in freshwater and terrestrial environments: Evaluating the current understanding to identify the knowledge gaps and future research priorities. *Science of the Total Environment*, 586, 127–141.

Huang, M., Zhu, Y., Chen, Y., & Liang, Y. (2023). Microplastics in soil ecosystems: soil fauna responses to field applications of conventional and biodegradable microplastics. *Journal of Hazardous Materials*, 441, 129943.

Hurley, R. R., & Nizzetto, L., (2018). Fate and occurrence of micro (nano) plastics in soils: Knowledge gaps and possible risks. *Current Opinion in Environmental Science & Health*, 1, 6–11.

Jambeck, J. R., Geyer, R., Wilcox, C., Siegler, T. R., Perryman, M., Andrady, A., …, & Law, K. L. (2015). Plastic waste inputs from land into the ocean. *Science*, 347(6223), 768–771.

Koelmans, A. A., Mohamed Nor, N. H., Hermsen, N., Kooi, M., Mintenig, S. M., De France, J., (2019). Microplastics in fresh waters and drinking water: Critical review and assessment of data quality. *Water Research*, 155, 410–422.

Lambert, S., Sinclair, C. J., & Boxall, A. B. (2014). Occurrence, degradation and effect of polymerbased materials in the environment. *Reviews of Environmental Contamination and Toxicology*, 227, 1–53. https://doi.org/10.1007/978-3-319-01327-5_1

Lau, W. W., Shiran, Y., Bailey, R. M., Cook, E., Stuchtey, M. R., Koskella, J., …, & Palardy, J. E. (2020). Evaluating scenarios toward zero plastic pollution. *Science*, 369(6510), 1455–1461.

Li, J., Song, Y., & Cai, Y. B., (2020). Focus topics on microplastics in soil: Analytical methods, occurrence, transport, and ecological risks. *Environmental Pollution*, 257, 113570.

Li, X., Wu, J., Cheng, X., Cai, Z., Wang, Z., & Zhou, J. (2024). Biodegradable microplastics reduce the effectiveness of biofertilizers by altering rhizospheric microecological functions. *Journal of Environmental Management*, 352, 120071.

Li, Y., Gu, P., Zhang, W., Sun, H., Wang, J., Wang, L., …, & Wang, L. (2023). Effects of biodegradable and non-biodegradable microplastics on bacterial community and PAHs natural attenuation in agricultural soils. *Journal of Hazardous Materials*, 449, 131001.

Liu, M., Lu, S., Song, Y., Lei, L., Hu, J., Lv, W., Zhou, W., Cao, C., Shi, H., Yang, X., & He, D. (2018). Microplastic and mesoplastic pollution in farmland soils in suburbs of Shanghai, China. *Environmental Pollution*, 242, 855–862.

Liu, T., Liu, Z., Gu, W., Zhang, J., Gong, S., & Li, J. (2024). Ultratough nacre-inspired soybean protein isolate/graphene nanocomposite with flame-retardant, thermal conductivity and recyclable. *Composites Part B: Engineering*, 291, 111998.

Lozano, Y. M., Lehnert, T., Linck, L. T., Lehmann, A., & Rillig, M. C. (2021). Microplastic shape, polymer type, and concentration affect soil properties and plant biomass. *Frontiers in Plant Science*, *12*, 616645.

Luo, H., Liu, C., He, D., Xu, J., Sun, J., Li, J., & Pan, X. (2022). Environmental behaviors of microplastics in aquatic systems: A systematic review on degradation, adsorption, toxicity and biofilm under aging conditions. *Journal of Hazardous Materials*, 423, 126915.

Mao, M., Ahrens, L., Luka, J., Contreras, F., Kurkina, T., Bienstein, M., …, & Schwaneberg, U. (2024). Material-specific binding peptides empower sustainable innovations in plant health, biocatalysis, medicine and microplastic quantification. *Chemical Society Reviews*. *53*, 6445–6510. https://doi.org/10.1039/D2CS00991A

Market Research Future. (2023). Bioplastics market analysis and forecast 2023–2027. (Report No. MRFR-1964, 111 pp.). Market Research Future.

Miles, C., DeVetter, L., Ghimire, S., & Hayes, D.G. (2017). Suitability of biodegradable plastic mulches for organic and sustainable agricultural production systems. *HortScience*, 52, 10–15.

Niaounakis, M. (2013). *Biopolymers: Reuse, Recycling, and Disposal*. Elsevier Inc., Oxford.

Nizzetto, L., Binda, G., Hurley, R., Baann, C., Selonen, S., Velmala, S., & van Gestel, C. A. (2024). Comments to "Degli-Innocenti, F. The pathology of hype, hyperbole and publication bias is creating an unwarranted concern towards biodegradable mulch films" [J. Hazard. Mater. 463 (2024) 132923]. *Journal of Hazardous Materials*, 471, 133690.

Qi, R., Jones, D. L., Tang, Y., Gao, F., Li, J., Chi, Y., & Yan, C. (2024). Regulatory path for soil microbial communities depends on the type and dose of microplastics. *Journal of Hazardous Materials*, 473, 134702.

Qi, Y. (2020). Microplastics in the Agroecosystem: Effects of Plastic Mulch Film Residues on the Soil-Plant system (Doctoral dissertation, Wageningen University and Research).

Qi, Y., Ossowicki, A., Yergeau, É., Vigani, G., Geissen, V., & Garbeva, P. (2022). Plastic mulch film residues in agriculture: Impact on soil suppressiveness, plant growth, and microbial communities. *FEMS Microbiology Ecology*, 98(2), fiac017.

Rillig, M. C. (2018). Microplastic disguising as soil carbon storage. *Environmental Science and Technology*, 52, 6079–6080.

Rillig, M. C., Ingraffia, R., & de Souza Machado, A. A., (2017). Microplastic incorporation into soil in agroecosystems. *Frontiers in Plant Science*, 8, 1805. https://doi.org/10.3389/fpls.2017.01805

Rillig, M. C., Kim, S. W., & Zhu, Y. G. (2024). The soil plastisphere. *Nature Reviews Microbiology*, 22(2), 64–74.

Rillig, M. C., Lehmann, A., de Souza Machado, A. A., & Yang, G., (2019). Microplastic effects on plants. *New Phytologist*, 223(3), 1066–1070. https://doi.org/10.1111/nph.15794

Rodriguez-Seijo, A., Lourenço, T. A. P., Costa, J. D., Duarte, A. C., Vala, H., & Pereira, R. (2017). Histopathological and molecular effects of micro plastics in *Eisenia andrei* Bouche. *Environmental Pollution*, 220, 495–503.

Rujnić-Sokele, M., & Pilipović, A. (2017). Challenges and opportunities of biodegradable plastics: A mini review. *Waste Management & Research*, 35(2), 132–140.

Shen, M. Y., Chu, C. Y., Sawatdeenarunat, C., & Bhuyar, P. (2023). Production, downstream processing, and characterization of polyhydroxyalkanoates (PHAs) boosted by pyruvate supplement using mixed microbial culture (MMC) and organic wastewater. *Biomass Conversion and Biorefinery*, 13(17), 15861–15869.

Shirai, T., et al. (2023). Field studies on bioplastic degradation in varied soil types. *European Journal of Soil Science*, in press.

Smith, M., Love, D. C., Rochman, C. M., & Neff, R. A. (2018). Microplastics in seafood and the implications for human health. *Current Environmental Health Reports*, 5, 375–386.

Tokiwa, Y., Calabia, B. P., Ugwu, C. U., & Aiba, S. (2009). Biodegradability of plastics. *International Journal of Molecular Sciences*, 10(9), 3722–3742.

Wang, W., Gao, H., Jin, S., Li, R., & Na, G. (2019). The ecotoxicological effects of microplastics on aquatic food web, from primary producer to human: A review. *Ecotoxicology and Environmental Safety*, 173, 110–117.

Yang, C., Huang, Y., Long, B., & Gao, X. (2022). Effects of biodegradable and polyethylene film mulches and their residues on soil bacterial communities. *Environmental Science and Pollution Research*, *29*(59), 89698–89711.

Yang, X., Bento, C. P. M., Chen, H., Zhang, H., Xue, S., Lwanga, E. H., Zomer, P., Ritsema, C. J., & Geissen, V., (2018). Influence of microplastic addition on glyphosate decay and soil microbial activities in Chinese loess soil. *Environmental Pollution*, 242(Pt A), 338–347.

Zhang, Z., Wang, W., Liu, J., & Wu, H. (2024). Discrepant responses of bacterial community and enzyme activities to conventional and biodegradable microplastics in paddy soil. *Science of the Total Environment*, 909, 168513.

Section III

Environmental Impact

10 Occurrence, Effects, and Analysis of Phthalate Esters in the Environment

Babafemi Raphael Babaniyi

10.1 INTRODUCTION

Phthalate esters (PE) are synthetic chemicals with diverse functions, widely incorporated into various household and industrial products (Perestrelo et al., 2021). They primarily serve as plasticizers or additives in plastic materials to enhance flexibility and durability. Typically, they are colorless liquids (Edjere et al., 2016). Annually, approximately 3 million metric tons of PE are produced globally (Perestrelo et al., 2021).

As plasticizers, PE are not chemically bonded to the polymer chains in plastics. This allows them to leach easily from plastic materials into the surrounding environment, leading to their widespread dispersion within the ecosystem. They are often regarded as significant man-made environmental pollutants (Zhang et al., 2015; Adelaja and Babaniyi, 2020). Studies have shown that phthalates are present in nearly all environmental compartments (soil, water, and air) which serve as pathways for their migration and exposure to plants, humans, and animals (Selyarai et al., 2015). Common phthalates used in the production of packaging materials and other plastic products include benzyl butyl phthalate (BBP), dibutyl phthalate (DBP), di(2-ethylhexyl) phthalate (DEHP), diethyl phthalate (DEP), di-isobutyl phthalate (DiBP), di-n-hexyl phthalate (DnHP), and di-isodecyl phthalate (DiDP). Chemically, phthalates are esters of phthalic acid or alkyl/aryl esters of 1,2-benzenedicarboxylic acid, with distinct chemical structures (Figure 10.1).

PE are synthetic chemicals with diverse functions and applications in both household and industrial products (Perestrelo et al., 2021). They are primarily used as plasticizers or additives to improve the flexibility and durability of plastic materials. Phthalates are typically colorless liquids (Edjere et al., 2016) and are industrially produced through the reaction of phthalic anhydride with ethanol in the presence of a concentrated sulfuric acid catalyst. Globally, approximately 6–8 million tons of phthalates are consumed annually (Seyoum and Pradhan, 2019; Zhang et al., 2021).

The physical properties of PE are largely unaffected by the polar carboxyl group, except when R and R' are lower alkyl groups such as methyl and ethyl (Zhang et al., 2021; Perestrelo et al., 2021). Differences in the physical and chemical compositions of phthalates determine their specific applications (Wang et al., 2019). For instance, their low melting points and high boiling points make them suitable as heat carriers in fluids (Giuliani et al., 2020). Linear PE are preferred for plastic manufacturing due to their efficient flexibility at lower temperatures and reduced volatilization compared to branch esters (Baloyi et al., 2021).

Phthalates are widely used in products such as food packaging, medical devices, vinyl flooring, wall coverings, furniture, toys, car interiors, paints, adhesives, cables, wires, hoses, and pharmaceuticals. They are also present in cosmetics, including nail polish, shampoos, deodorants, body washes, and hair products, even when not explicitly listed on product labels (Saeidnia, 2014; Ai et al., 2021).

DOI: 10.1201/9781003602828-13

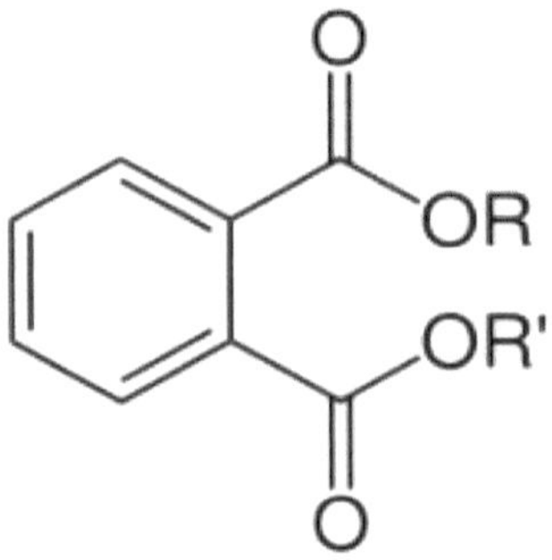

FIGURE 10.1 Phthalate ester.

Despite their versatility, the improper disposal of plastic waste, agricultural chemicals, industrial effluents, and leaching from solid waste into stormwater runoff are primary routes through which phthalates enter the environment (Adeleke et al., 2011; WHO, 2021). Studies reveal that phthalates are present in various environmental compartments, including soil, water, and air, facilitating their migration to plants, animals, and humans (Selyarai et al., 2015). The European Commission's Research Centre found phthalates in food products and human samples, indicating exposure through food consumption, water intake, inhalation, intravenous injection, and dermal absorption. Urinalysis in the United States has detected phthalate metabolites in the general population. The hydrophobicity of phthalates leads to their sorption onto organic phases in soils, slowing their entry into groundwater and aquatic ecosystems. However, their bioaccumulation and persistence in the environment remain significant concerns (Adegbola et al., 2012; Babaniyi et al., 2020).

In response to the toxic nature of certain phthalates, regulatory agencies worldwide have implemented guidelines to limit exposure. For instance, the European Food Safety Authority (EFSA) recommends tolerable daily intake levels of 0.01 mg/kg body weight per day for di-n-butyl phthalate (DBP), 0.5 mg/kg for benzyl butyl phthalate (BBP), and 0.05 mg/kg for DEHP based on toxicological studies (Alin, 2012). Previous research works were investigated in this chapter to highlight the impacts of PE on aquatic ecosystems, soils, plants, microbes, animals, and humans. It also examines the occurrence, extraction, and analytical approaches for investigating phthalates in various environmental matrices, providing a broader understanding of their spatial distribution and circulation. The study aims to propose better strategies to mitigate or prevent their environmental and health impacts.

10.2 PHTHALATE ESTERS IN AQUATIC ENVIRONMENT

The widespread use of phthalates in domestic and industrial applications has led to the detection of phthalate pollutants in various compartments of aquatic habitats, including sediments, plankton, biomass, water, and wastewater treatment plants (Cheng et al., 2019). Phthalates are frequently discharged during manufacturing processes or through improper disposal of waste, allowing them to leach unimpeded into aquatic environments (Henkel et al., 2019). As plastics age and break down over time, the release of PE accelerates, increasing their migration into water bodies (Das et al., 2021). Studies have reported significant concentrations of phthalates in aquatic ecosystems such as wetlands, lakes, river networks, and marine environments (Wang et al., 2017; Gao et al., 2019; Arfaeinia et al., 2019; Zacharia, 2019).

PE are found in aquatic systems at concentrations ranging from trace levels (μg/L) to more significant amounts (mg/L), depending on the degree of pollution. For instance, concentrations as high as 2,705 μg/L were detected in river water in Nigeria (Adeniyi et al., 2011), while levels ranging from 0.02 to 0.77 μg/L have been reported in aquatic regions of China (Liu et al., 2015). Similarly, surface waters in Iran have shown phthalate levels between undetectable limits and 0.43 μg/L (Abtahi et al., 2019). Areas near industries that use or manufacture phthalates are particularly vulnerable to heightened exposure, as water systems often act as receptors for a wide array of contaminants.

Products containing phthalates and associated waste are frequently introduced into aquatic systems through anthropogenic activities. This underscores the importance of understanding their prevalence, biodegradation pathways, bioaccumulation potential, persistence, and detrimental effects on aquatic life (He et al., 2020; Babaniyi et al., 2020b; Ai et al., 2021; Baloyi et al., 2021).

PE are known for their environmental stability and resistance to natural degradation processes. Research has shown that these compounds can bioaccumulate and biotransform within the food chain (Sun et al., 2015; Dong et al., 2022). However, biodegradation remains a primary mechanism for the breakdown of phthalates in the environment (Xu et al., 2020). Marine microalgae and various waterborne microbes have demonstrated the ability to degrade PE, although microbial degradation can be influenced by external environmental factors (Ren et al., 2018; Yu et al., 2020). A lack of efficient microbial strains capable of degrading phthalates may prolong their half-life in aquatic environments, potentially extending from several years to centuries (Fang et al., 2018). Recent research has identified promising microbial strains for phthalate degradation. Zhang et al. (2018a) reported that *Bacillus mojavensis B1811* efficiently degraded phthalates such as DEHP, DBP, BBP, and DPP at a concentration of 0.5 μg/L within four days in a mineral salt medium under optimal stirring conditions. Short-chain phthalates such as DMP and DEP were also degraded by B1811, although with lower efficiency (5.9 and 42.9%, respectively). Similarly, Xu et al. (2020) demonstrated that *Paracoccus kondratievae BJQ0001* effectively degraded mixtures of DMP, DEP, DBP, DIBP, and DEHP, albeit with a moderately reduced degradation rate.

Bioaccumulation studies have also highlighted variations in phthalate retention across organisms. He et al. (2020) investigated phthalate bioaccumulation in the muscle tissues of livestock (cattle, pigs, and chickens) and fish in China. They found that most phthalate congeners had bioaccumulation factors (BAFs) below 2, except for DEP, DMP, and BBP, which showed higher BAFs. Pigs demonstrated a greater ability to store phthalates compared to cattle and chickens, likely due to their higher fat content, as phthalates are lipophilic.

Table 10.1 shows the concentration of PE in relation to its effects on both living and non-living components in their locations which are discussed in details below under different headings.

10.2.2 Effect of Phthalate Esters in Aquatic Life

According to the reviewed articles, the percentage of PE in aquatic environments is approximately 25%, as shown in Figure 10.2. This significant presence poses a severe threat to aquatic ecosystems, particularly in lakes, dams, rivers, and wetlands located near farms, factories, residential areas, and urban communities. These areas are hotspots for phthalate release due to various anthropogenic activities (Li et al., 2021). Notably, under natural conditions, cyanobacteria and freshwater algae can produce phthalate derivatives such as monoethylhexyl phthalate (MEHP) or dibutyl phthalate (DBP), which are subsequently released into aquatic systems. Anthropogenic activities such as pesticide use in agriculture, industrial effluent discharge, wastewater from commercial processes, leachates from landfills, and improper disposal of household electronics and solid waste are major pathways for phthalates to enter aquatic ecosystems (Zhang et al., 2021).

Phthalates exhibit high partition coefficients (Kow/logKow values between 1.61 and 9.46) and low vapor pressures (Pa25) (ranging from 1.84×10^{-6} to 0.263), resulting in low volatility. These properties facilitate their migration into various water systems and their subsequent bioavailability to aquatic organisms (Das et al., 2021; Zhang et al., 2021). The high Kow values and longer alkyl chain lengths of certain phthalates contribute to their hydrophobic nature and higher molecular mass, increasing their sorption to organic matter (Zhang et al., 2021). Once phthalates enter aquatic ecosystems, they disrupt ecological niches, often causing ecological succession in unfavorable

TABLE 10.1
Toxic Effects of Some Phthalate Esters on Aquatic Organisms (Baloyi et al., 2021)

Phthalate Esters	Test Concentration Levels	Organisms	Toxicity
DEHP	0, 10, 100, 200, and 400 μg/L, 14 days	African sharptooth catfish (*Clarias gariepinus*)	Consistent and unique pattern of biphasic effect on steroidogenic enzyme genes, which increases at low concentration (10 μg/L), and affects the brain and head kidney
DEHP	109 μg/L, 10 days	Estuarine copepod (*Eurytemora affinis*)	Exposure to DEHP, especially from embryonic development, could results in endocrine disruption
DBP	2, 10, and 50 μg/L, 30 days	Abalone (*Haliotis diversicolor supertexta*)	Major metabolite changes related to metabolic energy, environmental stress, and osmotic regulation, with increase in intracellular metabolites the most obvious change
DEP, BBP, DBP	DEP: 0.08, 0.4, 2, and 10 mg/L; BBP: 0.02, 0.1, 0.5, and 2.5 mg/L; DBP: 0.0008, 0.004, 0.02, 0.1 mg/L, 14 days	Zebrafish embryos (*Danio rerio*)	Sex hormones disruption via modulation of key steroidogenic genes in male zebrafish over 14 days exposure period to DEP, BBP, and DBP.
BBP	10^{-3}, 10^{-2}, 10^{-1}, 1, and 10 μg/L, 48 hours	Larvae midge (*Chironomus riparius*)	Novel and significant dose and time-dependent alterations in expression profile of genes involved in cell processes such as energy metabolism, stress response, hormone pathways, and detoxication activities even at lower concentrations

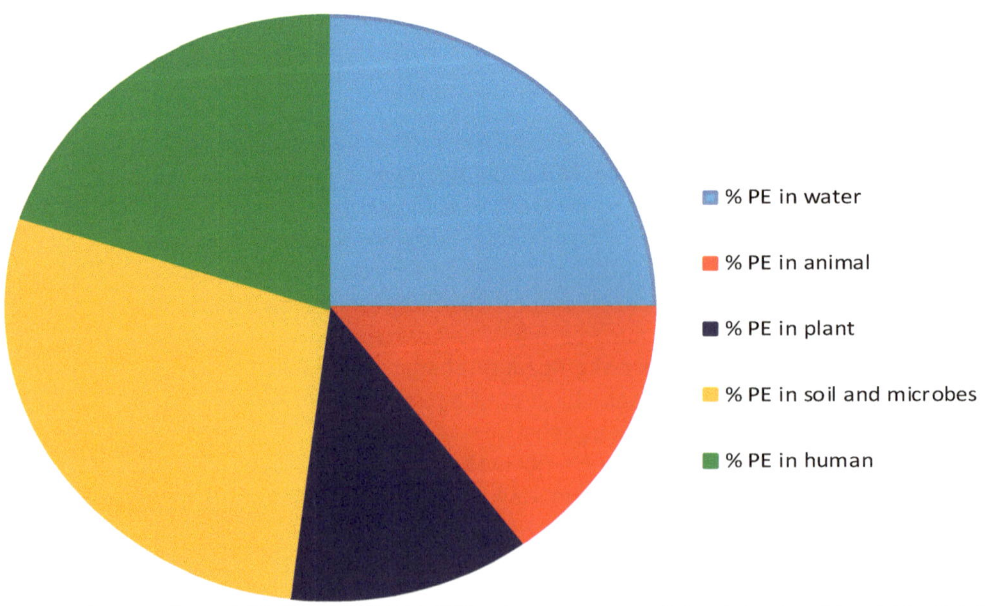

FIGURE 10.2 Distribution of phthalate esters according to the reviewed journals.

ways. Both lower and higher trophic-level aquatic organisms absorb or ingest phthalates, leading to contamination of the food chain. This bioaccumulation has far-reaching implications for aquatic life and the health of interconnected ecosystems.

The effects of PE on aquatic ecosystems are diverse and multifaceted. They evoke genotoxic, neurotoxic, immunotoxic, endocrine, and metabolic disorders while inducing toxin production in fish and other invertebrates. Common phthalates like BBP, DEHP, and DBP are particularly harmful, significantly impacting both living and non-living components of aquatic ecosystems (European Commission, 2008; Arambourou et al., 2019). Staples et al. (1997) reported that exposure to phthalates such as DMP, DEP, DAP, DBP, DIBP, and BBP can result in short- or long-term adverse effects on invertebrates, fish, and algae. Similarly, recent findings by Zhang et al. (2021) indicate manifestations in aquatic organisms such as necrosis, tortuous tails, cardiac edema, impaired tactile responses, egg damage, and embryonic death. Adult aquatic animals exposed to phthalates often exhibit reproductive system dysfunction, liver and kidney damage (Gao et al., 2018).

Phthalates with low molecular weight induce acute toxic effects and are chronically harmful at even low concentrations in aqueous environments. Toxicity generally increases with the alkyl chain length of the phthalate up to four carbon atoms (Staples et al., 1997). Ecotoxicological studies reveal that aquatic animals are particularly vulnerable to phthalates with longer alkyl chains, even at trace concentrations (ng/L to μg/L) commonly found in the environment (Zhang et al., 2021).

The chemical impacts on aquatic organisms can be quantified using median lethal dosage (LC50) measurements. Zhang et al. (2021) reported 96-hour LC50 values for common carp exposed to DEP, DBP, DOP, and DEHP as 34.8–53, 16.30–35, 7.95, and 37.95 mg/L, respectively. For juvenile fish, the toxicity is markedly higher, with LC50 values for DMP, DEP, DBP, and DEHP reduced to 3.29, 6.60, 2.08, and 5.41 mg/L, respectively. These levels are lethal to various aquatic species. DEP's LC50/EC50 values for algae and protozoans in marine environments are 3 and 132 mg/L, respectively, with the lowest No-Observed-Effect Concentrations (NOECs) ranging from 1.7 to 4 mg/L in algae, fish, and invertebrates (Net et al., 2015a). Interestingly, some aquatic organisms, such as microalgae, demonstrate the capacity to degrade PE instead of being adversely affected. For instance, benthic diatoms can facilitate the degradation of DEP and DBP (Li et al., 2015). While decomposition of phthalates with shorter carbon chains is feasible, many phthalates tend to bioaccumulate in biological specimens, leading to persistent environmental and ecological challenges (UNEP, 2020).

10.2.3 Effect of Phthalate Esters on Human Health

According to Figure 10.2, human exposure to PE is increasing daily. These compounds are ubiquitous in various domestic items, including cups, packaging materials, plates, cosmetics, curtains, toys, bottles, and electrical appliances. Consequently, human interaction with phthalates is virtually unavoidable, given the rising industrial production and widespread utilization of these products (Gong et al., 2018). Exposure to phthalates from aquatic sources occurs through various pathways, including dermal absorption, inhalation, and ingestion of contaminated food and water (Giuliani et al., 2020). The health effects of such exposure are significant and diverse, ranging from mutagenicity and teratogenicity to carcinogenicity. The toxic, mutagenic, teratogenic, and carcinogenic potential of phthalates is now a global concern (Zacharia, 2019; USEPA, 2021).

According to Table 10.2, exposure to PE, even at low doses, can be harmful to human health. Such exposure disrupts the endocrine system, damaging glands and impairing hormonal functions (De Toni et al., 2017; Zhu et al., 2019). Endocrine disruptors mimic or interfere with the body's natural hormones, leading to either an increase or decrease in hormone levels (Talia et al., 2021; You and Song, 2021). Recent investigations have highlighted the biological damage caused by endocrine disruptors in animals, raising concerns about similar threats to humans from low-dose phthalate ester exposure (Balaguer et al., 2019; USEPA, 2021). Studies and theories suggest that even low

TABLE 10.2
Toxicity Limits Uptake of Some Phthalate Esters in Humans

Phthalates	MRLs/TDI (μg/kg/day)	Toxicity Endpoint	References
DOP	3,000 (MRL-Acute)	Hepatic	ATSDR (2021)
	400 (MRL-Intermediate)		
DBP	500 (MRL-Acute)	Developmental	ATSDR (2021)
	10 (TDI)	Reproduction	
DEHP	100 (MRL-Intermediate)	Reproduction	ATSDR (2019)
	60 (MRL-Chronic)	Reproduction	ATSDR (2019)
	50 (TDI)	Reproduction	European Food Safety Authority (2019)
DEP	7,000 (MRL-Acute)	Reproduction	ATSDR (2021)
	6,000 (MRL-Intermediate)	Hepatic	ATSDR (2021)
BBP	500 (TDI)	Reproduction	European Food Safety Authority (2019)

concentrations of hormonally active substances like phthalates can interact with receptor hormones, disrupt reproductive systems, and interfere with growth and other hormonal functions (Xu and Yin, 2019; Okpara et al., 2022). Given that endogenous hormone concentrations in the body are naturally low, exposure to small amounts of exogenous phthalates could significantly disrupt endocrine functionality (Marty et al., 2011).

PE evoke negative endocrine effects at lower doses compared to other toxic substances. Biomonitoring studies, including analyses of blood plasma and human urine, indicate that phthalate diesters have a half-life of less than 24 hours in the body (Shin et al., 2020). Pharmacokinetics research shows that diesters rapidly hydrolyze to monoesters in the gastrointestinal tract (Frederiksen et al., 2007; Cequier et al., 2015). While Frederiksen et al. (2007) suggested that phthalates are eliminated through biliary excretion, enterohepatic circulation, and protein binding in blood plasma, Wang et al. (2019) emphasized sweat and urinary excretion as the primary pathways. Further studies revealed specific health risks associated with phthalate exposure in pregnant women, especially during the third trimester, including prolonged blood clotting times, reduced hemoglobin levels, and an elevated risk of anemia (Jiang et al., 2018). Timing and dosage of phthalate exposure are critical, as hormones play different roles during various developmental stages (UNEP, 2018). Concentrations of PE have been linked to hormonal imbalances, liver damage, testicular abnormalities, and birth defects (Hoffman et al., 2018; Sathyanarayana et al., 2014).

Prenatal exposure to phthalates can result in significant developmental abnormalities. For instance, Puerto Rican girls exposed to high phthalate levels experienced premature breast development, while boys born to mothers with elevated phthalate levels showed reduced anogenital distances (Colon et al., 2000; Swan et al., 2005). Additionally, prenatal exposure has been associated with miscarriages, impaired births, and reproductive organ abnormalities (Mu et al., 2015; Hoffman et al., 2018).

The health effects of phthalates extend beyond reproductive health. Endometriosis, a leading cause of infertility in women, has been linked to phthalate exposure, along with reduced birth weight, obesity, preterm births, diabetes, and autism spectrum disorders (Reddy et al., 2006; Stojanoska et al., 2017; Jeddi et al., 2016). Furthermore, phthalates have been implicated in hormonal variations, prostate and breast cancer, compromised immune systems, bronchial obstruction, and impaired neuropsychological development (Kim et al., 2018; Zhang et al., 2019). In males, phthalates disrupt testosterone levels, impairing reproductive functions and causing potentially irreversible damage to reproductive organs, especially during early development (Xie et al., 2019). Effective monitoring and regulation of phthalate exposure are crucial to mitigate these widespread health risks and protect human health from the detrimental effects of endocrine-disrupting chemicals (Chen et al., 2019).

10.2.4 Phthalate Esters in Plants

The trapping and bioaccumulation of phthalates in plants represent a critical pathway for the migration and transformation of PE in the environment, as illustrated in the Figure 10.3.

Plant uptake of these compounds is particularly concerning due to their ability to accumulate and cause long-term adverse effects (Feng et al., 2017; Ma et al., 2017).When PE are present in vegetables such as lettuce, they reduce carotenoid concentrations. Carotenoids play a vital role in protecting cells from mutagens, thereby preventing mutations in the body cells of consumers. Additionally, carotenoids capture free radicals, preventing damage to proteins, lipids, and DNA through lipid peroxidation (Ma et al., 2017). As potent antioxidants, carotenoids delay cellular aging, boost immunity, and help maintain the equilibrium of information transmission and physiological functions in cells (Yuan et al., 2015).

PE also inhibit the production of vitamin C (ascorbic acid) in plants. In higher plants, vitamin C prevents oxidative stress during cell division and elongation (Fridovich, 1995). It enhances plant resistance to diseases and other adverse conditions, regulates redox balance as an essential buffer,

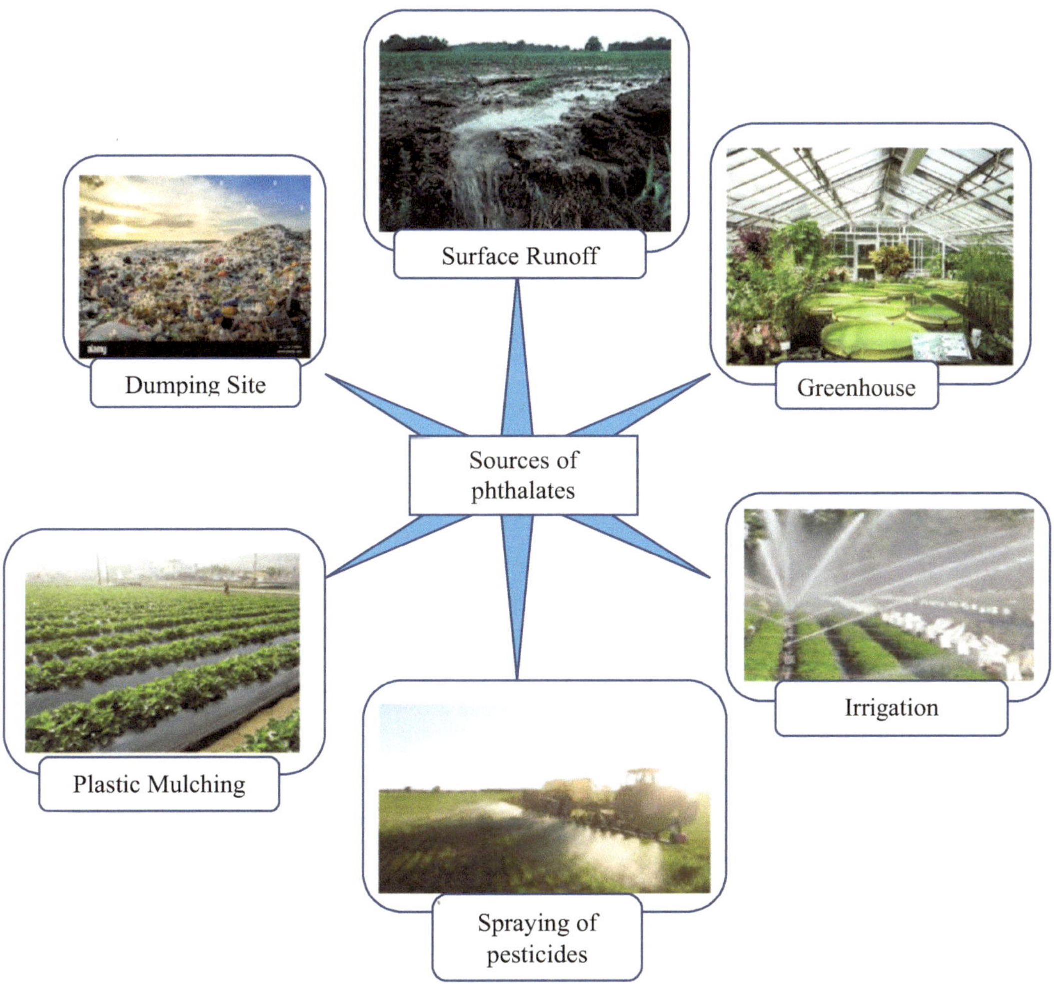

FIGURE 10.3 Some routes of phthalate exposure to plants.

and coordinates the transcription and translation of specific genes while functioning as a coenzyme (Ma et al., 2018). Therefore, the presence of PE compromises all these critical functions of vitamin C.

The metabolites of phthalates further exacerbate plant stress by decreasing leaf mass, inhibiting shoot elongation, and increasing proline bioaccumulation (Li et al., 2016). According to Ma et al. (2018), reduced chlorophyll content due to phthalates diminishes a plant's photosynthetic capacity. This hampers the plant's ability to synthesize organic matter, leading to reduced biomass and the inevitable decrease in dry matter accumulation. Roots, being the primary interface for water and nutrient absorption, are more sensitive to soil contaminants like phthalates than shoots. The direct contact of roots with the soil exposes them to these pollutants, leading to altered root morphology and disruptions in nutrient uptake. Consequently, these effects manifest as impaired photosynthesis and reduced plant growth. PE can inhibit seed germination, block normal germination, reduce biomass and nutrient content, disrupt cell division, and impair plant metabolism (Ma et al., 2018). PE exhibit low water solubility, preferentially binding to soil organic matter instead of dissolving in soil water pores. This persistence in the soil contributes to their toxicity (Chang et al., 2004). Furthermore, phthalates increase membrane permeability, genotoxicity, and oxidative damage in plants (Gao et al., 2017).

Aerobic metabolism is crucial for plant germination and development as it provides the necessary energy through oxygen reduction to form water. However, when this reduction process is incomplete, it leads to the production of reactive oxygen species (ROS), such as superoxide anions (O_2^-), hydroxyl radicals, singlet oxygen, and hydrogen peroxide. These ROS have strong oxidative properties that can cause structural membrane damage, disrupt chlorophyll, and impair photosynthesis (Liu et al., 2007). The presence of pollutants like phthalates exacerbates the production of ROS, including harmful by-products like hydroxyl radicals and hydrogen peroxide. These compounds accumulate and further hinder photosynthesis. Additionally, they aggravate membrane lipid peroxidation, leading to cellular membrane damage, functional degradation, and metabolic disturbances (Starratt & Lazarovits, 1999). Under normal conditions, plants possess mechanisms to neutralize ROS and protect cells from oxidative damage. However, in the presence of PE, the production of reactive oxygen far exceeds the plant's ability to mitigate these effects, leading to oxidative stress and cellular harm (Luo et al., 1998).

10.2.5 Effect of Phthalates on Soil Microorganisms

PE, when present in moderate amounts, stimulate the growth of various soil microorganisms. However, in higher concentrations, they disrupt microbial populations. A study by Xu et al. (2018) found that black soil polluted with di-n-butyl phthalate (DBP) led to significant changes in the soil's microbial community. Genera such as *Arthrobacter* and *Nocardioides* exhibited substantial proliferation. Conversely, the growth of several essential genera involved in nutrient cycling, such as *Nitrososphaera*, *Streptomyces*, *Frankia*, *Conexibacter*, *Pseudonocardia*, *Acidothermus*, *Nocardiopsis*, *Lactococcus*, *Mycobacterium*, *Pseudomonas*, and *Tetrasphaera*, was inhibited. These genera play critical roles in maintaining soil health and supporting plant growth (Briceno et al., 2016; Tanvir et al., 2016).

Genera like *Amycolatopsis* and *Intrasporangium*, known for their bioremediation potential, also declined in response to DBP pollution (Maqbool et al., 2015; Babaniyi & Adelaja, 2020). These shifts in the microbial community indicate that DBP pollution can disturb the soil ecosystem in the short term. Contamination of soil and microorganisms by DBP accelerates metabolic processes such as nitrogen cycling, glycolysis, the pentose phosphate pathway, sulfur metabolism, and the tricarboxylic acid (TCA) cycle (Xu et al., 2018). Nitrogen cycling, carbon metabolism, and sulfur metabolism are the three fundamental nutrient cycling processes in soil ecosystems (Grein et al., 2013; Liu et al., 2022).

In nitrogen cycling, key genes involved in dissimilatory nitrate reduction, assimilatory nitrate reduction, denitrification, and nitrification (*NarGHIJ*, *AlasAB*, *NxrAB*, *NirBD*, and *NorBC*) increase during DBP pollution. DBP treatment also accelerates carbon metabolism, enhancing the conversion of carbon in substrates to CO_2, making more carbon available for soil microorganisms. Moreover, DBP facilitates the formation of 3′-phosphoadenosine 5′-phosphosulfate (PAPS), a critical intermediate in sulfur metabolism, within soil microorganisms (Turchyn et al., 2016).

According to Fiore et al. (2019) and Cheng et al. (2018), DBP and bis(2-ethylhexyl) phthalate (DEHP) are the most common PE in soil. High residual concentrations of these compounds pose significant threats by reducing the functional diversity of the soil microbial community, microbial biomass, and soil basal respiration. They also inhibit carbon cycling and enzymatic activity (Xu et al., 2018; Babaniyi et al., 2021). Under diverse soil conditions, DBP and DEHP impact both vegetable crops and microbial assemblages in greenhouse soil (Ma et al., 2019).

Increased growth of cyanobacteria due to DBP can release toxins into surface water, contributing to eutrophication (Beghelli et al., 2016). *Dyella ginsengisoli* LA-4, capable of degrading several aromatic compounds, including biphenyl, shows promise as a novel bioresource for environmental bioremediation (Li et al., 2009, 2012). Similarly, rhizobacteria, a highly adaptive microbial group, significantly enhance crop productivity, nutrient cycling, and soil health. Their diverse properties suggest they could potentially replace chemical fertilizers (Yildirim et al., 2008; Upadhyay, 2009).

In summary, DBP pollution accelerates critical soil metabolic processes, including nitrogen cycling, carbon metabolism, and sulfur excretion, affecting the quality of nutrients within the soil. Xu et al. (2018) postulated that DBP pollution elevates the expression of key genes such as TCS genes, malate dehydrogenase, histidine kinase, and citryl-CoA lyase. These increases drive accelerated metabolism in black soil. Overall, DBP pollution alters the structure of the microbial community, enhances the abundance of genes associated with nutrient cycling, and promotes the degradation of DBP through microbial pathways.

10.2.6 Remedy to Exposure of Phthalate Esters

Evidence suggests that PE contribute to human health issues, including reproductive abnormalities and developmental deficiencies, as observed in animal studies at exposure levels comparable to those humans encounter. Consequently, scientists and regulatory bodies are working diligently to develop substitutes for phthalates as plasticizers (Lowell, 2011). Citrates, sebacates, adipates, and phosphates have been identified as potential alternatives. Ongoing studies are investigating their suitability for use in medical devices, toys, and food packaging. However, limited information is available on the health effects of these chemicals when adopted as replacements for phthalates. Additionally, some researchers advocate for the use of materials that do not require phthalates, such as polyurethane and polyethylene (Lowell, 2011). While complete eradication of phthalates would be ideal, current research has yet to provide universally suitable and readily available alternatives. Therefore, managing the negative impacts of phthalates remains critical until a long-term solution is found. To mitigate these effects, several measures should be implemented as follows:

1. Factories producing or utilizing phthalates should be located far from human settlements.
2. The quantity of phthalates used in products, especially those intended for medical or domestic use, should be minimized.
3. Public education campaigns should discourage the use of foods packaged in or wrapped with plastic materials and promote the consumption of balanced diets.
4. Individuals should be encouraged to practice regular handwashing and avoid cosmetic products containing phthalates, such as creams, powders, and deodorants (Braun, 2017; Wang et al., 2019).

A study by Chen et al. (2015) in Taiwan demonstrated the effectiveness of simple behavioral interventions in reducing phthalate exposure. Over a one-week period, urinary metabolites of phthalates were significantly reduced through frequent handwashing (p = 0.009) and avoiding the use of plastic containers for eating or drinking (p = 0.016). These strategies were particularly effective in decreasing mono-n-butyl phthalate (MBP) metabolites found in urine, sweat, and body moisture. Reducing reliance on untreated water, especially in areas with limited access to potable water, could further reduce exposure risks (Chen et al., 2015).

In agriculture, careful management is essential to prevent phthalate contamination of livestock and crops. Grazing fields, forage, water sources, and the siting and construction of animal pens or ranches should be monitored for phthalates. Similarly, agricultural farmlands should be tested for phthalates before being used to cultivate crops, particularly cereals, grains, and vegetables. These commonly consumed foods are often produced with chemicals that may contain phthalates and must be screened before distribution (He et al., 2020; Xu et al., 2020).

Government regulations play a pivotal role in ensuring environmental and public health safety. Many countries have enacted legislative measures to limit the use of phthalates. For example,

i. Denmark implemented a tax on phthalate-containing products, although this policy was eventually invalidated due to decreased production and usage of such materials.
ii. South Korea's Ministry of Food and Drug Safety banned the manufacture, import, and use of medical devices containing DBP, BBP, and DEHP.
iii. China's National Food Safety Standard prohibited the use of DMP, DBP, DOP, and DiDP in food packaging materials and restricted the use of DBP, BBP, DEHP, DOP, DiNP, and DiDP in children's textiles.
iv. Canada banned DEHP in cosmetics and medical tools, along with limiting BBP, DBP, DEHP, DOP, DiNP, and DiDP to less than 1 mg/kg in childcare materials (UNEP, 2020).

Globally, organizations such as the United Nations Environment Programme (UNEP), the United States Centers for Disease Control and Prevention (USCDC), the United States Environmental Protection Agency (USEPA), the Stockholm Convention, the EFSA, and the Agency for Toxic Substances and Disease Registry (ATSDR) have recognized the importance of monitoring phthalate use. These bodies are working to establish regulations and guidelines for safe toxicity levels in phthalate-containing products to mitigate their adverse environmental and health impacts.

10.3 EXTRACTION AND ANALYSIS APPROACHES TO PHTHALATE ESTERS

PE are lipophilic, meaning they bond with lipid-soluble compounds. They are commonly detected in the environment at concentrations ranging from parts per billion (ppb) to parts per million (ppm) (Lee et al., 2019; Dong et al., 2019; Annamalai & Vasudevan, 2020). The analysis of PE typically involves three main steps: extraction, column clean-up, and chromatographic quantification. Studies have outlined various methods for determining PE, with well-documented results. These studies employed a range of separation techniques, including sample collection, extraction of target compounds, washing of extracts, and both qualitative and quantitative analyses. The key aspects of these procedures are reviewed in Table 10.3.

10.3.1 Extraction of Phthalate Esters from Solid Phase

In this procedure, analytes are isolated from different types of matrices in a mixture based on their physical and chemical properties. This approach is commonly employed because it is flexible and does not consume excessive solvent or time (Falenas, 2019). With considerable success, the isolation of PE from river and tap waters has been achieved using the solid-phase extraction method

TABLE 10.3
Some of the Recent Extraction and Analysis Approaches Adopted on Aquatic Samples (Baloyi et al., 2021)

Environmental Samples	Extraction Method	Instrumental Analysis	Recovery Rate (%)	Detection Limit (μg/L)
Water (wastewater/pond)	SPE	HPLC-MS/MS	70–120	0.006–0.178
Water (marine)	SPME	GC-MS	68–114	0.0004–0.00032
Water (surface/river)	SPME	GC-FID	98–102	0.017–0.035
Sediments (river)	SE	GC-MS	68–119	0.00044–0.055
Sediments (wetlands)	USE	GC-MS	84–110	0.35–0.78
Sediments (river)	USE	GC-ECD	75–123	0.0001–0.075
Water (river/marine)	LLE	HPLC-UV	92–102	0.9976–0.9993

(Dominguez et al., 2014; Selvaraj et al., 2015). Recent advances have further enhanced the use and reliability of solid-phase extraction by incorporating organic covalent adsorbents (Santana et al., 2018; Tong et al., 2019). For instance, the successful development and use of COF-(TpBD)/Fe_3O_4, a quick magnetic isolation material that can be reused after cleaning with methanol, has demonstrated the ability to extract 15 PE. This method has shown benchmark recoveries of 79.3–121.8% at concentrations as low as 0.005 μg/L, with traceable points ranging from 0.01 to 0.28 μg/L and recovery rates of 80–112.8%. Additionally, the use of a core-shelled Fe_3O_4 graphitic sub-microcube has achieved traceable bands between 0.006 and 0.178 μg/L, with recoveries ranging from 70 to 120%. Similarly, Fe_3O_4 nanomaterials coated with reduced graphene oxide have shown promising results. However, these advances are fragile and capital-intensive (Pang et al., 2020; Santana et al., 2018).

10.3.2 Solid Phase-Microextraction (SPME)

This model involves the use of a smaller automated device to integrate samples, enrich, separate, purify, concentrate, and inject them into a single, efficient method that is more environmentally friendly (Yang et al., 2015). Furthermore, the incorporation of modulating substances in the synthesis of fibers greatly influences the efficient isolation of PE (Guo et al., 2019). The novel solid-phase microextraction (SPME) approach, with improved selective coatings, gained significant attention in the 21st century (Mehdinia and Aziz, 2013). In 2010, 16 phthalate ester compounds within oil matrices were analyzed using a unique fiber as a sorbent in SPME. This technique does not require manipulation of samples, thereby reducing the risk of contamination from glassware, solutions, and specimens. However, with slight adjustments in the process, many researchers have successfully applied this method to analyze PE in water samples. The results revealed relative standard deviations (RSDs) of less than 10% and better recoveries, with values ranging from 68–114, 55.4–110.9, and 75–116% for South African wastewater treatment plants (WWTPs) discharges (Zhang et al., 2017; Salaudeen et al., 2018).

Miriany et al. (2014) investigated the migration of PE in food by directly extracting samples using an 85 μm polyacrylate (PA) fiber. This study focused on the mobility of dibutyl phthalate (DBP) and benzyl butyl phthalate (BBP) in plastic food packaging materials using liquid food simulants. Extraction was performed using the solid-phase cold fiber microextraction technique. A cryogenic mechanism was employed, cooling the fiber with a copper tube (70 cm in length, 2.4 mm outer diameter, and 1.6 mm inner diameter) to transfer liquid nitrogen from a Dewar flask to the SPME device with an 85 μm PA fiber. The extraction process involved immersion of the fiber in 22 mL Pyrex vials, sealed with silicone and aluminum tops, containing the samples. In the study, various parameters were varied factorially, including extraction time (10 and 30 minutes), extraction temperature (25°C and 65°C), and salt concentration (0 and 10% NaCl, w/v). The optimized conditions were extraction time of 20 minutes, temperature of 45°C, and 5% NaCl concentration. Standard solutions were prepared at 100 μg/L per compound. The fiber was then analyzed by gas chromatography with a mass spectrometer equipped with an ion trap detector (Finnigan Trace GC/PolarisQ, Thermo, Austin, TX, USA). To assess phthalate migration in liquid food, ultra-treated water and 10% (v/v) ethanol were used as simulants in line with European guidelines (EU Regulation 10/2011), while Brazilian Health Ministry Document No. 26/MS/SVS supports water as a simulant for liquid foods with a pH higher than 5. The results revealed prominent peaks in the chromatographic analysis, and the interaction between time and temperature showed significant effects on the extraction of DBP and BBP, with prolonged extraction time (30 minutes) and increased temperature (65°C) corresponding to larger chromatographic signals (Miriany et al., 2014).

In a recent assessment, headspace solid-phase microextraction (HSSPME) coupled with gas chromatography-mass spectrometry (GC-MS) was optimized through multivariate analysis to examine four common phthalates—DIBP, butyl-benzyl phthalate (BBP), di-n-octyl phthalate (DOP), and 2,2,4,4-tetrabromodiphenyl (BDE)—in various food packaging materials. The extraction efficiency was recorded using polydimethylsiloxane/divinylbenzene (PDMS/DVB) fibers at 80°C for 30 minutes. The results showed good linearity and a RSD of less than 13%, with recoveries ranging from 90.2 to 111%. Detection limits ranged from 0.03 to 0.08 μg/L for some phthalates and between 0.10 and 0.24 μg/L for others. Among the food packaging materials analyzed, DIBP was the most commonly detected phthalate, occurring in 71.4% of samples with concentrations ranging from 3.61 to 10.7 μg/L. BBP was only quantified in one sample, and BDE was found in trace amounts (<LOQ) in two samples (Perestrelo et al., 2021).

10.3.3 Soxhlet Extraction

This approach is suitable for extracting non-evaporative organic compounds from solid phases such as soils, sediments, and sludges (Sibali, 2008). It is not capital-intensive as it only requires glassware and organic solvents. The technique is simple and convenient to operate. Once the process begins, it does not require further manipulation and provides efficient extraction. In an assessment of 16 PE in lakes and rivers in China, Soxhlet extraction achieved recaptures ranging from 61 to 112%, with average standard deviations less than 10% in water and sediment samples (Liu et al., 2014). The Soxhlet extraction apparatus was optimized to isolate PE from soil and sediment samples from an open dump in Muledane, Thohoyandou, Limpopo, with recapture rates between 88 and 90% (Adeniyi et al., 2008).

Live samples of *Cyprinus carpio* and *Clarias gariepinus* were harvested from upstream, midstream, and downstream locations of a river in Ibadan Metropolis, Nigeria. The samples were assumed to represent the concentrations of PE discharged at those locations. The scales, fins, and dorsal tissues of the fish were removed using a surgical blade for investigation. After drying, the specimens were homogenized, and 5.0 g of each sample was placed in a Soxhlet extractor for 8 hours of reflux extraction. The residues were filtered, and the filtrate was dried over anhydrous Na_2SO_4. The isolates were collected in hexane. Standard stock solutions of DEP, DMP, BBP, and DnBP were prepared (1 mg/mL) in hexane and stored at 4°C. Working solutions were prepared by diluting the stock solutions. Calibration curves for each standard were developed by running High-Performance Liquid Chromatography (HPLC). The recovery percentages were found to range between 52.2 and 91.13%. DBP was not detected in any of the samples. DMP, DPP, and DEP were found in higher concentrations at the three locations, with DMP being detected only in *C. gariepinus*. DEP was detected in a location with the lowest concentration in *C. gariepinus* (Adegbola et al., 2012).

In another study, the impact of fish storage conditions in ponds was investigated to assess changes in phthalic acid content, specifically di-n-butyl phthalate (DBP) and DEHP, in the carcasses of carp.

Fishes were harvested from two fish ponds in South Moravia and analyzed after being stored for seven weeks. The samples from both ponds were evaluated after the storage period. The DBP content increased from 0.56±0.39 to 0.83±0.26 mg/kg in one pond, and from 0.31±0.08 to 0.35±0.11 mg/kg in the other pond. For DEHP, the concentration in specimens from pond 1 increased from 0.26±0.23 to 0.72±0.16 mg/kg, while in pond 2, it decreased from 0.71±0.17 to 0.12±0.18 mg/kg (Alžbeta et al., 2011).

10.3.4 Ultrasonication Extraction (USE)

This method is applicable when high-powered, low-frequency ultrasound waves are coupled to a slurry containing bioactive material in a solvent. The ultrasonic probe processor is deeply immersed in the slurry, and high-powered ultrasound waves pass through the liquid, generating alternating high- and low-pressure cycles. These cycles lead to acoustic cavitation, which results in extreme local temperatures, pressure, rapid heating or cooling rates, and high shear forces within the medium. The implosion of cavitation bubbles over solid surfaces enhances the extraction process. This approach has been employed by Selvaraj et al. (2015) to isolate PE from sediments, with total recapture rates between 79 and 121% in India. Similarly, in China, a modified ultrasonication extraction technique was used to extract PE from sediments, achieving a recapture range of 69–123%, with respective standard deviations under 10% (Weizhen et al., 2020). Li et al. (2015) also adopted this method to assess the biodegradation of DEP and dibutyl PE in marine sediments, though recovery percentages were not specified.

Additionally, phthalate ester evaluation was conducted in soils of varying depths—0–10, 10–20, and 20–40 cm—across 37 greenhouses for vegetable cultivation on the Shandong Peninsula, East China (Chai et al., 2014). Following the procedure, the soil was spiked with 5.0 g of benzyl benzoate (as a surrogate standard) and isolated by ultrasonication for 15 minutes, repeated three times using 90 mL of acetone/petroleum ether (1:3, v:v). The isolates were combined, filtered, and concentrated to about 1 mL. The extracts were cleaned with 3 g of anhydrous sodium sulfate, and the column was cleaned with 10 mL of petroleum ether (10:0.4, v:v), followed by elution of PE with 90 mL of petroleum ether (10:3, v:v). The extracts were then reduced to 1.0 mL in hexane, and the internal standard, di-isophenyl phthalate, was introduced before instrumental analysis. GC-MS was used to identify and quantify 16 PE. Total recoveries for the 16 PE in the study areas ranged from 1.939 to 35.442 mg/kg, with an average of 6.748 mg/kg.

Notably, the highest mean and optimal concentrations of the 16 phthalates were found in the 0–10 cm soil depth in the Weifang area, which has a long history of vegetable farming, particularly in large greenhouses. While the concentration of the 16 PE varied, their composition remained consistent. DEHP, di-n-octyl phthalate (DnOP), di-n-butyl phthalate (DnBP), and DiBP were the most prevalent among the 16 investigated phthalates. These findings were compared to those from other agricultural soils in China, where vegetable greenhouse farming had led to elevated phthalate ester concentrations. Dimethyl phthalate (DMP), DEP, and DnBP were found to exceed the permissible levels of phthalates in soils, with over 90% of the samples exceeding the threshold, and DnOP exceeding 20%. The Shandong Peninsula was found to have particularly high concentrations of phthalates, leading to the conclusion that the area is severely contaminated with PE, making it unsuitable for vegetable farming (Chai et al., 2014).

10.3.5 Liquid–Liquid Extraction (LLE)

This technique, also referred to as "solvent extraction," relies on partitioning based on the traditional method of compound separation, which depends on the relative solubility of different immiscible liquids, such as water and organic solvents. The extraction method is employed to purify compounds or separate mixtures of compounds, particularly when isolating a product from a complex reaction mixture. It is also used to extract lipophilic compounds from a water solution, yielding excess water along with the organic components. The two phases are placed into a device called a separating funnel, where the compounds will distribute between the two phases before partitioning. The method offers several advantages, including simple phase dissociation, rapid equilibrium partitioning, and a safer process with a clearer understanding of subsequent analyses (Cai et al., 2007).

This procedure has been used in various studies, including the analysis of PE in aquatic environments (Sibali, 2008). Fatoki et al. (2010) optimized the Liquid–Liquid Extraction (LLE) method for phthalate ester analysis in South Africa, reporting recoveries in the range of 81–90%, with less than a 5% standard deviation. Similarly, Sibali (2008) slightly modified this technique and successfully isolated PE in samples from the Jukskei River catchment in Gauteng province, South Africa, with recoveries ranging from 95 to 122%, and a standard deviation of less than 10%. In a study by Nantaba et al. (2021) in Uganda, recovery rates ranged from 83 to 115% along Lake Victoria. PE were also measured in various bottled water products using air-assisted liquid–liquid microextraction, followed by identification with GC-MS. Additionally, a risk assessment for exposure to these compounds was conducted by measuring the potential cancer risk. The concentration of average PE (DEHP, DBP, DIBP, DEP, and DMP) after 5–15 days of incubation at 42°C, and after 15, 45, and 75 days of incubation at 25°C, showed initial concentrations of 7.28, 8.99, 1.78, 5.6, 6.45, and 8.55 ppb, respectively. The non-carcinogenic risk for consumers was minimal (HQ < 1), and the cancer risk due to di(2-ethylhexyl) phthalate (DEHP) in bottled water was considered negligible (Zabihollah et al., 2019).

10.3.6 Clean-Up and Analysis

Crude PE extracted from the environment are often found in compound forms, combined with various other organic substances, leading to impurities that can interfere with the accuracy of instrumental analysis. Therefore, it is essential to clean up the samples before instrumental analysis. Sample clean-up is typically performed using chromatography techniques, such as columns packed with aluminum oxide or silica gel, which trap unwanted compounds. If clean-up is not performed, the capillary column may become contaminated with extracts during the instrumental analysis.

The setup for this clean-up process involves using a glass column, which is lagged with glass wool. Typically, 1–5 g of silica gel or aluminum oxide slurry, mixed with an extracting solvent and anhydrous sodium sulfate, is placed above the glass wool (approximately 1 cm) to dry the extracts. Once the extracting solvent saturates the column, the extracts are added to the chamber, and excess extracting solvent is used to elute the compounds. Finally, about 1 mL of the eluate is concentrated before being analyzed by chromatography instrumentation (Lee et al., 2019; Santana et al., 2018).

The development of improved gas chromatography techniques, particularly fused-silica capillary columns, has significantly increased the separation efficiency for PE. This is due to the reduced molecular mass, decreased polarity, and improved thermal stability of the PE (Yang et al., 2015). The combination of GC with MS, electron capture detectors (ECD), or flame ionization detectors (FID) has recently become one of the most significant methods for identifying and quantifying PE in various environmental matrices (Weizhen et al., 2020; Li et al., 2020). However, the development of absorbents for covalent organic compounds using solid-phase extraction remains a tedious and time-consuming process. Meanwhile, SPME has emerged as an improved, solvent-free extraction technique that offers better recoveries.

10.4 CONCLUSION

The widespread presence of PE in the environment is primarily attributed to the increased production of chemicals and products containing these compounds. Contributing factors include rising population, improper disposal of domestic plastic waste, agricultural runoff, industrial effluent discharges, and leaching from solid waste into stormwater. Scientific investigations have revealed the harmful effects of phthalate exposure on humans, aquatic life, microbial biodiversity, soil, plants, and animals. In response to the toxicity of PE to living organisms and the environment, several policies have been formulated and implemented by regulatory agencies and nations to mitigate the identified risks. Some measures include ensuring wholesome livestock production, providing clean grazing fields, forage, and water, and properly siting ranches and pens to prevent exposure to phthalates. In certain countries, more extreme strategies have been adopted, such as imposing taxes on producers and users of phthalate-containing materials, highlighting the seriousness of the danger they pose.

Phthalate ester analysis typically involves a three-step procedure: extraction, column clean-up, and quantification through chromatography. Over the years, several analytical techniques have been developed, depending on the type or state of the samples, such as gaseous or liquid. Various separation techniques are employed, including sample collection, extraction of target compounds, clean-up of extracts, and both qualitative and quantitative instrumental analysis. In general, PE are harmful to humans, animals, microorganisms, plants, and ecosystems, and continuous efforts are necessary to remove them from water, farmlands, livestock, and the environment.

REFERENCES

Abtahi, M., Dobaradaran, S., Torabbeigi, M., Jorfi, S., Gholamnia, R., Koolivand, A., et al. (2019). Health Risk of Phthalates in Water Environment: Occurrence in Water Resources, Bottled Water, and Tap Water, and Burden of Disease from Exposure through Drinking Water in Tehran. *Iran. Environ. Res.*, 173, 469–479. doi: 10.1016/j.envres.2019.03.071.

Adegbola, R. A., Adewuyi, G. O., and Adekanmbi, A. I. (2012). Analysis of Fish Samples as a Biomarker of Levels of Pollutants In Ona River, Ibadan, Oyo State, Nigeria. *IOSR J. Appl. Chem.*, 2 (3), 32–37.

Adelaja, O. A., and Babaniyi, B. R. (2020). Mechanical and Morphological Characterization of Plasticized Low-Density Polyethylene (LDPE) Granules Blend with Polyhydroxybutyrate (PHB). *Int. J. Adv. Materials Res.*, 6 (2), 29–36.

Adeniyi, A., Dayomi, M., Siebe, P., and Okedeyi, O. (2008). An Assessment of the Levels of Phthalate Esters and Metals in the Muledane Open Dump, Thohoyandou, Limpopo Province. South Africa. *Chem. Cent. J.*, 2 (9), 1–9. doi: 10.1186/1752-153X-2-9.

Adeniyi, A. A., Okedeyi, O. O., and Yusuf, K. A. (2011). Flame Ionization Gas Chromatographic Determination of Phthalate Esters in Water, Surface Sediments and Fish Species in the Ogun River Catchments, Ketu, Lagos, Nigeria. *Environ. Monit. Assess*, 172, 561–569. doi: 10.1007/s10661-010-1354-2.

Ai, S., Gao, X., Wang, X., Li, J., Fan, B., Zhao, S., et al. (2021). Exposure and Tiered Ecological Risk Assessment of Phthalate Esters in the Surface Water of Poyang Lake. China. *Chemosphere*, 262, 127864. doi: 10.1016/j.chemosphere.2020.127864.

Alin, J. (2012). Migration from Plastic Food Packaging during Microwave Heating. American Chemical Society, *J. Agric. Food Chem.*, 59 (10), 1405–1415.

Annamalai, J., and Vasudevan, N. (2020). Detection of Phthalate Esters in PET Bottled Drinks and lake Water Using Eterase/PANI/CNT/CuNP Based Electrochemical Biosensor. *Anal. Chim. Acta*, 1135, 175–186. doi: 10.1016/j.aca.2020.09.041.

Arambourou, H., Planelló, R., Llorente, L., Fuertes, I., Barata, C., and Delorme, N., (2019). Chironomus Riparius Exposure to Field-Collected Contaminated Sediments: From Subcellular Effect to Whole-Organism Response. *Sci. Total Environ.*, 671, 874–882. doi: 10.1016/j.scitotenv.2019.03.384.

Arfaeinia, H., Fazlzadeh, M., Taghizadeh, F., Saeedi, R., Spitz, J., and Dobaradaran, S. (2019). Phthalate Acid Esters (PAEs) Accumulation in Coastal Sediments from Regions with Different Land Use Configuration along the Persian Gulf. *Ecotoxicol. Environ. Saf.*, 169, 496–506. doi: 10.1016/j.ecoenv.2018.11.033.

ATSDR. (2019). Minimal Risk Levels (MRLs) for Hazardous Substances. Available at: www.atsdr.cdc.gov/mrls/mrllist.asp (Accessed May 07, 2022).

ATSDR. (2021). Minimal Risk Levels (MRLs) for Hazardous Substances. Available at: wwwn.cdc.gov/TSP/MRLS/mrlsListing.aspx (Accessed May 07, 2022).

Babaniyi, B. R., Adebomi, J. I., Adejoro, F., and Babaniyi, E. E. (2021). *Bacillus Siamensis* KCTC 13613(T) Cultured in Plantain Peels Flour for the Production of Polyhydroxyalkanoate. *Int. J. Modern Dev. In Eng. Sci.*, 1 (1), 1–5.

Babaniyi, B. R., and Adelaja O. A. (2020). Production of Poly-β-Hydroxybutyric Acid (PHB) by *Bacillus cereus* on Pineapple Peels. *GSC Adv. Res. Rev.*, 4 (1), 24–30.

Babaniyi, B. R. Afolabi, F., and Obagunwa, M. P. (2020). Recycling of Used Polyethylene through Solvent Blending of Plasticized Polyhydroxybutyrate and Its Degradation Potential. *Elsevier J. Composite Part C*, 2 (2020), 100021.

Babaniyi, B. R., Chidozie, O. V., and Ademola, B.-O. (2020). Characterization and Blending of Polyhydroxyalkanoate (PHA) Produced by *Bacillus Safensis* FO-366(T) on Cassava Peels. *Int. J. Adv. Materials Res.*, 5 (2), 151–158.

Balaguer, P., Delfosse, V., and Bourguet, W. (2019). Mechanisms of Endocrine Disruption through Nuclear Receptors and Related Pathways. *Curr. Opin. Endocr. Metab. Res.*, 7, 1–8. doi: 10.1016/j.coemr.2019.04.008.

Baloyi, N. D., Tekere, M., Maphangwa, K. W., and Masindi, V. (2021). Insights into the Prevalence and Impacts of Phthalate Esters in Aquatic Ecosystems. *Front. Environ. Sci.*, 9, 684190. doi: 10.3389/fenvs.2021.684190.

Beghelli, F. G. D. S., Frascareli, D., Pompêo, M. L. M., and Moschini-Carlos, V. (2016). Trophic State Evolution over 15 Years in a Tropical Reservoir with Low Nitrogen Concentrations and Cyanobacteria Predominance. *Water Air Soil Pollut.*, 227, 1–15.

Braun, J. M. (2017). Early-Life Exposure to EDCs: Role in Childhood Obesity and Neurodevelopment. *Nat. Rev. Endocrinol.*, 13 (3), 161–173. doi: 10.1038/ nrendo.2016.186.

Briceno, G., et al., (2016). Use of Pure and Mixed Culture of Diazinon-Degrading Streptomyces to Remove Other Organophosphorus Pesticides. *Int. Biodeterior. Biodegradation*, 114, 193–201.

Cai, Y., Cai, Y., Shi, Y., Liu, J., Mou, S., and Lu, Y. (2007). A Liquid–Liquid Extraction Technique for Phthalate Esters with Water-Soluble Organic Solvents by Adding Inorganic Salts. *Microchim. Acta*, 157 (1–2), 73–79. doi: 10.1007/ s00604-006-0625-7.

Cequier, E., Sakhi, A. K., Marcé, R. M., Becher, G., and Thomsen, C. (2015). Human Exposure Pathways to Organophosphate Triesters – A Biomonitoring Study of Mother-Child Pairs. *Environ. Int.*, 75, 159–165. doi: 10.1016/ j.envint.2014.11.009.

Chai, C., Cheng, H., Ge, W., Ma, D., and Shi, Y. (2014). Phthalic Acid Esters in Soils from Vegetable Greenhouses in Shandong Peninsula, East China. *PLoS One*, 9 (4), e95701. doi: 10.1371/journal.pone.0095701.

Chang, B. V., Yang, C. M., Cheng, C. H., and Yuan, S. Y. (2004). Biodegradation of Phthalate Esters by Two Bacteria Strains. *Chemosphere*, 55, 533–538. doi: 10.1016/j.chemosphere.2003.11.057.

Chen, C.-Y., Chou, Y.-Y., Lin, S.-J., and Lee, C.-C. (2015). Developing an Intervention Strategy to Reduce Phthalate Exposure in Taiwanese Girls. *Sci. Total Environ.*, 517, 125–131. doi: 10.1016/j.scitotenv.2015.02.021.

Chen, Y., Jiang, L., Lu, S., Kang, L., Luo, X., Liu, G., et al. (2019). Organophosphate Ester and Phthalate Ester Metabolites in Urine from Primiparas in Shenzhen, China: Implications for Health Risks. *Environ. Pollut.*, 247, 944–952. doi: 10.1016/j.envpol.2019.01.107.

Cheng, Z., Li, H. H., Yu, L., Yang, Z. B., Xu, X. X., Wang, H. S., and Wong, M. H. (2018). Phthalate Esters Distribution in Coastal Mariculture of Hong Kong, China. *Environ. Sci. Pollut. Res.*, 25, 17321–17329.

Cheng, Z., Liu, J.-B., Gao, M., Shi, G.-Z., Fu, X.-J., Cai, P., et al. (2019). Occurrence and Distribution of Phthalate Esters in Freshwater Aquaculture Fish Ponds in Pearl River Delta. China. *Environ. Pollut.*, 245, 883–888. doi: 10.1016/ j. envpol.2018.11.085.

Colón, I., Caro, D., Bourdony, C. J., Rosario, O., and Rossario, O. (2000). Identification of Phthalate Esters in the Serum of Young Puerto Rican Girls with Premature Breast Development. *Environ. Health Perspect.*, 108 (9), 895–900, doi: 10.1289/ehp.00108895.

Das, M. T., Kumar, S. S., Ghosh, P., Shah, G., Malyan, S. K., Bajar, S., et al. (2021). Remediation Strategies for Mitigation of Phthalate Pollution: Challenges and Future Perspectives. *J. Hazard. Mater.*, 409, 124496. doi: 10.1016/ j. jhazmat.2020.124496.

De Toni, L., Tisato, F., Seraglia, R., Roverso, M., Gandin, V., Marzano, C., et al. (2017). Phthalates and Heavy Metals as Endocrine Disruptors in Food: A Study on Pre-Packed Coffee Products. *Toxicol. Rep.*, 4, 234–239. doi: 10.1016/ j.toxrep.2017.05.004.

Dong, L., Lin, L., Pan, X., Zhang, S., Lv, Z., and Mi, C. (2022). Distribution Dynamics of Phthalate Esters in Surface Water and Sediment of the Middle-Lower Hanjiang River, China. *Int. J. Environ. Res. Public Health*, 19, 2702. doi.org/10.3390/ ijerph19052702.

Dong, W., Guo, R., Sun, X., Li, H., Zhao, M., Zheng, F., et al. (2019). Assessment of Phthalate Ester Residues and Distribution Patterns in Baijiu Raw Materials and Baijiu. *Food Chem.*, 283, 508–516. doi: 10.1016/ j.foodchem.2019.01.069.

Edjere, O., Asibor, I., and Otolo, S. (2016). Evaluation of the Levels of phthalate Ester Plasticizers in Surface Water of Ethiope River System, Delta State, Nigeria. *J. Appl. Sci. Environ. Manage.*, 20 (3), 608–614.

European Commission. (2008). European Union Risk Assessment Report Bis(2Ethylhexyl) Phthalate (DEHP), CAS-No. 117-81-7. 1:80-85. EU R 233 84 EN. Luxembourg: Office for Official Publications of the European Communities.

European Food Safety Authority. (2019). Update of the Risk Assessment of Di-butyl Phthalate (DBP), Butyl-benzyl-phthalate (BBP), Bis(2-ethylhexyl) Phthalate (DEHP), Di-isononyl Phthalate (DINP) and Di-isodecyl Phthalate (DIDP) for Use in Food Contact Materials. *EFSA J.*, 17 (12), 1–85. doi: 10.2903/ j.efsa.2019.5838.

Falenas, F. (2019). Discussion: Automated Filtration or Really Just Solid Phase Extraction. *J. Chromatogr. A*, 1601, 388–389. doi: 10.1016/j.chroma.2019.03.052.

Fang, C., Chu, Y., Jiang, L., Wang, H., Long, Y., and Shen, D. (2018). Removal of Phthalic Acid Diesters through a Municipal Solid Waste Landfill Leachate Treatment Process. *J. Mater. Cycles Waste Manag.*, 20, 585–591. doi: 10.1007/ s10163-017-0625-1.

Fatoki, O. S., Bornman, M., Ravandhalala, L., Chimuka, L., Genthe, B., and Adeniyi, A. (2010). Phthalate Ester Plasticizers in Freshwater Systems of Venda, South Africa and Potential Health Effects. *Water SA*, 36 (1), 117–126. doi: 10.4314/ wsa.v36i1.50916.

Feng, Y. H., Zhang, Y., Zheng, L. P., Ying, R. R., Lin, Y. S., and Wang, G. Q. (2017). Distribution Characteristics of Phthalic Acid Esters in Soil and Vegetable under Greenhouse in Different Areas of Jiangsu Province, China. *J. Ecol. Rural Environ.*, 33, 308–316.

Fiore, M., Conti, G. O., Caltabiano, R., Buffone, A., Zuccarello, P., Cormaci, L., Cannizzaro, M. A., and Ferrante, M. (2019). Role of Emerging Environmental Risk Factors in Thyroid Cancer: A Brief Review. *Int. J. Environ. Res. Public Health*, 16, 1185.

Frederiksen, H., Skakkebaek, N. E., and Andersson, A.-M. (2007). Metabolism of Phthalates in Humans. *Mol. Nutr. Food Res.*, 51, 899–911. doi: 10.1002/ mnfr.200600243.

Fridovich, I. (1995). Superoxide Radical and Superoxide Dismutases. *Ann. Rev. Biochem.*, 64, 97–112, pmid:7574505.

Gao, M., Dong, Y., Zhang, Z., Song, W., and Qi, Y. (2017). Growth and Antioxidant Defence Responses of Wheat Seedlings to di-n-butyl phthalate and di (2-ethylhexyl) Phthalate Stress. *Chemosphere*, 172, 418–428.

Gao, D., Li, Z., Wang, H., and Liang, H. (2018). An Overview of Phthalate Acid Ester Pollution in China over the Last Decade: Environmental Occurrence and Human Exposure. *Sci. Total Environ.*, 645, 1400–1409. doi: 10.1016/ j.scitotenv.2018.07.093.

Gao, X., Li, J., Wang, X., Zhou, J., Fan, B., Li, W., et al. (2019). Exposure and Ecological Risk of Phthalate Esters in the Taihu Lake basin, China. *Ecotoxicol. Environ. Saf.*, 171, 564–570. doi: 10.1016/ j.ecoenv.2019.01.001.

Giuliani, A., Zuccarini, M., Cichelli, A., Khan, H., and Reale, M. (2020). Critical Review on the Presence of Phthalates in Food and Evidence of Their Biological Impact. *Int. J. Environ. Res. Public Health*, 17 (16), 5655. doi: 10.3390/ijerph17165655.

Gong, C.-B., Wei, Y.-B., Chen, M.-J., Liu, L.-T., Chow, C.-F., and Tang, Q. (2018). Double Imprinted Photoresponsive Polymer for Simultaneous Detection of Phthalate Esters in Plastics. *Eur. Polym. J.*, 108, 295–303. doi: 10.1016/ j.eurpolymj.2018.09.009.

Grein, F., Ramos, A. R., Venceslau, S. S., and Pereira, I. A. (2013). Unifying Concepts in Anaerobic Respiration: Insights from Dissimilatory Sulfur Metabolism. *Biochim. Biophys. Acta (BBA)-Bioenerg.*, 1827, 145–160.

Guo, H., Song, N., Wang, D., Ma, J., and Jia, Q. (2019). A Modulation Approach for Covalent Organic Frameworks: Application to Solid Phase Microextraction of Phthalate Esters. *Talanta*, 198, 277–283. doi: 10.1016/ j.talanta.2019.02.025.

He, M.-J., Lu, J.-F., Wang, j., Wei, S.-Q., and Hageman, K. J. (2020). Phthalate Esters in Biota, Air and Water in an Agricultural Area of Western China, with Emphasis on Bioaccumulation and Human Exposure. *Sci. Total Environ.*, 698, 134264. doi: 10.1016/j.scitotenv.2019.134264.

Henkel, C., Hüffer, T., and Hofmann, T. (2019). The Leaching of Phthalates from PVC Can Be Determined with an Infinite Sink Approach. *MethodsX*, 6, 2729–2734. doi: 10.1016/j. mex. 2019.10.026.

Hoffman, K., Stapleton, H. M., Lorenzo, A., Butt, C. M., Adair, L., Herring, A. H., et al. (2018). Prenatal Exposure to Organophosphates and Associations with Birthweight and Gestational Length. *Environ. Int.*, 116, 248–254. doi: 10.1016/ j.envint.2018.04.016.

Jarošová, A., Puškárová, L., and Stancová, V. (2011). di-2-Ethylhexyl Phthalate and di-n-butyl Phthalate in Tissues of Common Carp (*Cyprinus carpio L.*) after Harvest and after Storage in Fish Storage Tanks. *J. Microbiol. Biotechnol. Food Sci.*, 1 (3), 277–286.

Jeddi, M. Z., Janani, L., Memari, A. H., Akhondzadeh, S., and yunesian, M. (2016). The Role of Phthalate Esters in Autism Development: A Systematic Review. *Environ. Res.*, 151, 493–504. doi: 10.1016/ j.envres.2016.08.021.

Jiang, M., Li, Y., Zhang, B., Zhou, A., Zhu, Y., Li, J., et al. (2018). Urinary Concentrations of Phthalate Metabolites Associated with Changes in Clinical Hemostatic and Hematologic Parameters in Pregnant Women. *Environ. Int.*, 120, 34–42. doi: 10.1016/j.envint.2018.07.021.

Kim, Y.-M., Kim, J., Cheong, H.-K., Jeon, B.-H., and Ahn, K. (2018). Exposure to Phthalates Aggravates Pulmonary Function and Airway Inflammation in Asthmatic Children. *PLoS One*, 13 (12), e0208553. doi: 10.1371/ journal.pone.0208553.

Lee, Y.-M., Lee, J.-E., Choe, W., Kim, T., Lee, J.-Y., Kho, Y., et al. (2019). Distribution of Phthalate Esters in Air, Water, Sediments, and Fish in the Asan Lake of Korea. *Environ. Int.*, 126, 635–643. doi: 10.1016/ j.envint.2019.02.059.

Li, A., Qu, Y. Y., Pi, W. Q., Zhou, J. T., Gai, Z. H., and Xu, P. (2012). Metabolic Characterization and Genes for the Conversion of Biphenyl in Dyella Ginsengisoli La-4. *Biotechnol. Bioeng.*, 109, 609–613.

Li, A., Qu, Y. Y., Zhou, J., and Fang, M. (2009). Characterization of a Newly Isolated Biphenyl-Degrading Bacterium, Dyella ginsengisoli La-4. *Appl. Biochem. Biotechnol*, 159, 687–695.

Li, K. K., Ma, D., Wu, J., Chai, C., and Shi, Y. X. (2016). Distribution of Phthalate Esters in Agricultural Soil with Plastic Film Mulching in Shandong Peninsula, East China. *Chemosphere*, 164, 314–321. doi: 10.1016/j.chemosphere.2016.08.068

Li, J., Wang, Z., Wang, Q., Guo, L., Wang, C., Wang, Z., et al. (2021). Construction of Hyper Cross-Linked Polymers for High-Performance Solid Phase Microextraction of Phthalate Esters from Water Samples. *J. Chromatogr. A*, 1641, 461972. doi: 10.1016/j.chroma.2021.461972.

Li, Y., Gao, J., Meng, F., and Chi, J. (2015). Enhanced Biodegradation of Phthalate Acid Esters in Marine Sediments by Benthic Diatom Cylindrotheca Closterium. *Sci. Total Environ.*, 508, 251–257. doi: 10.1016/ j.scitotenv.2014.12.002.

Li, Y. Y., He, W., Liu, W. X., Yang, B., He, Q. S., Yang, C., et al. (2020). Impacts of Anthropogenic Activities on Spatial Variations of Phthalate Esters in Water and Suspended Particulate Matter from China's Lakes. *Sci. Total Environ.*, 724, 1–11. doi: 10.1016/j.scitotenv.2020.138281

Liu, H., Cui, K., Zeng, F., Chen, L., Cheng, Y., Li, H., et al. (2014). Occurrence and Distribution of Phthalate Esters in Riverine Sediments from the Pearl River Delta Region, South China. *Mar. Pollut. Bull.*, 83 (1), 358–365. doi: 10.1016/ j.marpolbul.2014.03.038.

Liu, H. R., Song, H. X., Liu, D. P., Guan, C. Y., Liu, Q., and Chen, S. Y. (2007). Dynamics Changes of Soluble Sugar and Free Amino Acid Contents in Stem and Leaf of Different Oilseed Rape Varieties. *Acta Agri. Boreali-occidentalis Sin.*, 16, 123–126.

Liu, J., Lv, M., Sun, A., Ding, J., Wang, Y., Chang, X., and Chen, L. (2022). Exposure to Microplastics Reduces the Bioaccumulation of Sulfamethoxazole but Enhances Its Effects on Gut Microbiota and the Antibiotic Resistome of Mice. *Chemosphere*, 294, 133810. doi: 10.1016/j.chemosphere.2022.133810.

Liu, X., Shi, j., Bo, T., Li, H., and Crittenden, J. C. (2015). Occurrence and Risk Assessment of Selected Phthalates in Drinking Water from Waterworks in China. *Environ. Sci. Pollut. Res.*, 22, 10690–10698. doi: 10.1007/s11356-0154253-9.

Lowell Center for Sustainable Production. (2011). Phthalates and Their Alternatives: Health and Environmental Concerns (p. 24). University of Massachusetts Lowell.

Luo, L. X., Sun, T. H., and Jin, Y. H. (1998). Accumulation of Superoxide Radical in Wheat Leaves under Cadmium Stress. *Acta Sci. Circumst.*, 18, 495–499.

Ma, T., Liu, L., Zhou, W., Chen, L., and Christie, P. (2019). Effects of Phthalate Esters on *Ipomoea aquatica* Forsk. Seedlings and the Soil Microbial Community Structure under Different Soil Conditions. *Int. J. Environ. Res. Public Health*, 16, 3489, doi: 10.3390/ijerph16183489.

Ma, T., Zhou, W., Chen, L., Wu, L., Christie, P., and Zhang, H. (2017). Toxicity Effects of di-(2-ethylhexyl) Phthalate to *Eisenia fetida* at Enzyme, Cellular and Genetic Levels. *PLoS One*, 12 (3), 0173957, pmid:28319143.

Ma, T. T., Zhou, W., Chen, L. K., Wu, L. H., Christie, P., and Liu, W. X. (2018). Toxicity of Phthalate Esters to Lettuce (*Lactuca sativa*) and the Soil Microbial Community under Different Soil Conditions. *PLoS One*, 13, 0208111.

Maqbool, Z. (2015). Isolating, Screening and Applying Chromium Reducing Bacteria to Promote Growth and Yield of Okra (*Hibiscus esculentus L.*) in Chromium Contaminated Soils. *Ecotoxicol. Environ. Saf.*, 114, 343–349.

Marty, M. S., Carney, E. W., and Rowlands, J. C. (2011). Endocrine Disruption: Historical Perspectives and Its Impact on the Future of Toxicology Testing. *Toxicol. Sci.*, 120 (1), S93–S108. doi: 10.1093/toxsci/kfq329.

Mehdinia, A., and Aziz-Zanjani, M. O. (2013). Advances for Sensitive, Rapid and Selective Extraction in Different Configurations of Solid-Phase Microextraction. *Trac. Trends Anal. Chem.*, 51, 13–22. doi: 10.1016/ j.trac.2013.05.013.

Moreira, M.A., André, L. C., and Cardeal, Z. L. (2014). Analysis of Phthalate Migration to Food Simulants in Plastic Containers during Microwave Operations. *Int. J. Environ. Res. Public Health*, 11, 507–526; doi: 10.3390/ijerph110100507.

Mu, D., Gao, F., Fan, Z., Shen, H., Peng, H., and Hu, J. (2015). Levels of Phthalate Metabolites in Urine of Pregnant Women and Risk of Clinical Pregnancy Loss. *Environ. Sci. Technol.*, 49, 10651–10657, doi: 10.1021/acs.est.5b02617.

Nantaba, F., Palm, W.-U., Wasswa, J., Bouwman, H., Kylin, H., and Kümmerer, K. (2021). Temporal Dynamics and Ecotoxicological Risk Assessment of Personal Care Products, Phthalate Ester Plasticizers, and Organophosphorus Flame Retardants in Water from Lake Victoria, Uganda. *Chemosphere*, 262, 127716. doi: 10.1016/j.chemosphere.2020.127716.

Net, S., Sempéré, R., Delmont, A., Paluselli, A., and Ouddane, B. (2015a). Occurrence, Fate, Behavior and Ecotoxicological State of Phthalates in Different Environmental Matrices. *Environ. Sci. Technol.*, 49 (7), 4019–4035. doi: 10.1021/es505233b.

Okpara, K. E., Phoungthong, K., Agbozu, I., Edwin-Isotu, E., and Techato, K. (2022). Phthalate Esters in Tap Water, Southern Thailand: Daily Exposure and Cumulative Health Risk in Infants, Lactating Mothers, Pregnant and Nonpregnant Women. *Int. J. Environ. Res. Public Health*, 19, 2187. doi.org/10.3390/ijerph19042187.

Pang, Y.-H., Yue, Q., Huang, Y.-Y., Yang, C., and Shen, X.-F. (2020). Facile Magnetization of Covalent Organic Framework for Solid-Phase Extraction of 15 Phthalate Esters in Beverage Samples. *Talanta*, 206, 120194. doi: 10.1016/j.talanta.2019.120194.

Perestrelo, R., Silva, C. L., Algarra, M., and Câmara, J. S. (2021). Evaluation of the Occurrence of Phthalates in Plastic Materials Used in Food Packaging. *Appl. Sci.*, 11, 2130. doi.org/10.3390/app11052130.

Qureshi, M. S., Yusoff, A. R. b. M., Wirzal, M. D. H., Barek, J., Afridi, H. I., Üstündag, Z., et al. (2016). Methods for the Determination of EndocrineDisrupting Phthalate Esters. *Crit. Rev. Anal. Chem.*, 46 (2), 146–159. doi: 10.1080/10408347.2015.1004157.

Reddy, B., Rozati, R., Reddy, B., and Raman, N. (2006). General Gynaecology: Association of Phthalate Esters with Endometriosis in Indian Women. *Int. J. Gynecol. Obstet.*, 113, 515–520. doi: 10.1111/j.1471-0528.2006.00925.x.

Ren, L., Lin, Z., Liu, H., and Hu, H. (2018). Bacteria-Mediated Phthalic Acid Esters Degradation and Related Molecular Mechanisms. *Appl. Microbiol. Biotechnol.*, 102, 1085–1096. doi: 10.1007/s00253-017-8687-5.

Rodgers, K., Rudel, R., Just, A., Snedeker, S. M. (ed.) (2014). *Toxicants in Food Packaging and Household Plastics, Molecular and Integrative Toxicology*, London: Springer-Verlag. doi: 10.1007/978-1-4471-6500-2_2.

Saeidnia, S. (2014). Phthalates. 928–933, doi: 10.1016/B978-0-12-386454-3.00963-5.

Salaudeen, T., Okoh, O., Agunbiade, F., and Okoh, A. (2018). Fate and Impact of Phthalates in Activated Sludge Treated Municipal Wastewater on the Water Bodies in the Eastern Cape, South Africa. *Chemosphere*, 203, 336–344. doi: 10.1016/j.chemosphere.2018.03.176.

Santana-Mayor, Á., Socas-Rodríguez, B., Afonso, M. d. M., Palenzuela-López, J. A., and Rodríguez-Delgado, M. Á. (2018). Reduced Graphene Oxide-Coated Magnetic-Nanoparticles as Sorbent for the Determination of Phthalates in Environmental Samples by Micro-Dispersive Solid-Phase Extraction Followed by Ultra-High-Performance Liquid Chromatography Tandem Mass Spectrometry. *J. Chromatogr.*, A1565, 36–47, doi: 10.1016/j.chroma.2018.06.031.

Sathyanarayana, S., Barrett, E., Butts, S., Wang, C., and Swan, S. H. (2014). Phthalate Exposure and Reproductive Hormone Concentrations in Pregnancy. *Hum. Reprod.*, 147, 401–409, doi: 10.1530/rep-13-0415.

Selvaraj, K. K., Sundaramoorthy, G., Ravichandran, P. K., Girijan, G. K., Sampath, S., and Ramaswamy, B. R. (2015). Phthalate Esters in Water and Sediments of the Kaveri River, India: Environmental Levels and Ecotoxicological Evaluations. *Environ. Geochem. Health*, 37 (1), 83–96. doi: 10.1007/s10653-014-9632-5.

Seyoum, A., and Pradhan, A. (2019). Effect of Phthalates on Development, Reproduction, Fat Metabolism and Lifespan in Daphnia Magna. *Sci. Total Environ.*, 654, 969–977. doi: 10.1016/j.scitotenv.2018.11.158.

Shin, H.-M., Dhar, U., Calafat, A. M., Nguyen, V., Schmidt, R. J., and HertzPicciotto, I. (2020). Temporal Trends of Exposure to Phthalates and Phthalate Alternatives in California Pregnant Women during 2007–2013: Comparison with Other Populations. *Environ. Sci. Technol.*, 54 (20), 13157–13166. doi: 10.1021/acs.est.0c03857.

Sibali, L. L. (2008). Determination of Endocrine Disrupting Compounds in Water and Sediments from Jukskei River Catchment Area. South Africa: Tshwane University of Technology [PhD thesis].

Staples, C. A., Adams, W. J., Parkerton, T. F., Gorsuch, J. W., Biddinger, G. R., and Reinert, K. H. (1997). Aquatic Toxicity of Eighteen Phthalate Esters. *Environ. Toxicol. Chem.*, 16 (5), 875–891. doi: 10.1002/etc.5620160507.

Starratt, A. N., and Lazarovits, G. (1999). Herbicide-Induced Disease Resistance and Associated Increases in Free Amino Acid Levels in Melon Plants. *Can. J. Plant. Pathol.*, 21, 33–36.

Stojanoska, M. M., Milosevic, N., Milic, N., and Abenavoli, L. (2017). The Influence of Phthalates and Bisphenol A on the Obesity Development and Glucose Metabolism Disorders. *Endocrine*, 55, 666–681. doi: 10.1007/s12020-016-1158-4.

Sun, J., Wu, X., and Gan, J. (2015). Uptake and Metabolism of Phthalate Esters by Edible Plants. *Environ. Sci. Technol.*, 49, 8471–8478. doi: 10.1021/acs.est.5b01233.

Swan, S. H., Main, K. M., Liu, F., Stewart, S. L., Kruse, R. L., Calafat, A. M., et al. (2005). Decrease in Anogenital Distance among Male Infants with Prenatal Phthalate Exposure. *Environ. Health Perspect.*, 113 (8), 1056–1061. doi: 10.1289/ ehp.8100.

Talia, C., Connolly, L., and Fowler, P. A. (2021). The Insulin-Like Growth Factor System: A Target for Endocrine Disruptors? *Environ. Int.*, 147, 106311. doi: 10.1016/j.envint.2020.106311.

Tanvir, R., Sajid, I., Hasnain, S., Kulik, A., and Grond, S. (2016). Rare *Actinomycetes Nocardia caishijiensis* and *Pseudonocardia carboxydivorans* as Endophytes, Their Bioactivity and Metabolites Evaluation. *Microbiol. Res.*, 185, 22–35.

Tong, Y., Liu, X., and Zhang, L. (2019). Green Construction of Fe_3O_4@GC Submicrocubes for Highly Sensitive Magnetic Dispersive Solid-Phase Extraction of Five Phthalate Esters in Beverages and Plastic Bottles. *Food Chem.*, 277, 579–585. doi: 10.1016/j.foodchem.2018.11.021.

Turchyn, A. V., Antler, G., Byrne, D., Miller, M., and Hodell, D. A. (2016). Microbial Sulfur Metabolism Evidenced from Pore Fluid Isotope Geochemistry at Site U1385. *Glob. Planet. Change*, 141, 82–90.

UNEP. (2020). An Assessment Report on Issues of Concern: Chemicals and Waste Issues Posing Risks to Human Health and the Environment. Available at: https://wedocs.unep.org/bitstream/handle/20.500.11822/33807/ARIC.pdf? Sequence=1 & is Allowed = y (Accessed November 21, 2021).

Upadhyay, S. K., Singh, D. P., and Saikia, R. (2009). Genetic Diversity of Plant Growth Promoting Rhizobacteria Isolated from Rhizospheric Soil of Wheat under Saline Condition. *Curr. Microbiol.*, 59, 489–496.

USEPA. (2018). America's Children and the Environment. Appendix A: Data Tables – Biomonitoring – Phthalates.

USEPA. (2021). Bis (2-ethylhexyl) Phthalate (DEHP). Available at: www. epa.gov/sites/production/files/2016-09/documents/bis-2-ethylhexyl-phthalate.pdf (Accessed November 11, 2021).

Wang, H., Liang, H., and Gao, D. (2017). Occurrence and Distribution of Phthalate Esters (PAEs) in Wetland Sediments. *J. For. Res.*, 28 (6), 1241–1248. doi: 10.1007/s11676-017-0371-1.

Wang, Y., Zhu, H., and Kannan, K. (2019). A Review of Biomonitoring of Phthalate Exposures. *Toxics*, 7 (2), 21. doi: 10.3390/toxics7020021.

Weizhen, Z., Xiaowei, Z., Peng, G., Ning, W., Zini, L., Jian, H., et al. (2020). Distribution and Risk Assessment of Phthalates in Water and Sediment of the Pearl River Delta. *Environ. Sci. Pollut. Res.*, 27, 12550–12565. doi: 10.1007/ s11356-019-06819-y.

WHO. (2021). Food Safety and Zoononess, Areas of Work Chemical Risks.

Xie, F., Chen, X., Weng, S., Xia, T., Sun, X., and Luo, T. (2019). Effects of Two Environmental Endocrine Disruptors Di-n-butyl Phthalate (DBP) and MonoN-Butyl Phthalate (MBP) on Human Sperm Functions In Vitro. *Reprod. Toxicol.*, 83, 1–7. doi: 10.1016/j.reprotox.2018.10.011.

Xu, T., and Yin, D. (2019). The Unlocking Neurobehavioral Effects of Environmental Endocrine-Disrupting Chemicals. *Curr. Opin. Endocr. Metab. Res.*, 7, 9–13. doi: 10.1016/j.coemr.2019.04.010.

Xu, W., You, Y., and Wang, Z. (2018). Dibutyl Phthalate Alters the Metabolic Pathways of Microbes in Black Soils. *Sci. Rep.*, 8, 2605–2617, doi: 10.1038/s41598-018-21030-8.

Xu, Y., Minhazul, K. A. H. M., Wang, X., Liu, X., Li, X., Meng, Q., et al. (2020). Biodegradation of Phthalate Esters by Paracoccus Kondratievae BJQ0001 Isolated from Jiuqu (Baijiu Fermentation Starter) and Identification of the Ester Bond Hydrolysis Enzyme. *Environ. Pollut.*, 263, 114506. doi: 10.1016/ j.envpol.2020.114506.

Yang, J., Li, Y., Wang, Y., Ruan, J., Zhang, J., and Sun, C. (2015). Recent Advances in Analysis of Phthalate Esters in Foods. *Trac. Trends Anal. Chem.*, 72, 10–26. doi: 10.1016/j.trac.2015.03.018.

Yildirim, E., Turan, M., and Donmez, M. F. (2008). Mitigation of Salt Stress in Radish (*Raphanus sativus* L.) by Plant Growth Promoting Rhizobacteria. *Roum. Biotechnol. Lett.*, 13, 3933–3943.

Yousefi, Z., Ala, A., Babanezhad, E., and Mohammadpour, R. A. (2019). Evaluation of Exposure to Phthalate Esters through the Use of Various Brands of Drinking Water Bottled in Polyethylene Terephthalate (PET) Containers under Different Storage Conditions. *Environ. Health Eng. Manage. J.*, 6 (4), 247–255, doi: 10.15171/EHEM.2019.28.

You, H. H., and Song, G. (2021). Review of Endocrine Disruptors on Male and Female Reproductive Systems. *Comp. Biochem. Physiol. C: Toxicol. Pharmacol.*, 244, 109002. doi: 10.1016/j.cbpc.2021.109002.

Yu, H., Wang, L., Lin, Y., Liu, W., Tuyiringire, D., Jiao, Y., et al. (2020). Complete Metabolic Study by Dibutyl Phthalate Degrading Pseudomonas Sp. DNB-S1. *Ecotoxicol. Environ. Saf.*, 194, 110378. doi: 10.1016/ j.ecoenv.2020.110378.

Yuan, L., Liu, X. G., and Tang, Y. (2015). The Comparison of Free Radical Scavenging Capacity of Carotenoids. *Pack. Food Mach.*, 33, 7–11.

Zacharia, J. T. (2019). Degradation Pathways of Persistent Organic Pollutants (POPs) in the Environment.

Zhang, B., Zhang, T., Duan, Y., Zhao, Z., Huang, X., Bai, X., et al. (2019). Human Exposure to Phthalate Esters Associated with E-Waste Dismantling: Exposure Levels, Sources, and Risk Assessment. *Environ. Int.*, 124, 1–9. doi: 10.1016/ j.envint.2018.12.035.

Zhang, J., Zhang, C., Zhu, Y., Li, J., and Li, X. (2018a). Biodegradation of Seven Phthalate Esters by *Bacillus Mojavensis* B1811. *Int. Biodeterior. Biodegradation*, 132, 200–207. doi: 10.1016/j.ibiod.2018.04.006.

Zhang, Y., Jiao, Y., Li, Z., Tao, Y., and Yang, Y. (2021). Hazards of Phthalates (PAEs) Exposure: A Review of Aquatic Animal Toxicology Studies. *Sci. Total Environ.*, 771, 145418. doi: 10.1016/j.scitotenv.2021.145418.

Zhang, Y., Wang, P. J., Wang, L., Sun, G. Q., Zhao, J. Y., and Zhang, H. (2015). The Influence of Facility Agriculture Production on Phthalate Esters Distribution in Black Soils of Northeast China. *Sci. Total Environ.*, 506, 118–125.

Zhang, Z.-M., Zhang, H.-H., Li, J.-L., and Yang, G.-P. (2017). Determination of Phthalic Acid Esters in Seawater and Sediment by Solid-phase Microextraction and Gas Chromatography-Mass Spectrometry. *Chin. J. Anal. Chem.*, 45 (3), 348–356. doi: 10.1016/s1872-2040(17)60999-x.

Zhu, M., Wu, J., Ma, X., Huang, C., Wu, R., Zhu, W. (2019). Butyl Benzyl Phthalate Promotes Prostate Cancer Cell Proliferation through miR-34a Downregulation. *Toxicol. Vitro*, 54, 82–88. doi: 10.1016/ j.tiv.2018.09.007.

11 Environmental Effects of Phenolic Compound Additives

Abdulmujeeb Bolaji Hamzat, Akintunde Adewole Adegbite, Taofeek Olawale Olabiyi, Moromade Abibat Lere-Aliu, and Adewumi Yetunde Oriji

11.1 INTRODUCTION TO PHENOLIC COMPOUNDS

Phenolic compounds, often referred to as phenolics, encompass a diverse class of chemical substances distinguished by the presence of one or more hydroxyl groups directly bonded to an aromatic hydrocarbon ring (benzene ring). These compounds are structurally defined by the phenol group and exhibit significant diversity, encompassing thousands of natural and synthetic derivatives (Rappoport, 2004). This structural variability underpins their extensive applications across industrial, pharmaceutical, and agricultural domains (Figure 11.1).

Simple Phenols: These are the most basic form; they are characterized by a single benzene ring with one or more hydroxyl (–OH) groups. These compounds represent the fundamental class of phenolic compounds and exhibit versatile chemical properties and applications. Examples include phenol itself, catechol, and hydroquinone. The simplest form, with one hydroxyl group directly attached to the benzene ring (phenol), is used extensively as a precursor in the production of resins, pharmaceuticals, and disinfectants. Catechol (1,2-benzenediol): This compound has two hydroxyl groups in adjacent positions (ortho configuration) and is widely used in photography, dye production, and as a chemical intermediate for antioxidants. Hydroquinone (1,4-benzenediol): With hydroxyl groups positioned opposite each other (para configuration), hydroquinone finds applications in photography, as a polymerization inhibitor, and in skin-lightening formulations. Simple phenols are noted for their antioxidant properties due to their ability to donate hydrogen atoms, stabilizing reactive free radicals. Catechol and hydroquinone are involved in enzymatic and oxidative pathways in plants and microbes, influencing biological processes such as lignin synthesis and biodegradation.

Polyphenols: Polyphenols are compounds characterized by the presence of multiple phenolic hydroxyl groups in their chemical structure. In fact, most phenolic compounds derived from plants are polyphenols, making them a dominant subgroup of plant phenolics. Polyphenols can be described as a diverse class of naturally occurring bioactive molecules containing multiple phenolic functionalities (Lattanzio et al., 2013). These compounds are predominantly found in higher plants and play a vital role in imparting key organoleptic properties to plant-based foods and beverages. Specifically, they contribute to the color, flavor, and taste profiles of these products. Their presence not only enhances sensory qualities but also supports various biological and antioxidant activities (Xie et al., 2023). Polyphenols are abundant in seeds, fruits, and various plant tissues, where they occur either in a free form or conjugated with sugars as glycosides and esters. These compounds constitute a diverse family of thousands of molecules found in a wide range of everyday foods

DOI: 10.1201/9781003602828-14

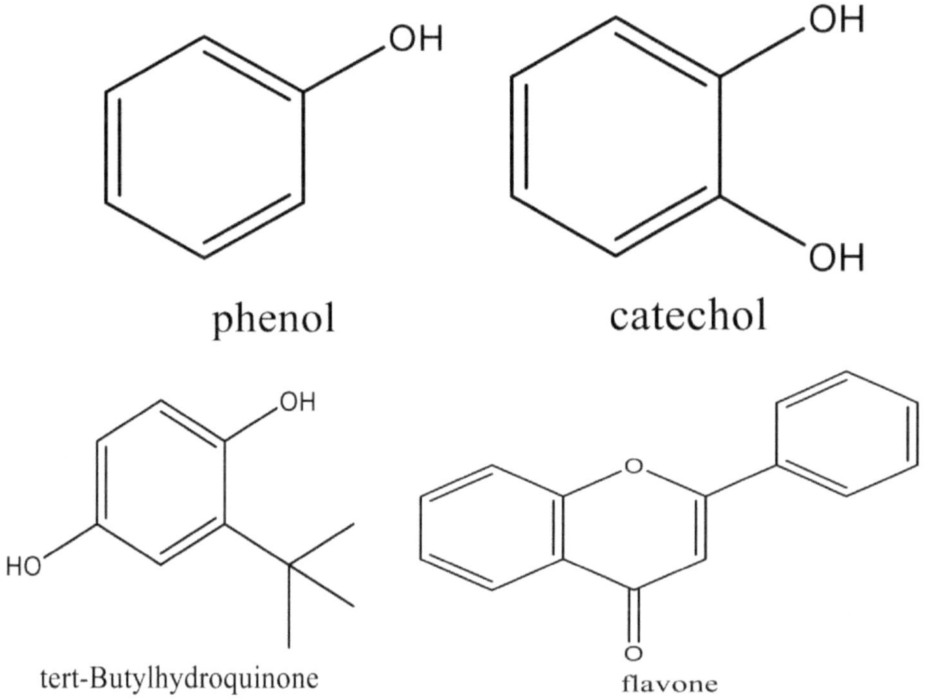

FIGURE 11.1 Chemical structures of some phenolic compound derivatives.

and beverages, including cereals, fruits, wine, tea, chocolate, and onions (Belščak-Cvitanović et al., 2018). The natural kingdom hosts three primary classes of polyphenols: flavonoids, lignin–carbohydrate complexes, and tannins. However, some classifications further subdivide these groups to reflect their structural and functional diversity.

Substituted Phenols: Substituted phenols, such as bisphenol A (BPA) and para-tert-butylphenol (PTBP), represent an important class of compounds due to their tailored reactivity and functionalities imparted by additional groups like alkyl, methoxy, or halogens. These substitutions influence their chemical properties and applications across industries. BPA, a widely used substituted phenol, serves as a precursor in the production of polycarbonates, epoxy resins, and other plastics. However, its endocrine-disrupting effects and potential toxicity have prompted significant research into sustainable alternatives. Recent advancements include renewable BPA substitutes derived from lignin-based phenolics, such as bisguaiacols. These bio-based alternatives maintain similar mechanical and thermal properties while offering reduced environmental and health impacts. They are synthesized via catalytic processes, often involving green chemistry approaches like mild conditions and renewable solvents, highlighting their eco-friendly potential. PTBP, known as 4-tert-butylphenol, is another prominent substituted phenol used in manufacturing phenolic resins, adhesives, and coatings. Its inclusion improves the binding properties and tackiness of rubber-based materials and inks. This compound is also applied in UV stabilizers and oil additives, demonstrating its versatility in enhancing material performance.

Phenolic compound additives are valued for their chemical stability and reactive properties. They are frequently used as starting materials for polymers, resins, and adhesives. The aromatic ring provides thermal stability, while the hydroxyl groups allow for reactivity, especially as antioxidants and UV stabilizers (Zhang et al., 2022). They serve as antioxidants, which is essential in industries like food preservation, cosmetics, and pharmaceuticals, where controlling oxidation is key to extending product shelf life.

11.1.2 Classification of Phenolic Additives

Phenolic additives can be broadly categorized based on their origin (natural or synthetic), chemical structure, and functional characteristics (Tsimogiannis and Oreopoulou, 2019). Each type is tailored to specific applications, especially in industrial settings where durability, oxidative stability, and bioactivity are important.

11.1.2.1 Natural Phenolic Additives

Phenolic compounds, naturally occurring in plants, fruits, vegetables, herbs, and certain microbes, are widely recognized for their antioxidant, antimicrobial, and health-promoting properties. These compounds play a critical role in plant defense mechanisms and contribute to the flavor, color, and nutritional quality of foods (Asif, 2015). With increasing interest in sustainable and eco-friendly solutions, phenolic compounds such as tocopherols (vitamin E) and flavonoids are gaining popularity as natural alternatives to synthetic additives. Tocopherols, derived from vegetable oils, are extensively used in food and cosmetics for their antioxidant properties. Other notable phenolic compounds include flavonoids, tannins, and phenolic acids like caffeic and gallic acids, which are extracted from various plant sources. Compounds such as rosmarinic acid and catechins, obtained from rosemary and tea, respectively, are used in food preservation and cosmetic formulations due to their antioxidant capabilities (Tsimogiannis and Oreopoulou, 2019). Studies by Lee and Zhang (2023) have highlighted their effectiveness in enhancing the oxidative stability in food oils, showcasing their potential as sustainable alternatives to synthetic antioxidants. The growing demand for phenolic compounds spans multiple industries, driven by the need for products that align with health and environmental standards. Continued research and innovation in phenolic additives are advancing food preservation, material science, and sustainability. Fruits, such as berries (e.g., strawberries, blueberries, blackberries), are rich in flavonoids, tannins, and phenolic acids like ellagic acid and anthocyanins, providing antioxidant benefits (Zhang et al., 2023b, while citrus fruits (e.g., oranges, lemons) contain flavonoids like hesperidin and naringenin known for their anti-inflammatory properties (Martínez-Zamora et al., 2023). Vegetables like leafy greens (e.g., spinach, kale) offer flavonoids like quercetin for cardiovascular health, and cruciferous vegetables (Delgado et al., 2023), such as broccoli, are high in phenolic acids like caffeic and ferulic acids, studied for potential anti-cancer effects (Melim et al., 2022). Herbs and spices, such as green tea (*Camellia sinensis*) and turmeric (*Curcuma longa*), contain polyphenols like catechins and curcumin, offering antioxidant and anti-inflammatory effects (Chen et al., 2023; Fuloria et al., 2022). Grains and legumes, including whole grains (e.g., oats, barley) and soybeans, contain phenolic acids like ferulic acid and isoflavones, promoting antioxidant benefits and heart health (Zhang et al., 2023b; Hu et al., 2020). Nuts and seeds, such as walnuts, almonds, and flaxseeds, are high in ellagic acid and polyphenols that support cardiovascular and brain health, while red wine contains resveratrol and olive oil is rich in hydroxytyrosol, both providing cardiovascular and anti-aging benefits (Kumar et al., 2024; Martínez Alvarez et al., 2023; Rodríguez et al., 2024). Medicinal plants like willow bark (*Salix* spp.) and ginger (*Zingiber officinale*) offer phenolic compounds like salicylic acid and gingerols, known for their analgesic, anti-inflammatory, and antioxidant effects (Aleman et al., 2023)

11.1.2.2 Synthetic Phenolic Additives

Synthetic Phenols: These compounds are produced via engineered chemical synthesis, including alkylphenols and phenolic resins, engineered to fulfill specific industrial requirements, such as enhancing stability, durability, and functionality, generating compounds like nonylphenol and octylphenol. Making them ideal for high-performance materials.

Monophenols: They contain a single-hydroxyl group attached to a benzene ring. Due to their simple structure, monophenols are primarily used in plastics, rubbers, and adhesives to prevent oxidative degradation. Examples include butylated hydroxytoluene (BHT) and butylated hydroxyanisole (BHA). BHT and BHA are particularly effective in stabilizing food products by preventing rancidity in oils and fats. Additionally, they are commonly added to packaging materials to extend product life.

Diphenols and polyphenols, with two or more hydroxyl groups, offer stronger antioxidant properties due to the greater number of hydroxyl sites available for neutralizing free radicals. A well-known polyphenolic additive is tert-butylhydroquinone (TBHQ), often added to vegetable oils and processed foods for enhanced stability. Polyphenols are also widely utilized in cosmetics and pharmaceuticals, where they help stabilize formulations and enhance product longevity.

i. Hindered phenols are modified phenols with bulky substituents (often alkyl groups) near the hydroxyl group, which increases their stability and effectiveness as antioxidants. These additives, such as Irganox 1010 and Irganox 1076, are widely used in the plastics and rubber industries to protect against oxidation under high temperatures. Their hindrance provides prolonged stability, which is especially important for polymers used in outdoor applications (they are often referred to as primary antioxidants or phenolic antioxidants).
ii. Phenolic phosphites combine phenolic structures with phosphite groups, enhancing both antioxidant and UV-stabilizing properties. These are commonly used in applications requiring protection from both thermal and photo-oxidative degradation, such as in polyolefins and polyesters. An example is tris(2,4-di-tert-butylphenyl) phosphite (Irgafos 168), often used in combination with hindered phenols to stabilize high-performance plastics. They are often referred to as secondary antioxidants and often combined with primary antioxidants, help decompose hydroperoxides into stable products. Compounds like phosphites or thioesters work synergistically with phenolic additives to further enhance oxidative stability. Recent research emphasizes the importance of using these combinations to increase stability in food and plastic industries, particularly where long-term storage is required (Genea et al., 2023).

11.1.3 Applications of Phenolic Compounds

The versatility of phenolic compounds has led to their widespread use across multiple industries. In each field, specific phenolic derivatives are selected for their unique properties, enabling them to fulfill diverse functional requirements (Albuquerque et al., 2021). Phenolic additive usage in various industries;

11.1.3.1 Industrial Uses

Plastics and Polymers: Phenolic additives are widely used in plastic production, especially for enhancing durability and thermal stability. By preventing oxidative degradation, they help maintain mechanical properties, color, and transparency of plastics, even when exposed to UV radiation or heat. Some examples include the following: BHT: Used in polyolefins, such as polyethylene and polypropylene, BHT protects against oxidation, preserving the plastic's structure and appearance. Phenolic resins, such as novolac and resol, are key components widely used in molded plastic parts for automotive and electronics applications. They offer high thermal resistance, flame retardancy, and chemical stability.

Coatings: Phenolic resins and additives are critical in coatings for both aesthetic and protective purposes. They help prevent corrosion and provide UV resistance, which is essential for coatings on outdoor structures, pipelines, and marine equipment. Some examples include BPA-based epoxy resins: These resins are widely used in coatings for metal cans, steel pipes, and marine applications due to their durability and resistance to chemicals and moisture.

Rubbers: In the rubber industry, phenolic additives improve resistance to oxidation and ozone cracking, extending the lifespan of rubber products like tires, seals, and hoses. Examples include alkylated phenols (e.g., tert-butylphenols): These compounds prevent oxidative degradation and hardening of rubber in harsh environments, making them suitable for automotive and industrial applications.

Lubricants: Phenolic antioxidants are used in lubricants to prevent oxidative breakdown, which can cause sludge formation and reduce lubricant effectiveness. This is particularly important in high-temperature applications, like engine oils and hydraulic fluids. BHA and BHT: These phenolic antioxidants stabilize lubricants, help in retaining their viscosity, and reduce the need for frequent replacements.

Agriculture: Phenolic compounds play a role in pesticide formulations and plant growth regulators. For example, phenolic herbicides disrupt photosynthesis in weeds. Additionally, certain phenolic compounds act as growth regulators, enhancing plant resilience to stress. In animal feed, antioxidants from phenolic sources are added to prevent the oxidation of fats, preserving feed quality. Lignin-derived phenolics, extracted from wood processing by-products, are compounds also emerging as sustainable antioxidants in bioplastics and material science applications.

Pharmaceuticals and Health Products: Many phenolic compounds exhibit antimicrobial and anti-inflammatory properties. This has prompted their use in pharmaceuticals, where they serve as both therapeutic agents and stabilizers in drug formulations. Some phenolic compounds, like salicylic acid, are also active ingredients with therapeutic roles, known for their efficacy in treating skin conditions and inflammation.

11.1.3.2 Antioxidant Properties

The antioxidant properties of phenolic additives are essential in applications that require prolonged stability and resistance to oxidative damage. Phenolics act by donating hydrogen atoms or electrons to neutralize free radicals, thereby slowing down oxidation and extending the shelf life of products.

Polymers: Phenolic antioxidants are key in polymer stabilization, preventing deterioration due to exposure to heat, light, and oxygen. This is crucial in plastics, where oxidative degradation can cause discoloration, brittleness, and loss of mechanical properties. BHT is often added to polyolefin-based packaging materials to prevent oxidation, helping to maintain the material's clarity and flexibility over time.

Food: In the food industry, phenolic antioxidants are used to delay rancidity in fat-containing foods. They help maintain flavor, color, and nutritional value, extending the product's shelf life. BHA is commonly used in processed foods, such as snacks and cereals, to prevent lipid oxidation, while natural phenolic compounds like rosemary extract are gaining popularity as natural alternatives. Flavonoids: Found in fruits, vegetables, and beverages like tea and wine, flavonoids are responsible for vibrant colors and exhibit antioxidant, anti-inflammatory, and preservative properties.

Cosmetics: Phenolic antioxidants are essential in cosmetics, where they prevent rancidity and degradation of ingredients. They protect formulations from UV and oxidative damage, extending product shelf life and ensuring stability. Vitamin E (a natural phenolic antioxidant) is widely used in skincare products to stabilize oils and reduce oxidation-induced skin irritation.

11.1.4 Structure–Function Relationship in Phenolic Additives

Phenolic additives are profoundly influenced by their functionality and structural variations, impacting their antioxidant efficacy, reactivity, solubility, and stability under various conditions (Rudrapal et al., 2024).

Hydroxyl Group Positioning: The position and number of hydroxyl groups (–OH) on the aromatic ring are crucial to the antioxidant activity of phenolic compounds. Hydroxyl groups act as electron donors, stabilizing free radicals through hydrogen donation and resonance stabilization (Chen et al., 2020). Example: Catechol, which has hydroxyl groups at ortho-positions, is more reactive and has stronger antioxidant properties than phenol itself, which only has one hydroxyl group.

Substituent Effects (Alkyl, Alkoxy Groups): Introducing alkyl groups (e.g., tert-butyl groups) increases the stability of phenolic antioxidants by making the molecule more resistant to degradation at high temperatures. Alkyl-substituted phenols are often chosen for polymer stabilization due to their enhanced oxidative resistance (Zhang et al., 2023a). Example: BHT and BHA have alkyl groups that not only increase lipid solubility (useful in food and cosmetic applications) but also enhance antioxidant stability by stabilizing phenoxy radicals.

Conjugation and Multiple Aromatic Rings: Polyphenols, with multiple aromatic rings and conjugated double bonds, have extensive electron delocalization, making them highly effective at neutralizing free radicals (Belščak-Cvitanović et al., 2018). This feature also increases their stability in UV-exposed environments. Example: Flavonoids (natural polyphenols) have conjugated structures that provide broad-spectrum antioxidant and UV-protective functions, useful in cosmetics and food packaging.

Polymeric or Network Structures: Phenolic resins, such as those derived from BPA or novolac, form crosslinked networks, providing superior durability, chemical resistance, and thermal stability. These features make them ideal for high-stress applications, such as automotive and aerospace industries (Asim et al., 2018). Example: BPA-based epoxy resins are widely used in protective coatings and electronic encapsulation because the network structure resists degradation and provides electrical insulation.

Steric Effects: Bulky substituents, such as tert-butyl groups, can shield the phenolic hydroxyl group, slowing down oxidative reactions (Jordan et al., 2022). These steric effects are leveraged in designing additives that need to provide long-term stability without readily reacting under mild conditions. Example: Tris(3,5-di-tert-butyl-4-hydroxybenzyl) isocyanurate has bulky tert-butyl groups that prevent rapid oxidation, providing prolonged stabilization in polyolefins.

11.2 MECHANISMS OF ACTION

11.2.1 How Phenolic Additives Function as Antioxidants and UV Stabilizers

Phenolic additives are widely used as antioxidants and UV stabilizers, essential for protecting materials from oxidative and photodegradative damage (Wang et al., 2022). Their functionality arises from their chemical structure, particularly with the presence of hydroxyl groups attached to an aromatic ring, which can stabilize reactive species and prevent chain reactions of degradation.

Antioxidants: Phenolic antioxidants are widely used in various industrial applications due to their ability to inhibit oxidative degradation, particularly in polymers, rubber, and lubricant products (Zhao et al., 2023). In food and cosmetics, phenolic antioxidants inhibit lipid oxidation, a process that leads to rancidity and degradation of quality. By scavenging free radicals, they preserve the freshness, flavor, and effectiveness of products and enhance the stability of materials that are exposed to heat, light, and oxygen (Abeyrathne et al., 2021). Recent studies highlight their growing importance, applications, and the development of newer, more efficient phenolic antioxidants due to ongoing concerns regarding environmental impact and human health.

11.2.2 Mechanism of Action

Phenolic antioxidants prevent oxidation primarily by scavenging free radicals. When materials like plastics, rubber, or oils are exposed to heat, light, or oxygen, they form free radicals, which lead to chain reactions that cause material degradation. Phenolic antioxidants, containing a hydroxyl group on the aromatic ring, donate a hydrogen atom to neutralize these free radicals, effectively breaking the oxidation chain reaction. This action is crucial for preventing discoloration, loss of mechanical strength, and brittleness in materials. These properties have led to their extensive use in the plastic and rubber industries, as well as in the stabilization of lubricants and other industrial fluids (Jiang et al., 2021).

Radical Scavenging Mechanism: The antioxidant property of phenolics is largely due to their ability to donate hydrogen atoms from their hydroxyl groups (–OH) to neutralize free radicals. When a phenolic compound encounters a free radical (R•), it donates a hydrogen atom to form a stable molecule (RH) and a phenoxyl radical (Ph-O•), which is more stable and less reactive due to resonance stabilization.

$$\text{Ph OH} + \text{R}\bullet \rightarrow \text{PhO}\bullet + \text{RH}$$

This donation mechanism effectively "quenches" the free radical, stopping it from propagating a chain reaction that would lead to further molecular breakdown.

Resonance Stabilization of Phenoxyl Radicals: After hydrogen donation, the resulting phenoxyl radical (Ph-O•) is stabilized by resonance across the aromatic ring. This stabilization reduces the likelihood of further reactions, making the phenoxyl radical less reactive compared to other radicals. Resonance is particularly strong in polyphenols, where multiple hydroxyl groups and aromatic rings contribute to electron delocalization, enhancing antioxidant effectiveness.

Electron Delocalization and Conjugation: Many phenolic compounds, especially polyphenols, exhibit conjugation (a series of alternating double bonds) within their structure. This conjugation allows for extensive electron delocalization, meaning the electrons are spread out over the structure, which further stabilizes the phenoxyl radical. Conjugated structures in phenolic compounds also enhance UV absorption, making them more effective as UV stabilizers. This is why certain phenolics are especially suited to high-UV environments, like outdoor coatings or packaging materials.

Quenching of Singlet Oxygen and Other Reactive Oxygen Species (ROS): Phenolic compounds can deactivate singlet oxygen (a high-energy form of oxygen that contributes to oxidative stress) through a process known as physical quenching, converting it into a more stable ground state without forming radicals. Phenolic additives can also scavenge other ROS, such as hydroxyl radicals (•OH) and superoxide anions (O2•–), which play a significant role in oxidative degradation, especially in food and cosmetic applications where the stability of sensitive ingredients is vital.

Steric Hindrance and Oxidative Stability: Many synthetic phenolics have bulky substituent groups, such as tert-butyl groups, near the hydroxyl site. These groups create steric hindrance, physically blocking reactive species from accessing the phenolic hydroxyl group. This steric effect allows the additive to remain effective over longer periods, which is particularly beneficial in high-temperature applications. Example: BHT has tert-butyl groups that protect its phenolic hydroxyl, making it a widely used antioxidant in polymers and lubricants where long-term oxidative resistance is required.

Synergistic Effects with Other Stabilizers: Phenolic antioxidants often work in conjunction with other stabilizers, such as phosphites and thioethers, to provide a "multi-layered" defense against degradation. Phosphites, for instance, act as secondary antioxidants by decomposing hydroperoxides (ROOH), which are formed during oxidation and can accelerate degradation. By breaking down hydroperoxides, phosphites regenerate the phenolic antioxidant, allowing it to continue neutralizing free radicals. Example: In polymer applications, the combination of a phenolic antioxidant and a phosphite (such as triphenyl phosphite) is common, providing effective stabilization under both processing and long-term conditions.

11.2.3 UV Stabilizers

UV light can break chemical bonds in polymers, generating free radicals that initiate degradation. Phenolic UV stabilizers absorb UV radiation, dissipating the energy as heat rather than allowing it to trigger chemical reactions in the material. Phenolic-based UV stabilizers play a critical role in protecting materials that are frequently exposed to sunlight, such as outdoor plastics, coatings, and paints (Rahman et al., 2022). Exposure to UV radiation leads to the breakdown of polymer chains, resulting in fading, brittleness, discoloration, and loss of mechanical strength. Phenolic UV stabilizers prevent this degradation by absorbing and dissipating harmful UV radiation before it can damage the material. Recent advancements in phenolic-based UV stabilizers have improved their efficiency, extending the durability and aesthetic appeal of materials exposed to sunlight. Example: Some phenolic compounds contain chromophores, which are groups that absorb UV light and protect the material by converting UV energy to less harmful forms, thus preserving the polymer's integrity.

11.2.4 Mechanism of UV Stabilization

Phenolic UV stabilizers function by absorbing UV light and dissipating it as low-level heat, which prevents the light from reaching the polymer matrix where it could initiate degradation. These stabilizers have hydroxyl groups on their aromatic ring structures that can absorb UV radiation and undergo non-destructive excitation and relaxation cycles, transforming the energy into harmless thermal energy. This unique property allows them to act as sacrificial agents that protect the polymer from photodegradation (Yañez-Cruz et al., 2022). Also, they are highly stable under prolonged UV exposure, allowing them to provide lasting protection for materials.

11.2.5 Types of Phenolic UV Stabilizers

Phenolic-based UV stabilizers can be categorized into several types based on their chemical structure and specific applications:

i. Hydroxyphenyl Benzotriazoles: Known for their high UV-absorbing capacity, these stabilizers are effective in a wide range of polymers, including polyvinyl chloride (PVC), polycarbonate, and polyethylene. They are especially popular in outdoor applications, such as garden furniture, car parts, and protective films, where prolonged sunlight exposure can significantly degrade material performance (Zhou et al., 2023).
ii. Hydroxybenzophenones: These UV absorbers work well in transparent plastics and are commonly used in packaging materials, coatings, and consumer products that need to maintain clarity. Hydroxybenzophenones effectively absorb and dissipate UV light, reducing discoloration and maintaining the visual appeal of products. They are often combined with other stabilizers to enhance UV resistance, particularly in polyolefins and styrenics, where they reduce the risk of yellowing and brittleness (Snedeker, 2014).
iii. Hindered Amine Light Stabilizers (HALS): While not strictly phenolic, HALS are often used in combination with phenolic UV stabilizers to improve light stability in polymers. HALS works by scavenging free radicals generated during photodegradation, thereby protecting the material from light-induced degradation. Their synergy with phenolic UV stabilizers is highly effective in polyolefin-based applications, including automotive parts, agricultural films, and outdoor furnishings (Deng et al., 2024).

11.2.2 Environmental and Human Health Implications

11.2.2.1 Human Health Impacts

Phenolic additives, including synthetic antioxidants like BHT, BHA, and BPA, have distinct toxicity profiles and varying levels of health risks. BHT and BHA are generally recognized as safe by the U.S. Food and Drug Administration (FDA) when used within established limits. However, high doses have been associated with potential carcinogenicity, liver toxicity, and endocrine disruption in animal studies, raising concerns for cumulative exposure in humans. Phenolic additives, such as BPA, nonylphenol, and other alkylphenols, are of growing concern due to their widespread use and potential negative effects on human health. These chemicals are often used in the production of plastics, resins, detergents, and other industrial products, leading to human exposure through various routes (**Food and Water**, **Dietary Intake**, **Dermal Exposure, Inhalation**). One of the most significant health impacts of phenolic additives, particularly BPA and alkylphenols like nonylphenol, is their ability to act as endocrine disruptors. These substances can mimic or interfere with the body's natural hormones, especially estrogen, leading to various health problems. By mimicking estrogen, phenolic additives can disrupt the normal hormonal balance, potentially leading to reproductive and developmental issues (Gałązka and Jankiewicz 2022).

Studies have shown that exposure to phenolic additives can impair fertility in both men and women. In women, BPA exposure has been associated with abnormalities in ovarian function and menstrual cycles. In men, reduced sperm quality and altered hormone levels have been observed with high exposure to phenolic compounds. Pregnant women exposed to phenolic additives may be at risk of passing these chemicals to their unborn children, potentially leading to developmental issues. Fetal exposure to BPA has been linked to abnormal development of the brain, reproductive organs, and other systems (Meli et al., 2020). BPA is a well-known endocrine disruptor, affecting hormonal balance and linked to reproductive, developmental, and metabolic disorders. Regulatory agencies in many countries have restricted BPA use in products like baby bottles and food packaging due to its potential impact on human health, particularly for vulnerable populations like infants. Some phenolic additives have been linked to cancer development. BPA, for example, has been studied for its potential role in increasing the risk of certain cancers, such as breast and prostate cancer, due to its estrogen-like activity. Prolonged exposure to these chemicals could disrupt normal cell growth and increase the likelihood of cancerous mutations. BPA has been shown to promote the growth of breast cancer cells in vitro by mimicking estrogen, suggesting a potential link between BPA exposure and increased breast cancer risk. Animal studies have shown that BPA exposure may contribute to the development of prostate cancer by disrupting normal hormone signaling in the prostate gland (Winz et al., 2023). Phenolic additives have been implicated in the development of metabolic disorders, including obesity and diabetes. The endocrine-disrupting properties of these compounds can interfere with metabolism, insulin sensitivity, and fat storage. BPA exposure has been associated with weight gain and fat accumulation, particularly in children. This may be due to BPA's ability to interfere with hormones that regulate appetite and fat metabolism. Several studies have shown a link between BPA exposure and the development of insulin resistance, which is a precursor to type 2 diabetes. BPA may disrupt the pancreas's ability to regulate blood sugar levels (Wade et al., 2020). There is emerging evidence that phenolic additives may contribute to cardiovascular diseases. BPA, in particular, has been linked to an increased risk of hypertension (high blood pressure) and cardiovascular inflammation. BPA exposure has been shown to cause a temporary increase in blood pressure, possibly due to its effects on blood vessel function and the regulation of heart rate. Long-term exposure to phenolic additives has been linked to an increased risk of heart disease, including atherosclerosis (hardening of the arteries), due to the chemicals' ability to cause oxidative stress and inflammation in blood vessels (Zhor et al., 2023). Exposure to phenolic additives during critical periods of development, such as pregnancy and early childhood, may have lasting effects on brain development and behavior. BPA exposure has been associated with cognitive deficits in children, such as problems with learning, memory, and attention. These effects may be due to BPA's disruption of normal brain development and neurotransmitter signaling. Studies have suggested a potential link between prenatal exposure to phenolic additives and behavioral disorders, such as attention-deficit hyperactivity disorder (ADHD) and increased aggression in children (Xiong et al., 2023). Phenolic additives may also affect the immune system, particularly through their ability to cause oxidative stress and inflammation. Chronic exposure to phenolic additives has been linked to increased levels of inflammation in the body, which can contribute to various chronic conditions, including autoimmune diseases. There is some evidence that phenolic additives may increase the risk of allergic reactions and asthma, particularly in children. This may be due to their ability to disrupt the immune system's normal response to allergens (Dębińska and Sozańska, 2023). Exposure to phenolic additives can also have toxic effects on the kidneys. Prolonged exposure to BPA and related compounds has been shown to result in kidney damage, including inflammation, fibrosis, and impaired kidney function. This can lead to chronic kidney disease over time. In developmental studies, exposure to phenolic compounds has been linked to abnormalities in kidney development, potentially increasing the risk of kidney disease later in life (Priego et al., 2021). Phenolic antioxidants like BHT and BHA may cause allergic reactions and skin sensitization in sensitive individuals. In cosmetics and personal care products, these additives are carefully regulated to minimize the risk of contact dermatitis and other allergic reactions. While some phenolic compounds, like BHA, have shown carcinogenic effects in animal studies, evidence of similar effects in humans is inconclusive. Nevertheless, regulatory agencies set strict usage limits, particularly in food and consumer products, to minimize potential long-term health risks. Similarly, exposure to certain phenolic additives, especially during prenatal and early-life stages, has been linked to developmental and neurological effects. For example, BPA exposure has been associated with neurobehavioral changes in children, prompting stricter regulations on BPA use in children's products.

11.2.2.2 Environmental Impacts

According to Wang et al. (2024), these additives are detected in varying proportions, and most are not chemically bonded to the polymer chain. As a result, these additives can progressively leach into the environment and infiltrate living organisms through the use, recycling, and deterioration of plastic products. This occurrence poses a significant hazard to both terrestrial and aquatic organisms as well as human health.

Bioaccumulation and Persistence: Many synthetic phenolic additives are resistant to biodegradation, persisting in ecosystems and accumulating in the food chain, such as BHT and BPA, which have limited biodegradability; this accumulation can disrupt ecosystems and impact wildlife health. BPA and nonylphenol are known to bioaccumulate in aquatic organisms, posing risks to marine ecosystems by disrupting endocrine functions in marine organisms and potentially affecting species higher in the food chain, including humans, resulting in hormonal imbalance.

Aquatic Toxicity: Phenolic compounds, especially alkyl phenols and bisphenols, are toxic to aquatic life, causing reproductive and developmental issues in fish and invertebrates. They disrupt hormonal balance, leading to endocrine disruption even at low concentrations, which affects species biodiversity and ecosystem stability.

Soil and Water Contamination: Due to their persistence, phenolic additives leach into soil and groundwater, potentially contaminating drinking water supplies. This contamination can impact both plant and animal life in terrestrial ecosystems, leading to bioaccumulation and toxicity in various organisms.

11.2.3 Safety Regulations

Safety regulations vary globally, but several major organizations and regulations govern the use of phenolic additives, particularly for food, cosmetics, and industrial products:

Food Safety Authorities: In the U.S., the FDA (2023) regulates phenolic additives in food, limiting their concentrations and requiring labeling. The European Food Safety Authority (EFSA) and Codex Alimentarius Commission also set maximum allowable levels for synthetic phenolics like BHT, BHA, and BPA in food products (Le Bloch et al., 2024).

Environmental and Chemical Safety Agencies: The U.S. Environmental Protection Agency (EPA) and the European Chemicals Agency (European Chemicals Agency (ECHA) 2023) assess phenolic additives for environmental safety. BPA, for instance, is classified as a substance of very high concern (SVHC) under Europe's REACH regulation, leading to significant restrictions in consumer products (Wang et al., 2021).

Cosmetic and Pharmaceutical Standards: In cosmetics, regulatory bodies such as the FDA (2023) and the European Commission impose restrictions on phenolic compounds due to skin irritation, sensitization risks, and potential endocrine disruption. The EU Cosmetics Regulation, for example, limits BHA concentrations and mandates labeling of known sensitizers (Sarkar et al., 2023).

Occupational Exposure Limits: Regulatory agencies, including OSHA (Occupational Safety and Health Administration) and NIOSH (National Institute for Occupational Safety and Health 2010), establish permissible exposure limits (PELs) for workers handling phenolic additives, focusing on ventilation, protective equipment, and training to mitigate risks (Crawford et al., 2008).

Natural Phenolic Additives: Natural phenolics from sources like rosemary and green tea are generally safer alternatives to synthetic phenolics, with antioxidant properties and minimal toxicity (Calderón-Oliver and Ponce-Alquicira 2021). However, excessive ingestion of concentrated natural phenolics may lead to adverse effects, such as gastrointestinal distress or liver toxicity, highlighting the importance of dose regulation.

11.2.4 Sustainable and Safer Alternatives

With increasing awareness of environmental impact, the use of phenolic additives, especially synthetic ones, is under scrutiny. Key concerns include biodegradability, sustainability, and eco-toxicity. While synthetic phenolics have robust performance characteristics, some may persist in the environment, leading to potential toxicity. In response to growing concerns, researchers and industries are focusing on developing safer, biodegradable phenolic alternatives, including natural antioxidants from plants like rosemary, thyme, and green tea. These natural phenolics exhibit antioxidant properties similar to synthetic compounds but with fewer toxicity risks. Other green chemistry approaches include designing phenolic additives with improved biodegradability and reduced toxicity profiles, helping to minimize their ecological footprint and support sustainable industrial practices. There is also growing interest in developing bio-based phenolic additives derived from renewable sources, such as plant extracts, to reduce reliance on petroleum-based chemicals. These alternatives are typically more biodegradable and less toxic. Tannins and lignin-derived phenolics from wood and agricultural waste are increasingly studied as eco-friendly alternatives to synthetic phenolics.

11.2.5 Analytical Techniques for Phenolic Additives

Phenolic additives in complex matrices for quantification and identification are essential for quality control, product optimization, and regulatory compliance. Various analytical techniques have been developed to detect and analyze phenolic additives in plastics, food, cosmetics, and environmental samples (Martins et al., 2019).

Fourier-Transform Infrared (FTIR) Spectroscopy: FTIR spectroscopy is widely used to identify the functional groups present in phenolic compounds by measuring the absorption of infrared radiation at specific wavelengths. Phenolic additives display characteristic absorbance peaks, particularly for hydroxyl (–OH) and aromatic (C=C) groups, making FTIR a quick and reliable technique for qualitative analysis. A non-destructive technique that requires minimal sample preparation, preserving the sample for further analysis and quick identification, for rapid screening of phenolic compounds based on their functional groups. Though it is less effective for quantification and may struggle with low concentrations of phenolics in complex matrices. They are widely used in quality control for polymer manufacturing to confirm the presence of phenolic stabilizers and detect any degradation products in the sample.

High-Performance Liquid Chromatography (HPLC): HPLC is a powerful technique for the separation, identification, and quantification of phenolic additives. It allows precise quantification of individual phenolic compounds, often using detectors such as UV-Vis or photodiode arrays (PDAs) for enhanced specificity. It is a highly sensitive instrument ideal for detecting trace amounts of phenolic compounds, compatible with various detectors, including mass spectrometry (LC–MS), which can provide additional structural information. It often requires solvent-based extraction and sample preparation, which may be time-consuming. This technique is widely used for quantifying phenolic antioxidants in foods, cosmetics, and polymers, as well as monitoring compliance with regulatory limits for phenolic residues.

Gas Chromatography–MS (GC–MS): GC–MS combines the separation capabilities of GC with the sensitive detection and identification capabilities of MS. Phenolic compounds can be detected as they volatilize and are ionized in the MS detector, making GC–MS particularly effective for volatile phenolic additives or those that can be derivatized. Excellent for trace analysis and complex mixtures, with high specificity, sensitivity, and structural elucidation, which provides molecular weight and fragmentation pattern data, enabling identification of unknown compounds. Phenolic compounds often require derivatization to improve volatility and thermal stability before GC analysis, which adds an extra step. GC–MS is widely used in environmental testing to detect and quantify phenolic compounds in water and soil samples, as well as in materials testing to identify phenolic degradation products in polymers.

According to Thakur and Kumar (2022), the need for faster, more sensitive, and eco-friendly analytical methods has spurred the development of novel techniques and modifications to traditional methods. These emerging techniques are focused on improving accuracy, reducing sample preparation times, and meeting stricter regulatory standards.

Ultra-HPLC (UHPLC): UHPLC is an advanced form of HPLC that operates at higher pressures, allowing for faster separation and-m MS (MS/MS) provides fragmentation data that enables more precise structural identification. Used extensively in regulatory compliance and environmental monitoring, LC–MS/MS can detect low concentrations of phenolic additives and their metabolites in biological and environmental samples, supporting eco-toxicological assessments.

Surface-Enhanced Raman Spectroscopy (SERS): SERS is a highly sensitive technique that enhances the Raman scattering of molecules adsorbed on nanostructured metal surfaces. For phenolic compounds, SERS enables detection at trace levels without extensive sample preparation. SERS is emerging in food safety testing, where trace levels of phenolic antioxidants or contaminants need to be detected. The technique's sensitivity and minimal sample preparation make it ideal for rapid testing in industrial settings.

Electrochemical Detection Methods: Electrochemical sensors and biosensors are gaining attention for phenolic compound analysis due to their portability, rapid response, and low cost. These sensors are based on the redox properties of phenolics, allowing them to detect compounds at low concentrations. Electrochemical methods are used in environmental monitoring and real-time quality control. For instance, biosensors with enzyme-based recognition elements (such as tyrosinase or laccase) are used to detect phenolic contaminants in water samples.

Green Analytical Chemistry Approaches: As sustainability becomes a priority, green chemistry approaches, such as solvent-free extraction and micro-extraction techniques (e.g., solid-phase microextraction or SPME), are being adopted. These methods reduce solvent use, lower waste production, and offer eco-friendly alternatives to traditional sample preparation. Green extraction methods are increasingly used in environmental analysis, where minimal chemical impact is prioritized. Microextraction techniques like SPME are particularly beneficial for fieldwork, allowing on-site sampling with minimal environmental disruption.

11.3 CONCLUSION

Phenolic additives are essential across various industries, providing exceptional antioxidant, UV stabilization, and functional properties that enhance the performance and durability of products, such as polymers, coatings, food, cosmetics, and pharmaceuticals. Their versatility stems from diverse structural variations, enabling applications in material stabilization, product preservation, and performance enhancement. By scavenging free radicals, inhibiting oxidation, and protecting against UV radiation, phenolic additives play a critical role in mitigating degradation and ensuring product stability. Their contribution to extending shelf life and improving material durability underscores their industrial importance. However, challenges surrounding toxicity, environmental persistence, and bioaccumulation have stimulated the search for safer, sustainable alternatives. Regulatory frameworks are vital for ensuring safe usage, while bio-based and biodegradable phenolic additives offer promising solutions to these concerns. Advanced analytical techniques, such as FTIR, HPLC, and GC–MS, are instrumental in quality control, regulatory compliance, and the development of innovative phenolic additives by enabling precise characterization and monitoring of their properties. The future of phenolic additives lies in sustainable innovation. Bio-based and biodegradable phenolic compounds derived from renewable resources are emerging as transformative solutions. Nanotechnology, functional modifications, and computational tools provide opportunities to enhance performance while addressing environmental and safety concerns. Interdisciplinary research and global collaboration will be key to developing safer, cost-effective, and high-performance phenolic additives, expanding their applications across industries. As a critical class of materials bridging functionality, innovation, and sustainability, phenolic additives remain indispensable. Ongoing research and development will unlock their full potential, balancing industrial demands with environmental stewardship and human health. By embracing green chemistry principles and leveraging emerging technologies, the future of phenolic additives promises to meet the needs of a rapidly evolving world in a sustainable and transformative manner.

REFERENCES

Abeyrathne, E.D.N.S., Nam, K., Ahn, D.U., 2021. Analytical methods for lipid oxidation and antioxidant capacity in food systems. *Antioxidants* 10(10), 1587.

Albuquerque, B.R., Heleno, S.A., Oliveira, M.B.P., Barros, L., Ferreira, I.C., 2021. Phenolic compounds: current industrial applications, limitations, and future challenges. *Food & Function* 12(1), 14–29.

Aleman, R.S., Marcia, J., Duque-Soto, C., Lozano-Sánchez, J., Montero-Fernández, I., Ruano, J.A., Hoskin, R.T., Moncada, M., 2023. Effect of microwave- and ultrasound-assisted extraction on the phytochemical and in vitro biological properties of willow (*Salix alba*) bark aqueous and ethanolic extracts. *Plants* 12(13), 2533.

Asif, M., 2015. Chemistry and antioxidant activity of plants containing some phenolic compounds. *Chemistry International* 1(1), 35–52.

Asim, M., Saba, N., Jawaid, M., Nasir, M., Pervaiz, M., Alothman, O.Y., 2018. A review on phenolic resin and its composites. *Current Analytical Chemistry* 14(3), 185–197.

Belščak-Cvitanović, A., Durgo, K., Huđek, A., Bačun-Družina, V., Komes, D., 2018. Overview of polyphenols and their properties. In *Polyphenols: Properties, Recovery, and Applications* (3–44). Woodhead Publishing.

Calderón-Oliver, M., Ponce-Alquicira, E., 2021. Environmentally friendly techniques and their comparison in the extraction of natural antioxidants from green tea, rosemary, clove, and oregano. *Molecules* 26(7), 1869.

Chen, J., Yang, J., Ma, L., Li, J., Shahzad, N., Kim, C.K., 2020. Structure-antioxidant activity relationship of methoxy, phenolic hydroxyl, and carboxylic acid groups of phenolic acids. *Scientific Reports* 10(1), 2611.

Chen, Y., Lai, L., You, Y., Gao, R., Xiang, J., Wang, G., Yu, W., 2023. Quantitative analysis of bioactive compounds in commercial teas: profiling catechin alkaloids, phenolic acids, and flavonols using targeted statistical approaches. *Foods* 12(16), 3098.

Crawford, J., Faroon, O., Llados, F., Wilson, J.D., 2008. Toxicological profile for phenol. Agency for Toxic Substances and Disease Registry (ATSDR), U.S. Department of Health and Human Services. Retrieved from https://www.atsdr.cdc.gov/ToxProfiles/tp115.pdf

Dębińska, A., Sozańska, B., 2023. Dietary polyphenols—natural bioactive compounds with potential for preventing and treating some allergic conditions. *Nutrients* 15(22), 4823.

Delgado, A., Gonçalves, S., Romano, A., 2023. Mediterranean diet: the role of phenolic compounds from aromatic plant foods. *Foods* 12(4), 840.

Deng, Q., He, B., Shen, M., Ge, J., Du, B., Zeng, L., 2024. First evidence of hindered amine light stabilizers as abundant, ubiquitous, emerging pollutants in dust and air particles: a new concern for human health. *Environmental Science & Technology* 58(2), 1349–1358.

European Chemicals Agency (ECHA), 2023. *Candidate list of substances of very high concern for authorisation*. European Chemicals Agency. Retrieved from https://echa.europa.eu/candidate-list-table

Fuloria, S., Mehta, J., Chandel, A., Sekar, M., Rani, N.N.I.M., Begum, M.Y., Subramaniyan, V., Chidambaram, K., Thangavelu, L., Nordin, R., Wu, Y.S., 2022. A comprehensive review on the therapeutic potential of *Curcuma longa* Linn. in relation to Its major active constituent curcumin. *Frontiers in Pharmacology* 13, 820806.

Gałązka, A., Jankiewicz, U., 2022. Endocrine-disrupting compounds (nonylphenol and bisphenol A): sources, harmfulness, and laccase-assisted degradation in the aquatic environment. *Microorganisms* 10(11), 2236.

Geana, E.I., Ciucure, C.T., Tamaian, R., Marinas, I.C., Gaboreanu, D.M., Stan, M., Chitescu, C.L., 2023. Antioxidant and wound healing bioactive potential of extracts obtained from bark and needles of softwood species. *Antioxidants* 12(7), 1383.

Hu, C., Wong, W.T., Wu, R., Lai, W.F., 2020. Biochemistry and use of soybean isoflavones in functional food development. *Critical Reviews in Food Science and Nutrition* 60(12), 2098–2112.

Jiang, S., Liu, Y., Zou, X., He, M., Zhang, K., Xu, G., Qin, S., 2021. Synthesis and application of new macromolecular hindered phenol antioxidants of polyamide 6. *Journal of Applied Polymer Science* 138(40), 51184.

Jordan, A., Hall, C.G., Thorp, L.R., Sneddon, H.F., 2022. Replacement of less-preferred dipolar aprotic and ethereal solvents in synthetic organic chemistry with more sustainable alternatives. *Chemical Reviews* 122(6), 6749–6794.

Kumar, H., Dhalaria, R., Guleria, S., Sharma, R., Kumar, D., Verma, R., Cruz-Martins, N., Dhanjal, D.S., Chopra, C., Kaur, T., Kumar, V., 2024. Non-edible fruit seeds: nutritional profile, clinical aspects, and enrichment in functional foods and feeds. *Critical Reviews in Food Science and Nutrition* 64(33), 13298–13317.

Lattanzio, V., 2013. Phenolic compounds: introduction 50. In *Natural Products* (1543–1580). Springer.

Le Bloch, J., Rouault, M., Iriantsoa, V., Michelet, O., 2024. Polyphenols in human nutrition: European regulations and potential classification as a novel food or food additive. *Journal of Agricultural and Food Chemistry*. 72(48), 26936–26942.

Lee, K., Zhang, H., 2023. Natural phenolic antioxidants as replacements for synthetic food preservatives: Efficacy and mechanisms. Food Chem. 416, 135878.

Martínez Álvarez, J.R., Lopez Jaen, A.B., Cavia-Saiz, M., Muñiz, P., Valls-Belles, V., 2023. Beneficial effects of olive oil enriched with lycopene on the plasma antioxidant and anti-inflammatory profile of hypercholesterolemic patients. *Antioxidants* 12(7), 1458.

Martínez-Zamora, L., Cano-Lamadrid, M., Artés-Hernández, F., Castillejo, N., 2023. Flavonoid extracts from lemon by-products as a functional ingredient for new foods: a systematic review. *Foods* 12(19), 3687.

Martins, F.C., Sentanin, M.A., De Souza, D., 2019. Analytical methods in food additives determination: compounds with functional applications. *Food Chemistry* 272, 732–750.

Meli, R., Monnolo, A., Annunziata, C., Pirozzi, C., Ferrante, M.C., 2020. Oxidative stress and BPA toxicity: an antioxidant approach for male and female reproductive dysfunction. *Antioxidants* 9(5), 405.

Melim, C., Lauro, M.R., Pires, I.M., Oliveira, P.J., Cabral, C., 2022. The role of glucosinolates from cruciferous vegetables (*Brassicaceae*) in gastrointestinal cancers: from prevention to therapeutics. *Pharmaceutics* 14(1), 190.

Priego, A.R., Parra, E.G., Mas, S., Morgado-Pascual, J.L., Ruiz-Ortega, M., Rayego-Mateos, S., 2021. Bisphenol A modulates autophagy and exacerbates chronic kidney damage in mice. *International Journal of Molecular Sciences* 22(13), 7189.

Rahman, M.S., Alom, J., Nitai, A.S., Hasan, M.S., Ahmed, M.B., Nam, S., Mondal, M.I.H., 2022. Ultraviolet-blocking protective textiles. In *Protective Textiles from Natural Resources* (395–444). Woodhead Publishing.

Rappoport, Z.V.I., 2004. *The Chemistry of Phenols*. Wiley.

Rodriguez, L., Muñoz-Bernal, Ó.A., Fuentes, E., Alvarez-Parrilla, E., Palomo, I., 2024. Antiplatelet activity of phenolic compounds-fortified merlot wine and pure phenolic compounds. *Applied Sciences* 14(13), 5707.

Rudrapal, M., Rakshit, G., Singh, R.P., Garse, S., Khan, J., Chakraborty, S., 2024. Dietary polyphenols: review on chemistry/sources, bioavailability/metabolism, antioxidant effects, and their role in disease management. *Antioxidants* 13(4), 429.

Sarkar, S., Pandey, A., Pant, A.B., 2023. Regulatory requirements for safety/toxicity assessment of cosmetics/nanocosmetic products: challenges and opportunities. In *Skin 3-D Models and Cosmetics Toxicity* (149–176). Elsevier.

Snedeker, S.M., 2014. Benzophenone UV-photoinitiators used in food packaging: potential for human exposure and health risk considerations. In *Toxicants in Food Packaging and Household Plastics: Exposure and Health Risks to Consumers*, 151-176. Springer-Verlag London

Thakur, A., Kumar, A., 2022. Recent advances on rapid detection and remediation of environmental pollutants utilizing nanomaterials-based (bio) sensors. *Science of the Total Environment* 834, 155219.

Tsimogiannis, D., Oreopoulou, V., 2019. Classification of phenolic compounds in plants. In *Polyphenols in Plants* (263–284). Academic Press.

U.S. Food and Drug Administration (FDA), 2023. Code of Federal Regulations – Title 21: Food additives. U.S. Government Publishing Office. Part 170 – Food Additives, Retrieved from https://www.ecfr.gov/current/title-21

Wade, M., Delawder, V., Reneau, P., Dos Santos, J.M., 2020. The effect of BPA exposure on insulin resistance and type 2 diabetes: the impact of muscle contraction. *Medical Hypotheses* 140, 109675.

Wang, T., Zhao, J., Yang, Z., Xiong, L., Li, L., Gu, Z., Li, Y., 2022. Polyphenolic sunscreens for photoprotection. *Green Chemistry* 24(9), 3605–3622.

Wang, W., Xiong, P., Zhang, H., Zhu, Q., Liao, C., Jiang, G., 2021. Analysis, occurrence, toxicity and environmental health risks of synthetic phenolic antioxidants: a review. *Environmental Research* 201, 111531.

Wang, Z., Liang, G., Jiang, S., Wang, F., Li, H., Li, B., Zhu, H., Lu, A., Gong, W., 2024. Understanding the environmental impact and risks of organic additives in plastics: a call for sustained research and sustainable solutions. *Emerging Contaminants*, 10, 100388.

Winz, C., Zong, W.X., Suh, N., 2023. Endocrine-disrupting compounds and metabolomic reprogramming in breast cancer. *Journal of Biochemical and Molecular Toxicology* 37(12), e23506.

Xie, A., Dong, Y., Liu, Z., Li, Z., Shao, J., Li, M., Yue, X., 2023. A review of plant-based drinks addressing nutrients, flavor, and processing technologies. *Foods* 12(21), 3952.

Xiong, R.G., et al., 2023. New insights into the protection of dietary components on anxiety, depression, and other mental disorders caused by contaminants and food additives. *Trends in Food Science & Technology* 138, 44–56.

Yañez-Cruz, M.G., Villanueva-Ibáñez, M., Méndez-Arriaga, F., Lucho-Constantino, C.A., Hernández-Pérez, M.D.L.Á., Ramírez-Vargas, M.D.R., Flores-González, M.A., 2022. Green route synthesis and characterization of β-Bi_2O_3/SiO_2 and β-$Bi_2O_3/Bi_2O_{2\cdot 75}/SiO_2$ using *Juglans regia* L. shell aqueous extract and photocatalytic properties for the degradation of RB-5. *Journal of Analytical Science and Technology* 13(1), 52.

Zhang, C., Xue, J., Yang, X., Ke, Y., Ou, R., Wang, Y., Madbouly, S.A., Wang, Q., 2022. From plant phenols to novel bio-based polymers. *Progress in Polymer Science*, 125, 101473.

Zhang, J., Ke, Q., Bai, J., Yang, M., 2023a. Synthesis of a novel organic-inorganic hindered phenol antioxidant derived from polyhedral oligomeric silsesquioxane and its anti-oxidative behavior in polypropylene. *Polymer Degradation and Stability* 218, 110550.

Zhang, M.Q., Zhang, J., Zhang, Y.T., Sun, J.Y., Prieto, M.A., Simal-Gandara, J., Putnik, P., Li, N.Y., Liu, C., 2023b. The link between the phenolic composition and the antioxidant activity in different small berries: a metabolomic approach. *LWT* 182, 114853.

Zhao, W., He, J., Yu, P., Jiang, X., Zhang, L., 2023. Recent progress in the rubber antioxidants: a review. *Polymer Degradation and Stability* 207, 110223.

Zhor, C., Wafaa, L., Ghzaiel, I., Kessas, K., Zarrouk, A., Ksila, M., Kharoubi, O., 2023. Effects of polyphenols and their metabolites on age-related diseases. *Biochem. Pharmacol.* 214, 115674

Zhou, H., Hu, X., Liu, M., Yin, D., 2023. Benzotriazole ultraviolet stabilizers in the environment: a review of analytical methods, occurrence, and human health impacts. *TrAC Trends in Analytical Chemistry*, **162,** 117170.

12 Bioplastic Additive in Marine Environment

*Abisola Abiodun Ibikunle**,
Daramola Oluwatosin Olaoluwa,
and Abdulmojeed Abdulsalam Adopisa

12.1 INTRODUCTION

The ecological and toxicological impacts of plastic in the marine environment are typically discussed or modeled as though they consist of "pure" polymers (Cole et al., 2015;; Yin et al., 2018). However, no plastic found in today's waste stream is produced without additives or as a "bare" formulation. Every plastic item is composed of a distinct combination of the base polymer, along with residual monomers or catalysts, and various chemical additives introduced during the manufacturing process (Hahladakis et al., 2018). A vast array of additives is used in plastic formulations, with each one serving a specific purpose in enhancing the functional properties, performance, or appearance of the plastic product (Marturano et al., 2016). Depending on the formulation, plastics can contain anywhere from less than 1% to over 50% by weight of plasticizers (Chaudhary et al., 2016; Marturano et al., 2016). Plasticizers, fillers, and flame retardants (FRs) are typically used in high weight fractions in plastic formulations, collectively making up about three-quarters of all additives produced. Other additives, including antioxidants and light stabilizers, are utilized in significantly smaller amounts. Contrary to the common belief that plastics are everlasting, they are organic materials that experience substantial degradation when subjected to processing or environmental factors such as high temperatures, ultraviolet (UV) radiation, atmospheric oxygen, and water. The expected durability and performance of thermoplastics would not be achievable without the deliberate addition of these chemical compounds. Since the mid-20th century, the production and use of plastics—and consequently plastic additives—have grown at an exponential rate (Binetti et al., 2008). At the current production rates, a cumulative total of 2,000 million metric tons of additives is projected to be produced by the end of 2050 (Geyer et al., 2017). This estimate assumes that plastics, on average, contain 7% additives by mass (Geyer et al., 2017; Lynch et al., 2022). Estimates suggest that at least 190 metric tons of additive chemicals entered the ocean in 2015 alone, with this rate projected to double by 2025 (De Frond et al., 2019). The plastics that have accumulated in the ocean over decades were produced without regard for their potential disposal in marine environments.

The majority of plastic additives identified in coastal and open ocean environments (Paluselli et al., 2018; Schmidt et al., 2019; Lynch et al., 2022) endocrine disruptors (Zhang et al., 2018) and may interact with marine living organisms through (i) release from ingested plastic debris, (ii) assimilation by contact with the chemicals in the dissolved water phase, or (iii) food web transfer (Barrios et al., 2018). Most plastic additives identified in coastal and open ocean environments (Cole et al., 2013), organophosphate esters (Net et al., 2015), and bioplastic (van der Veen & de Boer, 2012),

 DOI: 10.1201/9781003602828-15

as well as plastic leachate mixtures (Chen et al., 2016; Capolupo et al., 2020). These compounds are not chemically bound to the polymer and the release of their water-extractible fraction is thus considered to be primarily governed by polymer to water partitioning (Kpw) (Zimmermann et al., 2019). Furthermore, naturally occurring marine prokaryotes have been identified as a crucial parameter in influencing phtalate esters release from plastic to the aqueous phase (Fauvelle et al., 2021).

This chapter seeks to present an overview of bioplastic additives, emphasizing their chemistry and functionality, transport and fate, detection in marine environments, and associated toxicities. To simplify the extensive variety of additives, they can be categorized into three main groups: functional additives, colorants, and fillers/reinforcements. Table 12.1 highlights the most commonly used additives, providing details on typical loadings in polymers, key examples, and their intended effects.

TABLE 12.1
Type, Typical Loadings, Class in the Polymer, Key Examples, and Desired Effects of Additive

Type	Typical Loading (wt%)	Additive Class	Purpose	Example
Flame retardants	3–25	Functional	Preventing ignition or flame propagation	Poly (bromodiphenyl ethers), organophosphorus,
Plasticizer	10–70	Functional	Soften polymer and make more flexible	Phthalates, Tris(2-chloroethyl) phosphate, citrate esters, glycerol
Antioxidant	0.05–3	Functional	To prevent discoloration and degradation during processing, use, and weathering	Hindered phenols, alkylphenols, phosphates, lactones, hydroxylamines
Fillers	1–50	Fillers	To provide mechanical reinforcement or formulation cost reductions	Clays, silica, glass, chalk, talc, wood flour asbestos, alumina, carbon black,
Colorant	0.25–5	Colorant	To impart a desired color on the finished product	Cadmium, chromium, lead, and cobalt compounds, titanium dioxide, carbon black; organic dyes, natural dye and pigment.
Light stabilizer	0.05–3	Functional	To prevent degradation of the plastic upon exposure to light sources	Hindered amine light stabilizers, benzophenone light-absorbing compounds, calcium stearate, zinc stearate
Lubricant	0.01–3	Functional	Allows easier processing of the plastic (i.e., lower temperatures, faster processing times)	Waxes, oils, long-chain esters of polymeric alcohols
Impact modifier	0.5–30	Functional	To increase flexibility and impact strength to meet physical property requirements of rigid parts	Acrylonitrile butadiene styrene (ABS), acrylonitrile styrene acrylate (ASA)
Heat stabilizer	0.1–8	Functional	To prevent thermal degradation during processing	Dialkyl maleates or laureates and dialkyl mercaptides of tin in PVC formulations

12.2 ADDITIVES

Plastic additives are incorporated into polymers to enhance and modify specific properties. Among the most common are plasticizers, which are used to reduce a polymer's glass transition temperature (making it softer in some way), fillers to make it cheaper (usually chalk), oil composition to improve its rheology (Al-Malaika et al., 2017).

Plastic production involves over 13,000 chemicals used in a wide variety of applications. Among these, more than 3,200 substances—including monomers, additives, processing aids, and non-intentionally added substances—are potentially concerning due to their hazardous properties. On average, non-fiber plastics consist of approximately 93% polymer resin and 6–7% additives by mass (Geyer et al., 2017). However, the proportion varies based on the material's intended function; for instance, additives can make up as much as 80% of the total volume in polyvinyl chloride (PVC) (Hahladakis et al., 2018). Plasticizers, FRs, and fillers are the most commonly used plastic additives. Collectively, these additives account for 75% of the global annual production of additives (Geyer, 2020; Geyer et al., 2017). While data on the production and use of plastic additives is limited, it is estimated that 27 million tons of additives were produced in 2017, based on a plastic production volume of 438 million metric tons (Geyer, 2020). A significant challenge in identifying and quantifying additives in plastics lies in the vast number and diversity of substances used, coupled with issues of confidentiality and the commercial sensitivity of accessing such information (Maes et al., 2023).

12.3 ROUTE OF ADDITIVES INTO THE MARINE ECOSYSTEM

The release of plastic additives into marine ecosystems poses a significant and escalating environmental challenge, one that has become increasingly pervasive and persistent in the 21st century. While the broader issue of plastic waste in oceans is well documented (Unuofin, 2020), the introduction of plastic additives into marine environments begins well before the plastic itself reaches the ocean. Sources of plastic pollution include beach litter, improperly managed waste from urban and rural areas, and industrial discharges. Although much of the focus has been on visible, large plastic debris, the leaching of additives from plastics into marine ecosystems remains an often-overlooked issue.

Although alternative plastic production methods, such as bioplastics, have garnered attention, they have not completely resolved the issue. Bioplastics frequently rely on the same chemical additives used in conventional plastic production, addressing the challenge of plastic degradation but leaving the environmental impacts of these additives unmitigated. Additives in plastics can enter marine environments through various pathways, often causing significant ecological harm.

Plastics do not biodegrade in the conventional sense but instead break down into smaller fragments through physical and chemical processes such as UV radiation, thermal degradation, and mechanical wear. As plastics degrade, they become increasingly prone to the leaching of additives. This process eventually leads to the formation of microplastics, which not only retain chemicals from the original plastic additives but also absorb other toxic substances from the surrounding water, amplifying the environmental threat (Jambeck et al., 2015). Even when plastics do not fragment into microplastics, larger items like fishing nets, bottles, and food packaging can persist in marine environments for decades or longer, continuing to leach harmful chemicals into ecosystems (Pawar et al., 2016).

Industrial discharges represent another major pathway for the release of plastic additives into marine environments. Plastics used in manufacturing processes often contain additives that can enter the environment through industrial runoff or improper disposal of waste. In some cases, plastics designed with specific additives are directly discarded into the ocean (Chen et al., 2021). Such direct releases exacerbate the growing problem of plastic pollution and its associated environmental impacts.

In addition to primary sources, plastics in the ocean interact with other pollutants by absorbing toxic substances from seawater, including pesticides, heavy metals, and persistent organic pollutants (POPs). When ingested by marine organisms, these contaminants can leach into their systems, posing a heightened threat to both ecosystems and human health (Okoye et al., 2022; Ighalo et al., 2022). This interaction creates a cumulative toxic burden that is far more harmful than individual pollutants in isolation.

Despite increasing awareness of plastic waste accumulation in the oceans, the transport and fate of plastic additives in marine environments remain poorly understood (Tian et al., 2020). Plastic additives can be released into air, water, and soil throughout a product's life cycle, reaching marine ecosystems through various interconnected pathways. While plastics are known to leach additives into marine environments (De Frond et al., 2019; Koelmans et al., 2014; Paluselli et al., 2018; Pereao et al., 2020; Sun et al., 2019; Teuten et al., 2009), other significant sources include direct wastewater effluents, atmospheric deposition, and river transport. For example, in 2019, U.S. industries reported releasing 14,742 metric tons of styrene into the environment, independent of its leaching from plastic products, according to the U.S. Environmental Protection Agency's Toxic Release Inventory. Additionally, wastewater treatment plant effluents, particularly those near plastic production facilities, are major contributors to the discharge of pre-leached plastic additives into aquatic environments (Kim et al., 2021; Liu et al., 2020; Peng et al., 2007; Staples et al., 1997; Zhang et al., 2024, 2018).

While much of the focus on plastic additives in the ocean has emphasized plastics as the primary carriers of these chemicals, they are not the only significant transport mechanism. Koelmans et al. (2014) argue that plastics retain only a small fraction of hydrophobic organic compounds, including plastic additives, compared to the amounts present in other environmental media such as seawater, atmosphere, sediment, and biota. Consequently, plastic-mediated transport plays a relatively minor role in the overall distribution of these compounds.

The fate of plastic additives in marine environments is complex and often differs from that of the plastics themselves. Additives that leach from plastics can bioaccumulate in marine organisms, undergo degradation, or be re-adsorbed by other plastic particles that initially lacked these chemicals. The ongoing processes of sorption and desorption between various environmental compartments—plastic debris, water, sediment, air, and organic matter—are driven by chemical equilibrium kinetics.

As research into the environmental transport and fate of plastic additives advances, it is becoming increasingly evident that these chemicals pose a significant yet often underappreciated threat to marine ecosystems. Addressing plastic pollution requires moving beyond the visible plastic debris to fully account for the impacts of additives, including their leaching, bioaccumulation, and interactions with other pollutants in marine environments.

12.4 IMPACTS OF ADDITIVES IN MARINE ENVIRONMENT

12.4.1 Effects on Water Quality

When plastics degrade or are exposed to environmental conditions, the additives they contain can leach into surrounding water bodies, adversely affecting water quality. This issue, particularly concerning microplastics and their associated chemical additives, has become a pressing environmental concern with implications for both aquatic ecosystems and human health. Of particular concern is the release of plasticizers such as phthalates, which are used to enhance the flexibility and durability of plastics but are known endocrine disruptors. Studies show that phthalates can migrate from plastics into water, interfering with the reproductive and developmental processes of aquatic organisms. For example, exposure to phthalates has been linked to altered hormone levels and developmental abnormalities in fish and amphibians, potentially leading to population declines. These chemicals have been detected in both freshwater and marine environments (Rochman, 2015; Maddela et al., 2023).

Another category of hazardous plastic additives includes FRs like polybrominated diphenyl ethers (PBDEs). These substances are added to plastics to reduce flammability, but they are highly persistent in the environment and can accumulate in aquatic organisms. PBDEs have been detected in water samples worldwide, with concentrations notably higher in areas with significant industrial activity or plastic waste. The buildup of these compounds within aquatic food chains presents long-term toxicity risks, affecting not only the organisms exposed but also humans who consume contaminated seafood (Koelmans et al., 2014).

Plastic stabilizers, such as organotin compounds, are another class of additives that can leach into water from degraded plastics, particularly marine debris. While these stabilizers help prolong the lifespan of plastics by preventing UV degradation, they are highly toxic to aquatic life, especially mollusks and crustaceans. Research has shown that even low levels of organotin compounds can impair growth, reproduction, and behavior in marine species. This toxicity threatens biodiversity and disrupts crucial ecosystem services, such as water filtration and nutrient cycling (Bridson et al., 2021). Additionally, as plastics degrade into microplastic particles smaller than 5 mm, the environmental risks posed by plastic additives are amplified. Microplastics can adsorb a range of harmful chemicals, including pesticides and heavy metals, on their surfaces. Aquatic organisms may ingest these particles along with their associated toxins, leading to internal contamination and potential toxicity. Recent studies have emphasized that microplastic ingestion can cause physical harm to aquatic animals, while also carrying chemical contaminants that may lead to long-term health effects (Asmara et al., 2024).

The leaching of plastic additives into water bodies not only threatens wildlife but also poses significant risks to human health. Many chemicals present in plastics are persistent, bioaccumulative, and toxic. For instance, phthalates have been linked to reproductive and developmental issues in humans, while prolonged exposure to FRs and stabilizers has been associated with cancer and neurodevelopmental disorders. Given the widespread presence of plastic pollution in both freshwater and marine environments, humans are likely exposed to these harmful chemicals through drinking water and seafood consumption (Maddela et al., 2023). In addition to direct toxicity concerns, plastic additives also contribute to the overall degradation of water quality. As plastics break down in the environment, they release various hazardous substances into the water, altering its physical and chemical properties. This can impact oxygen levels, pH balance, and the overall health of aquatic ecosystems, making it more difficult to maintain water quality and ensure the sustainability of water resources, particularly in areas heavily burdened by plastic waste (Bridson et al., 2021).

12.4.2 Effect on Aquatic Plant

The presence of additives in aquatic environments can significantly impact aquatic plants. These chemicals, often leached from plastics in marine and freshwater ecosystems, pose various risks to plant life due to their potential toxicity and ability to disrupt biological processes.

Phthalates, for example, are known endocrine disruptors, and their presence in aquatic environments can interfere with the normal hormonal functions of aquatic plants, potentially affecting growth and reproduction (Godswill & Godspell, 2019). Research has shown that phthalates can impair photosynthetic activity in aquatic plants, reducing their ability to capture energy, which is crucial for their survival and growth (Lyche, 2017). Additionally, phthalates may alter the composition of plant tissues by disrupting cellular structures such as membranes and organelles, leading to weakened structural integrity. This disruption can hinder the plant's ability to absorb nutrients effectively, reducing its overall health and vitality. In some cases, exposure to phthalates has been linked to changes in cell division rates and plant development, further compromising the plant's ability to thrive in polluted environments (Koelmans et al., 2014).

Calcium stearate, a fatty acid salt commonly used as a stabilizer in plastics, also leaches into aquatic ecosystems, where it can negatively affect aquatic plants (Zughaibi et al., 2022). While calcium is an essential nutrient for plant growth, excessive amounts of calcium stearate can cause nutrient imbalances, interfering with the plant's uptake of other critical minerals like potassium and magnesium (Paluselli et al., 2018). This can result in stunted growth and poor development, as plants are unable to effectively utilize the nutrients available in their environment. Calcium stearate can also interact with other pollutants in the water, forming insoluble complexes that hinder nutrient absorption and potentially cause toxic effects. Moreover, the accumulation of calcium stearate particles on the plant's surface can block light penetration, further limiting photosynthesis and reducing the plant's ability to produce energy from sunlight (Sridharan et al., 2022).

Phosphates, while essential for plant growth in small amounts, can be highly harmful when present in excess. Elevated phosphate levels, resulting from industrial discharges or agricultural runoff, can lead to eutrophication in aquatic environments (Höfer & Hinrichs, 2010). This process triggers excessive algae growth, which depletes oxygen levels in the water and reduces the amount of sunlight reaching aquatic plants. Without sufficient sunlight, photosynthetic plants struggle to generate the energy needed for survival, resulting in poor growth, chlorosis (yellowing of leaves), and, in severe cases, plant death (Cairns et al., 2020). The excess phosphate disrupts the nutrient balance within aquatic plants, causing an accumulation of toxic substances and a decline in overall plant health. Phosphate-driven eutrophication has also been shown to alter plant community structures, favoring species that can tolerate low light and oxygen conditions, thus reducing biodiversity and ecosystem stability (Zhang et al., 2024).

These additives present complex and diverse threats to aquatic plants. While each chemical can directly affect plant physiology and growth, they can also interact with one another and with other pollutants, magnifying their harmful effects. The cumulative impact of these additives on aquatic plant communities can lead to significant disruptions in ecosystem functions, including nutrient cycling, oxygen production, and the provision of habitats for other organisms (Häder et al., 2015).

12.4.3 Effect on Aquatic Animals

Additives used in the formulation of bioplastics, which are often leached from plastics in the ocean, can disrupt biological processes and result in various harmful effects on marine life. A key concern is the role of plastic additives as endocrine disruptors. Phthalates and other organic plasticizers are known to interfere with the endocrine systems of marine organisms. When marine animals ingest microplastic particles or consume prey contaminated with these additives, they may experience hormonal imbalances that disrupt reproductive and developmental processes. For instance, a study by researchers at the University of Exeter found that chemicals leached from plastics caused deformities in sea urchin embryos, including delayed development and malformations in the skeleton and nervous system (Rendell-Bhatti et al., 2021). Phthalates have also been shown to impair fish reproduction by causing abnormalities in gonadal development and altering hormone levels (Delbes et al., 2022). These endocrine disruptions can lead to reduced fertility, altered growth rates, and weakened immune function, all of which threaten the health and survival of marine populations (Golshan & Alavi, 2019).

Coral reefs are particularly susceptible to the effects of plastic additives. Research from Tel Aviv University revealed that these chemicals negatively impact the reproductive processes and larval development of corals and other reef organisms. The study found that certain additives inhibited larval settlement and fertilization, crucial processes for maintaining healthy coral populations (Vered & Shenkar, 2022). Given the importance of coral reefs for biodiversity and ecosystem stability, any disruption to their reproductive success poses a significant threat to marine biodiversity and overall ecosystem health.

When these chemicals leach into the marine environment, they can be toxic to aquatic organisms (Al Mamun et al., 2023). For example, FRs have been shown to interfere with the nervous systems of marine animals. Fish exposed to these chemicals may exhibit behavioral changes, including impaired locomotion and altered feeding behavior, which can hinder their ability to survive and reproduce (Zhang et al., 2018). The toxicity of these additives is especially concerning for organisms at lower trophic levels, such as plankton and small fish, which are more likely to absorb these substances and transfer them through the food chain.

Bioplastic additives can also have indirect effects on marine animals by altering their habitats. As bioplastics degrade, the leaching of additives can change the chemical composition of the water, affecting nutrient availability and phytoplankton growth, which is critical for the marine food web (Laranjeiro et al., 2024). Disruptions at this foundational level can have cascading effects, reducing the food supply for higher trophic levels and ultimately leading to a decline in biodiversity. In this way, the effects of bioplastic additives can harm not only individual species but also entire marine ecosystems (Okeke et al., 2022). The presence of plastic additives in marine environments represents a complex threat to aquatic life, as the combined chemical toxicity from leached additives creates a challenging situation for marine organisms.

12.4.4 Effect on Microbes

The marine microbiome, which includes a diverse array of microorganisms such as bacteria, fungi, algae, and viruses, plays a crucial role in maintaining ecosystem health, facilitating nutrient cycling, and supporting the broader food web. However, the introduction of plastic additives into marine environments can disrupt these delicate microbial communities, leading to significant ecological consequences (Hermabessiere et al., 2017). When plastic debris, especially those containing additives, enters the ocean, the leaching of these chemicals into the water can create a toxic environment for microorganisms. Additives like plasticizers, stabilizers, and FRs can alter the chemical composition of seawater, making it less hospitable for certain microbial species. Research has shown that these substances can hinder microbial growth and diversity by interfering with critical cellular processes such as energy production and nutrient uptake. Since microorganisms are typically the first to encounter these pollutants, they may experience more immediate effects than larger marine organisms (Almroth & Eggert, 2019).

A major consequence of exposure to plastic additives is the disruption of microbial biodiversity. Certain additives can selectively inhibit the growth of specific microbial groups, leading to an imbalance within the microbial community. This imbalance reduces the resilience of marine ecosystems, as diverse microbial populations are essential for processes such as bioremediation, nutrient cycling, and carbon fixation. A loss in microbial diversity can also increase vulnerability to other stressors, such as disease outbreaks or fluctuations in environmental factors like temperature and salinity (Tanentzap, 2023). For example, toxic plasticizers may favor resistant microbial species while suppressing others, leading to shifts toward less efficient or beneficial microbial communities. In addition to affecting microbial populations, plastic additives can influence microbial activity in marine ecosystems (Zhang et al., 2024). Interactions between microorganisms and microplastic particles carrying additives can alter metabolic activities. Studies suggest that certain plastic additives, particularly those with antimicrobial properties, can directly affect how microbes decompose organic matter in oceanic environments. This disruption can slow nutrient cycling, limiting essential nutrients available for other marine organisms and disrupting the entire food web.

Furthermore, the accumulation of plastic debris provides surfaces for biofilm growth—thin layers of microorganisms on plastic surfaces—where additives can modify both the composition and function of these communities, potentially promoting pathogenic bacteria or harmful algal blooms. The effects of plastic additives on the microbiome extend beyond microorganisms themselves, impacting higher trophic levels in the food web. Microbial communities are vital for the health and survival of larger marine organisms, such as fish and marine mammals. Changes in microbial populations or their activities can reduce food availability for these animals. Additionally, some additives may bioaccumulate in microorganisms and biomagnify through the food chain, ultimately affecting larger organisms, including humans who consume seafood.

Moreover, interactions between plastic additives and other marine pollutants, such as heavy metals or POPs, can intensify their effects on the microbiome. These additives may act as carriers for additional toxic substances, enhancing their bioavailability to microorganisms. As a result, marine microbes could absorb not only plastic additives but also other contaminants adsorbed onto plastic surfaces. This combined exposure complicates the ability of these organisms to perform their essential ecological functions.

12.5 THE BIOACCUMULATION OF BIOPLASTIC ADDITIVES

The bioaccumulation of bioplastic additives presents significant ecological and health risks, particularly as marine animals ingest contaminated food or microplastics carrying these chemicals. Over time, these additives accumulate in the tissues of marine organisms, leading to prolonged exposure to potentially harmful substances. This accumulation impacts not only the health of individual animals but also poses risks to entire populations and ecosystems (Ribeiro et al., 2019). For example, studies have shown that FRs and plasticizers, commonly found in bioplastics, can accumulate in the tissues of fish and shellfish. This bioaccumulation endangers not only predators, such as marine mammals, that consume these organisms but also humans who eat seafood (Teuten et al., 2009). The presence of these substances in the food chain raises concerns for both ecological balance and public health. The bioaccumulation process is facilitated by the chemical properties of bioplastic additives, which are often lipophilic, enabling them to integrate easily into the fatty tissues of marine animals (Khan et al., 2022). As these animals consume contaminated prey, the concentration of harmful additives in their bodies increases over time. This process is worsened by the fact that many bioplastic additives are not chemically bound within the plastic matrix and can leach into the environment, making them more available for uptake by marine organisms (Yu et al., 2023).

Moreover, the impact of bioaccumulated additives extends beyond individual species. The disruption of predator–prey dynamics due to contaminated food sources can lead to shifts in population structures and community dynamics within marine ecosystems (Alava et al., 2017). For instance, if certain fish species accumulate higher levels of toxic additives, their predators may experience reduced reproductive success or increased mortality rates, further destabilizing the ecosystem (Satkar et al., 2024). In addition to the direct health impacts on marine life, bioaccumulation poses broader risks through biomagnification, the process by which toxic substances increase in concentration at each trophic level of the food web. As larger predators consume smaller contaminated prey, they may accumulate even higher levels of harmful substances. This phenomenon threatens biodiversity and raises serious concerns for human health, as these toxins can enter the human food chain through seafood consumption (Sonone et al., 2020).

12.6 MITIGATING RISK OF ADDITIVES IN MARINE

The marine microbiome, which includes a diverse array of microorganisms such as bacteria, fungi, algae, and viruses, plays a crucial role in maintaining ecosystem health, facilitating nutrient cycling, and supporting the broader food web. However, the introduction of plastic additives into marine environments can disrupt these delicate microbial communities, leading to significant ecological consequences (Hermabessiere et al., 2017). When plastic debris, especially those containing additives, enters the ocean, the leaching of these chemicals into the water can create a toxic environment for microorganisms. Additives like plasticizers, stabilizers, and FRs can alter the chemical composition of seawater, making it less hospitable for certain microbial species. Research has shown that these substances can hinder microbial growth and diversity by interfering with critical cellular processes such as energy production and nutrient uptake. Since microorganisms are typically the first to encounter these pollutants, they may experience more immediate effects than larger marine organisms (Almroth & Eggert, 2019).

A major consequence of exposure to plastic additives is the disruption of microbial biodiversity. Certain additives can selectively inhibit the growth of specific microbial groups, leading to an imbalance within the microbial community. This imbalance reduces the resilience of marine ecosystems, as diverse microbial populations are essential for processes such as bioremediation, nutrient cycling, and carbon fixation. A loss in microbial diversity can also increase vulnerability to other stressors, such as disease outbreaks or fluctuations in environmental factors like temperature and salinity (Tanentzap, 2023). For example, toxic plasticizers may favor resistant microbial species while suppressing others, leading to shifts toward less efficient or beneficial microbial communities. In addition to affecting microbial populations, plastic additives can influence microbial activity in marine ecosystems (Zhang et al., 2024). Interactions between microorganisms and microplastic particles carrying additives can alter metabolic activities. Studies suggest that certain plastic additives, particularly those with antimicrobial properties, can directly affect how microbes decompose organic matter in oceanic environments. This disruption can slow nutrient cycling, limiting essential nutrients available for other marine organisms and disrupting the entire food web.

Furthermore, the accumulation of plastic debris provides surfaces for biofilm growth thin layers of microorganisms on plastic surfaces, where additives can modify both the composition and function of these communities, potentially promoting pathogenic bacteria or harmful algal blooms. The effects of plastic additives on the microbiome extend beyond microorganisms themselves, impacting higher trophic levels in the food web. Microbial communities are vital for the health and survival of larger marine organisms, such as fish and marine mammals. Changes in microbial populations or their activities can reduce food availability for these animals. Additionally, some additives may bioaccumulate in microorganisms and biomagnify through the food chain, ultimately affecting larger organisms, including humans who consume seafood. Moreover, interactions between plastic additives and other marine pollutants, such as heavy metals or POPs, can intensify their effects on the microbiome. These additives may act as carriers for additional toxic substances, enhancing their bioavailability to microorganisms. As a result, marine microbes could absorb not only plastic additives but also other contaminants adsorbed onto plastic surfaces. This combined exposure complicates the ability of these organisms to perform their essential ecological functions.

12.7 CONCLUSION

The marine ecosystem is highly sensitive to pollutants, making it essential to adopt a comprehensive approach to mitigate the risks posed by additives. This includes implementing stricter regulations on harmful chemicals from industries such as shipping and agriculture, with bans on heavy metals and toxic antifouling agents. Rigorous testing of chemicals for environmental impact before approval is important. Real-time monitoring systems in sensitive marine areas can track pollution levels and provide early warnings. Reducing plastic pollution is vital, achievable through bans on single-use plastics, improved waste management, and promoting biodegradable alternatives. Efforts should also focus on controlling microplastics from household and industrial products. Managing nutrient runoff is critical to prevent eutrophication and harmful algal blooms. This can be addressed by reducing chemical fertilizers and adopting sustainable farming practices, along with establishing buffer zones to filter excess nutrients before they reach the ocean. Restoring marine ecosystems through protected areas and habitat restoration enhances resilience against pollutants. In the shipping industry, adopting cleaner technologies—such as low-sulfur fuels and renewable energy—can minimize harmful additives. Stronger regulations are needed to ensure compliance with environmental standards. Public awareness and education are vital in reducing the risk of additives entering marine environments. Encouraging responsible consumer behavior and community involvement in conservation efforts can make a significant difference. Research into safer alternatives for materials and additives is essential for reducing environmental footprints. Continued investigation into the long-term impacts of additives on marine life will improve risk assessments and inform regulations. International cooperation is necessary to address marine pollution globally, as pollutants often cross borders. Strengthening agreements like the MARPOL Convention for ship-generated pollution is crucial to minimize the transboundary movement of harmful additives. By implementing these strategies, we can significantly reduce the risks of additives in marine ecosystems, preserving biodiversity and protecting ocean health for future generations.

REFERENCES

Alava, J. J., Cheung, W. W., Ross, P. S., & Sumaila, U. R. (2017). Climate change–contaminant interactions in marine food webs: Toward a conceptual framework. *Global Change Biology*, *23*(10), 3984–4001.

Ali, H., Khan, E., & Ilahi, I. (2019). Environmental chemistry and ecotoxicology of hazardous heavy metals: Environmental persistence, toxicity, and bioaccumulation. *Journal of Chemistry*, *2019*(1), 6730305.

Al-Malaika, S., Axtell, F., Rothon, R., & Gilbert, M. (2017). Additives for plastics. In *Brydson's Plastics Materials* (pp. 127–168). Butterworth-Heinemann.

Al Mamun, A., Prasetya, T. A. E., Dewi, I. R., & Ahmad, M. (2023). Microplastics in human food chains: Food becoming a threat to health safety. *Science of the Total Environment*, *858*, 159834.

Almroth, B. C., & Eggert, H. (2019). Marine plastic pollution: Sources, impacts, and policy issues. *Review of Environmental Economics and Policy*, *13*. https://doi.org/10.1093/reep/rez012.

Asmara Iftikhar, Z. Q., Sarfraz, W., Ejaz, U., Aqeel, M., Rizvi, Z. F., & Khalid, N. (2024). Understanding the leaching of plastic additives and subsequent risks to ecosystems. https://doi.org/10.20517/wecn.2023.58

Barrios-Estrada, C., de Jesús Rostro-Alanis, M., Muñoz-Gutiérrez, B. D., Iqbal, H. M., Kannan, S., & Parra-Saldívar, R. (2018). Emergent contaminants: Endocrine disruptors and their laccase-assisted degradation – A review. *Science of the Total Environment*, *612*, 1516–1531.

Binetti, R., Costamagna, F. M., Ceccarelli, F., D'angiolini, A., Fabri, A., Riva, G., Satalia, S., Marcello, I. 2008. The ISS National Inventory of Chemical Substances (INSC). *Annali dell'Istituto Superiore di Sanità*, *44*(1), 16–26.

Bridson, J. H., Gaugler, E. C., Smith, D. A., Northcott, G. L., & Gaw, S. (2021). Leaching and extraction of additives from plastic pollution to inform environmental risk: A multidisciplinary review of analytical approaches. *Journal of Hazardous Materials*, *414*, 125571.

Cairns, J., Niederlehner, B. R., & Smith, E. R. (2020). Ecosystem effects: Functional end points. In *Fundamentals of Aquatic Toxicology* (pp. 589–607). CRC Press.

Capolupo, M., Sørensen, L., Jayasena, K. D. R., Booth, A. M. & Fabbri, E. (2020). Chemical composition and ecotoxicity of plastic and car tire rubber leachates to aquatic organisms. *Water Research*, *169*, 115270.

Chaudhary, B. I., Liotta, C. L., Cogen, J. M., & Gilbert, M. (2016). Plasticized PVC. In *Reference Module in Materials Science and Materials Engineering*. Amsterdam, Elsevier.

Chen, D., Kannan, K., Tan, H., Zheng, Z., Feng, Y. L., Wu, Y., & Widelka, M. (2016). Bisphenol analogues other than BPA: Environmental occurrence, human exposure, and toxicity A review. *Environmental Science & Technology*, *50*(11), 5438–5453.

Chen, Y., Awasthi, A. K., Wei, F., Tan, Q., & Li, J. (2021). Single-use plastics: Production, usage, disposal, and adverse impacts. *Science of the Total Environment*, *752*, 141772.

Cole, M., Lindeque, P., Fileman, E., Halsband, C., & Galloway, T. S. (2015). The impact of polystyrene microplastics on feeding, function and fecundity in the marine copepod Calanus helgolandicus. *Environmental Science & Technology*, *49*(2), 1130–1137.

Cole, M., Lindeque, P., Fileman, E., Halsband, C., Goodhead, R., Moger, J., & Galloway, T. S. (2013). Microplastic ingestion by zooplankton. *Environmental Science & Technology*, *47*(12), 6646–6655.

De Frond, H. L., van Sebille, E., Parnis, J. M., Diamond, M. L., Mallos, N., Kingsbury, T., Rochman, C. M. 2019. Estimating the mass of chemicals associated with ocean plastic pollution to inform mitigation efforts. *Integrated Environmental Assessment and Management*, 15(4), 596–606.

Delbes, G., Blázquez, M., Fernandino, J. I., Grigorova, P., Hales, B. F., Metcalfe, C., …, & Marlatt, V. (2022). Effects of endocrine disrupting chemicals on gonad development: Mechanistic insights from fish and mammals. *Environmental Research*, *204*, 112040.

Fauvelle, V., Garel, M., Tamburini, C., Nerini, D., Castro-Jiménez, J., Schmidt, N., …, & Sempéré, R. (2021). Organic additive release from plastic to seawater is lower under deep-sea conditions. *Nature Communications*, *12*(1), 4426. https://doi.org/10.1038/s41467-021-24738-w

Geyer, R. (2020). Production, use, and fate of synthetic polymers. In *Plastic Waste and Recycling* (pp. 13–32). Academic Press. https:// doi.org/10.1016/B978-0-12-817880-5.00002-5

Geyer, R., Jambeck, J. R., & Law, K. L. (2017). Production, use, and fate of all plastics ever made. *Science Advances*, *3*(7): e1700782. https://doi.org/10.1126/sciadv.1700782

Godswill, A. C., & Godspel, A. C. (2019). Physiological effects of plastic wastes on the endocrine system (Bisphenol A, Phthalates, Bisphenol S, PBDEs, TBBPA). *International Journal of Bioinformatics and Computational Biology*, *4*(2), 11–29.

Golshan, M., & Alavi, S. M. H. (2019). Androgen signaling in male fishes: Examples of anti-androgenic chemicals that cause reproductive disorders. *Theriogenology*, *139*, 58–71.

Häder, D. P., Williamson, C. E., Wängberg, S. Å., Rautio, M., Rose, K. C., Gao, K., …, & Worrest, R. (2015). Effects of UV radiation on aquatic ecosystems and interactions with other environmental factors. *Photochemical & Photobiological Sciences*, *14*(1), 108–126.

Hahladakis, J. N., Velis, C. A., Weber, R., Iacovidou, E., & Purnell, P. (2018). An overview of chemical additives present in plastics: Migration, release, fate and environmental impact during their use, disposal and recycling. *Journal of Hazardous Materials*, *344*, 179–199. https://doi.org/10.1016/j.jhazmat.2017.10.014. ISSN 0304-3894

Höfer, R., & Hinrichs, K. (2010). Additives for the manufacture and processing of polymers. In *Polymers-Opportunities and Risks II: Sustainability, Product Design and Processing* (pp. 97–145). Berlin, Heidelberg: Springer Berlin Heidelberg.

Ighalo, J. O., Yap, P. S., Iwuozor, K. O., Aniagor, C. O., Liu, T., Dulta, K., …, & Rangabhashiyam, S. (2022). Adsorption of persistent organic pollutants (POPs) from the aqueous environment by nano-adsorbents: A review. *Environmental Research*, *212*, 113123.

Jambeck, J. R., Geyer, R., Wilcox, C., Siegler, T. R., Perryman, M., Andrady, A., …, & Law, K. L. (2015). Plastic waste inputs from land into the ocean. *Science*, *347*(6223), 768–771.

Khan, F. R., Patsiou, D., & Catarino, A. I. (2022). Pollutants bioavailability and toxicological risk from microplastics. In *Handbook of Microplastics in the Environment* (pp. 697–736). Springer International Publishing.

Kim, S., Kim, Y., & Moon, H.B. (2021). Contamination and historical trends of legacy and emerging plasticizers in sediment from highly industrialized bays of Korea. *Science of the Total Environment*, 765, 142751.

Koelmans, A. A., Besseling, E., & Foekema, E. M. (2014). Leaching of plastic additives to marine organisms. *Environmental Pollution (Barking, Essex: 1987)*, *187*, 49–54. https://doi.org/10.1016/j.envpol.2013.12.013

Laranjeiro, F., Rotander, A., López-Ibáñez, S., Vilas, A., Seilitz, F. S., Clerandeau, C., …, & Beiras, R. (2024). Comparative assessment of the acute toxicity of commercial bio-based polymer leachates on marine plankton. *Science of the Total Environment*, *946*, 174403.

Liu, R., & Mabury, S. A. (2020). Synthetic phenolic antioxidants: A review of environmental occurrence, fate, human exposure, and toxicity. *Environmental Science & Technology*, *54*(19), 11706–11719.

Lyche, J. L. (2017). Phthalates. In *Reproductive and developmental toxicology* (pp. 829–856). Academic Press.

Lynch, J. M., Knauer, K., & Shaw, K. R. (2022). Plastic additives in the ocean. *Plastics and the Ocean: Origin, Characterization, Fate, and Impacts*, 43–76.

Maddela, N. R., Kakarla, D., Venkateswarlu, K., & Megharaj, M. (2023). Additives of plastics: Entry into the environment and potential risks to human and ecological health. *Journal of Environmental Management*, *348*, 119364. https://doi.org/10.1016/j.jenvman.2023.119364

Maes, T., Preston-Whyte, F., Lavelle, S., Gomiero, A., Booth, A. M., Belzunce-Segarra, M. J., …, & De Witte, B. (2023). A recipe for plastic: Expert insights on plastic additives in the marine environment. *Marine Pollution Bulletin*, *196*, 115633.

Marturano, V., Cerruti, P., & Ambrogi, V. (2016). Polymer additives. *Physical Sciences Reviews*, *2*(6), 1–20.

Massos, A., & Turner, A. (2017). Cadmium, lead and bromine in beached microplastics. *Environmental Pollution*, *227*, 139–145.

Mészáros, E., Bodor, A., Kovács, E., Papp, S., Kovács, K., Perei, K., & Feigl, G. (2023). Impacts of plastics on plant development: Recent advances and future research directions. *Plants (Basel, Switzerland)*, *12*(18), 3282. https://doi.org/10.3390/plants12183282

Net, S., Sempéré, R., Delmont, A., Paluselli, A., & Ouddane, B. (2015). Occurrence, fate, behavior and ecotoxicological state of phthalates in different environmental matrices. *Environmental Science & Technology*, *49*(7), 4019–4035.

Okeke, E. S., Olagbaju, O. A., Okoye, C. O., Addey, C. I., Chukwudozie, K. I., Okoro, J. O., ..., & Abesa, S. (2022). Microplastic burden in Africa: A review of occurrence, impacts, and sustainability potential of bioplastics. *Chemical Engineering Journal Advances*, *12*, 100402.

Okoye, C. O., Addey, C. I., Oderinde, O., Okoro, J. O., Uwamungu, J. Y., Ikechukwu, C. K., ..., & Odii, E. C. (2022). Toxic chemicals and persistent organic pollutants associated with micro-and nanoplastics pollution. *Chemical Engineering Journal Advances*, *11*, 100310.

Paluselli, A., Fauvelle, V., Schmidt, N., Galgani, F., Net, S., & Sempere, R. (2018). Distribution of phthalates in marseille bay (NW Mediterranean Sea). *Science of the Total Environment*, *621*, 578–587.

Pawar, P. R., Shirgaonkar, S. S., & Patil, R. B. (2016). Plastic marine debris: Sources, distribution and impacts on coastal and ocean biodiversity. *Pencil Publication of Biological Sciences*, *3*(1), 40–54.

Peng, X., Wang, Z., Mai, B., Chen, F., Chen, S., Tan, J., Yu, Y., Tang, C., Li, K., Zhang, G., & Yang, C. (2007). Temporal trends of nonylphenol and bisphenol A contamination in the Pearl River Estuary and the adjacent South China Sea recorded by dated sedimentary cores. *Science and the Total Environment*, *384*(1–3), 393–400.

Pereao, O., Opeolu, B., & Fatoki, O. (2020). Microplastics in aquatic environment: characterization, ecotoxicological effect, implications for ecosystems and developments in South Africa. *Environmental Science and Pollution Research*, *27*(18), 22271–22291.

Rendell-Bhatti, F., Paganos, P., Pouch, A., Mitchell, C., D'Aniello, S., Godley, B. J., ..., & Jimenez-Guri, E. (2021). Developmental toxicity of plastic leachates on the sea urchin *Paracentrotus lividus*. *Environmental Pollution*, *269*, 115744.

Ribeiro, F., O'Brien, J. W., Galloway, T., & Thomas, K. V. (2019). Accumulation and fate of nano-and microplastics and associated contaminants in organisms. *TrAC Trends in Analytical Chemistry*, *111*, 139–147.

Rochman, C. M. (2015). The complex mixture, fate and toxicity of chemicals associated with plastic debris in the marine environment. *Marine Anthropogenic Litter*, 117–140.

Satkar, S. G., Kumar, A., Bhusare, S., Sahu, A., & Gautam, R. K. (2024). Impact of pollution and toxic stress on fish health: Mechanisms, consequences, and mitigation strategies. *Uttar Pradesh Journal of Zoology*, *45*(6), 29–45.

Schmidt, N., Fauvelle, V., Ody, A., Castro-Jiménez, J., Jouanno, J., Changeux, T., ..., & Sempere, R. (2019). The Amazon River: A major source of organic plastic additives to the tropical North Atlantic. *Environmental Science & Technology*, *53*(13), 7513–7521.

Sonone, S. S., Jadhav, S., Sankhla, M. S., & Kumar, R. (2020). Water contamination by heavy metals and their toxic effect on aquaculture and human health through food Chain. *Letters in Applied NanoBioScience*, *10*(2), 2148–2166.

Sridharan, S., Kumar, M., Saha, M., Kirkham, M. B., Singh, L., & Bolan, N. S. (2022). The polymers and their additives in particulate plastics: What makes them hazardous to the fauna?. *Science of the Total Environment*, *824*, 153828.

Staples, C. A., Adams, W. J., Parkerton, T. F., Gorsuch, J. W., Biddinger, G. R., & Reinert, K. H. (1997). Aquatic toxicity of eighteen phthalate esters. *Environmental Toxicology and Chemistry*, *16*(5): 875–891.

Suhrhoff, T. J., & Scholz-Böttcher, B. M. (2016). Qualitative impact of salinity, UV radiation and turbulence on leaching of organic plastic additives from four common plastics – A lab experiment. *Marine Pollution Bulletin*, *102*, 84–94.

Sun, B., Hu, Y., Cheng, H., &Tao, S. (2019). Releases of brominated flame retardants (BFRs) from microplastics in aqueous medium: Kinetics and molecular-size dependence of diffusion. *Water Research*, *151*, 215–225.

Tanentzap, A. J. (2023). Plastics as non-toxic disruptors of aquatic ecosystems. *Limnology and Oceanography Bulletin*, *32*(4), 125–128.

Teuten, E. L., Saquing, J. M., Knappe, D. R., Barlaz, M. A., Jonsson, S., Bjorn, A., Rowland, S. J., Thompson, R. C., Galloway, T. S., Yamashita, R., Ochi, D., Watanuki, Y., Moore, C., Viet, P. H., Tana, T. S., Prudente, M., Boonyatumanond, R., Zakaria, M. P., Akkhavong, K., Ogata, Y., Hirai, H., Iwasa, S., Mizukawa,

K., Hagino, Y., Imamura, A., Saha, M., & Takada, H. (2009). Transport and release of chemicals from plastics to the environment and to wildlife. *Philosophical Transactions of the Royal Society B: Biological Sciences*, *364*(1526), 2027–2045.

Tian, K., & Bilal, M. (2020). Research progress of biodegradable materials in reducing environmental pollution. *Abatement of environmental pollutants*, 313–330. https://doi.org/10.1016/B978-0-12-818095-2.00015-1

Unuofin, J. O. (2020). Garbage in garbage out: The contribution of our industrial advancement to wastewater degeneration. *Environmental Science and Pollution Research*, *27*(18), 22319–22335.

van der Veen, I., & de Boer, J. (2012). Phosphorus flame retardants: properties, production, environmental occurrence, toxicity and analysis. *Chemosphere*, 88, 1119–1153.

Vered, G., & Shenkar, N. (2022). Limited effects of environmentally-relevant concentrations in seawater of dibutyl phthalate, dimethyl phthalate, bisphenol A, and 4-nonylphenol on the reproductive products of coral-reef organisms. *Environmental Pollution*, *314*, 120285.

Yu, Y., Kumar, M., Bolan, S., Padhye, L. P., Bolan, N., Li, S., …, & Li, Y. (2023). Various additive release from microplastics and their toxicity in aquatic environments. *Environmental Pollution*, 123219.

Zhang, S., Hou, R., Sun, C., Huang, Q., Lin, L., Li, H., …, & Xu, X. (2024). Metabolic activity of gut microbial enrichment cultures from different marine species and their transformation abilities to plastic additives. *Environment International*, *190*, 108882.

Zhang, Z. M., Zhang, H. H., Zhang, J., Wang, Q. W., & Yang, G. P. (2018). Occurrence, distribution, and ecological risks of phthalate esters in the seawater and sediment of Changjiang River Estuary and its adjacent area. *Science of the Total Environment*, *619–620*, 93–102.

Zimmermann, L., Dierkes, G., Ternes, T. A., Völker, C., & Wagner, M. (2019). Benchmarking the in vitro toxicity and chemical composition of plastic consumer products. *Environmental Science & Technology*, *53*, 11467–11477.

Zughaibi, T. A., Sheikh, I. A., & Beg, M. A. (2022). Insights into the endocrine disrupting activity of emerging non-phthalate alternate plasticizers against thyroid hormone receptor: A structural perspective. *Toxics*, *10*(5), 263.

13 Biocompatibility and Neurotoxicity of Bioplastic Additives

Adedamola Aminat Bayo-Olugbami,
Tolulope Timothy Arogundade, Abraham Olufemi Asuku,
and Aminat Omolola Imam-Fulani

13.1 INTRODUCTION

There is no doubting the fact that our daily lives revolve around many plastic materials. They are used in toys, electronics, textiles, healthcare goods, and packaging, making them relatively inescapable. These plastics are typically made using petroleum and are readily available because of their multifunctionality. However, the application of these plastics has been heavily criticized during the last decade, primarily because of the environmental issue of plastic pollution caused by the industrial-age exploitation of its characteristics (Barnes, 2019; Zimmermann et al., 2020).

Slowly but surely, the increasing awareness of sustainability has encouraged the use of bioplastics as a better alternative to conventional petroleum-derived plastics. According to a report by ResearchandMarkets.com, the global bioplastics market is projected to witness rapid growth to \$20.48 billion in 2028 at a compound annual growth rate (CAGR) of 17.1%. This potential forecast is attributed to an increase in consumer demand for sustainable products and regulatory initiatives aimed at reducing plastic waste (Shen et al., 2020). Bioplastics are made using starch blends, polyhydroxyalkanoates (PHAs), polylactic acid (PLA), and cellulose, which are all considered renewable biological sources. Broadly, there are two types of bioplastics: biodegradable and non-biodegradable. Micro- or macroscopic organisms interact with biodegradable bioplastics in such a way that only water, CO_2, and biomass are left behind, and no permanent environmental change occurs. With these, it is said that reliance on fossil resources would gradually decrease, all while producing a lesser amount of carbon in the atmosphere. However, like traditional plastics, bioplastics often require additives to enhance their functionality, stability, and aesthetics. These additives include plasticizers that increase flexibility and reduce brittleness; stabilizers that enhance resistance to heat and ultraviolet (UV) rays; colorants that provide aesthetic appeal; and flame retardants that improve safety by reducing flammability (Boccalon & Gorrasi, 2022; Narancic et al., 2020).

Interestingly, while the environmental benefits of bioplastics are widely touted, the safety of their additives is a growing concern. Additives, whether derived from natural or synthetic sources, can interact with biological systems in complex ways, potentially posing risks to human health and the environment (Barrick et al., 2021). Chief among the health concerns is the risk of neurotoxicity – the structural change or functionally adverse response of the nervous system to a chemical, biological, or physical agent – which has profound implications for both acute and chronic health outcomes (Prüst et al., 2020; Reyes & Medina, 2024). The nervous system's complexity and limited regenerative capacity make it especially susceptible to toxicants, with even low-level exposures potentially leading to long-term behavioral, cognitive, and developmental disruptions that may or may not be reversible (Wallace & Djordjevic, 2020). This concern is heightened by the widespread use of

DOI: 10.1201/9781003602828-16

bioplastic additives, which may inadvertently contribute to these risks if not adequately tested and regulated. Therefore, given the nervous system's intricate and highly sensitive nature and particularly vulnerable nature to chemical exposures, there is a need to explore the potential ways bioplastic additives may influence neurological health.

In this chapter, we aimed to bridge this knowledge gap by critically examining the biocompatibility and neurotoxicity of bioplastic additives, examples as earlier mentioned. We also review the biological mechanisms of action to provide a comprehensive understanding of their impacts. Next, we explore how these substances interact with the adult and developing nervous systems (using examples from animal and human studies), the exposure pathways, and their potential short- and long-term health implications. Furthermore, we highlight the importance of balancing the benefits of bioplastics with the imperative to safeguard human health and the environment.

13.2 BIOCOMPATIBILITY OF BIOPLASTIC ADDITIVES

The ability of a material to elicit its desired function without triggering adverse biological reactions is known as biocompatibility, and it is considered one of the pillars of the safety assessment of different molecules and compounds, including bioplastic additives (Black & Black, 2005; Boisseaux et al., 2023; Marin et al., 2020). With the increased use of bioplastics, here, we emphasize the importance of understanding physicochemical changes and biocompatibility of these materials and their additives, considering their application in food packaging, personal care products, and medical devices. Typically, biocompatibility has multiple dimensions and includes the assessment of cytocompatibility, immunogenicity, and the absence of toxic effects on specific organ systems, particularly over prolonged exposure periods (Basu, 2017; Marin et al., 2020). We briefly describe these methods in the next subsections.

13.2.1 Assessing the Biocompatibility of Bioplastics

In order to investigate the biocompatibility of bioplastics, a combination of *in vitro* and *in vivo* methods is usually employed. To start with, *in vitro* tests typically involve exposing cultured cells to the bioplastic material or its leachates to evaluate cytotoxicity, proliferation, and metabolic activity. For instance, the MTT (3-[4,5-dimethylthiazol-2-yl]-2,5 diphenyl tetrazolium bromide) assay is commonly used to measure mitochondrial activity that can indicate the viability of cells, while complementary tests, like those assessing oxidative stress or genotoxicity with enzyme-linked immunosorbent assay kits, provide a more comprehensive picture of cellular responses (Akoueson et al., 2023; Boccalon & Gorrasi, 2022; Folino et al., 2023; Savva et al., 2022).

Conversely, *in vivo* studies expose whole organisms, from rodents to aquatic species and nematodes, to additives to examine systemic effects like inflammation, immune response, and histopathology (Folino et al., 2023). Such studies are important for understanding the complex interactions between material systems and living systems, especially when additives are metabolized or bioaccumulated over time. For example, additives like phthalates and bisphenols are well-documented to bioaccumulate in various body tissues and have metabolic and endocrine-disrupting properties (Meeker et al., 2009; Wang et al., 2021; Xu et al., 2024).

13.2.2 Interaction with Biological Systems

At this point, it is vital to emphasize that bioplastic additives interact with biological systems at multiple levels of structural organization, depending on their chemical structure, concentration, and duration of exposure. At the cellular level, additives may induce oxidative stress by generating reactive oxygen species (ROS), leading to lipid peroxidation, DNA damage, and apoptosis. Specifically, it has been reported that plasticizers cause disruptions in mitochondrial integrity and altered calcium signaling pathways, leading to an impairment in energy metabolism and cellular homeostasis (Cui et al., 2024; Trebuňová et al., 2023).

Additionally, immune reactions varying from mild inflammation to severe hypersensitivity can be triggered by additives. The immunogenic potential of stabilizers and colorants that contain heavy metals or aromatic compounds is especially well-known (Burgos-Aceves et al., 2021; Campanale et al., 2020; Kumar et al., 2022). Such reactions are of concern in bioplastics used for medical applications, where immune compatibility is paramount.

Beyond localized cellular interactions, additives can enter the systemic circulation and accumulate in specific tissues. The ability of certain compounds to cross biological barriers, such as the blood–brain barrier (BBB) or placental barrier, highlights the importance of evaluating their biocompatibility in vulnerable populations, including pregnant women and children. For instance, bisphenols have been experimentally shown to be significantly teratogenic in *Xenopus laevis*, causing aberrant development of body tissues (Gu et al., 2022; Gyimah et al., 2021; Metruccio et al., 2024; Prüst et al., 2020).

13.2.3 Long-Term Considerations

Despite the fact that short-term studies provide valuable initial insights, the long-term biocompatibility of bioplastic additives is still an area of active research. The continuous exposure to low doses of some additives can increase the likelihood of suffering from chronic health conditions such as neurodegenerative diseases, endocrine dysregulation, or even cancer. For instance, overexposure to organophosphate flame retardants is associated with severe developmental neurotoxicity and hormonal disruptions in babies, as shown through experimental studies (Chen et al., 2021; Patisaul et al., 2021; Sun et al., 2016a, 2016b). Given this, it is not unreasonable to postulate that some so-called adult diseases might be latent expressions from developmental exposures to these additives. For an in-depth analysis of the above, we refer readers to detailed reviews on the Developmental Origins of Health and Disease (DOHaD) hypothesis (Lacagnina, 2020; Safi-Stibler & Gabory, 2020).

In light of the aforementioned, there is a biocompatibility challenge as it pertains to bioplastic additives, and this necessitates the development of safer alternatives, probably through green chemistry innovations. According to the US Environmental Protection Agency (EPA), designing chemical products and processes that minimize or do away with the use or production of hazardous compounds is known as "green chemistry." The scope of green chemistry includes but is not limited to a chemical product's life cycle, design, production, use, and eventual disposal (Zimmerman et al., 2020). Interestingly, some recent studies have been investigating the potential of natural additives, such as plant-derived plasticizers or antioxidants as promising alternatives, but these still require rigorous evaluation to ensure their efficacy and safety (Begum et al., 2024; Edayadulla et al., 2024; Jarwar et al., 2024; Mekonnen et al., 2013; Vieira et al., 2022). More so, there is a need to develop regulatory frameworks that prioritize biocompatibility testing and encourage transparency in additive composition.

Since bioplastics are increasingly becoming important across the board, it will be paramount to ensure that their additives are biocompatible as, this way, they are capable of fulfilling their promise of being safe and healthy materials. In the subsequent section, the probable neurotoxicity of bioplastic additives is examined, mainly the relevant routes of exposure, mechanisms/molecular targets of the materials, and biological effects on the neurodevelopmental trajectory.

13.3 NEUROTOXICITY OF BIOPLASTIC ADDITIVES

The nervous system is complex and highly specialized. It is a network of organ systems that comprises the brain, spinal cord, and peripheral nerves and is responsible for nearly all bodily activities, including movement, sensory processing, cognition, and emotion. The system serves an important purpose while it is also very fragile as it has a high metabolic demand, tightly regulated signaling pathways, and limited self-repair potential, all of which leave it open to external insults (Heyer & Meredith, 2017; Wallace & Djordjevic, 2020). Recent research has raised questions regarding some of the additives used in bioplastics. For example, leachates from both plastics and bioplastics were observed to increase neurotoxicity levels while reducing locomotion and lifespan in *Caenorhabditis elegans* (Reyes & Medina, 2024). For that reason, although bioplastics are claimed to be an environmentally friendly option, they potentially leach more chemicals than conventional plastics. Similarly, it has been demonstrated that some bioplastic materials, for instance, polyhydroxybutyrate (PHB), do have a slightly toxic effect on sea urchin larvae (Uribe-Echeverría & Beiras, 2022). Furthermore, Savva et al. (2022) showed that bioplastic particles and extracts can cause oxidative stress in human cell cultures. Additionally, certain bioplastic additives, especially isocyanates and maleic anhydrides, have proven highly toxic to *C. elegans* by causing death, impeding development, and causing oxidative damage (Li et al., 2022). Considering these facts, it becomes evident that the neurotoxic effects of bioplastic additives should be critically investigated due to their widespread use in daily life.

13.3.1 Mechanisms of Neurotoxicity

It is important to mention that not all chemicals have the same degree of toxicity; some could potentially be harmful at low concentrations, while others may not have the slightest effect, even at very high concentrations. For example, Buhler and Williams (1988) stated that a chemical would have a toxic effect if it is absorbed into the body, reaches the target tissue site(s), and remains in sufficient concentration for a period. The possible means of exposure include the following: lungs (breathing/inhalation), skin (dermal absorption), or gastrointestinal tract (ingestion/eating). From there, it gets absorbed, usually in the liver, where it becomes biochemically activated and, once fed from the blood, settles in various organs (Rahmati et al., 2020; Tilson et al., 1998; Zimmermann et al., 2019).

As with many other organs, there are some potentially harmful substances that could breach the BBB and directly affect the functioning of the brain, and these effects are determined by the particular functional zone that has sustained the damage. Such compounds, it should be noted, affect anatomical reserve, blood supply, or both. The term "anatomical reserve" may be understood to mean structural changes in neuronal or glial cells and their specialized subcellular elements like myelin, cytoskeleton (microfilaments, intermediate filaments, and microtubules), organelles such as mitochondria and endoplasmic reticulum (ER), which in turn imply oxidative stress on the system. Furthermore, an agent might, at the molecular level, target very precise cellular processes involving nucleic acid or protein biosynthesis and thereby disrupt some critical metabolic pathways. Some other agents may modify the production and secretion of trophic factors or specific neurotransmitters and perturb the system by overstimulating receptors, blocking some synaptic vesicles from fusing to the membrane in order to release the neurotransmitter, impairing the metabolic breakdown of the transmitting substance, or inhibiting the intracellular synthesis of precursors that could also lead to changes in the neurotransmitter systems that would in turn affect synaptic plasticity (Carmona et al., 2021; Ceccatelli, 2013; Tilson et al., 1998; Wang et al., 2017).

Importantly, neurotoxic agents are the most threatening during early neurodevelopmental windows (which begins early on *in utero* and is not complete until approximately after puberty). Even a minor disturbance can be immediately deleterious in outcome or have hidden (latent) ramifications concerning the trajectory of the system. The development of the mammalian nervous system is a highly intricate process because it consists of many stages that are meticulously controlled by molecular actors. Therefore, exposure to chemicals during pregnancy is capable of resulting in a plethora of effects which range from an unviable conceptus, gross structural abnormalities, and retarded growth to less apparent effects (Heyer & Meredith, 2017; Rock & Patisaul, 2018).

For context, different bioplastic additives can potentially have neurotoxic effects, and their mechanisms for neurotoxicity are likely to be just as broad as their applications. Paths like atmospheric exposure during construction (inhalation of volatile organic compounds (VOCs)), skin absorption of consumer products, or even drinking contaminated water can be surprisingly extensive than expected (Ahmad et al., 2023; Gupta et al., 2024; Hahladakis et al., 2018; Le et al., 2023). Additives, in an effort to serve their purpose, may affect the nervous system through several mechanistic pathways, which in turn have neurotoxic effects. These substances are often used to enhance the material properties of bioplastics to make them flexible, more durable, or resistant to heat; thus, their interaction with the nervous system can be far-reaching. For example, phthalates and organophosphates, as evidenced in the publications by Sun et al. (2016b) and Wang et al. (2021), can induce the production of ROS that basically leads to oxidative damage of neurons and glial cells. This ROS aggravates oxidative stress, disrupting mitochondrial function, impairing energy production, and triggering apoptotic pathways. Also, certain additives can provoke chronic glial activation, releasing pro-inflammatory cytokines that damage neuronal networks and alter synaptic plasticity (Carmona et al., 2021; Wang et al., 2021; Yin et al., 2020). In the same vein, additives may interact with biogenic amines and interfere with their synthesis, release, or reuptake, disrupting the delicate balance required for proper neuronal communication. For instance, organophosphate flame retardants inhibit acetylcholinesterase, leading to an accumulation of acetylcholine and subsequent cholinergic overstimulation and causing negative consequences for the neural milieu (Kassab et al., 2019; Wopara et al., 2021) (Figure 13.1).

This schematic representation highlights the primary routes through which bioplastic additives can enter the body, including inhalation, ingestion, dermal absorption, and maternal–fetal transfer. Upon systemic circulation, these additives may disrupt the BBB – depending on the type of additive, route, and duration of exposure, leading to their accumulation in the brain. Key neurotoxic mechanisms include oxidative stress (mediated by factors such as Nrf2 and antioxidant enzymes), mitochondrial and ER dysfunction, apoptosis, neurotransmitter dysregulation, glial homeostasis disturbances, excitotoxicity, and neuroinflammation. The diagram stresses the multifaceted impact of these additives on neural health and their potential contributions to neurological disorders.

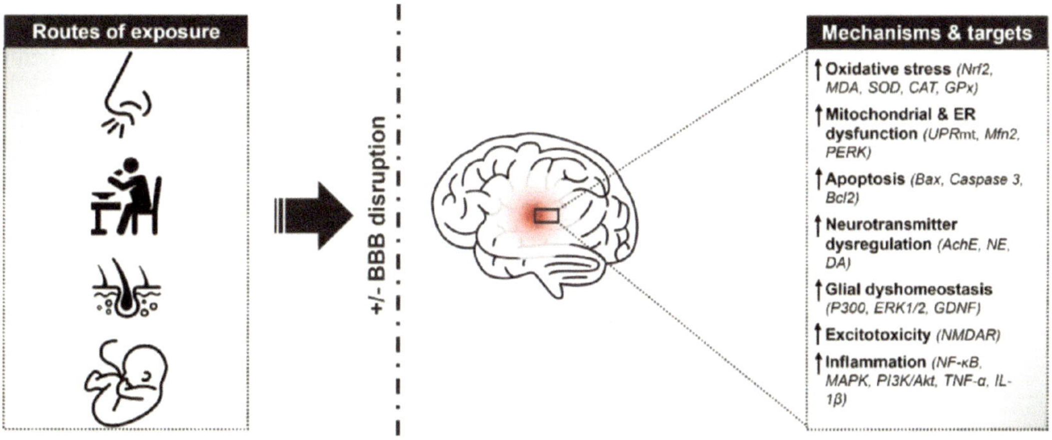

FIGURE 13.1 Pathways of exposure and neurotoxic mechanisms of bioplastic additives.

Abbreviations:

Nrf2	Nuclear factor erythroid 2-related factor 2;
MDA	Malondialdehyde;
SOD	Superoxide dismutase;
CAT	Catalase;
GPx	Glutathione peroxidase;
UPRmt	Unfolded protein response in mitochondria;
Mfn2	Mitofusin-2;
PERK	Protein kinase RNA-like endoplasmic reticulum kinase;
Bax	Bcl-2-associated X protein;
Caspase 3	Cysteine-aspartic protease 3;
Bcl2	B-cell lymphoma 2;
AchE	Acetylcholinesterase;
NE	Norepinephrine;
DA	Dopamine;
P300	Transcriptional co-activator p300;
ERK1/2	Extracellular signal-regulated kinases 1/2;
GDNF	Glial cell line-derived neurotrophic factor;
NMDAR	N-Methyl-D-aspartate receptor;
NF-κB	Nuclear factor kappa-light-chain-enhancer of activated B cells;
MAPK	Mitogen-activated protein kinase;
PI3K/Akt	Phosphatidylinositol 3-kinase/Protein kinase B pathway;
TNF-α	Tumor necrosis factor-alpha;
IL-1β	Interleukin-1 beta.

As earlier alluded to, the risks posed by bioplastic additives are particularly concerning during developmental windows of vulnerability, where the nervous system's dynamic growth and specialization heighten its sensitivity to external insults. Recent studies suggest that additives in bioplastics and 3D-printed plastics may pose developmental neurotoxicity risks. Isocyanates and maleic anhydrides used in biodegradable plastics showed prominent toxicity in *C. elegans* – it caused high lethality in the nematodes at very low concentrations while inhibiting growth and lifespan and instigating oxidative damage and mitochondrial dysfunction (Li et al., 2022). Similarly, leachates from 3D-printed plastics induced developmental abnormalities and neurobehavioral alterations in zebrafish embryos, with toxic effects evidenced by upregulated ROS, lipid and protein damage, and acetylcholinesterase inhibition (Walpitagama et al., 2019). On the other hand, Uribe-Echeverría and Beiras (2022) showed that while some bioplastics like PLA cups and PLA/PHA filaments appeared innocuous, others like PHB resin contained potentially harmful additives. Organophosphate esters (OPEs), used as flame retardants and plasticizers in various products, have been shown to alter neurodevelopmental processes through non-cholinergic pathways – perturbation of glutamate, GABA, and neuroendocrine signaling pathways – at environmentally relevant doses, potentially affecting cognitive and social skill development in children (Patisaul et al., 2021). Notably, a human study observed the potential adverse effects of prenatal phthalate exposure on child neurodevelopment, influenced by genetic vulnerability to oxidative stress. The researchers analyzed data from a birth cohort of 1,064 women in Australia, measuring maternal urinary phthalate levels at 36 weeks of pregnancy and assessing neurodevelopmental outcomes in their children. Higher prenatal phthalate levels and a genetic predisposition to oxidative stress were both linked to an increased risk of autism spectrum disorder (ASD), cognitive impairments, and behavioral issues such as inattention and hyperactivity. Specific oxidative stress-related genetic variations amplified the effects of prenatal phthalate exposure. Additionally, other oxidative stress factors, such as prenatal smoking, further elevated the risks (Ponsonby et al., 2020).

Experimental studies have also highlighted the potential for compounds like microplastics, bisphenol A (BPA), and its alternatives like bisphenol S (BPS), to influence neurodevelopment by modulating receptor activity or altering dopamine-related pathways (Gyimah et al., 2021; Solleiro-Villavicencio et al., 2020; Welch & Mulligan, 2022; Zhou et al., 2017). Interestingly, exposure to BPA and BPB during the teratogenicity and neurobehavioral windows in *X. laevis* led to lethality, malformations, and behavioral deficits, respectively, with BPB being about three times more potent than BPA (Metruccio et al., 2024). These findings warrant concern and further investigation into the developmental neurotoxicity of bioplastic additives, especially in light of their increasing prevalence in consumer products and environmental exposure.

In light of the foregoing, a comprehensive understanding of the toxicokinetics and toxicodynamics of these additives is pertinent. Factors such as dose, duration of exposure, bioavailability, and the developmental stage could modulate the neurotoxic potential. Moreover, additives may not act in isolation; their interactions with other environmental chemicals could exacerbate neurotoxicity or introduce synergistic effects. This interplay demands detailed evaluation to ensure the acclaimed safety of bioplastic materials. Thus, while bioplastics offer a sustainable alternative to conventional plastics, the potential neurotoxic effects of their additives present a complex challenge, and adopting a precautionary approach that integrates rigorous testing protocols, regulatory oversight, and innovations in green chemistry may aid in identifying and developing safer additive alternatives (Table 13.1).

TABLE 13.1
Neurotoxic Effects of Common Bioplastic Additives

Additive Type	Example Compounds	Primary Neurotoxic Mechanisms	Relevant Studies
Plasticizers	Phthalates (e.g., DEHP, DBP)	Induction of oxidative stress, mitochondrial dysfunction, and disruptions in calcium signaling.	Kassab et al. (2019), Mohammadi and Ashari (2021), Nahla et al. (2024), Tseng et al. (2013), Zhao et al. (2024)
Flame Retardants	Organophosphates, PBDEs	Acetylcholinesterase inhibition, neurotransmitter imbalance, and developmental neurotoxicity.	Patisaul et al. (2021), Sun et al. (2016a, 2016b)
Stabilizers	Lead- and cadmium-based agents	Impairment of synaptic plasticity, oxidative damage to neurons, and cognitive deficits.	Lee et al. (2022), Reyes and Medina (2024), Santa Maria et al. (2019), Turna Demir et al. (2022)
Colorants	Azo dyes, heavy metals	Triggering of neuroinflammation and disruption of neurotransmitter systems.	Biswas et al. (2023, 2024), Haridevamuthu et al. (2024), Rubalajyothi et al. (2022)
Antimicrobial Agents	Triclosan, silver nanoparticles	Disruption of neuronal ion channels, oxidative stress, and microglial activation.	De la Rosa et al. (2022), Dziendzikowska et al. (2022), Sun et al. (2016c), Szychowski et al. (2019)
Degradation Byproducts	Bisphenol A (BPA), Bisphenol S (BPS), microplastics	Neurotransmitter dysregulation, oxidative stress, endoplasmic-mitochondrial dyshomeostasis, synaptic impairment, cytoskeletal derangement, and excitotoxicity.	Bi et al. (2022), Li et al. (2022), Wang et al. (2023a, 2023b, 2024), Yin et al. (2020)

13.4 INTERACTION WITH THE ENVIRONMENT AND SECONDARY NEUROTOXICITY

In the previous section, we showed the direct neurotoxic impacts of bioplastic additives; however, the interaction between them and the environment is also a critical factor influencing their potential for secondary neurotoxicity. Interestingly, these additives undergo various physical, chemical, and biological transformations that can modify their toxicity profiles once released into the environment – whether during manufacturing, usage, or disposal. These interactions can result in the formation of secondary toxicants, bioaccumulate in ecosystems, and eventually re-enter human systems through air, water, and food pathways, and in a sense, perpetuate a vicious cycle of toxicity (Hahladakis et al., 2018; Le et al., 2023).

13.4.1 Environmental Degradation of Additives

Bioplastic additives are often not chemically bonded to the polymer matrix and can leach into the surrounding environment. Exposure to environmental factors such as UV radiation, heat, and microbial activity can further degrade these additives, creating byproducts with altered toxicity profiles. For instance, the photodegradation of phthalates can produce more bioactive metabolites capable of disrupting hormonal and neurological processes (Hahladakis et al., 2018; Janzadeh et al., 2022). Furthermore, the persistence of certain additives in soil and water systems raises concerns about their bioavailability to aquatic and terrestrial organisms. These additives can accumulate in biota, leading to biomagnification across trophic levels and indirectly exposing humans to neurotoxic risks through dietary intake (Barboza et al., 2018; Pan et al., 2023).

13.4.2 Pathways to Secondary Neurotoxicity

Secondary neurotoxicity occurs when environmental interactions amplify or extend the adverse effects of bioplastic additives on the nervous system. A summary of the key pathways includes the following:

1. **Bioaccumulation and Biomagnification:** Additives are able to bioaccumulate in aquatic creatures like fish and shellfish, which is a source of food for dozens of people. By eating these particular seafoods, people likewise come in contact with these neurotoxicants that are able to traverse biological barriers like BBB and interfere with neural homeostasis (Mamun et al., 2023; Okeke et al., 2022).
2. **Soil and Water Contamination:** Additives that infiltrate the subsoil and subaqueous surroundings are able to bring alteration to microbial populations that serve as stabilizers to the ecosystem. If such changes occur in the microbial ecosystem, it has an add-on effect on humans, animals, and even plants. Soil microbiology alterations, for instance, may have effect on agricultural productivity and also possibly incorporate neurotoxic residues in the food chain (Barrick et al., 2021; Brown et al., 2023; Gao et al., 2023; Liu et al., 2020).
3. **Airborne Transmission:** VOCs that occur during the bioplastic additives breakdown may contaminate air supplies. This method of contact with neurotoxins is rather under-researched but it does emphasize the fact that the neurotoxin exposure can come from breathing the volatiles and in heavily industrialized places or places with poor ventilation, a common occurrence in employees or even the people surrounding them (Wang et al., 2017).
4. **Microplastic and Nanoplastic Vehicles:** Micro- and nanoplastics, which contain leftover additives and environmental contaminants, can be created as bioplastics break down. In the previous section, we illuminated how these particles can enter biological systems and produce localized inflammation and oxidative stress, worsening neurotoxic consequences. More significantly, micro/nanoplastics can affect neurotoxicity by interacting with other additives, such as bisphenols (Gao et al., 2023; Prüst et al., 2020; Sofield et al., 2024).

13.4.3 Synergistic Effects of Additives with Environmental Stressors

It is important to note that the additives currently being studied exist in a variety of other products including the surface coatings, detergents, and air fresheners which means bioplastic additives are likely to resume contact with other pollutants in the environment, including heavy metals, pesticides, and industrial chemicals, leading to the synergistic or additive-additive toxic effects/ co-toxicity (Maddela et al., 2023; Maity et al., 2021; Menéndez-Pedriza & Jaumot, 2020). For example, animals co-exposed to phthalates and titanium dioxide nanoparticles showed exacerbated neurodevelopmental deficits. This experimental observation highlights the complexity of secondary neurotoxicity pathways (Chen et al., 2024).

The implication of the aforestated is far-reaching as the interplay between environmental factors and human vulnerability would further amplify the risk of toxicity. Populations in areas with high environmental contamination or occupational exposure to bioplastics are at particular risk, as are vulnerable groups such as children and pregnant individuals, who are more sensitive to chemical insults. For example, the significant consumption and mismanagement of plastics in Africa ranks it second globally after Asia. It is also worth noting that over 70% of African plastic waste is mismanaged and thus creates microplastic pollution from tourism, fishing, and water transport. The consequences of attitude include metal toxicity in aquatic and terrestrial organisms, disruption of ecosystems, and health challenges for humans consuming contaminated seafood (Okeke et al., 2022).

13.4.4 Mitigation Strategies

Having considered the interaction of bioplastics with the environment and the potential for secondary neurotoxicity, attempts to mitigate such occurrences can employ a blend of approaches. Firstly, there must be comprehensive tracking of additive leaching, degradation products, and their environmental impacts to provide critical data for risk assessment (Barrick et al., 2021). Secondly, innovations in additive chemistry should focus on inventing materials that degrade into non-toxic byproducts during the environmental process (Barnes, 2019). Additionally, the placement of strict limits on the environmental release of high-risk additives, in conjunction with incentives for low-risk alternatives, can minimize their ecological footprint (Boisseaux et al., 2023). Lastly, public relations campaigns that highlight the association between environmental contamination and secondary neurotoxicity can empower communities to challenge existing practices.

Collectively, the interactions between bioplastic additives and the environment reveal a complex web of pathways through which neurotoxicity can manifest. As bioplastics continue to stand in for conventional plastics, it is vital to learn more about these processes so as to preserve ecosystem health and human health. The next part summarizes the recent developments, barriers, and future outlook in the area of bioplastic materials and additives regulation and safety regulations.

13.5 REGULATORY AND SAFETY CONSIDERATIONS

As various strategies are developed to enhance these bioplastics and their additives, they also bring with them the need to be safety-ensured. From the available evidence, bioplastics are indeed more favorable than traditional plastics, but the potential risks accompanying their use concern their biocompatibility and the potential neurotoxicity of the additives. There is a need to have established practices for regulations and adequate assessment of safety. Proper regulation enhances the safety of humans as well as the environment and increases the confidence of the consumers in the industry and the integrity of the industry itself (Cruz et al., 2022; Painuli et al., 2024).

Currently, the regulation of bioplastic additives varies from one region to another and is predominantly a consequence of differing levels of awareness, scientific evidence, and political priorities. For instance, in the European Union (EU), bioplastic chemicals are governed by regulatory standards such as REACH (Registration, Evaluation, Authorisation, and Restriction of Chemicals) (Coria, 2018). However, these regulations often target known risks and do not effectively manage the use of new additives or degradation products. Similarly, the US Food and Drug Administration (FDA) oversees the safety of bioplastics used in food packaging, while the EPA evaluates environmental impacts. It is conceivable that the fragmented nature of oversight may make enforcement more difficult, particularly if cross-talks between both agencies are minimal or non-existent. In contrast, however, regulatory oversight is rare in developing countries, leaving significant room for the unregulated use of potentially hazardous additives.

Despite these laudable efforts, no comprehensive global framework exists to address the unique challenges posed by bioplastic additives, particularly their long-term biocompatibility and neurotoxic potential. More so, the existing regulatory frameworks often fail to fully account for the complexity of bioplastic additives and are, therefore, fraught with limitations (Moshood et al., 2021; Silva et al., 2024). For instance, many regulations focus on acute toxicity or specific endpoints, neglecting chronic exposure, developmental neurotoxicity, or additive interactions. Furthermore, there is also the issue of absent transparency – manufacturers are not always required to disclose the complete composition of bioplastic additives, limiting independent evaluation and public awareness (Fletcher et al., 2021). Even then, the available regulations rarely address the environmental fate of additives, including their degradation into secondary toxicants, coupled with the fact that divergent regional standards pose an obstacle to global trade and uniform safety assurance (Rosenboom et al., 2022).

Therefore, regulatory bodies must adopt a more comprehensive and proactive approach to address these gaps (Cucina, 2023). These strategies may entail regulatory agencies mandating studies that evaluate chronic exposure, developmental effects, and cumulative impacts over time; combining *in vitro*, *in vivo*, and *in-silico* methods to enhance the depth and efficiency of safety evaluations while minimizing ethical concerns associated with animal testing and consideration of the synergistic effects of multiple additives, as these mixtures may exhibit enhanced toxicity that is not captured by individual assessments.

Additionally, international cooperation is essential given the global nature of bioplastic production and distribution. Harmonized standards would ensure consistent safety requirements, streamline manufacturers' compliance, and reduce trade barriers. Organizations such as the World Health Organization (WHO) and the United Nations Environment Programme (UNEP) could play pivotal roles in facilitating these efforts. Notably, regulation alone cannot achieve safety; active participation from the bioplastics industry is essential. Manufacturers must adopt transparent practices, including full disclosure of additive compositions and collaboration with independent researchers to validate safety claims. Voluntary certification programs like eco-labeling initiatives can incentivize safer additive development and promote consumer confidence. Moreover, consumers and advocacy groups are vital in shaping regulatory priorities. Public demand for safer and more sustainable bioplastics can drive stricter regulations and accelerate the transition to non-toxic additives. Educational campaigns highlighting the potential risks of unsafe additives and the benefits of rigorous safety standards are essential for fostering informed decision-making. Excellent publications are available for more details on policies, regulatory aspects, and barriers to the uptake of bioplastics and biodegradable materials (Fletcher et al., 2021; Raj & Matche, 2011).

13.6 FUTURE DIRECTIONS

Bioplastics boom should provide extra impetus for research and development of new materials, which have less toxic potential and better biocompatibility. As far as understanding the biological activity of the said compounds, much has been done, yet several important issues and scope for advancement continue to exist. Against this backdrop, there is no denying the fact that future works would require integration and collaboration across different fields – material sciences, toxicology, and health science, among others (Boccalon & Gorrasi, 2022; Boisseaux et al., 2023; Painuli et al., 2024; Raj & Matche, 2011).

To start with, the design and development of more appropriate bioplastic additives could be one of the main directions in the future. Efforts should be made to ensure that the new plasticizers, stabilizers, and other functional agents which will be used in the future will not be *bio-incompatible* (Edayadulla et al., 2024; Vieira et al., 2022). Furthermore, research into additive-free bioplastics, where polymer properties are enhanced through intrinsic modifications rather than external chemicals, could eliminate several safety concerns. Such advancements would represent a paradigm shift in materials science, paving the way for a new generation of non-toxic, high-performance bioplastics.

Importantly, future studies should assess the cumulative effects of low-dose, long-term exposure to additives, particularly during critical periods of neurodevelopment (Heyer & Meredith, 2017; Welch & Mulligan, 2022). They should also examine the combined effects of multiple additives and their environmental degradation products, which may exhibit synergistic toxicity. These studies could incorporate organ-on-chip systems and computational toxicology tools to simulate human-specific responses and reduce reliance on animal models.

Still, given the complexity of additive interactions with environmental systems, future research must prioritize understanding how these compounds degrade, bioaccumulate, and impact ecosystems. This includes evaluating their potential to generate secondary neurotoxicants and exploring mechanisms of environmental remediation. Equally, as bioplastics gain widespread use, existing regulatory frameworks must evolve to address their unique challenges. Policies should mandate comprehensive biocompatibility and neurotoxicity testing for all additives, including their degradation products. Additionally, international collaboration and public and stakeholder engagement will be essential to harmonizing safety standards and ensuring consistent enforcement across regions. Regulatory agencies can also incentivize safer additive development by providing funding, tax breaks, or certification programs for manufacturers prioritizing sustainability and health (Ashiwaju et al., 2023; Constance Obiuto et al., 2024; Fletcher et al., 2021; Rosenboom et al., 2022).

13.7 CONCLUSION

As society comes to terms with bioplastics as a viable substitute for traditional plastics, the safety and environmental impact of the additives used to improve their functionality becomes more important. Bioplastic additives contribute to the service lifespan and biodegradability of these plastics, but their biocompatibility and potential for neurotoxicity require in-depth investigation. The evidence presented in this chapter emphasizes how these additives mutually interact with biological and ecological systems, from systemic neurotoxic effects to cellular-level impacts like oxidative stress and immune modulation. In addition, their decomposition and chemical bonding with other environmental toxins might further generate secondary toxicants, which adds risks more than anticipated.

We must stress that a multifaceted strategy is needed to overcome the challenges we have highlighted above. Developing safer and greener additives that minimize the negative impacts on people and ecosystems should be the top priority for green chemistry advancements. At the same time, regulatory structures need to change to guarantee worldwide harmonization, transparency, and thorough safety testing. It is impossible to overestimate the contribution that consumers, industry stakeholders, researchers, and legislators have made to this change (Fletcher et al., 2021).

Bioplastics have much potential as an environmentally mindful material, but their success depends on our ability to balance responsibility and innovation. We can steer toward a future where sustainability and safety coexist and create a healthier world for future generations by recognizing and resolving the possible dangers associated with bioplastic additives.

REFERENCES

Ahmad, M., Chen, J., Khan, M. T., Yu, Q., Phairuang, W., Furuuchi, M., Ali, S. W., Nawab, A., & Panyametheekul, S. (2023). Sources, analysis, and health implications of atmospheric microplastics. *Emerging Contaminants*, *9*(3), 100233. https://doi.org/10.1016/j.emcon.2023.100233

Akoueson, F., Paul-Pont, I., Tallec, K., Huvet, A., Doyen, P., Dehaut, A., & Duflos, G. (2023). Additives in polypropylene and polylactic acid food packaging: Chemical analysis and bioassays provide complementary tools for risk assessment. *Science of the Total Environment*, *857*, 159318. https://doi.org/10.1016/j.scitotenv.2022.159318

Al Mamun, A., Prasetya, T. A. E., Dewi, I. R., & Ahmad, M. (2023). Microplastics in human food chains: Food becoming a threat to health safety. *Science of The Total Environment*, *858*, 159834. https://doi.org/10.1016/j.scitotenv.2022.159834

Ashiwaju, B. I., Orikpete, O. F., Fawole, A. A., Alade, E. Y., & Odogwu, C. (2023). A step toward sustainability: A review of biodegradable packaging in the pharmaceutical industry. *Matrix Science Pharma*, *7*(3), 73–84. https://doi.org/10.4103/mtsp.mtsp_22_23

Barboza, L. G. A., Vieira, L. R., Branco, V., Figueiredo, N., Carvalho, F., Carvalho, C., & Guilhermino, L. (2018). Microplastics cause neurotoxicity, oxidative damage and energy-related changes and interact with the bioaccumulation of mercury in the European seabass, *Dicentrarchus labrax* (Linnaeus, 1758). *Aquatic Toxicology*, *195*, 49–57. https://doi.org/10.1016/j.aquatox.2017.12.008

Barnes, S. J. (2019). Understanding plastics pollution: The role of economic development and technological research. *Environmental Pollution*, *249*, 812–821. https://doi.org/10.1016/j.envpol.2019.03.108

Barrick, A., Champeau, O., Chatel, A., Manier, N., Northcott, G., & Tremblay, L. A. (2021). Plastic additives: Challenges in ecotox hazard assessment. *PeerJ*, *9*, e11300. https://doi.org/10.7717/peerj.11300

Basu, B. (2017). *Probing Toxicity of Biomaterials and Biocompatibility Assessment* (pp. 291–351). https://doi.org/10.1007/978-981-10-3059-8_9

Begum, S. S., Divakaran, D., Suyambulingam, I., Senthamaraikannan, P., Handayani, M., Murali, A., & Han, S. S. (2024). Comprehensive characterization of bioplasticizer from the *Murraya koenigii* leaves: From biomass to biomaterial for polymer composite applications. *Industrial Crops and Products*, *222*, 119950. https://doi.org/10.1016/j.indcrop.2024.119950

Bi, N., Gu, X., Fan, A., Li, D., Wang, M., Zhou, R., Sun, Q.-C., & Wang, H.-L. (2022). Bisphenol-A exposure leads to neurotoxicity through upregulating the expression of histone deacetylase 2 in vivo and in vitro. *Toxicology*, *465*, 153052. https://doi.org/10.1016/j.tox.2021.153052

Biswas, P., Jain, J., Hasan, W., Bose, D., & Yadav, R. S. (2023). Azo food dye neurotoxicity in rats: A neurobehavioral, biochemical, and histopathological study. *Food and Chemical Toxicology*, *181*, 114067. https://doi.org/10.1016/j.fct.2023.114067

Biswas, P., Jain, J., Hasan, W., Bose, D., & Yadav, R. S. (2024). Biochemical alterations and motor dysfunctions in corpus striatum of rats brain exposed to azo dyes. *Toxicology Research*, *13*(6). https://doi.org/10.1093/toxres/tfae216

Black, J., & Black, J. (2005). *Biological Performance of Materials*. CRC Press. https://doi.org/10.1201/9781420057843

Boccalon, E., & Gorrasi, G. (2022). Functional bioplastics from food residual: Potentiality and safety issues. *Comprehensive Reviews in Food Science and Food Safety*, *21*(4), 3177–3204. https://doi.org/10.1111/1541-4337.12986

Boisseaux, P., Hopkinson, P., Santillo, D., Smith, C., Garmulewicz, A., Powell, Z., & Galloway, T. (2023). Environmental safety of second and third generation bioplastics in the context of the circular economy. *Ecotoxicology and Environmental Safety*, *256*, 114835. https://doi.org/10.1016/j.ecoenv.2023.114835

Brown, R. W., Chadwick, D. R., Zang, H., Graf, M., Liu, X., Wang, K., Greenfield, L. M., & Jones, D. L. (2023). Bioplastic (PHBV) addition to soil alters microbial community structure and negatively affects plant-microbial metabolic functioning in maize. *Journal of Hazardous Materials*, *441*, 129959. https://doi.org/10.1016/j.jhazmat.2022.129959

Buhler, D. R., & Williams, D. E. (1988). The role of biotransformation in the toxicity of chemicals. *Aquatic Toxicology*, *11*(1–2), 19–28. https://doi.org/10.1016/0166-445X(88)90004-5

Burgos-Aceves, M. A., Abo-Al-Ela, H. G., & Faggio, C. (2021). Physiological and metabolic approach of plastic additive effects: Immune cells responses. *Journal of Hazardous Materials*, *404*, 124114. https://doi.org/10.1016/j.jhazmat.2020.124114

Campanale, C., Massarelli, C., Savino, I., Locaputo, V., & Uricchio, V. F. (2020). A detailed review study on potential effects of microplastics and additives of concern on human health. *International Journal of Environmental Research and Public Health*, *17*(4), 1212. https://doi.org/10.3390/ijerph17041212

Carmona, A., Roudeau, S., & Ortega, R. (2021). Molecular mechanisms of environmental metal neurotoxicity: A focus on the interactions of metals with synapse structure and function. *Toxics*, *9*(9), 198. https://doi.org/10.3390/toxics9090198

Ceccatelli, S. (2013). Mechanisms of neurotoxicity and implications for neurological disorders. *Journal of Internal Medicine*, *273*(5), 426–428. https://doi.org/10.1111/joim.12053

Chen, H., Wang, Y., & Liang, H. (2024). The combined neurotoxicity of DBP and nano-TiO_2 in embryonic zebrafish (*Danio rerio*) revealed by oxidative activity, neuro-development genes expression and metabolomics changes. *Aquatic Toxicology*, *269*, 106881. https://doi.org/10.1016/j.aquatox.2024.106881

Chen, X., Guo, W., Lei, L., Guo, Y., Yang, L., Han, J., & Zhou, B. (2021). Bioconcentration and developmental neurotoxicity of novel brominated flame retardants, hexabromobenzene and pentabromobenzene in zebrafish. *Environmental Pollution*, *268*, 115895. https://doi.org/10.1016/j.envpol.2020.115895

Constance Obiuto, N., Ninduwezuor-Ehiobu, N., Chigozie Ani, E., Andrew Olu-lawal, K., & David Ugwuanyi, E. (2024). Implementing circular economy principles to enhance safety and environmental sustainability in manufacturing. *International Journal of Advanced Multidisciplinary Research and Studies*, *4*(2), 22–29. https://doi.org/10.62225/2583049X.2024.4.2.2432

Coria, J. (2018). Policy monitor—The economics of toxic substance control and the REACH directive. *Review of Environmental Economics and Policy*, *12*(2), 342–358. https://doi.org/10.1093/reep/rey003

Cruz, R. M. S., Krauter, V., Krauter, S., Agriopoulou, S., Weinrich, R., Herbes, C., Scholten, P. B. V., Uysal-Unalan, I., Sogut, E., Kopacic, S., Lahti, J., Rutkaite, R., & Varzakas, T. (2022). Bioplastics for food packaging: Environmental Impact, trends and regulatory aspects. *Foods*, *11*(19), 3087. https://doi.org/10.3390/foods11193087

Cucina, M. (2023). The lesser of two evils: Enhancing biodegradable bioplastics use to fight plastic pollution requires policy makers interventions in Europe. *Environmental Impact Assessment Review*, *103*, 107230. https://doi.org/10.1016/j.eiar.2023.107230

Cui, J.-G., Zhang, H., Chen, M.-S., Wang, J.-X., Zhao, Y., & Li, J.-L. (2024). Calcium homeostasis imbalance mediates DEHP induced mitochondrial damage in cerebellum and the antagonistic effect of lycopene. *Science of the Total Environment*, *954*, 176351. https://doi.org/10.1016/j.scitotenv.2024.176351

De la Rosa, V., Guzmán-Hernández, M. L., & Carrillo, E. (2022). Triclosan is a KCNQ3 potassium channel activator. *Pflügers Archiv – European Journal of Physiology*, *474*(7), 721–732. https://doi.org/10.1007/s00424-022-02692-w

Dziendzikowska, K., Wilczak, J., Grodzicki, W., Gromadzka-Ostrowska, J., Węsierska, M., & Kruszewski, M. (2022). Coating-dependent neurotoxicity of ilver nanoparticles—An in vivo study on hippocampal oxidative stress and neurosteroids. *International Journal of Molecular Sciences*, *23*(3), 1365. https://doi.org/10.3390/ijms23031365

Edayadulla, N., Divakaran, D., Chandraraj, S. S., Sriariyanun, M., Suyambulingam, I., Sanjay, M. R., & Siengchin, S. (2024). Suitability study of novel bio-plasticizer from *Agave sisalana* leaf for biofilm applications: A biomass to biomaterial approach. *Biomass Conversion and Biorefinery*, *14*(16), 19675–19691. https://doi.org/10.1007/s13399-023-04172-2

Fletcher, C. A., Niemenoja, K., Hunt, R., Adams, J., Dempsey, A., & Banks, C. E. (2021). Addressing stakeholder concerns regarding the effective use of bio-based and biodegradable plastics. *Resources*, *10*(10), 95. https://doi.org/10.3390/resources10100095

Folino, A., Pangallo, D., & Calabrò, P. S. (2023). Assessing bioplastics biodegradability by standard and research methods: Current trends and open issues. *Journal of Environmental Chemical Engineering*, *11*(2), 109424. https://doi.org/10.1016/j.jece.2023.109424

Gao, D., Junaid, M., Chen, X., Liao, H., Chen, G., & Wang, J. (2023). Interaction of micro(nano)plastics and bisphenols in the environment: A recent perspective on adsorption mechanisms, influencing factors and ecotoxic impacts. *TrAC Trends in Analytical Chemistry*, *165*, 117132. https://doi.org/10.1016/j.trac.2023.117132

Gu, J., Guo, M., Yin, X., Huang, C., Qian, L., Zhou, L., Wang, Z., Wang, L., Shi, L., & Ji, G. (2022). A systematic comparison of neurotoxicity of bisphenol A and its derivatives in zebrafish. *Science of the Total Environment*, *805*, 150210. https://doi.org/10.1016/j.scitotenv.2021.150210

Gupta, R. K., Pipliya, S., Karunanithi, S., Eswaran U, G. M., Kumar, S., Mandliya, S., Srivastav, P. P., Suthar, T., Shaikh, A. M., Harsányi, E., & Kovács, B. (2024). Migration of chemical compounds from packaging materials into packaged foods: Interaction, mechanism, assessment, and regulations. *Foods*, *13*(19), 3125. https://doi.org/10.3390/foods13193125

Gyimah, E., Xu, H., Dong, X., Qiu, X., Zhang, Z., Bu, Y., & Akoto, O. (2021). Developmental neurotoxicity of low concentrations of bisphenol A and S exposure in zebrafish. *Chemosphere*, *262*, 128045. https://doi.org/10.1016/j.chemosphere.2020.128045

Hahladakis, J. N., Velis, C. A., Weber, R., Iacovidou, E., & Purnell, P. (2018). An overview of chemical additives present in plastics: Migration, release, fate and environmental impact during their use, disposal and recycling. *Journal of Hazardous Materials*, *344*, 179–199. https://doi.org/10.1016/j.jhazmat.2017.10.014

Haridevamuthu, B., Murugan, R., Seenivasan, B., Meenatchi, R., Pachaiappan, R., Almutairi, B. O., Arokiyaraj, S., Kathiravan, M., K., & Arockiaraj, J. (2024). Synthetic azo-dye, Tartrazine induces neurodevelopmental toxicity via mitochondria-mediated apoptosis in zebrafish embryos. *Journal of Hazardous Materials*, *461*, 132524. https://doi.org/10.1016/j.jhazmat.2023.132524

Heyer, D. B., & Meredith, R. M. (2017). Environmental toxicology: Sensitive periods of development and neurodevelopmental disorders. *NeuroToxicology*, *58*, 23–41. https://doi.org/10.1016/j.neuro.2016.10.017

Janzadeh, A., Behroozi, Z., saliminia, F., Janzadeh, N., Arzani, H., Tanha, K., Hamblin, M. R., & Ramezani, F. (2022). Neurotoxicity of silver nanoparticles in the animal brain: A systematic review and meta-analysis. *Forensic Toxicology*, *40*(1), 49–63. https://doi.org/10.1007/s11419-021-00589-4

Jarwar, A. I., Ahmed, F., Iqbal, M., Shah, A. K., Usto, A., Jatoi, A. S., & Mubarak, N. M. (2024). Bio-based additives for food packaging. In *Application of Bio-Additives for the Food Industry* (pp. 67–88). Springer Nature Switzerland. https://doi.org/10.1007/978-3-031-71131-2_5

Kassab, R. B., Lokman, M. S., & Essawy, E. A. (2019). Neurochemical alterations following the exposure to di-n-butyl phthalate in rats. *Metabolic Brain Disease*, *34*(1), 235–244. https://doi.org/10.1007/s11011-018-0341-0

Kumar, R., Manna, C., Padha, S., Verma, A., Sharma, P., Dhar, A., Ghosh, A., & Bhattacharya, P. (2022). Micro(nano)plastics pollution and human health: How plastics can induce carcinogenesis to humans? *Chemosphere*, *298*, 134267. https://doi.org/10.1016/j.chemosphere.2022.134267

Lacagnina, S. (2020). The developmental origins of health and disease (DOHaD). *American Journal of Lifestyle Medicine*, *14*(1), 47–50. https://doi.org/10.1177/1559827619879694

Le, V.-G., Nguyen, M.-K., Nguyen, H.-L., Lin, C., Hadi, M., Hung, N. T. Q., Hoang, H.-G., Nguyen, K. N., Tran, H.-T., Hou, D., Zhang, T., & Bolan, N. S. (2023). A comprehensive review of micro- and nanoplastics in the atmosphere: Occurrence, fate, toxicity, and strategies for risk reduction. *Science of the Total Environment*, *904*, 166649. https://doi.org/10.1016/j.scitotenv.2023.166649

Lee, H., Lee, M. W., Warren, J. R., & Ferrie, J. (2022). Childhood lead exposure is associated with lower cognitive functioning at older ages. *Science Advances*, *8*(45). https://doi.org/10.1126/sciadv.abn5164

Li, X., Chen, Y., Gao, W., Mo, A., Zhang, Y., Jiang, J., & He, D. (2022). Prominent toxicity of isocyanates and maleic anhydrides to caenorhabditis elegans: Multilevel assay for typical organic additives of biodegradable plastics. *SSRN Electronic Journal*. https://doi.org/10.2139/ssrn.4159400

Li, Y.-Z., Wu, Z.-Y., Zhu, B.-Q., Wang, Y.-X., Kan, Y.-Q., & Zeng, H.-C. (2022). The BDNF–TrkB–CREB signalling pathway is involved in bisphenol S-induced neurotoxicity in male mice by regulating methylation. *Toxics*, *10*(8), 413. https://doi.org/10.3390/toxics10080413

Liu, W., Zhao, Y., Shi, Z., Li, Z., & Liang, X. (2020). Ecotoxicoproteomic assessment of microplastics and plastic additives in aquatic organisms: A review. *Comparative Biochemistry and Physiology Part D: Genomics and Proteomics*, *36*, 100713. https://doi.org/10.1016/j.cbd.2020.100713

Maddela, N. R., Kakarla, D., Venkateswarlu, K., & Megharaj, M. (2023). Additives of plastics: Entry into the environment and potential risks to human and ecological health. *Journal of Environmental Management*, *348*, 119364. https://doi.org/10.1016/j.jenvman.2023.119364

Maity, S., Biswas, C., Banerjee, S., Guchhait, R., Adhikari, M., Chatterjee, A., & Pramanick, K. (2021). Interaction of plastic particles with heavy metals and the resulting toxicological impacts: A review. *Environmental Science and Pollution Research*, *28*(43), 60291–60307. https://doi.org/10.1007/s11356-021-16448-z

Marin, E., Boschetto, F., & Pezzotti, G. (2020). Biomaterials and biocompatibility: An historical overview. *Journal of Biomedical Materials Research Part A*, *108*(8), 1617–1633. https://doi.org/10.1002/jbm.a.36930

Meeker, J. D., Sathyanarayana, S., & Swan, S. H. (2009). Phthalates and other additives in plastics: human exposure and associated health outcomes. *Philosophical Transactions of the Royal Society B: Biological Sciences*, *364*(1526), 2097–2113. https://doi.org/10.1098/rstb.2008.0268

Mekonnen, T., Mussone, P., Khalil, H., & Bressler, D. (2013). Progress in bio-based plastics and plasticizing modifications. *Journal of Materials Chemistry A*, *1*(43), 13379. https://doi.org/10.1039/c3ta12555f

Menéndez-Pedriza, A., & Jaumot, J. (2020). Interaction of environmental pollutants with microplastics: A critical review of sorption factors, bioaccumulation and ecotoxicological effects. *Toxics*, *8*(2), 40. https://doi.org/10.3390/toxics8020040

Metruccio, F., Battistoni, M., Di Renzo, F., Bacchetta, R., Santo, N., & Menegola, E. (2024). Teratogenic and neuro-behavioural toxic effects of bisphenol A (BPA) and B (BPB) on *Xenopus laevis* development. *Reproductive Toxicology*, *123*, 108496. https://doi.org/10.1016/j.reprotox.2023.108496

Mohammadi, H., & Ashari, S. (2021). Mechanistic insight into toxicity of phthalates, the involved receptors, and the role of Nrf2, NF-κB, and PI3K/AKT signaling pathways. *Environmental Science and Pollution Research*, *28*(27), 35488–35527. https://doi.org/10.1007/s11356-021-14466-5

Moshood, T., Nawanir, G., Mahmud, F., Mohamad, F., Ahmad, M., & Abdul Ghani, A. (2021). Expanding policy for biodegradable plastic products and market dynamics of bio-based plastics: Challenges and opportunities. *Sustainability*, *13*(11), 6170. https://doi.org/10.3390/su13116170

Nahla, E., Arya, P., Maneesha, P., & Chitra, K. C. (2024). Exposure to the plasticizer dibutyl phthalate causes oxidative stress and neurotoxicity in brain tissue. *Environmental Science and Pollution Research*, *31*(14), 21399–21414. https://doi.org/10.1007/s11356-024-32604-7

Narancic, T., Cerrone, F., Beagan, N., & O'Connor, K. E. (2020). Recent advances in bioplastics: Application and biodegradation. *Polymers*, *12*(4), 920. https://doi.org/10.3390/polym12040920

Okeke, E. S., Olagbaju, O. A., Okoye, C. O., Addey, C. I., Chukwudozie, K. I., Okoro, J. O., Deme, G. G., Ewusi-Mensah, D., Igun, E., Ejeromedoghene, O., Odii, E. C., Oderinde, O., Iloh, V. C., & Abesa, S. (2022). Microplastic burden in Africa: A review of occurrence, impacts, and sustainability potential of bioplastics. *Chemical Engineering Journal Advances*, *12*, 100402. https://doi.org/10.1016/j.ceja.2022.100402

Painuli, R., Raghav, S., & Kumar, D. (2024). Standards issues toward bioplastics. In *Bioplastics for Sustainability* (pp. 143–159). Elsevier. https://doi.org/10.1016/B978-0-323-95199-9.00010-X

Pan, I., Umapathy, S., Issac, P. K., Rahman, M. M., Guru, A., & Arockiaraj, J. (2023). The bioaccessibility of adsorped heavy metals on biofilm-coated microplastics and their implication for the progression of neurodegenerative diseases. *Environmental Monitoring and Assessment*, *195*(11), 1264. https://doi.org/10.1007/s10661-023-11890-7

Patisaul, H. B., Behl, M., Birnbaum, L. S., Blum, A., Diamond, M. L., Rojello Fernández, S., Hogberg, H. T., Kwiatkowski, C. F., Page, J. D., Soehl, A., & Stapleton, H. M. (2021). Beyond cholinesterase inhibition: Developmental neurotoxicity of organophosphate ester flame retardants and plasticizers. *Environmental Health Perspectives*, *129*(10). https://doi.org/10.1289/EHP9285

Ponsonby, A.-L., Symeonides, C., Saffery, R., Mueller, J. F., O'Hely, M., Sly, P. D., Wardrop, N., Pezic, A., Mansell, T., Collier, F., Burgner, D., Thompson, K., Vijayasarathy, S., Sugeng, E. J., Dwyer, T., Ranganathan, S., Anderson, P. J., Anderson, V., & Vuillermin, P. (2020). Prenatal phthalate exposure, oxidative stress-related genetic vulnerability and early life neurodevelopment: A birth cohort study. *NeuroToxicology*, *80*, 20–28. https://doi.org/10.1016/j.neuro.2020.05.006

Prüst, M., Meijer, J., & Westerink, R. H. S. (2020). The plastic brain: Neurotoxicity of micro- and nanoplastics. *Particle and Fibre Toxicology*, *17*(1), 24. https://doi.org/10.1186/s12989-020-00358-y

Rahmati, M., Silva, E. A., Reseland, J. E., A. Heyward, C., & Haugen, H. J. (2020). Biological responses to physicochemical properties of biomaterial surface. *Chemical Society Reviews*, *49*(15), 5178–5224. https://doi.org/10.1039/D0CS00103A

Raj, B., & Matche, R. S. (2011). Safety and regulatory aspects of plastics as food packaging materials. In *Multifunctional and Nanoreinforced Polymers for Food Packaging* (pp. 669–691). Elsevier. https://doi.org/10.1533/9780857092786.4.669

Reyes, M. S. S., & Medina, P. M. B. (2024). Leachates from plastics and bioplastics reduce lifespan, decrease locomotion, and induce neurotoxicity in *Caenorhabditis elegans*. *Environmental Pollution*, *357*, 124428. https://doi.org/10.1016/j.envpol.2024.124428

Rock, K. D., & Patisaul, H. B. (2018). Environmental mechanisms of neurodevelopmental toxicity. *Current Environmental Health Reports*, *5*(1), 145–157. https://doi.org/10.1007/s40572-018-0185-0

Rosenboom, J.-G., Langer, R., & Traverso, G. (2022). Bioplastics for a circular economy. *Nature Reviews Materials*, *7*(2), 117–137. https://doi.org/10.1038/s41578-021-00407-8

Rubalajyothi, P., Rajendran, A., Gangadhar, L., & Pandiyan, V. (2022). Chronic neurological effects and photocatalytic investigation of AZO dyes. *Neuroscience Informatics*, *2*(3), 100049. https://doi.org/10.1016/j.neuri.2022.100049

Safi-Stibler, S., & Gabory, A. (2020). Epigenetics and the developmental origins of health and disease: Parental environment signalling to the epigenome, critical time windows and sculpting the adult phenotype. *Seminars in Cell & Developmental Biology*, *97*, 172–180. https://doi.org/10.1016/j.semcdb.2019.09.008

Santa Maria, M. P., Hill, B. D., & Kline, J. (2019). Lead (Pb) neurotoxicology and cognition. *Applied Neuropsychology: Child*, *8*(3), 272–293. https://doi.org/10.1080/21622965.2018.1428803

Savva, K., Borrell, X., Moreno, T., Pérez-Pomeda, I., Barata, C., Llorca, M., & Farre, M. (2022). Cytotoxicity assessment and suspected screening of plastic additives in bioplastics of single-use household items. *SSRN Electronic Journal*. https://doi.org/10.2139/ssrn.4188439

Shen, M., Song, B., Zeng, G., Zhang, Y., Huang, W., Wen, X., & Tang, W. (2020). Are biodegradable plastics a promising solution to solve the global plastic pollution? *Environmental Pollution*, *263*, 114469. https://doi.org/10.1016/j.envpol.2020.114469

Silva, V. L., de Alvarenga Freire, M. T., De Almeida Oroski, F., Trentini, F., Costa, L. O., & de Batista, V. G. T. (2024). Bioplastics and the role of institutions in the design of sustainable post-consumer solutions. *Sustainability*, *16*(12), 5029. https://doi.org/10.3390/su16125029

Sofield, C. E., Anderton, R. S., & Gorecki, A. M. (2024). Mind over microplastics: Exploring microplastic-induced gut disruption and gut-brain-axis consequences. *Current Issues in Molecular Biology*, *46*(5), 4186–4202. https://doi.org/10.3390/cimb46050256

Solleiro-Villavicencio, H., Gomez-De León, C. T., Del Río-Araiza, V. H., & Morales-Montor, J. (2020). The detrimental effect of microplastics on critical periods of development in the neuroendocrine system. *Birth Defects Research*, *112*(17), 1326–1340. https://doi.org/10.1002/bdr2.1776

Sun, C., Yin, N., Wen, R., Liu, W., Jia, Y., Hu, L., Zhou, Q., & Jiang, G. (2016a). Silver nanoparticles induced neurotoxicity through oxidative stress in rat cerebral astrocytes is distinct from the effects of silver ions. *NeuroToxicology*, *52*, 210–221. https://doi.org/10.1016/j.neuro.2015.09.007

Sun, L., Tan, H., Peng, T., Wang, S., Xu, W., Qian, H., Jin, Y., & Fu, Z. (2016b). Developmental neurotoxicity of organophosphate flame retardants in early life stages of Japanese medaka (*Oryzias latipes*). *Environmental Toxicology and Chemistry*, *35*(12), 2931–2940. https://doi.org/10.1002/etc.3477

Sun, L., Xu, W., Peng, T., Chen, H., Ren, L., Tan, H., Xiao, D., Qian, H., & Fu, Z. (2016c). Developmental exposure of zebrafish larvae to organophosphate flame retardants causes neurotoxicity. *Neurotoxicology and Teratology*, *55*, 16–22. https://doi.org/10.1016/j.ntt.2016.03.003

Szychowski, K. A., Wnuk, A., Rzemieniec, J., Kajta, M., Leszczyńska, T., & Wójtowicz, A. K. (2019). Triclosan-evoked neurotoxicity involves NMDAR subunits with the specific role of GluN2A in caspase-3-dependent apoptosis. *Molecular Neurobiology*, *56*(1), 1–12. https://doi.org/10.1007/s12035-018-1083-z

Tilson, H. A., Kodavanti, P. R. S., Mundy, W. R., & Bushnell, P. J. (1998). Neurotoxicity of environmental chemicals and their mechanism of action. *Toxicology Letters*, *102–103*, 631–635. https://doi.org/10.1016/S0378-4274(98)00271-9

Trebuňová, M., Petroušková, P., Balogová, A. F., Ižaríková, G., Horňak, P., Bačenková, D., Demeterová, J., & Živčák, J. (2023). Evaluation of biocompatibility of PLA/PHB/TPS polymer scaffolds with different additives of ATBC and OLA plasticizers. *Journal of Functional Biomaterials*, *14*(8), 412. https://doi.org/10.3390/jfb14080412

Tseng, I.-L., Yang, Y.-F., Yu, C.-W., Li, W.-H., & Liao, V. H.-C. (2013). Phthalates induce neurotoxicity affecting locomotor and thermotactic behaviors and AFD neurons through oxidative stress in *Caenorhabditis elegans*. *PLoS One*, *8*(12), e82657. https://doi.org/10.1371/journal.pone.0082657

Turna Demir, F., Akkoyunlu, G., & Demir, E. (2022). Interactions of ingested polystyrene microplastics with heavy metals (cadmium or silver) as environmental pollutants: A comprehensive in vivo study using *Drosophila melanogaster*. *Biology*, *11*(10), 1470. https://doi.org/10.3390/biology11101470

Uribe-Echeverría, T., & Beiras, R. (2022). Acute toxicity of bioplastic leachates to *Paracentrotus lividus* sea urchin larvae. *Marine Environmental Research*, *176*, 105605. https://doi.org/10.1016/j.marenvres.2022.105605

Vieira, I. R. S., de de Carvalho, A. P. A., & Conte-Junior, C. A. (2022). Recent advances in biobased and biodegradable polymer nanocomposites, nanoparticles, and natural antioxidants for antibacterial and antioxidant food packaging applications. *Comprehensive Reviews in Food Science and Food Safety*, *21*(4), 3673–3716. https://doi.org/10.1111/1541-4337.12990

Wallace, D. R., & Djordjevic, A. B. (2020). Nervous system. In *An Introduction to Interdisciplinary Toxicology* (pp. 261–282). Elsevier. https://doi.org/10.1016/B978-0-12-813602-7.00020-X

Walpitagama, M., Carve, M., Douek, A. M., Trestrail, C., Bai, Y., Kaslin, J., & Wlodkowic, D. (2019). Additives migrating from 3D-printed plastic induce developmental toxicity and neuro-behavioural alterations in early life zebrafish (*Danio rerio*). *Aquatic Toxicology*, *213*, 105227. https://doi.org/10.1016/j.aquatox.2019.105227

Wang, J., Li, X., Wang, C., Li, Y., Wang, J., Fang, R., Wang, J., Chen, J., & Dong, J. (2021). Exposure to di-(2-ethylhexyl) phthalate reduces secretion of GDNF via interfering with estrogen pathway and downregulating ERK/c-fos signaling pathway in astrocytes. *Food and Chemical Toxicology*, *158*, 112592. https://doi.org/10.1016/j.fct.2021.112592

Wang, W., Li, Z., Yuan, S., Du, Z., Li, J., Peng, H., & Ru, S. (2024). A potential neurotoxic mechanism: Bisphenol S-induced inhibition of glucose transporter 1 leads to ATP excitotoxicity in the Zebrafish brain. *Environmental Science & Technology*, *58*(35), 15463–15474. https://doi.org/10.1021/acs.est.4c03870

Wang, W., Li, Z., Zhang, X., Zhang, J., & Ru, S. (2023a). Bisphenol S impairs behaviors through disturbing endoplasmic reticulum function and reducing lipid levels in the brain of Zebrafish. *Environmental Science & Technology*, *57*(1), 582–594. https://doi.org/10.1021/acs.est.2c07828

Wang, Y., Gai, T., Zhang, L., Chen, L., Wang, S., Ye, T., & Zhang, W. (2023b). Neurotoxicity of bisphenol A exposure on *Caenorhabditis elegans* induced by disturbance of neurotransmitter and oxidative damage. *Ecotoxicology and Environmental Safety*, *252*, 114617. https://doi.org/10.1016/j.ecoenv.2023.114617

Wang, Y., Xiong, L., & Tang, M. (2017). Toxicity of inhaled particulate matter on the central nervous system: Neuroinflammation, neuropsychological effects and neurodegenerative disease. *Journal of Applied Toxicology*, *37*(6), 644–667. https://doi.org/10.1002/jat.3451

Welch, C., & Mulligan, K. (2022). Does bisphenol A confer risk of neurodevelopmental disorders? What we have learned from developmental neurotoxicity studies in animal models. *International Journal of Molecular Sciences*, *23*(5), 2894. https://doi.org/10.3390/ijms23052894

Wopara, I., Modo, E. U., Adebayo, O. G., Mobisson, S. K., Nwigwe, J. O., Ogbu, P. I., Nwankwo, V. U., & Ejeawa, C. U. (2021). Anxiogenic and memory impairment effect of food color exposure: Upregulation of oxido-neuroinflammatory markers and acetyl-cholinestrase activity in the prefrontal cortex and hippocampus. *Heliyon*, *7*(3), e06378. https://doi.org/10.1016/j.heliyon.2021.e06378

Xu, Y., Nie, J., Lu, C., Hu, C., Chen, Y., Ma, Y., Huang, Y., & Lu, L. (2024). Effects and mechanisms of bisphenols exposure on neurodegenerative diseases risk: A systemic review. *Science of the Total Environment*, *919*, 170670. https://doi.org/10.1016/j.scitotenv.2024.170670

Yin, Z., Hua, L., Chen, L., Hu, D., Li, J., An, Z., Tian, T., Ning, H., & Ge, Y. (2020). Bisphenol-A exposure induced neurotoxicity and associated with synapse and cytoskeleton in Neuro-2a cells. *Toxicology in Vitro*, *67*, 104911. https://doi.org/10.1016/j.tiv.2020.104911

Zhao, Y., Chang, Y.-H., Ren, H.-R., Lou, M., Jiang, F.-W., Wang, J.-X., Chen, M.-S., Liu, S., Shi, Y.-S., Zhu, H.-M., & Li, J.-L. (2024). Phthalates induce neurotoxicity by disrupting the Mfn2-PERK axis-mediated endoplasmic reticulum–mitochondria interaction. *Journal of Agricultural and Food Chemistry*, *72*(13), 7411–7422. https://doi.org/10.1021/acs.jafc.3c07752

Zhou, Y., Wang, Z., Xia, M., Zhuang, S., Gong, X., Pan, J., Li, C., Fan, R., Pang, Q., & Lu, S. (2017). Neurotoxicity of low bisphenol A (BPA) exposure for young male mice: Implications for children exposed to environmental levels of BPA. *Environmental Pollution*, *229*, 40–48. https://doi.org/10.1016/j.envpol.2017.05.043

Zimmerman, J. B., Anastas, P. T., Erythropel, H. C., & Leitner, W. (2020). Designing for a green chemistry future. *Science*, *367*(6476), 397–400. https://doi.org/10.1126/science.aay3060

Zimmermann, L., Dierkes, G., Ternes, T. A., Völker, C., & Wagner, M. (2019). Benchmarking the in vitro toxicity and chemical composition of plastic consumer products. *Environmental Science & Technology*, *53*(19), 11467–11477. https://doi.org/10.1021/acs.est.9b02293

Zimmermann, L., Dombrowski, A., Völker, C., & Wagner, M. (2020). Are bioplastics and plant-based materials safer than conventional plastics? In vitro toxicity and chemical composition. *Environment International*, *145*, 106066. https://doi.org/10.1016/j.envint.2020.106066

14 Environmental Impact Assessment of Bioplastic

*Abdulhakeem Idris Abdulhakeem,
Innocent Ojeba Musa,
Peter Olabisi Abioye, Helen Shnada Auta,
Mustapha Abdulsalam, Job Oloruntoba Samuel,
Miracle Uwa Livinus, Muhammed Abdulazeez,
Asmau M. Maude, Fatima Adamu Yahaya,
Shehu-Alimi Elelu, Mustapha Adamu,
and Auwal Sagir Muhammad*

14.1 INTRODUCTION

Bioplastics are derived from renewable biological sources and have emerged as a promising substitute to the conventional plastics as a result of their reduced environmental footprint. The growing concern over plastic pollution globally and climate change has driven researchers to explore sustainable solutions, leading to increased attention on bioplastics. Most organic plastics are made from petrochemicals, which are majorly responsible for greenhouse gas (GHG) emissions. However, bioplastics can be manufactured using plant materials, thus lowering the carbon emissions as well as minimizing the damage to the environment (Nandakumar *et al.*, 2021). Moreover, bioplastics can help solve the environmental concern caused by plastic in the first place by reducing non-renewable resources and providing a better alternative waste management (WM) system (Narancic *et al.*, 2020).

Biodegradability is one of the most important features of bioplastics and the trait that sets it apart from the traditional plastic. In optimal conditions, certain bioplastics, such as the polylactic acid (PLA) as well as the polyhydroxyalkanoates (PHAs), would only take a few years to degrade, which will greatly diminish plastic waste within ecosystems (Narancic *et al.*, 2020). Additionally, bioplastics can assist in the sequestration of carbon dioxide (CO_2) emissions during the bioplastic's life cycle. Because plants grow and absorb CO_2 from the air, the net carbon emissions are reduced, and this contributes to the fight against climate change (Yadav & Nikalje, 2024). However, the time required for the biodegradation of bioplastics is different, and the behaviour of bioplastics in their natural conditions is still being studied (Song *et al.*, 2025). Even with these ambiguities, bioplastics can effectively be utilized to lessen the burden of plastic pollution in landfills and oceans. If bioplastics have clear advantages, such as biodegradability, there remain obstacles in the recycling processes. Exploring the reuse of bioplastics has borne attempts to utilize mechanical and chemical recycling techniques. Even so, material contamination from conventional plastics and the poor quality of recycled plastic that is produced are major problems (Huang *et al.*, 2025). In addition, chemical recycling processes like pyrolysis and gasification require large amounts of energy, which may counter the environmental benefits of bioplastics (Emadian *et al.*, 2017). Thus, because bioplastics have the potential to mitigate adverse environmental impacts, there is a dire need to develop bioplastic recycling technologies so as to ensure a sustainable and circular economy (Lamberti *et al.*, 2020).

DOI: 10.1201/9781003602828-17

14.2 BENEFITS OF BIOPLASTICS OVER CONVENTIONAL PLASTICS

Bioplastics are viewed as eco-friendly substitutes for standard plastic (Nandakumar *et al.*, 2021). Bioplastics also have certain positive influences on the environment compared to common plastics. Prima facie, plastic manufacturing processes driven by fossil fuel combustions have been tagged awful with the emissions of GHGs that, in one way or another, aggravate food security issues and increase the frequency of wildfires and harsh weather conditions, among other vices. As a result, less consumption of fossil fuel resource is a notable benefit that would accrue with the adoption of bioplastic. In addition, bioplastics seem to offer advantages for recovery of the end-product and the energy contained in it, which shows that they will be a sustainable renewable resource (Landis *et al.*, 2018). Even though separation of bioplastics and conventional plastics is often necessary, the literature reports that bioplastics are only in trace amounts in most traditional plastic recycling streams and so doesn't impair the quality of the materials being recycled (Landis *et al.*, 2018). Bioplastics can go through take-back processes for feedstock recovery. For instance, feedstock recovery of biopolymers can be used for producing syngas when fused with municipal waste. Such syngas, from different sources, can be used for different purposes, such as the generation of electricity and the synthesis of chemicals like ethanol and/or methanol, or even synthetic petroleum. It is worth mentioning that the normal practice in most nations is to burn plastics in order to produce heat energy; however, the shift towards bioplastics allows for better utilization of biogenic carbon. These approaches further help to minimize landfill space requirements over a period of time and, thus, make the land available for other uses (Nandakumar *et al.*, 2021). Bioplastics can also lessen the net emission of CO_2. Plants grown for the bioplastics production can sequester CO_2 which would be released during the biodegradation of bioplastics (Zhao *et al.*, 2023). In one study, it was shown that synthesized PLA in a landfill experiences 70% less GHG during degradation (Benavides *et al.*, 2020). One other major advantage for many bioplastics is the biodegradability property of the material with numerous bioplastic fabrication under optimal settings. When compared to conventional plastics, biodegradable plastics, including PLA or PHA, on the other hand, have a much shorter time frame of degradation, ranging into a few years, making them ideal for curbing plastic waste in landfills or sea over time (Narancic *et al.*, 2020). Partial degradation of bioplastics in terrestrial systems might generate microplastics (MPs) that could be advantageous to the soil understory and carbon sinks by contributing to organic carbon accumulation. Some of the different types of MPs that were analysed for the study, Polybutylene Adipate Terephthalate Microplastics were observed to have a very stimulating effect on soil respiration. This strengthens the argument that edible plastics can promote microbial processes and carbon turnover in soil, allowing higher respiration rates within the soil (Song *et al.*, 2025). Such benefits of biodegradable materials on soil carbon cycles reinforce the prospects of employing eco-friendly materials in agricultural and soil engineering practices. Efforts to create a bioplastic material and to replace conventional plastic materials that can mitigate the impact of plastic pollution support the sustainability objective.

14.3 RECYCLING OPTIONS FOR BIOPLASTIC

Most synthetic biodegradable materials readily available in the market, biodegradable bioplastics disintegrate in nature slowly, even in the presence of microbiota, a fact that hurts biodegradation, which is usually regarded as a good way of disposing of waste bacteria (Emadian *et al.*, 2017). Biodegradation assessment involves a lot of parameters, such as change in mechanical properties of the bioplastics, change in chemical structural composition of the plastic polymer, reduction in mass of the substrate, and the measurement of CO_2 emissions (Monira *et al.*, 2023). Such low degradation could meaningfully contribute to a misunderstanding of the prospects of bioplastics to avert pollution (Lamberti *et al.*, 2020). Still, after this repeated use, followed by some limited recycling, the commercial bioplastics would eventually get biodegraded in some particular composting facilities. This is however considered to be the final phase of their life span. When bioplastics are thrown away, bio-based raw materials lactic acid for Polylactic Acid or PLA, for instance, are lost. Bioplastics could be re-used by mechanical recycling, but only if the material is good enough (Emadian *et al.*, 2017). Otherwise, thermal scholars explain, bioplastics could be destroyed during recycling. Chemical recycling, including solvolysis, gasification, and pyrolysis are able to convert waste bioplastics into monomers as well as intermediate compounds, provided the quality of the material is below a certain level (Lamberti *et al.*, 2020). This approach may help reduce the environmental burden posed by bioplastics and conserve primary renewable resources. However, this increasingly appears to be yet another disadvantage of chemical recycling.

14.3.1 Mechanical Recycling

There is another recycling definition that considers mechanical recycling as a physical recycling of waste whereby it's held that mechanical recycling is the main method of plastics' recovery. This is a relatively cheaper method than chemical recycling, and its environmental impact is not very terrible (Huang *et al.*, 2025). These are well known as the primary and secondary recycling processes. Biodegradable plastic waste with good characteristics that is reused for secondary use recycling where material obtained can provide total functionality similar to the virgin materials. On the other hand, the secondary recycling is utilized when the demand for application is low because the recycled plastic has lower properties (Emadian *et al.*, 2017). The impact strength and tensile strength were reduced by 20% and 5% to the initial level, respectively, as mentioned by Finnerty *et al.* (2023). Due to this, there are constraints that hinder the actualization of the traditional mechanical recycling process for the petrochemical plastics included. The presence of unfavourable polymers in blends contributes to a low mechanical characteristic of the recycled product. Nonetheless, the phenomenon of the different melting points and viscosities indicated in the case of the different polymers can complicate the whole process. The use of near-infrared spectroscopy (NIR) can solve this issue, and its result, with a percentage of accuracy above 97%, is effective in sorting polymers (Zhu *et al.*, 2019). There is an indication that density separation, air separation, and dye tracers are even other techniques for separation. bioplastics (Lamberti *et al.*, 2020). However, there are challenges because there are multiple layers of different types of material and different types of additives. Consequently, many different recycling techniques of bioplastics have been compared through life cycle assessments (LCAs). Mechanical recycling is associated with Global Warming Potential (GWP) and emissions. However, according to Van *et al.* (2022), it was found to have a negative GWP, with values of 0.04 kg CO_2/kg PLA and -0.6 kg CO_2/kg TPS, compared to the results from landfilling and industrial composting. The comparative form of PLA disposal that Fonseca *et al.* (2023) established includes mechanical recycling of the three.

Some of the traditional techniques are pyrolysis, chemical recycling, and composting. Hydrolysis and composting tests were also performed at the lab scale, and information about the mechanical recycling developed in these assays was integrated with data collected in an industrial recycling plant for plastics. Such analysis revealed that mechanical recycling led to 0.3 kg CO_2eq/kg PLA of climatic impact, 0.08 kg 1,4-DBeq/kg PLA of human toxicity, and 0.1 kg oil/kg PLA of fossil depletion, outranking other end-of-life (EOL) options. Concerning biodegradable containers made of PLA, PHA, and PBS, Islam *et al.* (2024) addressed EOL issues based on the E/E indicator, such as mechanical recycling, landfilling, crushing and composting. According to the study, biodegradable take-away boxes achieve the highest E/E expansion rate of 100% in mechanical recycling. Mechanical recycling is more preferred than chemical recycling, considering the properties such as energy use, GWP, cost, eutrophication potential (EP) and acidification potential (AP) (Finnerty *et al.*, 2023). For instance, mechanical recycling requires 1.6 MJ of fossil energy to recycle the 0.96 kg of PLA, while chemical recycling uses half the energy required by mechanical recycling but is five times more plentiful to produce same quantity of Polylactic Acid. None of the indicated studies have considered that with mechanical processing, product quality decreases, and no set methods may be employed to assess the quality of the final outcome (Fonseca *et al.*, 2023). Thus, mechanical recycling can serve as a strained long-term strategy, as far as the LCA research approves it, while other post-consumption EOL scenarios can be worthwhile after several mechanical recycling. Consequently, only mechanical recycling can be considered preferable in the short run in terms of LCA studies, while other options linked with EOL can be effective in the long run or after a few cycles of mechanical recycling.

14.3.2 Chemical Recycling

Chemical recycling employs both thermal and chemical methods in order to bring about the process of depolymerization. Pyrolysis, gasification, and solution-driven catalytic depolymerization including both alcoholysis and hydrolysis, fall under this broad category of state of art methodology. In these processes, heat is used to transform mixed plastics into intermediate compounds; pyrolysis proceeds anaerobically, whereas gasification uses steam or atmospheric oxygen (Li *et al.*, 2024). These methods also provide monomers that return back to plastics with different useful co-products, such as hydrocarbons, coke, and hydrogen (Merchan *et al.*, 2022). When we try to compare this chemical recycling with the mechanical recycling system, however, we very quickly come to the realization that chemical recycling tends to have greater environmental consequences, particularly regarding human toxicity, climate change, and fossil fuel depletion, as a result of its intensive energy requirements for the power for the breakers and other requirements for chemical synthesis (Fonseca *et al.*, 2023). However, gasification and pyrolysis processes require relatively cleaner feedstocks; this could be a bottleneck as feedstock requires further purification processes. The impurities produced by feedstocks include undesired by-products and emissions that could affect the quality of air and be hazardous to health. Also, with impurities in the feedstock, feed tends to foul and corrode the equipment downstream, hence the reduction in efficiency and increased maintenance cost. To date, limited research has been conducted in comparing the chemical recycling of bioplastics with other recycling practices, and so a cost and benefit analysis, performance analysis, and possibilities of application have not been made. Also, none of the above studies have been performed at industrially relevant conditions. Another factor may be the general cost of the catalysts, the accessibility of the secondary solvents, and primary conversion into a limited number of high-demand products - these all may influence overall feasibility for the large-scale application.

14.3.3 Incineration

The second most popular method of utilizing plastic waste is by burning the waste in incinerators after it has been landfilled. This method minimizes the generation of mixed solid waste or MSW and at the same time is capable of recovering the calorific value of wastes, that may be harnessed to produce electricity or heat (Zhang *et al.*, 2024). For example, in 2018, about 40% of the plastic waste generated in Europe was either dumped into the MSWI facilities, while 60% ended up in landfill sites. Bioplastics have characteristics like high energy content as well as burnability, which makes it possible to dispose of them in general plastic waste burning plants.

According to Negrete-Bolagay *et al.*, 2024, as with wood, starch-based bioplastics offer an equivalent energy storage regardless of their marginally lower gross calorific value than coal (Islam *et al.*, 2024). However, as these materials increasingly gain acceptance as structures in the soil, environment, and construction industry, it has been observed that instead of decomposing naturally, they burnt it in this process; it is considered that it is wasting nutrients which could have been used in our farms as manure. When burning, toxins are released, and although burning is prevalent in getting rid of plastics, it is usually associated with various health risks (Sheriff *et al.*, 2025). Municipal Solid Waste Incineration (MSWI) reduces the waste that needs to be managed at the same time, the ash it generates may contain health hazardous substances and MP, which requires special attention.

14.3.4 Compost

In the course of composting, organic matter is decomposed by a body of microorganisms to give out humus and CO_2. Compostable plastics have many definitions, but the most common ones emphasize the criteria related to the decrease of the organic mass over the set period of time, such as the one introduced in EN 13432 (European Bioplastics, 2024; Kim *et al.*, 2023). Such plastics are mostly engineered to decompose more rapidly into CO_2, water, and biomass. It has been stated that industry-level composting is usually done in a warm, humid region, and the temperature used should be between 60-70 °C and humidity should be 60% (Gu, 2021). Water content should be sufficient in supporting aerobic biodegradation, as this helps in the growth of microorganisms (Springle *et al.*, 2022). Besides, the physiochemical characteristics and composition of compost are also very crucial because they provide the nutritional needs of the microorganisms. For composting, a carbon-to-nitrogen (C:N) ratio. The proportion of moths to Lis dock (N) is preferred to be 14-43 among 47 when consuming castor meal (Springle *et al.*, 2022). A high Convolutional Neural Network (CNN) ratio is not conducive to biodegradation because nitrogen is present as the limiting reagent, while low CNN may cause a high conversion rate of nitrogen to ammonia (NH_3) (Springle *et al.*, 2022). Composting in real life may not work out as the experimental studies have depicted it to work out. The fermented municipal solid waste fraction/organic fraction of municipal solid waste (OFMSW) was blended with PLA water bottles in the industrial composting study performed by Islam *et al.* (2024) in real-world-like conditions. The study unveiled that because of PLAs bio-decomposition process, lactic acid accumulated and caused a decrease in the pH of the compost, thus hindering birth seed germination. Those apprehensions notwithstanding, composting continues to be the typical EOL procedure in LCA and biodegradation investigations of bioplastics. Nevertheless, it plays a minor role in the global WM industry. Almost 56% of the global MSW is organic; only 5% is recycled through composting, and the remaining is dumped into landfills (Rosenboom *et al.*, 2022). Chronic disposal of waste in the countryside contributes to the emission of GHGs, in addition to wasting a rare resource that could be useful in the preparation of fertilizers to reduce the use of fertilizer, which is made synthetically. Therefore, this approach, normal to other organic waste disposal, may not be the best to optimize resources for the bioplastics. Polyhydroxybutyrate (PHB), polylactic acid (PLA), and thermoplastic starch (TPS) are among the bioplastics found to be compostable under controlled laboratory settings; however, additional studies are required to evaluate their influence on the overall quality of the resulting compost.

14.3.5 Anaerobic Digestion (AD)

Anaerobic digestion (AD) offers a chance of energy and nutrient recovery in bioplastics when other mechanical recycling is impossible. The four processes of AD include hydrolysis, acidogenesis, genesis, and methanogenesis, as noted by Bátori *et al.* (2018). Each of them needs a special bacterial strain for the process to happen. It has been reported that the above processes should produce biogas with a methane content of about 50-60% with a CO_2 content of about 25-50%. In Europe, an example was given that 1 tonne of biowaste can generate the power production of 180 kWh, depending on the biogas produced. Methane from this biogas can in this case be flared for this other use as an energy source, as pointed out by Kim *et al.*, 2023. As stated earlier, considerations about waste management include at least the following part: It is expected that 50% of the final products can be transformed into biogas within 2 months (Batori *et al.*, 2018). The funding process of the anaerobic digestion that involves biodegradable plastics and food waste is one possible approach to waste management. At least 50% of the total products should be converted into biogas in the year 2000 in two months (Batori *et al.*, 2018). was described before, one potential WM solution is through the funding process of the AD of biodegradable plastics with food waste. The beauty of this method is that disposal of this for the consumers is easy since they are relieved of the duty of cleaning and sorting. Urban sludge waste mostly exhibits a specific carbon-to-nitrogen (C:N) balance (Kang *et al.*, 2022). Enhancing the carbon content can improve biomethane production: Besides, the incorporation of carbon-rich bioplastics aligned with standard food waste streams can also greatly reduce the N ratio. Previous studies have discovered that biomethane that originates from the municipal wastewater recovery systems can be boosted using biodegradable polymer materials within digestate. However, an excessive carbon-to-nitrogen input ratio might result in an accumulation of volatile fatty acids and consequently inhibit biogas production (Oduor *et al.*, 2022). Furthermore, the methanotrophic bacteria have potential to transform the methane formed from the AD into PHB; the advanced bio-material PHB can also be subjected to AD again. In consequence, the processes of applying AD for bioplastics' life cycle will create cyclic life cycles (Adnan *et al.*, 2022). From the waste, therefore, the cost of the expensive carbon sources which are used to feed the bacteria in the production process of PHB is lowered since they contribute to half of the total cost of synthesizing the polymer (Martínez *et al.*, 2024). Nonetheless, due to the composite structure of PHB and the process of its extraction from cells, the cost of PHB generated from biogas is somewhat high (McAdam *et al.*, 2020). Of great significance is that food waste and bioplastics, if fed into the AD systems, do not add to the landfill's burden, yet petroleum plastics are still capable of bringing about dangerous contamination. In current WM practices, waste fed to AD systems is processed and pre-treated in order to eliminate contaminants from conventional plastics and to produce soluble, high-value digested product known as digestate. At the same time, it is impossible to define the criterion for the identification of petro-plastics and bioplastics practically and at the same time have it be cheap. Possible solutions regarding this contamination problem may include the use of flexible food packaging material from the geared biodegradable bioplastics or finding a way to differentiate between the two plastics. Also for safety reasons, the industry standards recommend that biogas should take at least two months to biodegrade, while there are several anaerobic digesters that are working at very low HRTs. Hence, it does not align with other industrial practices. Eliminating such standards may decrease the level of AD contamination, but PLA and other biopolymers can be rejected for AD more easily. From the study conducted by Gu (2021), further research should be done on identifying the ways to increase the breakdown of bioplastics in AD systems since it is in high demand in industries.

14.3.6 Landfill

Landfilling remains the method of choice for disposal of Municipal Solid Waste (MSW), where it is estimated that 61% of it is dealt with through landfill gas emissions (Rosenboom *et al.*, 2022). In 2023, the EPA estimates it emitted 15% of the total methane emissions, which makes the industry the United States' third-leading contributor to anthropogenic methane emissions. Equally, landfills have adverse effects on human health, ecosystems and the soil and water resources. For instance, runoff liquids from landfill locations contain higher Chemical Oxygen Demand (COD), Biochemical Oxygen Demand (BOD), and heavy metals which affect the quality of the adjacent soil or water percolation (Xiang *et al.*, 2025). Moreover, this leachate contains bacteria like coliform, which may raise its toxic impact on plants in the soil (Kooch *et al.*, 2024). Asthma and Bronchitis have been among the health issues that individuals experience living near landfills, according to several research works (Kooch *et al.*, 2024). In addition, when throwing these bioplastics into the dump, the chance to melt them and reuse the materials to set up a new cycle is impossible. Defining the rates and amounts of methane produced in a landfill is extremely complicated by the continuous deposition of waste mass, the interphase interference, and the interaction of all biodegradation processes with the thermal, mechanical, and hydraulic actions. Quantitative and chemical properties of wastes and the landfill architectural design constitute the internal influences, while external factors include temperature, humidity, and geographical location (Yeşiller *et al.*, 2022). Nevertheless, recent studies

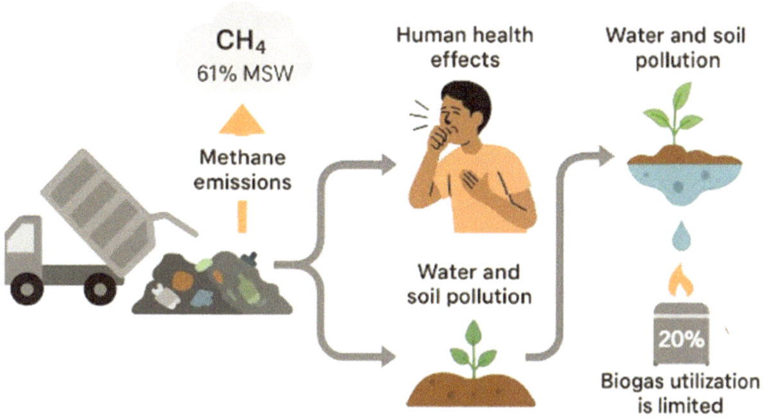

FIGURE 14.1 Environmental impact of landfill disposal of municipal solid waste.

have also provided models which utilize data of weather conditions and waste composition to forecast methane gas produced in a landfill in real-time (Yeşiller *et al.*, 2022). Some of these models have thus managed to explain some of the stochastic behaviours within a landfill system (Malmir *et al.*, 2023). Nevertheless, these last have been used only in a limited number of case histories and are not widely used in production industries (Figure 14.1).

Dumping of plastics continues to be widely used (EPA, 2024). However, when bioplastics are buried together with other hydrocarbon plastics, collection of biogas requires advanced technologies to reduce the emission of GHGs. Although biogas can substitute natural gas and has further use apart from the manufacturing of bioplastics, according to recent surveys, only about 20% of the landfills in the United States have facilities for methane extraction or flaring because of the expenses (Scharff *et al.*, 2023). However, the international energy agency (IEA) expects capture and use of biogas to increase by 50% by the year 2040 due to legislation on reduction of greenhouse gas (GHG) emissions (IEA, 2020); also, future advancement in technology holds high promise for improved efficiency in the capture of biogas, as new systems developed have an estimated capacity of 94% of methane released.

14.4 ENVIRONMENTAL EFFECTS

The evaluation of the environmental effects of bioplastics is still debatable, even though bioplastics are normally assumed to be superior to traditional bioplastics. The common kinds of bioplastics still had an impact on the environment and ecological systems all through the method, beginning from the production process of bioplastics, in the course of their use and even after disposal by somehow creating pollutants and adverse health hazards. Bioplastics are derived from plants, which means that, like most other products originating from plants, adverse land consumption is inevitable, thus straining the biodiversity of the planet. In addition, eutrophication and acidification may be linked to the extensive application of fertilizers and pesticides during the growth of crops for bioplastics production. Furthermore, bioplastics contribute to climate impacts throughout its life cycle and can break apart into MPs/nano-plastics (NPs) due to wrong biodegradation, which can also result from the wrong disposal of bioplastic products. Nevertheless, in the majority of the studies, authors pay attention to the cradle-to-gate and do not take into account the contribution of bioplastics during their usage and final disposal. In totality, attention should be drawn to the major impacts of bioplastics on the environment systematically.

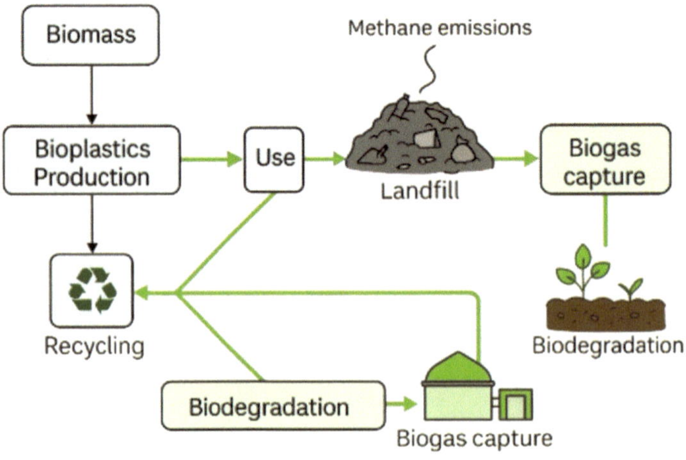

FIGURE 14.2 Bioplastics production, use, and end-of-life options.

14.4.1 Adverse Land Use Changes

Even though biomass can be sourced from a renewable resource, there is danger that the biomass may be overexploited like what could be observed as the impact of excessive use of fertilizer, water, and land erosion, and reduced number of available lands and different impacts on the level of biodiversity, as pointed out by Landis *et al.* (2018). Managing biomass supply among different consumer groups - energy, food and feed, biofuel, and materials/carbon - can be a complicated process because of the increased competition for the product (Anli Dino *et al.*, 2025). Due to direct land use change (LUC) in agricultural areas and rainforests, plant usage in the preparation of bioplastic has sparked some environmental issues (Landis *et al.*, 2018). For instance, where there is an extensive and wrong way of practising agriculture to enhance the production of bioplastic, there is a high possibility of enhancing soil degradation through either direct erosion or compaction (Song *et al.*, 2025). Biomass resources could be obtained from unavoidable forestry or agricultural residuals, and thus the plants and the agricultural lands needed for the production of food supplies do not come under threat due to movement towards and increased use of bioplastics (Taheripour & Tyner, 2013). These are the LUC and the land occupation (LO) mentally employed during the exploitation of the land for the production of bioplastics. Some of the pertinent pieces of work have also been conducted to determine and compare the LO and LUC of different categories of bioplastics. For example, Rezvani *et al.* (2021) calculated the amount of the LO for producing 1 kg PLA as 0.35 m^2 of organic cultivable land; in another example, Rezvani *et al.* (2021) estimated one of the LOS as 1.69 m^2 or m^2 for the same quantity of Polylactic Acid (PLA), respectively. Nevertheless, the estimated value arrived at by following the computational framework was still fairly broad. However, in the said studies, what was provided was information on the area utilized for production activities, and still missing were data as to the total land needed for the growth of crops used in the production inputs. What Tecorralco *et al.* (2024) too have reported, as stated earlier, is that for every kilogram of bio-HDPE produced in Brazil which is then exported to Europe, it takes 4.30 m^2 of land. Hence, by assuming the calculation of the LO measure only in m^2a, they only reveal how many acres of land are out of use temporally; they do not consider fluctuations in the quality of the soil in the occupied period whatsoever. Seasonal use of compartments for, among other uses, may have primary impact on the quality of the soil and other by-product impacts. It should be understood that direct effects which include compaction, erosion, chemical pollution of the soil reduces fertility, loss of species, and other ecosystem services associated with the soil. Otherwise, soil quality can be affected by LO by changes in land use, invasions of natural ecosystems, changes in the water regime, and the introduction of valuable species. Second, such indirect effects might further increase the intensification of each one of the mentioned processes of soil degradation and have long-term consequences for the soil capacity and its performance. Therefore, a higher Lobbes of assessment over and above that represented by LO might be required for turfing the effect of LO on the state of the soil. Tecorralco *et al.* (2024) included Land Use Change (LUC) emissions in their assessment of partially bio-based PET production, which is synthesised from two key monomers: purified terephthalic acid (PTA) and mono-ethylene glycol. Among these, PTA plays a more critical role in the formation of PET. The only product that utilized sugarcane ethanol was the monomer Monoethylene Glycol (MEG), a raw material input that accounted for 28% of the total mass. On the other side, an LCA can be performed without the potential effects of LUC, and the results could be misleading. For instance, biomaterials and/or bioenergy, associated with environmental advantages, generate increased CO_2 emissions in actual scenarios. Although the absence of a perceptible consensus framework for LUC analysis is impressive, to analyse the results an expanding range of subsequent metrics is present (De Rosa, 2018). Moreover, assessing the secondary consequences of this phenomenon is one of the most complex steps in LUC analysis. To estimate LUC quantitatively, it is critical for assessing relationships between the demand for products and changes in the utilization of global land, along with assessments of extent, nature and emission in indirectly affected areas (De Rosa, 2018).

14.5 CLIMATE CHANGE

Speaking of energy saving, the production of bioplastics might imply a high consumption of additional biomass resources; therefore, the CO_2 emission reduction is not significant. Bioplastics need more Non-Renewable Energy Use (NREU) in comparison to petro-plastics, as determined by Benavides *et al.* (2020), and GHG is emitted in the atmosphere, as reported by Atiwesh *et al.* (2021). Furthermore, emissions of SO_2 from NREU pollute water, and the emission of other chemicals also leads to acid rain (Marson *et al.*, 2023). But there is a solution to this, and it is the replacement of non-renewable energy with environmentally friendly alternatives. PLA thermoform boxes have a greater environmental effect than polystyrene plastics; indirect emissions from LUC contributed 81-91% of the impact category (Yadav & Nikalje, 2024). As a result, the use of renewable biological feedstock instead of the fossil one in bioplastics manufacturing might decrease CO_2 emissions during manufacturing and partially free the market from non-renewable resources, but at the same time, it might minimize plant photosynthesis periods and thus global CO_2 uptake. Moreover, analysing bio-based bioplastics in the context of global carbon content, it becomes mandatory to conclude that their use contributes to the overall rise in global carbon emissions. It also affects the current global carbon cycle in a similar manner to what the manufacture of the plastic derived from fossil fuels does: Liberating massive amounts of CO_2 into the atmosphere Numerous research articles show the unfair impact of bioplastics on climate change due to the absorption of the CO_2 by the biomass in the processing stage. Nevertheless, excluding the EOL CO_2 emission in the LCA of 'cradle to gate' can make it subjective and could indirectly misrepresent the actual picture (Huang *et al.*, 2025). It's also possible to link climate change with the EOL choices of bioplastics in some way. Since CO_2 produced from burning bioplastics - which is made basically of C/H/O - does not add carbon back into the cycle, it is said that it contributes no net carbon back into the cycle; the carbon carried in CO_2 in incineration plants is utilized for the creation of new biomass through photosynthesis (Rosenboom *et al.*, 2022). On the other hand, N/S/CI, when incorporated in bioplastics, may be oxidized into NO_x, SO_x, and HCl individually. Similarly, combustion of a number of constituents of bioplastic leads to a potential emission of toxic components which need costly measures to capture and treat (De Sadeleer *et al.*, 2024). The greater concern of landfilling is that methane emissions, which have 20 times more GHG potential than CO_2 is uncontrollably released; space occupancy and leachate production are also issues (Kwon *et al.*, 2023), Although landfill takes 58% of plastics, the incineration discharges more GHG than landfill (Kwon *et al.*, 2023). Despite such findings, the author established that the captivation of this methane gas from landfills is still low globally (Basta & Szewczyk, 2024). The filled marker represents cradle-to-gate, while the open marker represents cradle-to-grave. It is challenging to compare fossil-derived plastics with bioplastics primarily because spread data are provided in the majority of cases. Bio-based polymers and fossil-based polymers had comparable data dispersion in categories linked to climate change. In all the cases, bioplastic had a higher Global Warming Potential (GWP) than conventional plastics, except for the fossil-based PET, which had 742 kg CO_2eq/kg higher GWP than PLA. Thus, the overall conventional plastics were overshadowed by the bioplastic. However, when comparing at the cradle-to-gate level of analysis, PHA had the highest global warming potential, whereas when cradle-to-grave analysis had been done, PBS had the highest of all, making its total potency as high as PHA. However, there are two factors that might render this approach disputable and potentially inaccurate: (i) EOL emissions in the scope of cradle-to-gate analyses and (ii) CO_2-fixing biomass that may metamorphose into the fossil source for petro-plastics (Rosenboom *et al.*, 2022).

14.6 EUTROPHICATION AND ACIDIFICATION

Acidification is the process by which water becomes more acidic than it naturally is, while eutrophication is the process of an increased buildup of nutrients in water.

Fertilizers assist in the growth of crops used for bioplastics, which comes at a cost of high consumption that contributes to eutrophication and acidification. Research showed that the crop cultivation stages are the worst as far as environmental harm is concerned, more so as regards acidification and eutrophication. For example, those who performed cradle-to-gate analysis, De Sadeleer *et al.* (2024) found out that in the comparison of bio- and fossil-based PHB, the former had the most times more EP. There are also studies which demonstrated together with De Sadeleer *et al.* (2024) that bioplastics have an environmental impact 1.5-2.2 times greater than that of landfilling of conventional plastic because of Acidification Potential (AP) and Eutrophication Potential (EP) (Huang *et al.*, 2025). The Life Cycle Assessment (LCA) of the cradle-to-grave and the cradle-to-gate strategies revealed possibilities of acidification and eutrophication. Generally, bio-based polymers had more potential for acidification than fossil-based polymers. This was in particular seen in bio-based Polyethylene Terephthalate (PET), where it was established that it was 60% less efficient than the highest inefficient fossil polymer. Meanwhile, in eutrophication, similar to the scenario of Carbon Disclosure Project (CDP), bio-originated polymers were revealed to be less effective than fossil-originated polymers, including PLA. Due to the fact that conventional plastics are made from fossil fuel, the possible result should be no eutrophication impact (Huang *et al.*, 2025).

14.7 MICRO/NANO-PLASTIC POLLUTION

In biodegradation, bioplastics find themselves undergoing the fragmentation processes which, as a result, make them convert into macro/micro/nano sizes, and they might pose a threat to the environment in the future. The problem of ecotoxicity of these particles and their possible penetration into the food chain is becoming urgent. Biodegradable MPs should dissolve in the environment faster than ordinary plastics, but the threats they pose may remain the same (Yang *et al.*, 2025) and become more severe when the size of the particles is significantly smaller. Wang *et al.* (2024) observed that while in such a scenario, the rate at which bioplastics generate MPs is greater than that of conventional petroleum-based plastics. Biodegradation depends on parameters such as the polymer and chemical type, the type of organism, and chemical composition (Gola *et al.*, 2021). Bioplastics can only degrade in a controlled environment such as that provided by most industrial composting facilities. But a natural environment may not afford optimum conditions for such a favourable breakdown. The extent of biodegradation relies on the following properties of the bioplastic; they include crystallinity, molecular weight, and functional groups, predominantly because they are determinants of ease of access to degrading agents. For example, the existence of crystalline areas may slow the degradation process. Furthermore, the bioplastics' constituents affect degradation; for instance, Polybutylene Adipate Terephthalate (PBAT) bioplastics experienced more MP formation in seawater than typical plastics, according to Gola *et al.* (2021). However, these bioplastic materials of plastics degrade more rapidly than conventional plastics within 3-6 months (Akhlaq *et al.*, 2023); this phenomenon, if not controlled, leads to the formation of MPs. Further studies have shown that in several biodegradation media the material did not degrade completely and thus turned into MPs and NPs (Withana *et al.*, 2025). Besides, the samples and environment of the soil could contain more MPs than that of the sea and marine (Li *et al.*, 2023a). Some research studies indicated that the concentration of MPs in the industrial area can be higher for the road dust and storm water samples containing 2410 items per kg and 35 items per L in the industrial area; for the commercial area, 2130 items per kg and 27 items per L; and for the residential area, 1970 per kg and 24 per L, respectively (Kabir *et al.*, 2024). All these MPs are collected at the roadside as well as adjacent areas before they are washed off through surface runoff during rains and transported to aquatic settings (Monira *et al.*, 2023). A hypothesis was tested to calculate amounts of MPs released from bioplastic and conventional materials by using a formula. Published work showed that the assembly of five segments, each 5 cm in length, of conventional plastics provided an average of 17-52 MPs during composting, while in a controlled scenario, PLA samples had an average count of only four particles. Based on such data, the author concluded that the amount of MPs that decompose certainly depends on the type of plastics used as well as the thickness of the samples. A separate study reported that nanoparticle concentrations increased by 28-fold at an energy density of 32 kJ/L in water, with the smallest detected particle measuring 72.4 nm, originating from the degradation of weathered polystyrene initially sized at 106 μm (Monira *et al.*, 2023). They can alter the growth, reproduction, species composition and health of marine organisms (micro- and nano-sized plastics). The researchers have also presented data from Jewett *et al.* (2022), which indicate that PLA MPs affect the marine environment and species, changing bacteria and their activity. Owing to their nonpolar nature and large surface area, MPs/NPs can adsorb heavy dyes, micropollutants of complex matrices, anxiolytics, antidepressants, biocides, plasticizers, hormones, artificial sweeteners, and polycyclic aromatic hydrocarbons. As a result, solution species take up these particles and cause progressive tissue, organ, and intestinal tract injury; by the same token, these particles 'concentrate' in the food web (Li *et al.*, 2023b). Thus, the intake and elimination of freshwater amphipods were equal. Badea *et al.* (2023) identified different biological effects of MP derived from petro-plastic and bioplastic. Another effect that MP pollution has is that it can alter dynamics of the nature of soil carbon that has implications for the atmospheric concentration of GHGs (Yang *et al.*, 2021). This influence is subject to changes in environment, soil type, and MPs' properties defined by Sunil *et al.* (2024). Moreover, types of MP are also accountable for the impacts of dissolved organic carbon (DOC). For instance, Li *et al.* (2023a) stated to find a rich relationship with findings of higher impact on DOC in biodegradable Polybutylene Succinate (PBS) and Polylactic Acid (PLA) microplastics (MPs) than non-biodegradable Polyethylene (PE) and Polystyrene (PS) microplastics. This increase can be occasioned by incidental inclusion of biodegradable plastics in the soil environment (Figure 14.2). However, biodegradable Polybutylene Adipate Terephthalate Microplastics (PBAT-MPs) had a relatively smaller impact on the change in DOC than the non-biodegradable PE-MPs (Withana *et al.*, 2025). DOC was identified as being higher for biodegradable MPs as compared to non-biodegradable MPs (Figure 14.3). In so doing, this suggests that bioplastics could possibly enhance the leaching of organic carbon in soils, thereby impacting nutrient turnover and microbial functions (Song *et al.*, 2025).

However, bio-MPs remain in the environment for long, and this information has not been universally affirmed, and there is more research being done. Regarding the consequences and the fate, there remains only conjecture, and there is no proof (Wang *et al.*, 2021). Therefore, researchers

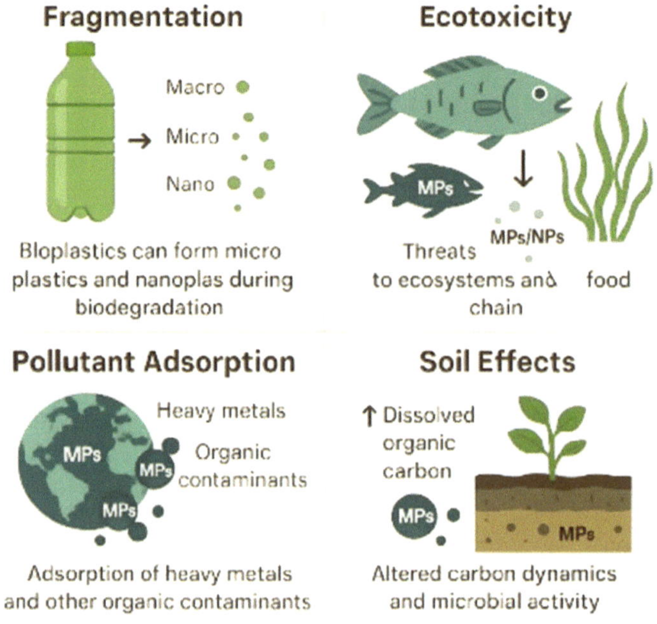

FIGURE 14.3 Environmental Impact of Micro/Nano-Plastic.

should aim at the formation and the possible consequences of the MPs and NPs rather than concentrating on only the bioplastics degradation on macro level.

14.8 CONCLUSION

Bioplastics, considered a new generation of plastics, can partially solve the problem of using fossil fuel, decreasing emissions of GHGs, and possessing biodegradable characteristics if disposed of under certain conditions. Based on this aspect, their capability to degrade into organic constituents, recycle carbon in the soil, and stimulate microbial action in agricultural applications makes them useful to sustainable development. The biodegradable nature of bioplastics can also be mechanically or chemically recycled and, in addition, contain energy recovery. However, if the wastes generated are to be appropriately managed, then the benefits that are postulated will have to accompany proper WM systems, and more so if a recycling or composting infrastructure needs to be set.

However, some bars have remained to be hurdles. Bioplastics appear to break down more slowly in natural ecosystems and environments, especially where conditions favourable for their biodegradation are absent, putting MP pollution and waste into focus. Secondly, chemical recycling processes are energy intensive, and the current recycling streams are being diluted with petrochemical plastics. This is important given that the benefits of bioplastics are not entirely clear given their ability to degrade into MPs and NPs. For bioplastics to capture their full environmental worth, enhanced approaches to production, disposal, and recycling procedures need to be implemented all over the world.

14.9 FUTURE RECOMMENDATION

The effect of bioplastics on the environment is still unclear since there is not adequate information to work with. This means that the increased demand for bioplastic may, in turn, mean a new demand for large pieces of land and water bodies, which may, in turn, have some impacts on biodiversity. Further, there are several other indirect effects of environmental pollution that have not been well quantified and are not well incorporated in today's Life Cycle Assessment (LCA) tools: The release of toxic substances during the microbial decomposition of bioplastics remains poorly understood, as does the formation of microplastics and nanoplastics resulting from their partial degradation. Secondly, LCA readers may not find relevant information on MP-specific data differences due to the limited available data on this topic, although 117,000 tons of MPs are released annually to aquatic systems globally. In LCAs, important policy hot spots within the MP life cycle can be delineated in the future and protective strategies can be implemented to minimize negative environmental impacts. The use of dynamic LCA methodology has the potential for including spatial factors in the assessment, expanding knowledge of MP concentrations and distribution over time. A correct assessment of the effects on the environment entails consideration of the emissions generated in the process of production as well as biodegradation of bioplastics. However, arguments continue to exist concerning the selection of impact categories, the extent of the study, and functional units appropriate to analyse the Life Cycle Inventory (LCI) data related to bioplastics. Therefore, more research should be conducted in the future to establish the adequate LCI database and a methodological framework for bioplastic products and, in addition, compare the life cycle implications of bioplastics and petrochemical plastics if analysed through the 'cradle-to-grave' model. To evaluate the environmental impacts of bioplastics production, it is essential to first develop a life cycle inventory covering feedstock cultivation, its conversion into polymers, the manufacturing process, and transportation. This enables estimation of factors such as Global Warming Potential (GWP), acidification, eutrophication, energy consumption, and Land Use Change (LUC). In addition, some emissions during Waste Management (WM) and other End of Life (EOL) choices shall be incorporated in order to compile a complete 'cradle-to-grave' Life Cycle Inventory (LCI) which significant studies lack. Future studies will give a clearer and broader picture to know how far its application is viable to replace the conventional plastics, which will result in a drastic reduction of fossil fuel utilization, GHG emissions and MP emissions.

REFERENCES

Adnan, M., Siddiqui, A. J., Ashraf, S. A., Snoussi, M., Badraoui, R., Alreshidi, M., Elasbali, A. M., Al-Soud, W. A., Alharethi, S. H., Sachidanandan, M., & Patel, M. (2022). Polyhydroxybutyrate (PHB)-based biodegradable polymer from *Agromyces indicus*: Enhanced production, characterization, and optimization. *Polymers, 14*(19), 3982. https://doi.org/10.3390/polym14193982

Akhlaq, S., Singh, D., Mittal, N., & others. (2023). A review on biodegradation of bioplastics in different environmental conditions. *Polymer Science Series B, 65*, 733-745. https://doi.org/10.1134/S1560090424600128

Anli Dino, A., Kishore, G., Rokhum, S. L., & Baskar, G. (2025). Bioplastic production using waste macroalgal biomass: A holistic review on challenges, prospects, economic viability and sustainability analysis. *Journal of Environmental Chemical Engineering, 13*(2), 116108. https://doi.org/10.1016/j.jece.2025.116108

Atiwesh, G., Mikhael, A., Parrish, C. C., Banoub, J., & Le, T.-A. T. (2021). Environmental impact of bioplastic use: A review. *Heliyon, 7*(9), e07918. https://doi.org/10.1016/j.heliyon.2021.e07918

Badea, M. A., Balas, M., & Dinischiotu, A. (2023). Microplastics in freshwaters: Implications for aquatic autotrophic organisms and fauna health. *Microplastics, 2*(1), 39-59. https://doi.org/10.3390/microplastics2010003

Basta, E., & Szewczyk, P. (2024). The use of methane from landfill gas to generate energy and its management at the plant as a way to reduce climate change. *Rocznik Ochrona Środowiska, 26*, 236-250. https://doi.org/10.54740/ros.2024.024

Bátori, V., Åkesson, D., Zamani, A., Taherzadeh, M. J., & Sárvári Horváth, I. (2018). Anaerobic degradation of bioplastics: A review. *Waste Management, 80*, 406-413. https://doi.org/10.1016/j.wasman.2018.09.040

Benavides, P. T., Lee, U., & Zarè-Mehrjerdi, O. (2020). Life cycle greenhouse gas emissions and energy use of polylactic acid, bio-derived polyethylene, and fossil-derived polyethylene. *Journal of Cleaner Production, 277*, 124010. https://doi.org/10.1016/j.jclepro.2020.124010

De Rosa, M. (2018). Land use and land-use changes in life cycle assessment: Green modelling or black boxing? *Ecological Economics, 144*, 73-81. https://doi.org/10.1016/j.ecolecon.2017.07.017

de Sadeleer, I., & Woodhouse, A. (2024). Environmental impact of biodegradable and non-biodegradable agricultural mulch film: A case study for Nordic conditions. *International Journal of Life Cycle Assessment, 29*, 275-290. https://doi.org/10.1007/s11367-023-02253-y

Emadian, S. M., Onay, T. T., & Demirel, B. (2017). Biodegradation of bioplastics in natural environments. *Waste Management, 59*, 526-536. https://doi.org/10.1016/j.wasman.2016.10.006

Environmental Protection Agency. (2024). Research climate data table dev. https://www.epa.ie/our-services/research/epa-funded-research/epa-funded-projects/research-climate-data-table-dev/

European Bioplastics. (2024). *Glossary*. https://www.european-bioplastics.org/glossary/

Finnerty, J., Rowe, S., Howard, T., Connolly, S., Doran, C., Devine, D. M., Gately, N. M., Chyzna, V., Portela, A., Bezerra, G. S. N., McDonald, P., & Colbert, D. M. (2023). Effect of mechanical recycling on the mechanical properties of PLA-based natural fiber-reinforced composites. *Journal of Composites Science, 7*(4), 141. https://doi.org/10.3390/jcs7040141

Fonseca, A., Ramalho, E., Gouveia, A., Figueiredo, F., & Nunes, J. (2023). Life cycle assessment of PLA products: A systematic literature review. *Sustainability, 15*(16), 12470. https://doi.org/10.3390/su151612470

Fredi, G., & Dorigato, A. (2021). Recycling of bioplastic waste: A review. *Advanced Industrial and Engineering Polymer Research, 4*(3), 159-177. https://doi.org/10.1016/j.aiepr.2021.06.006

Gola, D., Tyagi, P. K., Arya, A., Chauhan, N., Agarwal, M., Singh, S. K., & Gola, S. (2021). The impact of microplastics on marine environment: A review. *Environmental Nanotechnology, Monitoring & Management, 16*, 100552. https://doi.org/10.1016/j.enmm.2021.100552

Gu, J.D. (2021). Biodegradability of plastics: The issues, recent advances, and future perspectives. Environmental Science and Pollution Research, *28*(2), 1278-1282. https://doi.org/10.1007/s11356-020-11501-9

Huang, S., Dong, Q., Che, S., Li, R., & Tang, K. H. D. (2025). Bioplastics and biodegradable plastics: A review of recent advances, feasibility and cleaner production. *Science of The Total Environment, 969*, 178911. https://doi.org/10.1016/j.scitotenv.2025.178911

International Energy Agency (IEA). (2020). *The outlook for biogas and biomethane to 2040 - Outlook for biogas and biomethane: Prospects for organic growth - Analysis*. IEA. https://www.iea.org/reports/outlook-for-biogas-and-biomethane-prospects-for-organic-growth/the-outlook-for-biogas-and-biomethane-to-2040

Islam, M., Xayachak, T., Haque, N., Lau, D., Bhuiyan, M., & Pramanik, B. K. (2024). Impact of bioplastics on environment from its production to end-of-life. *Process Safety and Environmental Protection, 188*, 151-166. https://doi.org/10.1016/j.psep.2024.05.113

Jewett, E., Arnott, G., Connolly, L., Vasudevan, N., & Kevei, E. (2022). Microplastics and their impact on reproduction—Can we learn from the *C. elegans* model? *Frontiers in Toxicology, 4*, 748912. https://doi.org/10.3389/ftox.2022.748912

Kabir, S. M. A., Bhuiyan, M. A., Zhang, G., & Pramanik, B. K. (2024). Microplastic distribution and ecological risks: Investigating road dust and stormwater runoff across land uses. *Environmental Science: Advances, 3*(1), 62-75. https://doi.org/10.1039/D3VA00128H

Kang, J.H., Kang, S.W., Kim, W.J., Kim, D.H., & Im, S.W. (2022). Anaerobic co-digestion of bioplastics and food waste under mesophilic and thermophilic conditions: Synergistic effect and biodegradation. *Fermentation, 8*(11), 638. https://doi.org/10.3390/fermentation8110638

Kim, M. S., Chang, H., Zheng, L., Yan, Q., Pfleger, B. F., Klier, J., Nelson, K., Majumder, E. L. W., & Huber, G. W. (2023). A review of biodegradable plastics: Chemistry, applications, properties, and future research needs. *Chemical Reviews, 123*(16), 9915-9939. https://doi.org/10.1021/acs.chemrev.2c00876

Kooch, Y., Nouraei, A., Wang, L., Wang, X., Wu, D., Francaviglia, R., Frouz, J., & Parsapour, M. K. (2024). Long-term landfill leachate pollution suppresses soil health indicators in natural ecosystems of a semi-arid environment. *Chemosphere, 367*, 143647. https://doi.org/10.1016/j.chemosphere.2024.143647

Kwon, Y., Choi, K., & Jang, Y.C. (2023). Greenhouse gas emissions from incineration of municipal solid waste in Seoul, South Korea. *Energies, 16*(12), 4791. https://doi.org/10.3390/en16124791

Lamberti, F. M., Román-Ramírez, L. A., & Wood, J. (2020). Recycling of bioplastics: Routes and benefits. *Journal of Polymers and the Environment, 28*, 2551-2571. https://doi.org/10.1007/s10924-020-01795-8

Landis, D. A., Gratton, C., Jackson, R. D., Gross, K. L., Duncan, D. S., Liang, C., Meehan, T. D., Robertson, B. A., Schmidt, T. M., Stahlheber, K. A., Tiedje, J. M., & Werling, B. P. (2018). Biomass and biofuel crop effects on biodiversity and ecosystem services in the North Central US. *Biomass and Bioenergy, 114*, 18-29. https://doi.org/10.1016/j.biombioe.2017.02.003

Li, X., Chen, Y., Zhang, S., Dong, Y., Pang, Q., Lynch, I., Xie, C., Guo, Z., & Zhang, P. (2023). From marine to freshwater environment: A review of the ecotoxicological effects of microplastics. *Ecotoxicology and Environmental Safety, 251*, 114564. https://doi.org/10.1016/j.ecoenv.2023.114564

Li, Y., Gu, P., Zhang, W., Sun, H., Wang, J., Wang, L., Li, B., & Wang, L. (2023). Effects of biodegradable and non-biodegradable microplastics on bacterial community and PAHs natural attenuation in agricultural soils. *Journal of Hazardous Materials, 449*, 131001. https://doi.org/10.1016/j.jhazmat.2023.131001

Li, Y., Wang, S., Qian, S., Liu, Z., Weng, Y., & Zhang, Y. (2024). Depolymerization and re/upcycling of biodegradable PLA plastics. *ACS Omega, 9*(12), 13509-13521. https://doi.org/10.1021/acsomega.3c08674

Li, Z., Yang, Y., Chen, X., He, Y., Bolan, N., Rinklebe, J., Lam, S. S., Peng, W., & Sonne, C. (2023). A discussion of microplastics in soil and risks for ecosystems and food chains. *Chemosphere, 313*, 137637. https://doi.org/10.1016/j.chemosphere.2022.137637

Malmir, T., Lagos, D., & Eicker, U. (2023). Optimization of landfill gas generation based on a modified first-order decay model: A case study in the province of Quebec, Canada. *Environmental Systems Research, 12*, 6. https://doi.org/10.1186/s40068-023-00292-w

Marson, A., Piron, M., Zuliani, F., Fedele, A., & Manzardo, A. (2023). Comparative life cycle assessment in the plastic sector: A systematic literature review. *Cleaner Environmental Systems, 9*, 100119. https://doi.org/10.1016/j.cesys.2023.100119

Martínez, A. L. M., Yepes-Pérez, M., Contreras, K. A. C., & Moreno, P. E. Z. (2024). Production of poly(3-hydroxybutyrate-co-3-hydroxyvalerate) by *Bacillus megaterium* LVN01 using biogas digestate. *Applied Microbiology, 4*(3), 1057-1078. https://doi.org/10.3390/applmicrobiol4030072

McAdam, B., Fournet, M. B., McDonald, P., & Mojicevic, M. (2020). Production of polyhydroxybutyrate (PHB) and factors impacting its chemical and mechanical characteristics. *Polymers, 12*(12), 2908. https://doi.org/10.3390/polym12122908

Merchan, A. L., Fischöder, T., Hee, J., Lehnertz, M. S., Osterthun, O., Pielsticker, S., Schleier, J., Tiso, T., Blank, L. M., Klankermayer, J., Kneer, R., Quicker, P., Walther, G., & Palkovits, R. (2022). Chemical recycling of bioplastics: Technical opportunities to preserve chemical functionality as path towards a circular economy. *Green Chemistry, 24*(24), 9428-9449. https://doi.org/10.1039/D2GC02244C

Monira, S., Roychand, R., Hai, F. I., Bhuiyan, M., Dhar, B. R., & Pramanik, B. K. (2023). Nano and microplastics occurrence in wastewater treatment plants: A comprehensive understanding of microplastics fragmentation and their removal. Chemosphere, *334*, 139011. https://doi.org/10.1016/j.chemosphere.2023.139011

Nandakumar, A., Chuah, J. A., & Sudesh, K. (2021). Bioplastics: A boon or bane? Renewable and Sustainable Energy Reviews, *147*, 111237. https://doi.org/10.1016/j.rser.2021.111237

Narancic, T., Cerrone, F., Beagan, N., & O'Connor, K. E. (2020). Recent Advances in Bioplastics: Application and Biodegradation. Polymers, *12*(4), 920. https://doi.org/10.3390/polym12040920

Negrete-Bolagay, D., & Guerrero, V. H. (2024). Opportunities and challenges in the application of bioplastics: Perspectives from formulation, processing, and performance. *Polymers, 16*(18), 2561. https://doi.org/10.3390/polym16182561

Oduor, W. W., Wandera, S. M., Murunga, S. I., & Raude, J. M. (2022). Enhancement of anaerobic digestion by co-digesting food waste and water hyacinth in improving treatment of organic waste and bio-methane recovery. *Heliyon, 8*(9), e10580. https://doi.org/10.101/1016/j.heliyon.2022.e10580

Rezvani Ghomi, E., Khosravi, F., Saedi Ardahaei, A., Dai, Y., Neisiany, R. E., Foroughi, F., Wu, M., Das, O., & Ramakrishna, S. (2021). The life cycle assessment for polylactic acid (PLA) to make it a low-carbon material. *Polymers, 13*(11), 1854. https://doi.org/10.3390/polym13111854

Rosenboom, J. G., Langer, R., & Traverso, G. (2022). Bioplastics for a circular economy. *Nature Reviews Materials, 7*, 117-137. https://doi.org/10.1038/s41578-021-00407-8

Scharff, H., Soon, H. Y., Rwabwehare Taremwa, S., & others. (2023). The impact of landfill management approaches on methane emissions. *Waste Management & Research, 42*(11), 1052-1064. https://doi.org/10.1177/0734242X231200742

Senila, L., Kovacs, E., Resz, M. A., Senila, M., Becze, A., & Roman, C. (2024). Life cycle assessment (LCA) of bioplastics production from lignocellulosic waste (study case: PLA and PHB). *Polymers, 16*(23), 3330. https://doi.org/10.3390/polym16233330

Sheriff, S. S., Yusuf, A. A., Akiyode, O. O., Hallie, E. F., Odoma, S., Yambasu, R. A., Thompson-Williams, K., Asumana, C., Gono, S. Z., & Kamara, M. A. (2025). A comprehensive review on exposure to toxins and health risks from plastic waste: Challenges, mitigation measures, and policy interventions. *Waste Management Bulletin, 3*(3), 100204. https://doi.org/10.1016/j.wmb.2025.100204

Song, D., Jin, G., Su, Z., Ge, C., Fan, H., & Yao, H. (2025). Influence of biodegradable microplastics on soil carbon cycling: Insights from soil respiration, enzyme activity, carbon use efficiency and microbial community. *Environmental Research, 266*, 120558. https://doi.org/10.1016/j.envres.2024.120558

Springle, N., Li, B., Soma, T., & Shulman, T. (2022). The complex role of single-use compostable bioplastic food packaging and foodservice ware in a circular economy: Findings from a social innovation lab. *Sustainable Production and Consumption, 33*, 664-673. https://doi.org/10.1016/j.spc.2022.08.006

Sunil, S., Bhagwat, G., Vincent, S. G. T., & Palanisami, T. (2024). Microplastics and climate change: The global impacts of a tiny driver. *Science of The Total Environment, 946*, 174160. https://doi.org/10.1016/j.scitotenv.2024.174160

Taheripour, F., & Tyner, W. E. (2013). Biofuels and land use change: Applying recent evidence to model estimates. *Applied Sciences, 3*(1), 14-38. https://doi.org/10.3390/app3010014

Tecorralco-Bobadilla, A. L., Vázquez-Morillas, A., González-Cardoso, G., & Sotelo-Navarro, P. X. (2024). Life cycle assessment (LCA) of bioplastics. In Inamuddin, T. Altalhi, & V. C. Fernandes (Eds.), *Toxic effects of micro- and nanoplastics* (Chapter 2). https://doi.org/10.1002/9781394238163.ch2

U.S. Environmental Protection Agency (2023), Office of Research and Development.. *Quantifying Methane Emissions from Landfilled Food Waste*. Retrieved from https://www.epa.gov/land-research/quantifying-methane-emissions-landfilled-food-waste

Van, Y., Čuček, L., Klemeš, J. J., Vujanović, A., & Varbanov, P. S. (2022). Life cycle assessment approaches of plastic recycling with multiple cycles: Mini review. *Chemical Engineering Transactions, 94*, Article 014. https://doi.org/10.3303/CET2294014

Walker, S., & Rothman, R. (2020). Life cycle assessment of bio-based and fossil-based plastic: A review. Journal of Cleaner Production, *261*, 121158. https://doi.org/10.1016/j.jclepro.2020.121158

Wang, C., Yu, J., Lu, Y., Hua, D., Wang, X., & Zou, X. (2021). Biodegradable microplastics (BMPs): A new cause for concern? Environmental Science and Pollution Research, *28*(47), 66511-66518. https://doi.org/10.1007/s11356-021-16435-4

Wang, J., Jia, M., Zhang, L., Li, X., Zhang, X., & Wang, Z. (2024). Biodegradable microplastics pose greater risks than conventional microplastics to soil properties, microbial community and plant growth, especially under flooded conditions. *Science of The Total Environment, 931*, 172949. https://doi.org/10.1016/j.scitotenv.2024.172949

Withana, P. A., Yuan, X., Im, D., Choi, Y., Bank, M. S., Lin, C. S. K., Hwang, S. Y., & Ok, Y. S. (2025). Biodegradable plastics in soils: Sources, degradation, and effects. *Environmental Science: Processes & Impacts*. Advance online publication. https://doi.org/10.1039/D4EM00754A

Xiang, R., Wei, W., Mei, T., Wei, Z., Yang, X., Liang, J., & Zhu, J. (2025). A review on landfill leachate treatment technologies: Comparative analysis of methods and process innovation. *Applied Sciences, 15*(7), 3878. https://doi.org/10.3390/app15073878

Yadav, K., & Nikalje, G. C. (2024). Comprehensive analysis of bioplastics: Life cycle assessment, waste management, biodiversity impact, and sustainable mitigation strategies. *PeerJ, 12*, e18013. https://doi.org/10.7717/peerj.18013

Yang, H., Niu, S., Guo, M., & Xue, Y. (2025). A critical review of the ecotoxic effects of microplastics on aquatic, soil and atmospheric ecosystems and current research challenges. *Environmental Research, 274*, 121361. https://doi.org/10.1016/j.envres.2025.121361

Yang, L., Zhang, Y., Kang, S., Wang, Z., & Wu, C. (2021). Microplastics in soil: A review on methods, occurrence, sources, and potential risk. *Science of The Total Environment, 780*, 146546. https://doi.org/10.1016/j.scitotenv.2021.146546

Yeşiller, N., Hanson, J. L., Manheim, D. C., Newman, S., & Guha, A. (2022). Assessment of methane emissions from a California landfill using concurrent experimental, inventory, and modeling approaches. *Waste Management, 154*, 146-159. https://doi.org/10.1016/j.wasman.2022.09.024

Zhang, Z., Chen, Z., Zhang, J., Liu, Y., Chen, L., Yang, M., Osman, A. I., Farghali, M., Liu, E., Hassan, D., Ihara, I., Lu, K., Rooney, D. W., & Yap, P. S. (2024). Municipal solid waste management challenges in developing regions: A comprehensive review and future perspectives for Asia and Africa. *Science of The Total Environment, 930*, 172794. https://doi.org/10.1016/j.scitotenv.2024.172794

Zhao, X., Wang, Y., Chen, X., Yu, X., Li, W., Zhang, S., Meng, X., Zhao, Z. M., Dong, T., Anderson, A., Aiyedun, A., Li, Y., Webb, E., Wu, Z., Kunc, V., Ragauskas, A., Ozcan, S., & Zhu, H. (2023). Sustainable bioplastics derived from renewable natural resources for food packaging. *Matter, 6*(1), 97-127. https://doi.org/10.1016/j.matt.2022.11.006

Zhu, S., Chen, H., Wang, M., Guo, X., Lei, Y., & Jin, G. (2019). Plastic solid waste identification system based on near infrared spectroscopy in combination with support vector machine. *Advanced Industrial and Engineering Polymer Research, 2*(2), 77-81. https://doi.org/10.1016/j.aiepr.2019.04.001

Section IV

Management and Societal Impact

15 Effect of Bioplastic Additives on Packaging and Storage of Food

Eridiong Ogbonna Onyenweaku and Swapnil Godbharle

15.1 INTRODUCTION

15.1.1 General Overview of Food Packaging and Storage

Food packaging, a vital aspect of food processing, plays a key role in preserving products by preventing spoilage and contamination while extending shelf life. It serves multiple functions, including containment (holding the product), protection (ensuring quality, safety, and freshness), communication (through graphics and labels), and convenience of use (The Society of the Plastics Industry, Washington, DC). Beyond these primary roles, packaging offers additional benefits to both manufacturers and consumers. It can extend shelf life by modifying the atmosphere, enhance marketing efforts by conveying messages, improve security with tamper-evident features, and provide portion control to address "portion distortion." Packaging also safeguards food by acting as a barrier against external hazards such as light, oxygen, and moisture, which could compromise the product. It facilitates ease of transport, ensures adequate storage, provides essential information to consumers, and indicates potential tampering (Vaclavik et al., 2021). By fulfilling these diverse functions, packaging is integral to maintaining the integrity and usability of both raw and processed foods.

Adequate packaging is a critical technique in the food industry, used alongside additives and preservation methods to slow or prevent spoilage that would otherwise compromise taste, texture, and nutritional value. Proper packaging ensures both plant-based and animal-derived products are market-ready with extended shelf lives. Various materials, such as films and oxygen-controlling solutions, protect food from air exposure, while additional features like dating or doneness indicators help maintain time-sensitive products. Packaging also serves as a promotional tool, enhancing product appeal on retail shelves. Common packaging materials include metal, glass, paper, plastics, foil, wood crates, cotton, and burlap (jute). Advanced methods such as vacuum sealing, controlled or modified atmospheric packaging, and aseptic packaging further preserve food quality. Manufacturers must comply with Food and Drug Administration (FDA) regulations governing packaging methods and materials to ensure safety and effectiveness. Consumer demand has driven the popularity of convenient options, such as microwaveable packages, single-serve containers, resealable pouches, and tamper-evident designs, reflecting the need for functionality that meets the requirements of both consumers and food companies (Risch, 2009).

In recent years, the food packaging industry has seen a notable transition towards sustainable materials, largely fuelled by increasing environmental concerns and consumer demand for eco-friendly options. Bioplastics, made from renewable biological sources, have emerged as a promising solution to address these challenges. As bioplastics gain traction, the food packaging industry is also witnessing shifts in market dynamics and regulatory landscapes. Companies are increasingly investing in research and development to create high-performance bioplastics that meet industry

DOI: 10.1201/9781003602828-19

standards while being cost-competitive. This trend is supported by governmental policies and incentives aimed at promoting sustainable practices, pushing manufacturers to adopt bioplastics as a viable alternative to traditional materials (Keränen et al., 2021) for the packaging of food and other materials. As the world struggles with the environmental challenges posed by plastic pollution, bioplastics seem to offer a pathway towards a more sustainable future, aligning with both consumer preferences and the global imperative for responsible resource management.

Shlush and Davidovich-Pinhas (2022) assert that the growing role of bioplastics in food packaging not only reflects the industry's commitment to sustainability but also highlights the importance of ongoing research and collaboration among stakeholders to harness the full potential of these materials. Food packaging must meet certain standards to ensure quality and safety for consumers. Different packaging materials are more suitable for specific types of food products. According to Sarkar and Aparna (2020), food packaging has undergone over 300 years of evolution to reach its current form. Various packaging materials have been developed during this time, each serving a specific role in protecting and preserving food. Innovations such as biodegradable and edible packaging address the global demand for environmentally friendly and natural food solutions, reflecting growing concerns about sustainability. The food packaging industry has seen significant advancements with technologies like active, aseptic, smart, bioactive, and edible packaging, all of which represent current research trends.

The primary functions of food packaging are protection, preservation, and promotion. Its main purpose is to shield food from biological, chemical, and mechanical damage, maintain its integrity, and attract consumers by conveying essential information such as ingredients and nutritional content. These efforts help minimize food waste and ensure consumer health. Secondary functions, including convenience, traceability, and tamper or pilferage indication, are increasingly important (Sarkar & Aparna, 2020). Novel packaging technologies have not only extended food shelf life but also improved safety, quality, and consumer convenience. As suggested by Fellows (2000), the major types of packaging are as follows:

1. Primary Packaging: Comes in direct contact with the contents packed; smallest unit.
2. Secondary Packaging: Material present outside the primary package; also used to group primary packages.
3. Tertiary Packaging: The material which grasps secondary packages; used for bulk packaging to ease handling, transportation, distribution, shipping, or storage.

Examples of materials used in packaging in the food industry include paper/paper board, glass, metals (aluminium foil), tin cans, plastics (polyolefins – polyethene, polypropylene), polyesters, wood, cloth and vegetable fibres (cotton, jute, linen, sisal, etc.).

Generally, food packaging is carried out as a final process where food products are packed for storage or transport to consumers.

15.2 EFFECT OF BIOPLASTIC ADDITIVES ON FOOD QUALITY

Bioplastic additives which have been identified as substances used to enhance the properties and performance of bioplastics, are derived from renewable biomass sources such as plants, algae, or microorganisms. Some common bioplastic additives are plasticizers, impact modifiers, UV stabilizers, antimicrobial agents, fillers, reinforcing agents, lubricants, and flame retardants. These additives can affect the quality of food products packaged using bioplastics – either positively or negatively. Food quality encompasses the attributes of food that ensure wholesomeness and make it acceptable to consumers. Tanner (2016) defines food quality as a combination of characteristics that determine a product's acceptability, including external factors like appearance (size, shape, colour, gloss, and consistency), texture, and flavour, as well as internal factors such as chemical, physical, and microbial composition. It also involves product traceability, enabling recalls when necessary, and accurate labelling to provide correct ingredient and nutritional information. As a fundamental consumer expectation, food quality requires rigorous quality control measures to maintain standards (Hurst, 2010). It represents the degree of excellence in a food product, integrating attributes such as taste, appearance, nutritional content, and bacteriological stability (Vaclavik & Christian, 2008). While quality is somewhat challenging to define precisely, it represents the features that make food acceptable to consumers. Food quality can also be described as the sum of features that satisfy customer needs and influence both intended and unintended effects on stakeholders. In contrast, food safety specifically addresses the presence or absence of food-borne hazards at the point of consumption. Yang et al. (2019) argue that definitions of food quality and food safety indicate that "food safety in essence is one of the features and characteristics that compose food quality, despite that food safety is usually referred independent of food quality due to its importance to human health". For food to be considered safe and wholesome, it should be devoid of hazards that can cause harm to consumers. One major source of concern is usually chemicals that come in contact with food during processing and packaging. For the safety of consumers, it must be ensured that such chemicals do not contaminate the food and if at all they pass onto the food products, they are present in safe/tolerable limits which are acceptable by regulatory bodies such as the Codex Alimentarius and Food and Agricultural Organisation (FAO).

In the case of packaging materials, chemical additives are sometimes incorporated to boost the physical properties (mainly) of the materials such as plastics. Oluwasina and Awonyemi (2021) assert that an excellent food packaging plastic must be able to keep its stored product with an acceptable moisture content that prevents microbial growth and spoilage. Commonly used plastics in food packaging are polyolefins, polyester, polystyrene, and nylon polycarbonate. These plastics possess high tensile strength and excellent mechanical reliability, thereby meeting the basic requirements for food packaging applications (Behera et al., 2022). In order to maximize the quality of these plastics used in both bioplastic and conventional plastic packaging materials, some researchers have developed strategies to improve them. For example, Behera et al. (2022) found that a higher bentonite concentration (1.5 wt%) can effectively improve yam starch-based bioplastic properties, making it a potential replacement for petrochemical plastic in food packaging applications. Ongoing studies on bioplastic for food packaging applications are often focused on mechanical properties. However, to promote its application and ensure consumer safety, other tests are also needed; such as oxygen barrier properties, food safety, biocompatibility, and water vapour permeability (to prevent moisture transmission from the environment to the food).

Below are some bioplastic additives and the possible ways they could affect food quality/safety:

1. Plasticizers, Fillers, and Reinforcing Agents: These improve flexibility and processability (e.g., citrate esters and succinate esters). They enhance mechanical properties and help to reduce costs. These may not directly affect the quality of packaged food but may enhance its storage conditions and make it difficult for moisture or spoilage organisms to permeate the packaging material into the product. The use of plasticizers in bioplastics can help preserve the nutritional quality of food products by modifying the material's properties. Plasticizer concentration and type significantly influence bioplastics' characteristics. According to Abe et al. (2021), plasticizers increase the interstitial volume within the polymer matrix, reducing polymer-polymer bonds and altering mechanical (Sanyang et al., 2015), thermal (Lopez et al., 2011), and water vapour barrier properties. Variations in mechanical and barrier properties arise from the differing plasticizing effects of each substance (Abe et al., 2021). For instance, glycerol, due to its smaller molecular size, exhibits greater plasticizing capacity compared to sorbitol. The selection of plasticizers depends on their compatibility with the polymer matrix and the intended application, allowing for the safe use of different plasticizers in bioplastics (Tyagi & Bhattacharya, 2019; Arifa et al., 2021).

2. Impact Modifiers: Enhance toughness and impact resistance (e.g., biodegradable polymers, natural rubber). These may positively affect the quality of food by preventing mechanical damage and unnecessary loss of content/unwanted introduction of moisture.
3. UV Stabilizers: Protect against ultraviolet light degradation (e.g., plant-derived UV stabilizers). These will go a long way in preserving the photo-sensitive components of packaged food and hence promote food quality. For example, food items like milk can get easily degraded by light and must be covered from direct light rays in order to prevent spoilage and quality reduction.
4. Antimicrobial Agents: Inhibit the growth of microorganisms (e.g., essential oils and plant extracts). These ones go a long way to directly prevent microbial spoilage in foods especially those with high moisture content. The issue of concern here is adding these antimicrobial agents in concentrations that are safe for consumers peradventure they 'leach' into the food.
5. Lubricants: Reduce friction, and improve processability (e.g., plant-derived lubricants). Food-grade lubricants are typically manufactured at high temperatures to minimize the presence of living microorganisms. Additionally, their chemical composition inherently inhibits microbial growth, offering a significant advantage in maintaining hygiene and safety standards.

Furthermore, in the food industry's attempts to minimize contamination of food products during packaging and at the same time maintain maximum levels of effectiveness and safety, ecolabel lubricants, which are also known as green or environmentally friendly lubricants have been introduced (Caine, 2024). In this regard, manufacturers have worked hard to ensure that ecolabel lubricants either match or surpass the requirements set by their conventional counterparts. Additionally, because of their resistance to oxidation and stability at different temperatures, ecolabel lubricants can be used in various food packaging procedures. Since they provide better safety profiles than conventional choices, eco-label lubricants have become a crucial answer in this area. The strict standards required by regulatory institutions like the Food and Drug Administration (FDA) and the U.S. Department of Agriculture (USDA) – are met by many ecolabel lubricants, which are designed to be food-grade. This classification serves to mitigate the danger to consumer health in the unlikely event of inadvertent interaction with food goods. As an added safety measure, the non-toxic nature of these lubricants lowers the possibility of hazardous contamination during the packaging process (Caine, 2024).

15.3 IMPACT OF BIOPLASTIC ADDITIVES ON FOOD SHELF LIFE

The early development of bioplastics encountered challenges such as high production costs and inferior mechanical and barrier properties compared to traditional plastics. To overcome these limitations, recent advancements have focused on incorporating additives with antifungal, antimicrobial, and antioxidant properties (Abang et al., 2023). These additives significantly enhance the functional characteristics of bioplastics, broadening their applications across various industries. For example, the inclusion of antibacterial agents in bioplastics has facilitated the creation of active food packaging that inhibits spoilage-causing microorganisms, thereby extending the shelf life of food products (Abang et al., 2023). Current research highlights a variety of additives and fillers used to improve bioplastic performance, such as inorganic nanoparticles like zinc oxide and titanium dioxide (Suryanegara et al., 2021), natural fibres like pineapple leaf fibre (Armynah et al., 2022), and plant waste materials, including orange peels and avocado seeds (Quilez-Molina et al., 2022).

Antimicrobial agents like natamycin are commonly employed in bioplastic films to prevent bacterial growth, enhancing both the shelf life and safety of food products (Romero et al., 2016). Similarly, antioxidants play a vital role in protecting fatty foods by reducing lipid oxidation, which slows spoilage and helps minimize food waste (Merino et al., 2021; Romani et al., 2015). *Boswellia serrata* (BS), a natural filler, has also emerged as an effective additive for improving the properties of biopolymers. Known for its antifungal and antibacterial characteristics, BS enhances the durability of bioplastics and extends the shelf life of food products during storage (Abu Ubaidah et al., 2024). These advancements underscore the growing potential of bioplastics in addressing food preservation challenges while contributing to sustainability.

The inclusion of bioplastic additives also poses numerous challenges. One such challenge is the migration of these additives from packaging materials to food items or the adjacent environment. Factors such as temperature, duration of storage, and the type of food can impact this migration (Kudłak et al., 2019). The migration of certain additives can change the sensory qualities (taste, smell, texture) of the food and present potential health hazards if the levels of migrated substances surpass allowable limits. A number of these additives are categorized as endocrine disruptors (EDs) and may have harmful impacts on the reproductive system or possess carcinogenic properties (Hahladakis et al., 2018).

From the food industry's perspective, addressing various challenges associated with incorporating new additives is crucial. Essential factors like the safety and potential hazards linked to these substances require thorough testing and assessment. It is also important to adhere to current regulations to ensure that any new additives meet the strict safety standards set by regulatory agencies (Ghasemlou et al., 2024; Valdés et al., 2014).

Moreover, the effectiveness of bioplastic additives in extending the shelf life of food also depends on the specific type of food and its composition. Different food products possess unique compositions that can affect the performance of the additives. For instance, perishable items may benefit more from bioplastic formulations that create a barrier to moisture and gas transfer. Thus, it is important to understand both the characteristics of the food and the properties of the bioplastic additives in order to extend shelf life and ensure food safety (Romani et al., 2015). Also, the conditions in which the food is processed and the type of bioplastic used can also impact the efficacy of these additives.

To sum up, additives used in the plastic industries are vital in altering the properties of these materials. Nonetheless, it is important to cautiously assess the possible effects of these additives on human health and the environment. While bioplastic additives are crucial for enhancing bioplastic technology, their responsible use and comprehensive testing are vital.

15.4 EFFECTS OF BIOPLASTIC ADDITIVES ON PACKAGING DURABILITY

Additives enhance the durability and performance of bioplastic used for packaging by improving mechanical strength, barrier properties, and resistance to environmental degradation (Armynah et al., 2022). This enhancement improves food preservation and promotes sustainability by decreasing the frequency of packaging replacements, thereby reducing reliance on conventional plastics. Various bioplastic additives have been extensively explored for their possible uses in food packaging. Studies have concentrated on the application of oxidized starch, dialdehyde starch, and silica to improve the physico-mechanical properties of starch-based bioplastic films (Dintcheva & La Mantia, 2007). Oxidized starch, created through a controlled oxidation process, can improve the flexibility and strength of bioplastics. Similarly, dialdehyde starch enhances the barrier properties, which are essential for preserving food freshness by minimizing gas and moisture transfer. Additionally, silica is known for its strengthening capabilities and can significantly enhance the mechanical strength and thermal stability of starch-based bioplastics (Oluwasina et al., 2019, 2021).

Plasticizers are vital additives in bioplastics, used to improve flexibility, strength, and processability. Their primary role is to lower the viscosity of the bioplastic mixture, facilitating easier handling and manufacturing. This reduction in viscosity simplifies the moulding and shaping processes during production and allows for the creation of more intricate designs and shapes. As the material becomes more pliable, it can be processed at lower temperatures and with increased efficiency, ultimately resulting in energy savings during manufacturing. Glycerol and polyvinyl alcohol are two widely used plasticizers (Charles et al., 2022; Navasingh et al., 2023).

Compatibilizers are another type of additive designed to improve compatibility and adhesion among different polymers or polymer blends. These additives can enhance the mechanical strength, thermal stability, and overall performance of bioplastics by promoting better dispersion of polymers, reducing phase separation, and improving interfacial adhesion. Maleic anhydride is the most used compatibilizer. Additionally, fillers such as nanocrystalline cellulose have been employed as reinforcing agents in bioplastics production to enhance their mechanical properties, while also serving as bulk additives (Abu Ubaidah et al., 2024).

The addition of bioplastic additives has greatly improved the effectiveness of bioplastics for use in food packaging. The use of these chemical additives within standard regulatory concentrations will go a long way to protect consumers from any risk of harm.

15.5 ARTIFICIAL ADDITIVES IN FOOD PACKAGING AND CONSUMER SAFETY

The incorporation of artificial additives in packaging materials has raised concerns regarding the food safety of the packaged products. The wholesomeness of food which pertains to its quality must not be compromised in any way if food is going to be fit for consumption by the consumers. Consumer health has to do with the well-being of the people who end up consuming the food products. It comprises the totality of the health and safety of consumers including sanitary issues. Certain considerations need to be put in place when designing a packaging product in order to meet the standards and expectations of consumers. These include:

- Packaging should safeguard food from microbial contamination and be tamper-resistant.
- It should protect food from physical damage, such as breakage, leakage, or pilferage during transit and storage.
- Packaging must prevent the permeation of water vapour, gases (e.g., O_2, CO_2, N_2, ethylene), and other volatile compounds that could impact odour, both into and out of the package.
- It should enable compression and reduce the bulkiness of certain commodities.
- The material should be non-toxic and prevent the transfer of compounds between the food and the packaging.
- Packaging should support clear printability or labelling for product information.
- It should facilitate easy handling, storage, and transportation.
- Packaging should aid in product differentiation, branding, and advertising, as seen with examples like AMUL butter, KISSAN Jam, and MAGGI noodles.
- It should be visually appealing to attract consumers and encourage them to try the product.
- Packaging should be cost-effective, reusable, or easily disposable.

Meeting the above conditions will go a long way in ensuring food safety and product acceptability by consumers. Consumer needs regarding food packaging are summarized in Table 15.1.

Food safety regulations are essential for protecting consumers and ensuring that food products are safe for consumption, encompassing a broad spectrum of standards and practices. In the United States, the FDA regulates food additives, including bioplastics, under the Federal Food, Drug, and Cosmetic Act. The FDA mandates comprehensive testing of any substance intended to become a food component to verify its safety (U.S. FDA, 2024).

The leaching of certain substances from packaging materials can pose significant health risks. For example, some additives act as EDs, interfering with hormonal systems and potentially causing reproductive or developmental issues (Li et al., 2024). Furthermore, the migration of additives is affected by factors such as temperature, contact duration, and the nature of the food interacting with the packaging (Li et al., 2024). To mitigate these risks, regulations such as the FDA's guidelines on food contact substances require manufacturers to provide evidence that any leaching of additives remains below established safety thresholds (U.S. FDA, 2018). However, the evolving nature of

TABLE 15.1
Consumer Needs and Wants of Food Packaging

Requirement	Description
Quality	Processing and packaging in preserving and enhancing the flavour, nutrition, texture, colour, freshness, and overall acceptability of food products, etc.
Information	Product information, legibility, brand, use, etc.
Convenience	Ease of access, opening and disposal; shelf life, microwaveable, etc.
Product availability	Product available at all times
Variety	A wide range of products in a variety of pack sizes, designs and pack types
Health	Safety e.g., Enables the provision of extended shelf-life foods, without the use of preservatives
Safety	The prevention of product contamination and tampering
Environment	Environmental compatibility

bioplastic formulations means that continuous monitoring and assessment are necessary to safeguard public health. Compliance with global food packaging standards is essential for ensuring that food products remain safe for consumption and free from harmful contaminants. Various organizations and regulatory bodies set these standards, which include guidelines for the safety and testing of food packaging materials.

Global Standards for Food Packaging

1. European Union Regulations: The EU has established comprehensive regulations governing food contact materials, primarily outlined in Regulation (EC) No. 1935/2004. According to this rule, materials cannot add ingredients to food in amounts that might jeopardize human health cause an unacceptably altered composition of food, or deteriorate the organoleptic (taste and smell) properties of food (European Commission, 2004).
2. U.S. Standards: In the United States, the FDA regulates food contact substances under the Federal Food, Drug, and Cosmetic Act. The agency has a premarket notification system for food contact substances, requiring manufacturers to demonstrate that their products comply with safety standards before they can be marketed (U.S. FDA, 2018).
3. Global Harmonization: The Codex Alimentarius, a collection of internationally recognized standards, guidelines, and codes of practice, provides a framework for food safety and quality, including packaging materials. While the Codex does not set binding regulations, it serves as a reference for member countries in developing their own standards (FAO/WHO, 2003).

15.6 BIOPLASTIC ADDITIVES AND FOOD SAFETY CONSIDERATIONS

Negrete-Bolagay and Guerrero (2024) assert that the performance and safety of bioplastics depend significantly on the additives incorporated during production and the specific storage conditions of the foods they protect. Furthermore, storage conditions such as temperature, humidity, and light exposure, play a pivotal role in determining the performance of not only bioplastics but conventional plastics generally. Biodegradable polymers like polylactic acid (PLA) and starch-based materials (used in these packaging plastic materials) are sensitive to moisture and temperature fluctuations, which can affect their structural integrity and barrier properties. High humidity levels can weaken starch-based bioplastics due to their hydrophilic nature, compromising their ability to protect food effectively. Temperature is another critical factor. For example, PLA has a relatively low glass transition temperature, making it unsuitable for hot food applications unless modified with heat-resistant additives. Cold storage conditions, on the other hand, can lead to brittleness in some bioplastic formulations, necessitating the inclusion of plasticizers or copolymer blends to maintain flexibility (Cruz et al., 2022; Shah et al., 2024).

The transfer of additives from bioplastics into food products poses a significant concern, especially in extreme storage conditions. Studies have shown that migration rates vary depending on the bioplastic type, additive composition, and food matrix. PLA and starch-based polymers generally exhibit low migration rates, making them safer alternatives for food contact materials compared to conventional plastics containing toxic monomers like bisphenol A (BPA) (Negrete-Bolagay & Guerrero, 2024; Shah et al., 2024). However, certain additives used in bioplastics, such as degradation aids or nanomaterials, may pose cytotoxic risks (i.e. poisonous to living cells). For instance, bio-nanocomposites with nano-clays have shown varying cytotoxic effects, emphasizing the need for standardized testing protocols to ensure safety. Regulatory frameworks like the European Commission's standards for bioplastics help address these concerns by setting limits on additive migration and toxicity (Geueke, 2014).

As Cruz et al. (2022) rightly noted, the rising interest in bioplastics for food packaging stems from their perceived environmental gains and capacity to replace traditional plastics. However, their safety and impact on food quality are closely tied to the additives incorporated during production. Additives such as plasticizers, stabilizers, antimicrobial agents, and fillers enhance most plastic functionality, but their interactions with food and consumer health require scrutiny. The safety of these additives is influenced by their migration rates, which depend on factors like polymer structure, food type, and storage conditions. The following health concerns have been identified:

1. Chemical Migration: Additives such as plasticizers and antimicrobial agents may leach into food, particularly under high-temperature or acidic conditions. For instance, the migration of phthalate substitutes in biodegradable plastics can mimic endocrine-disrupting effects, posing risks to reproductive health (Muzeza et al., 2023).
2. Nanotoxicity: Nanoparticles used in bio-nanocomposites may cross biological barriers, leading to oxidative stress, inflammation, and potential DNA damage. Although not all nanoparticles exhibit these effects, the lack of standardized safety assessments complicates risk evaluation (Muzeza et al., 2023).
3. Toxicity from Stabilizers: UV stabilizers like benzophenones, although effective in protecting packaging materials, have shown potential carcinogenic and endocrine-disrupting effects. Their migration is of particular concern in fatty or oily foods (Gupta et al., 2024).
4. Effect on Organoleptic Properties of Food: Bioplastic additives can also influence the sensory (organoleptic) and nutritional quality of food. Changes in taste, odour, and texture may occur due to the interaction between food components and migrating additives. For example, high levels of certain plasticizers can impart an off-taste to packaged food, reducing its acceptability. Moreover, antimicrobial additives, while inhibiting microbial spoilage, may affect the natural microbial balance of certain foods, impacting their fermentation or ageing processes. Inadequate additive formulations could also lead to premature degradation of the packaging, exposing food to environmental contaminants (Shaikh et al., 2021). Chemical additives used in plastics significantly influence food's sensory and nutritional quality, leading to various concerns regarding their effects on taste, odour, texture, and overall food integrity. The interactions between food components and migrating additives can result in several notable impacts:

A. Sensory Changes

- Taste and Odour: High levels of certain plasticizers used in bioplastics can impart undesirable tastes and odours to packaged food. This can reduce consumer acceptability and overall satisfaction with the product. For instance, transporting plasticizers into food can lead to off-flavours, which may deter consumers from purchasing or consuming these products (Cruz et al., 2022).

- Texture: The interaction between food components and bioplastic additives can alter the texture of food items. Changes in texture may affect the mouthfeel and overall eating experience, potentially leading to negative perceptions of quality (Navasingh et al., 2023; Shaikh et al., 2021).

B. Nutritional Quality

- Microbial Balance: Antimicrobial additives are often included in bioplastics to inhibit microbial spoilage. However, their use can disrupt the natural microbial balance in certain foods, which is crucial for processes like fermentation and ageing. This disruption may affect not only the flavour profile but also the nutritional benefits derived from naturally occurring microbes (Cruz et al., 2022; Shaikh et al., 2021).
- Degradation of Nutrients: Inadequate formulations of bioplastic additives could lead to premature degradation of the packaging material itself. This degradation can expose food to environmental contaminants, potentially compromising its quality and safety (Cruz et al., 2022).

C. Environmental Contaminants

- If bioplastic packaging degrades too quickly or inadequately, it may release harmful substances into the packaged food. These contaminants can negatively impact food safety and quality, raising concerns about the long-term health effects of consuming contaminated products (Negrete-Bolagay & Guerrero, 2024).

15.7 REGULATORY AND SAFETY MEASURES

To mitigate risks, regulatory frameworks such as those by the European Food Safety Authority (EFSA) and the U.S. FDA evaluate additives for bioplastics intended for food contact. To establish safety standards, Cruz et al. (2022) state that these agencies assess migration limits, toxicological profiles, and exposure scenarios.

Recent advancements aim to reduce reliance on synthetic additives by exploring natural or bio-derived options. For instance, essential oils and plant extracts are being investigated as safer alternatives to conventional antimicrobial agents. Similarly, bio-nanocomposites using naturally occurring nanomaterials may offer improved safety profiles (Negrete-Bolagay & Guerrero, 2024).

In conclusion, bioplastic additives (like most other artificial compounds) present a double-edged sword: their environmental benefits are offset by the challenges associated with additive safety and food quality. Balancing these aspects is vital for the broader adoption of bioplastics in food packaging, ensuring consumer health and maintaining food integrity. Further research, coupled with stringent regulation, will drive innovations that prioritize both sustainability and safety.

15.8 CONCLUSION

Bioplastic additives are now being used in the manufacture of both conventional plastics and bioplastics. However, the complexities of biodegradability, resource use, food safety, waste management, and potential ecological impacts necessitate a cautious approach. Bioplastic additives are key in enhancing the shelf-life of food products. Additives in bioplastics play a crucial role in extending the shelf life of food products. By improving barrier properties, providing antimicrobial protection, and reducing oxidation, these additives contribute to minimizing food waste and ensuring food safety. However, challenges such as regulatory compliance, environmental concerns, and cost implications must be addressed. As technology advances, the development of more sustainable and effective bioplastic additives will be essential for the future of food packaging. As awareness of sustainability continues to rise, bioplastics will likely play an integral role in the future of packaging. By focusing on innovation, education, and sustainability, the food packaging industry can capitalize on these trends and contribute to a more safe and sustainable packaging landscape. Further research and compliance with safety regulations are essential to ensure consumer protection with the use of these chemical additives in the production of food packaging plastics. Regulatory bodies can also be leveraged to ensure that bioplastic additives are used within safe limits to prevent any adverse or unwanted effects on consumers.

REFERENCES

Abang, S., Wong, F., Sarbatly, R., Sariau, J., Baini, R., & Besar, N. A. (2023). Bioplastic classifications and innovations in antibacterial, antifungal, and antioxidant applications. *Journal of Bioresources and Bioproducts*, 8(4), 361–387. https://doi.org/10.1016/j.jobab.2023.06.005

Abe, M. M., Martins, J. R., Sanvezzo, P. B., Macedo, J. V., Branciforti, M. C., Halley, P., Botaro, V. R., & Brienzo, M. (2021). Advantages and disadvantages of bioplastics production from starch and lignocellulosic components. *Polymers*, 13(15), 2484. https://doi.org/10.3390/polym13152484

Abu Ubaidah, A. S., Mohamad Puad, N. I., Hamzah, F., Azmi, A. S., & Ahmad Nor, Y. (2024). Synthesis of bioplastics and the effects of additives on the mechanical, thermal and biodegradable properties. *Polymers from Renewable Resources*, 15(4), 430–490. https://doi.org/10.1177/20412479241292148

Arifa, S., Al-Zaqri, N., Tahir, A., & Alsalme, A. (2021). Synthesis and characterization of starch based bioplatics using varying plant-based ingredients, plasticizers and natural fillers. *Saudi Journal of Biological Sciences*, 28, 1739–1749.

Armynah, B., Anugrahwidya, R., & Tahir, D. (2022). Composite cassava starch/chitosan/Pineapple Leaf Fibre (PALF)/Zinc Oxide (ZnO): Bioplastics with high mechanical properties and faster degradation in soil and seawater. *International Journal of Biological Macromolecules*, 213, 814–823. https://doi.org/10.1016/j.ijbiomac.2022.06.038

Behera, L., Mohanta, M., & Thirugnanam, A. (2022). Intensification of yam-starch based biodegradable bioplastic film with bentonite for food packaging application. *Environmental Technology & Innovation*, 25, 102180.

Caine, M. (2024). What Effects Do Ecolabel Lubricants Have on the Food Packaging Sector? Readers Country PR. Retrieved from https://readerscountry.com

Charles, A. L., Motsa, N., & Abdillah, A. A. (2022). A comprehensive characterization of biodegradable edible films based on potato peel starch plasticized with glycerol. *Polymers*, 14(17), 3462. https://doi.org/10.3390/polym14173462

Costache, M.-A., Stancu, M., & Tudoroiu, E.-E., Ioniţă-Mîndrican, C.-B., Ziani, K., Negrei, C., & Magdalena, M. (2022). Food packaging and effects on consumer health. *Farmacist.ro*, 1(204), 30. https://doi.org/10.26416/Farm.204.1.2022.6165

Cruz, R. M. S., Krauter, V., Krauter, S., Agriopoulou, S., Weinrich, R., Herbes, C., Scholten, P. B. V., Uysal-Unalan, I., Sogut, E., Kopacic, S., Lahti, J., Rutkaite, R., & Varzakas, T. (2022). Bioplastics for food packaging: Environmental impact, trends and regulatory aspects. *Foods*, *11*(19), 1–39. https://doi.org/10.3390/foods11193087

Dintcheva, N. T., & La Mantia, F. P. (2007). Durability of a starch-based biodegradable polymer. *Polymer Degradation and Stability*, 92(4), 630–634. https://doi.org/10.1016/j.polymdegradstab.2007.01.003

European Commission (2004). Regulation (EC) No 1935/2004 of the European Parliament and of the Council of 27 October 2004 on materials and articles intended to come into contact with food and repealing Directives 80/590/EEC and 89/109/EEC.

Fellows, P. (2000). *Food Processing Technology*. Second Edition. CRS Press. Woodhead Publishing Ltd.

Food and Agriculture Organization of the United Nations (FAO). (2003). Basic texts on food hygiene – Third edition. Codex Alimentarius Commission." https://openknowledge.fao.org/server/api/core/bitstreams/a325d7a9-f624-4f7b-ac24-1f31b7ce9e45/content/y5307e00.htm

Geueke, B. (2014). Bioplastics as food contact materials. *Food Packaging Forum*, April, 1–8. https://doi.org/10.5281/zenodo.33517

Ghasemlou, M., Barrow, C. J., & Adhikari, B. (2024). The future of bioplastics in food packaging: An industrial perspective. *Food Packaging and Shelf Life*, 43, 101279. https://doi.org/10.1016/j.fpsl.2024.101279

Gupta, R. K., Pipliya, S., Karunanithi, S., Eswaran, M. G., Kumar, S., Mandliya, S., Srivastav, P. P., Suthar, T., Shaikh, A. M., Harsanyi, E., & Kovacs, B. (2024). Migration of chemical compounds from packaging materials into packaged foods: Interaction, mechanism, assessment, and regulations. *Foods*, 13, 1–32.

Hahladakis, J. N., Velis, C. A., Weber, R., Iacovidou, E., & Purnell, P. (2018). An overview of chemical additives present in plastics: Migration, release, fate and environmental impact during their use, disposal and recycling. *Journal of Hazardous Materials*, 344, 179–199. https://doi.org/10.1016/j.jhazmat.2017.10.014

Hurst, W. C., Tybor, P. T., Reynolds, A. E., and Schuller, G. A. (2010). Quality control: A model program for the food industry. *University of Georgia Cooperative Extension Bulletin*, 997, 1–15. Retrieved from https://athenaeum.libs.uga.edu/bitstream/handle/10724/12251/B997.pdf

Keränen, O., Komulainen, H., Lehtimäki, T., & Ulkuniemi, P. (2021). Restructuring existing value networks to diffuse sustainable innovations in food packaging. *Industrial Marketing Management*, 93, 509–519.

Kudłak, B., Jatkowska, N., Kubica, P., Yotova, G., & Tsakovski, S. (2019). Influence of storage time and temperature on the toxicity, endocrine potential, and migration of epoxy resin precursors in extracts of food packaging materials. *Molecules*, 24(23), 4396. https://doi.org/10.3390/molecules24234396

Li, Y., Liu, C., Yang, H., He, W., Li, B., Zhu, X., Liu, S., Jia, S., Li, R., & Tang, K. H. D. (2024). Leaching of chemicals from microplastics: A review of chemical types, leaching mechanisms and influencing factors. *Science of the Total Environment*, 906, 167666, https://doi.org/10.1016/j.scitotenv.2023.167666

López, O. V., Lecot, C. J., Zaritzky, N. E., & García, M. A. (2011). Biodegradable packages development from starch-based heat sealable films. *Journal of Food Engineering*, 105, 254–263.

Merino, D., Bertolacci, L., Paul, U. C., Simonutti, R., & Athanassiou, A. (2021). Avocado peels and seeds: Processing strategies for the development of highly antioxidant bioplastic films. *ACS Applied Materials & Interfaces*, 13(32), 38688–38699. https://doi.org/10.1021/acsami.1c09433

Muzeza, C., Ngole-Jeme, V., & Msagati, T. A. M. (2023). The mechanisms of plastic food-packaging monomers' migration into food matrix and the implications on human health. *Foods*, *12*(18). https://doi.org/10.3390/foods12183364

Navasingh, R. J. H., Gurunathan, M. K., Nikolova, M. P., & Królczyk, J. B. (2023). Sustainable bioplastics for food packaging produced from renewable natural sources. *Polymers*, 15(18), 3760. https://doi.org/10.3390/polym15183760

Negrete-Bolagay, D., & Guerrero, V. H. (2024). Opportunities and challenges in the application of bioplastics: Perspectives from formulation, processing, and performance. *Polymers*, 16(18), 2561. https://doi.org/10.3390/polym16182561

Oluwasina, O. O., Akinyele, B. P., Olusegun, S. J., Oluwasina, O. O., & Mohallem, N. D. S. (2021). Evaluation of the effects of additives on the properties of starch-based bioplastic film. *SN Applied Sciences*, 3(4), 421. https://doi.org/10.1007/s42452-021-04433-7

Oluwasina, O. O., & Awonyemi, I. O. (2021). Citrus peel extract starch-based bioplastic: Effect of extract concentration on packed fish and bioplastic properties. *Journal of Polymers and the Environment*, 29, 1706–1716.

Oluwasina, O. O., Olaleye, F. K., Olusegun, S. J., Oluwasina, O. O., & Mohallem, N. D. S. (2019). Influence of oxidized starch on physicomechanical, thermal properties, and atomic force micrographs of cassava starch bioplastic film. *International Journal of Biological Macromolecules*, 135, 282–293. https://doi.org/10.1016/j.ijbiomac.2019.05.150

Quilez-Molina, A. I., Mazzon, G., Athanassiou, A., & Perotto, G. (2022). A novel approach to fabricate edible and heat sealable bio-based films from vegetable biomass rich in pectin. *Materials Today Communications*, 32, 103871. https://doi.org/10.1016/j.mtcomm.2022.103871

Risch, S. J. (2009). Food packaging history and innovations. *Journal of Agricultural & Food Chemistry*, 57(18), 8089–8092.

Romani, S., Tappi, S., Balestra, F., Rodriguez Estrada, M. T., Siracusa, V., Rocculi, P., & Dalla Rosa, M. (2015). Effect of different new packaging materials on biscuit quality during accelerated storage. *Journal of the Science of Food and Agriculture*, 95(8), 1736–1746. https://doi.org/10.1002/jsfa.6888

Romero, V., Borneo, R., Passalacqua, N., & Aguirre, A. (2016). Biodegradable films obtained from triticale (x *Triticosecale Wittmack*) flour activated with natamycin for cheese packaging. *Food Packaging and Shelf Life*, 10, 54–59. https://doi.org/10.1016/j.fpsl.2016.09.003

Sanyang, M. L., Sapuan, S., Jawaid, M., Ishak, M., & Sahari, J. (2015). Effect of plasticizer type and concentration on tensile, thermal and barrier properties of biodegradable films based on sugar palm (*Arenga pinnata*) starch. *Polymers*, 7, 1106–1124.

Sarkar, S., & Aparna, K. (2020). Food Packaging and Storage. https://doi.org/10.22271/ed.book.959

Shah, Y. A., Bhatia, S., Al-Harrasi, A., & Khan, T. S. (2024). Advancements in the biopolymer films for food packaging applications: A short review. *Biotechnology for Sustainable Materials*, 1(1), 1–13. https://doi.org/10.1186/s44316-024-00002-1

Shaikh, S., Yaqoob, M., & Aggarwal, P. (2021). An overview of biodegradable packaging in food industry. *Current Research in Food Science*, 4(July), 503–520. https://doi.org/10.1016/j.crfs.2021.07.005

Shlush, E., & Davidovich-Pinhas, M. (2022). Bioplastics for food packaging. *Trends in Food Science & Technology*, 125, 66–80. https://doi.org/10.1016/j.tifs.2022.04.026

Suryanegara, L., Fatriasari, W., Zulfiana, D., Anita, S. H., Masruchin, N., Gutari, S., & Kemala, T. (2021). Novel antimicrobial bioplastic based on PLA-chitosan by addition of TiO_2 and ZnO. *Journal of Environmental Health Science and Engineering*, 19(1), 415–425. https://doi.org/10.1007/s40201-021-00614-z

Tanner, D. (2016). Food Quality, Storage, and Transport. In *Reference Module in Food Science* (pp. 1–5). Elsevier.

Tyagi, V., & Bhattacharya, B. (2019). Role of plasticizers in bioplastics. *MOJ Food Processing & Technology*, 7, 128–130.

U.S. FDA (2018). Determining the Regulatory Status of a Food Ingredient. www.fda.gov/food/food-ingredients-packaging/determining-regulatory-status-food-ingredient

U.S. FDA (2024). Understanding How the FDA Regulates Food Additives and GRAS Ingredients. www.fda.gov

Vaclavik, V. A., and Christian, E. W. (2008). *Essentials of Food Science*. Third Edition. Springer.

Vaclavik, V. A., Christian, E. W. & Campbell, T. (2021). Chapter 18: Food Packaging. In *Essentials of Food Science*, Food Science Text Series. Fifth Edition. Springer Nature, (pp. 367–373). https://doi.org/10.1007/978-3-030-46814-9_18

Valdés, A., Mellinas, A. C., Ramos, M., Garrigós, M. C., & Jiménez, A. (2014). Natural additives and agricultural wastes in biopolymer formulations for food packaging. *Frontiers in Chemistry*, 2. https://doi.org/10.3389/fchem.2014.00006

Yang, Y., Wei, L., and Pei, J. (2019). Application of Bayesian modelling to assess food quality and safety status and identify risky food in China market. *Food Control*, 100, 111–116.

16 Plastic Waste Incineration and Additive-Derived Pollution

Feyisayo Elizabeth Akinjoko, Adunbi John Olarewaju, and Olawoore Ayooluwa Samuel

16.1 INTRODUCTION

Plastics have revolutionized modern society influencing our ecosystem and becoming indispensable both in homes and in industries finding use in packaging, healthcare, and technology due to their versatility, durability, and cost-effectiveness. Their benefits, however, are marred by their environmental implications. As David Attenborough stated, "*The plastic problem has become a calamity we cannot ignore*." While plastics are essential to contemporary life, their widespread use has led to a mounting crisis: plastic waste. Since the advent of synthetic polymers in the early 20th century, plastic production has grown exponentially. By the 1950s, plastics became an integral part of everyday life, and today, annual global production exceeds 400 million metric tons (Geyer et al., 2017). Their properties—particularly their resistance to degradation—contribute to their persistence in the environment, creating significant waste management challenges.

Globally, the issue of plastic waste is alarming, of the 8.3 billion metric tons of plastics produced since the mid-20th century, approximately 79% accumulates in landfills or the natural environment, and only 9% has been recycled (Geyer et al., 2017). The remaining plastics are incinerated, contributing to emissions of harmful pollutants. These waste plastics, ranging from single-use items to industrial-grade materials, often litter landscapes, clog waterways, and disrupt ecosystems. Additionally, microplastics—tiny fragments formed by the breakdown of larger plastics—have infiltrated soil, water bodies, and even the food chain, raising concerns about their impact on human and environmental health.

Alternatively, efforts to address plastic waste have led to the development of various waste management strategies. Plastic waste management methods can be categorized into traditional and improved approaches. The traditional methods include landfilling, open burning, and mechanical recycling. Improved Methods include modern incinerators, chemical recycling, and circular economy models.

Landfilling, the most common method is cost-effective but unsustainable. Landfills require extensive land, often emit methane—a potent greenhouse gas—and pose risks of soil and groundwater contamination due to leachate (Denison, 1997; Jambeck et al., 2015). In developing regions, open burning of plastics is prevalent, releasing toxic pollutants like dioxins and furans into the atmosphere (Nagy & Kuti, 2016). Recycling, while environmentally friendly, is limited by challenges such as contamination, high processing costs, and difficulties in handling multi-layered or degraded plastics (Hopewell et al., 2009). Composting applies primarily to biodegradable polymers and does little to address the bulk of synthetic plastics.

In recent years, incineration has emerged as a viable option for managing non-recyclable plastic waste. This process reduces waste volume by up to 90% while generating energy in the form of heat

DOI: 10.1201/9781003602828-20

or electricity (Arena, 2012). Countries such as Sweden and Denmark have successfully integrated waste-to-energy (WTE) plants into their waste management systems, highlighting the potential benefits of incineration. However, this method is not without its drawbacks. Incineration of plastics, particularly those containing chemical additives, such as plasticizers, brominated flame retardants (BFRs), and stabilizers, become hazardous during combustion, releasing harmful substances such as dioxins, furans, volatile organic compounds (VOCs), and heavy metals. These emissions pose severe risks to human health and the environment, undermining the potential benefits of energy recovery. For example, the incineration of polyvinyl chloride (PVC) plastics produces dioxins and furans—persistent organic pollutants (POPs) known for their toxicity and ability to bioaccumulate in the food chain (Nagy & Kuti, 2016). Heavy metals such as lead, cadmium, and mercury, commonly found in plastic pigments and stabilizers, are also released into the air during burning, contaminating nearby soil and water resources.

The health impacts of these emissions are profound. Dioxins and furans, for instance, are linked to cancers, hormonal imbalances, and immune suppression. Heavy metals like cadmium and lead contribute to respiratory illnesses, neurological disorders, and developmental delays in children (Teuten et al., 2009). Furthermore, airborne particulate matter (PM) from incineration exacerbates cardiovascular and pulmonary diseases, particularly in vulnerable populations living near incineration facilities. Environmental consequences are equally concerning. Residual ash from incineration often contains concentrated levels of heavy metals, which require secure disposal to prevent leaching into groundwater. Additionally, the release of acid gases such as hydrogen chloride (HCl) and hydrogen bromide (HBr) contributes to air pollution and acid rain, further degrading ecosystems.

These challenges signaled the importance of developing sustainable and integrated waste management solutions. While incineration provides an immediate means of reducing waste and recovering energy, its long-term viability depends on mitigating its environmental and health impacts. Advanced technologies, such as high-temperature incinerators equipped with scrubbers and catalytic converters, have shown promise in reducing emissions. Similarly, waste segregation at the source can prevent hazardous plastics from entering incineration streams, thereby minimizing the release of toxins.

This study looked into the specific challenges of plastic waste incineration, focusing on the additive-derived air pollutants released during this process and their implications for health and the environment. By examining the sources, mechanisms, and impacts of these emissions, it aims to highlight the urgent need for technological advancements and policy interventions to mitigate their effects.

16.1.1 Importance of Plastic Waste Incineration as a Disposal Method

Incineration is a thermal treatment process used to dispose of waste by burning it at high temperatures, producing ash, flue gases, and energy. Denison (1997) defines incineration as the controlled combustion of waste materials aimed at reducing volume and recovering energy. According to Nagy and Kuti (2016), incineration is described as the decomposition of plastics into small molecules during combustion, with pyrolysis and charring as integral steps. PlasticsEurope (2022) highlights its application in energy recovery, though it is often criticized for undermining recycling efforts. These different opinions on incineration underscore its dual role as a waste reduction and energy recovery tool.

As global waste generation continues to increase, especially in urban areas, incineration presents a viable solution to the mounting waste management crisis in regions with limited landfill capacity (Geyer et al., 2017; Nagy & Kuti, 2016). Plastic waste incineration plays a significant role in modern waste management systems by addressing challenges associated with waste volume, energy recovery, and the treatment of non-recyclable materials.

1. Reduction of Waste Volume: One of the most notable benefits of incineration is its ability to reduce the volume of waste significantly. Incineration can decrease waste volume by up to 90%, making it especially beneficial in urban areas with limited landfill space (Geyer et al., 2017). The process produces residual ash, which occupies a fraction of the space required for untreated waste. Additionally, this ash can often be repurposed in construction materials, further contributing to sustainability efforts (Arena, 2012).
2. Energy Recovery and Resource Optimization: Modern WTE facilities capitalize on incineration's thermal process to generate electricity and heat. This dual-purpose approach reduces dependence on fossil fuels while addressing waste disposal needs. For instance, Sweden and Denmark successfully incorporated incineration into their energy grids, using waste-derived energy to supply power and heat to urban areas (Plastic Pollution Coalition, n.d.).
3. Compared to landfills, which emit methane, incineration primarily releases carbon dioxide—a less potent greenhouse gas. When combined with advanced pollution control technologies, such as scrubbers and filters, the environmental impact of incineration can be minimized (Zheng & Suh, 2019). Furthermore, materials like metals that survive the incineration process can be recovered and recycled, enhancing resource efficiency.
4. Treatment of Non-Recyclable Plastics: A significant portion of plastic waste, including contaminated and multi-layered plastics, is unsuitable for recycling due to technical or economic constraints. Incineration offers an effective alternative for managing these materials, converting them into energy rather than allowing them to accumulate in landfills. Nagy and Kuti (2016) highlight the role of incineration in handling plastics with high additive content, which are challenging to process through conventional recycling methods.
5. Mitigation of Open Burning: In regions with inadequate waste management systems, open burning remains a common disposal method, despite its severe health and environmental consequences. This practice releases dioxins, furans, and PM into the atmosphere, contributing to air pollution and respiratory illnesses. Incineration, when equipped with modern emission control technologies, provides a safer alternative, significantly reducing the release of harmful pollutants (Wiedinmyer et al., 2014).
6. Support for Urban Waste Management: Urbanization has led to a dramatic increase in waste generation, putting immense pressure on existing waste management systems. Incineration addresses this challenge by integrating seamlessly into urban waste disposal networks. High-capacity incineration plants can process large volumes of municipal waste, alleviating the burden on landfills. Additionally, the energy generated from these plants can be redirected to support urban infrastructure, such as powering public facilities and transportation systems (Jambeck et al., 2015).
7. Economic and Social Benefits: Despite the high initial investment required for incineration facilities, the long-term economic advantages are significant. Revenue generated from energy production and material recovery can offset operational costs. Moreover, these facilities create employment opportunities in their construction, operation, and maintenance. In regions transitioning from informal waste disposal methods, incineration facilities contribute to improved public health outcomes by reducing exposure to hazardous pollutants (Wilson et al., 2006).

16.2 ADDITIVES EMISSION OF AIRBORNE INCINERATION

The incineration of plastics releases an extensive variety of chemical compounds, a lot of which might be derived from the components used at some point in the manufacturing of plastic materials. These emissions pose massive environmental and fitness worries because of their toxicity and endurance within the surroundings. Understanding the sources, types, and influences of those emissions is crucial for growing techniques to mitigate their outcomes.

Plastic components are included at some point in production to improve their properties along with flexibility, durability, and resistance to warmth or UV radiation. Common components encompass plasticizers (e.g., phthalates), flame retardants (e.g., brominated compounds), stabilizers, pigments, and fillers (Lithner et al., 2011). While those components enhance the capability of plastics, they frequently emerge as environmental dangers while subjected to excessive temperatures at some point of incineration. PVC and polyethylene are mainly noteworthy as they include chlorine and different halogens, that may produce incredibly poisonous by-merchandise while burned (Teuten et al., 2009). Additionally, heavy metal-primarily based totally pigments and stabilizers, along with the ones containing cadmium or lead, in addition make contributions to unsafe emissions.

16.2.1 Types of Components Emission at Some Point of Plastic Incineration

16.2.1.1 Dioxins and Furans

Chlorinated components in plastics are the main participants in the formation of dioxins and furans at some point of incineration. They are classified as polychlorinated dibenzo-p-dioxins (PCDDs) and polychlorinated dibenzofurans (PCDFs), which are environmental contaminants of significant concern due to their toxicological profiles and persistence. Both compounds have two aromatic rings connected by oxygen atoms, with chlorine substitutions at various positions, usually at positions 2, 3, 7, and 8. These positions earn them their highest toxicity. Dioxins and Furans compounds are continual natural pollution (POPs) recognized for excessive toxicity. According to the World Health Organization (WHO), their chemical stability and lipophilicity contribute to their bioaccumulation in fatty tissues, making them long-lasting in biological and environmental systems (WHO, 2023; Lithner et al., 2011). They are labeled as carcinogenic and may result in intense endocrine and immune gadget disruptions even at low concentrations.

According to Nagy and Kuti (2016), dioxins and furans are shaped via incomplete combustion while chlorine-containing plastics, along with PVC, are burned. Among the 419 dioxin-like compounds identified, TCDD (2,3,7,8-tetrachlorodibenzo-p-dioxin) is considered the most toxic. Its flat aromatic configuration facilitates high-affinity binding to the aryl hydrocarbon receptor (AhR), a cytoplasmic protein involved in xenobiotic metabolism and gene regulation. Studies have shown that this interaction is responsible for the cascade of toxic effects observed in exposed organisms (WHO, 2023; Teuten et al., 2009).

16.2.1.1.1 Key Properties of Dioxins and Furans Include

i. High Toxicity: The WHO acknowledges dioxins as a number of the maximum poisonous chemical substances to people and wildlife. Even at low concentrations, they could cause extreme fitness issues.
ii. Environmental Persistence: Dioxins and furans face up to degradation, letting them stay within the surroundings for decades.
iii. Bioaccumulation: These compounds pay attention to better trophic degrees, posing sizeable dangers to people and different apex predators.

In biological activities, the AhR is a ligand-activated transcription factor that mediates the toxicological effects of dioxins. When exposed, TCDD reactively binds to AhR with high affinity, to form a ligand-receptor complex that moves to the nucleus. This complex formed dimerizes with the aryl hydrocarbon receptor nuclear translocator (ARNT) and forms a cleavage to xenobiotic response elements (XREs) in the DNA. The result of this action is altered expression of several genes, including those involved in xenobiotic metabolism (e.g., cytochrome P450 enzymes), cell proliferation, and apoptosis (WHO, 2002; FAO/WHO, 2006).

Recent studies have demonstrated that prolonged AhR activation is responsible for induced oxidative stress, inflammation, and disruption of normal cellular signaling, leading to pathological conditions. Some of these pathological conditions include immunosuppression, endocrine disruption, and carcinogenesis. Another study has shown that dioxins are not directly genotoxic but promote tumorigenesis through receptor-mediated pathways, including increased cell proliferation and reduced apoptosis (IARC, 2012; Nagy & Kuti, 2016).

Dioxins and furans have the ability to disrupt cellular homeostasis. Their functional groups, which are the chlorinated aromatic rings, confer chemical stability and enable strong binding to AhR.

Those following the Seveso disaster have linked epidemiological studies confirming that TCDD exposure increases risks of cancers, including non-Hodgkin lymphoma, soft tissue sarcoma, and lung cancer. This is because the reaction of chlorophenols under oxidative conditions can form TCDD.

$$2C_6H_4ClOH + O_2 \rightarrow CH_4Cl_4O_2 + H_2O$$

According to WHO, these effects are dose-dependent and reflect the long half-life of dioxins in biological systems, estimated to be 7–11 years (WHO, 2023; Geyer et al., 2017).

16.2.1.1.2 Health Effects of Airborne Dioxins and Furans

The emissions from plastic incineration have profound environmental consequences. Persistent pollution along with dioxins and furans gather within side the meal chain, affecting the natural world and humans. For example, Geyer et al. (2017) highlighted that dioxins launched into the ecosystem can settle onto agricultural lands, contaminating plants and livestock.

The incineration of plastic waste outcomes results in the launch of dioxins and furans, which might be many of the maximum poisonous materials recognized by science. These compounds, labeled as continual natural pollution (POPs), pose massive dangers to human fitness and the surroundings because of their toxicity, lengthy half-life, and capacity to bioaccumulate in meal chains.

Humans are exposed to dioxins and furans primarily through:

1. Inhalation of Contaminated Air: Incineration centers can launch airborne dioxins and furans into the atmosphere, wherein they may be inhaled with the aid of using close by populations.

   These compounds are released into the atmosphere during the combustion of organic materials containing chlorine, such as plastics and other municipal solid waste. The high-temperature conditions of incineration promote the formation of PCDDs and PCDFs, which are then emitted as fine PM or gaseous pollutants (Nagy & Kuti, 2016).

   Airborne dioxins are highly stable and can remain suspended in the atmosphere for extended periods. Wind currents disperse them over large distances, where they can settle on soil, water, and vegetation. Studies have demonstrated elevated concentrations of airborne dioxins near waste incineration plants, with proximity to such facilities linked to increased respiratory exposure and health risks. For instance, residents near incinerators in Japan and Taiwan showed higher serum dioxin levels compared to distant populations, correlating with increased rates of lung and liver diseases (WHO, 2023).

   The inhalation of dioxins is particularly harmful because these compounds bind strongly to PM (PM2.5 and PM10), which can penetrate deep into the lungs and enter the bloodstream. Once in the body, dioxins activate the AhR, leading to toxic effects such as immune suppression, endocrine disruption, and cancer. Studies on workers exposed to high levels of dioxins in industrial accidents or incineration facilities have reported an increased risk of respiratory illnesses and cancers, further illustrating the dangers of this exposure pathway (Teuten et al., 2009; Nagy & Kuti, 2016).

2. Dietary Intake: Over 90% of human publicity happens via the intake of infected animal merchandise which includes meat, fish, and dairy. These meals gather dioxins from environmental deposition onto plants and in aquatic ecosystems (Teuten et al., 2009).

   Dietary intake accounts for over 90% of human exposure to dioxins and furans, primarily through the consumption of contaminated animal products such as meat, fish, dairy, and eggs. These compounds bioaccumulate in the fatty tissues of animals through environmental deposition on crops, soil, and aquatic ecosystems. Dioxins, being lipophilic, persist in animal fat and magnify through the food chain, making humans at the top of the chain highly susceptible to their toxic effects (Lithner et al., 2011).

   Marine ecosystems are particularly vulnerable to dioxin contamination. Industrial runoff and atmospheric deposition introduce dioxins into water bodies, where they accumulate in sediments and are ingested by fish and shellfish. Studies in Baltic Sea fisheries have revealed high levels of dioxins in fatty fish like salmon and herring, posing risks to consumers in the region (Teuten et al., 2009). Similarly, contamination of agricultural products occurs when dioxins settle on crops or enter animal feed, transferring the toxins to livestock.

   Populations with diets rich in animal fats are disproportionately exposed to dioxins. For instance, rural communities relying on locally sourced food have shown higher serum dioxin levels compared to urban populations with diverse diets. The Seveso disaster in Italy (1976) highlighted the dietary risks of dioxins, where milk and dairy products from contaminated farmlands contributed significantly to local exposure. Chronic dietary intake of dioxins has been linked to cancers, reproductive disorders, and developmental delays, emphasizing the need for stringent monitoring and food safety regulations (WHO, 2023).

3. Dermal Contact: Direct touch with infected soil or ash from incineration centers can introduce those materials into the body.

   Dermal contact is another significant exposure pathway, particularly in occupational settings or areas with heavy environmental contamination. Direct skin contact with contaminated soil, ash, or water introduces dioxins and furans into the body. While dermal absorption rates are lower compared to inhalation and dietary intake, repeated or prolonged exposure can result in substantial accumulation, especially in industrial workers or those involved in waste management (Arena, 2012).

   The lipophilic nature of dioxins allows them to penetrate the skin barrier and accumulate in subcutaneous fat. For example, individuals handling fly ash or bottom ash from incinerators without proper protective equipment are at risk. Studies have shown that ash from WTE plants contains significant concentrations of PCDDs and PCDFs, which can leach onto the skin during handling.

   Environmental contamination also exacerbates dermal exposure risks. Soil and water in areas near incinerators or industrial sites can retain dioxins for decades, prolonging exposure risks for local populations. For instance, residents near contaminated sites in Vietnam, resulting from the use of dioxin-laden herbicides like Agent Orange, continue to exhibit high dioxin levels due to persistent environmental contamination and direct skin contact during agricultural activities (WHO, 2002).

   Additionally, dermal exposure can synergize with other pathways, such as inhalation or dietary intake, amplifying the overall toxic burden. Chronic exposure through the skin has been linked to conditions such as chloracne, a hallmark symptom of acute dioxin toxicity, as well as systemic effects like hormonal imbalances and immune suppression. Proper safety protocols and remediation measures are essential to minimize dermal contact risks, particularly for at-risk populations and workers handling dioxin-contaminated materials.

Health Impacts

The fitness consequences of dioxins and furans are extreme and multi-faceted, affecting a couple of organ systems. The maximum super effects include the following:

Carcinogenicity

The International Agency for Research on Cancer (IARC) classifies dioxins, especially 2,3,7,8-tetrachlorodibenzo-p-dioxin (TCDD), as a Group 1 carcinogen. Long-term period publicity will increase the hazard of cancers, inclusive of tender tissue sarcoma, non-Hodgkin lymphoma, and lung cancer (Lithner et al., 2011).

Endocrine Disruption

Dioxins intrude with hormone regulation, mimicking or blocking off the motion of herbal hormones. This disruption can result in developmental delays, reproductive disorders, and changes in thyroid function. Women exposed to excessive degrees of dioxins frequently revel in decreased fertility and menstrual irregularities (Nagy & Kuti, 2016).

Immune System Suppression

Prolonged publicity of dioxins suppresses immune responses, growing susceptibility to infections and decreasing the efficacy of vaccines. Studies suggest that even low degrees of publicity can result in a measurable decline in immune function, especially in children (Teuten et al., 2009).

Developmental and Reproductive Effects

Prenatal and early-lifestyle publicity to dioxins is connected to developmental problems, consisting of cognitive impairments, behavioral changes, and boom delays. High maternal publicity can bring about delivery defects and decreased delivery weights. Additionally, dioxins disrupt reproductive fitness by changing sperm excellent in men and increasing the danger of endometriosis in females (Lithner et al., 2011).

Metabolic Disorders

Evidence indicates a correlation between dioxin publicity and metabolic problems which include diabetes and obesity. Dioxins can disrupt lipid metabolism and insulin sensitivity, contributing to those conditions.

Populations at Risk

Certain corporations are disproportionately suffering from dioxin publicity because of their proximity to incineration facilities, occupation, or nutritional habits. Significantly, groups counting on conventional diets wealthy in fish and marine mammals are specifically susceptible to bioaccumulated dioxins.

16.2.1.1 Volatile Organic Compounds (VOCs)

The thermal degradation of plastic components can launch a whole lot of VOCs, together with benzene, toluene, and styrene. These compounds contribute to ground-stage ozone formation and respiration problems. Wiedinmyer et al. (2014) mentioned that VOCs launched at some point of open burning are a massive contributor to city air pollution.

16.2.1.3 Heavy Metals

Additives containing heavy metals, along with cadmium, lead, and mercury, vaporize at some point of incineration, contaminating air, soil, and water. Cadmium and lead are mainly harmful, with lengthy time periods of publicity connected to neurological harm, kidney dysfunction, and developmental troubles in children (Lithner et al., 2011).

16.2.1.3.11 Heavy Metals Released during Plastic Waste Incineration

Plastic waste incineration effects in the launch of several poisonous substances, consisting of heavy metals which might be through-merchandise of incomplete combustion or thermal degradation of plastics and their additives. Heavy Metals Released include the following:

Lead (Pb): Found in stabilizers and pigments. Lead interferes with enzyme features through cleavage to sulfhydryl groups, changing critical metals which include calcium and zinc in organic systems. Their danger includes disruptions to biosynthesis, main to anemia, and impairs neuron improvement in children.

Cadmium (Cd): Used in colorants and stabilizers. Cadmium induces oxidative pressure by using productive reactive oxygen species (ROS) and depleting antioxidant defenses. It accumulates within the kidneys, inflicting renal dysfunctionality and skeletal deformities. And it has the capacity to disrupt DNA (Nagy & Kuti, 2016).

Mercury (Hg): Present in positive styles of plastic waste, mercury damages the valuable anxious machine and bioaccumulates in aquatic ecosystems. Mercury's excessive volatility makes it a great environmental pollutant. It crosses the blood-mind barrier as methylmercury, interfering with neuronal features and main to cognitive and developmental disorders.

Chromium (Cr): Often utilized in pigments, hexavalent chromium (Cr(VI)) is exceedingly poisonous because of its capacity to penetrate mobile membranes and generate ROS, inflicting DNA harm, inflicting breathing, pores and skin problems and being carcinogenic.

Arsenic (As): Used in a few flame retardants, arsenic contributes to most cancers and cardiovascular problems upon publicity. Arsenic disrupts cell breathing with the aid of using inhibiting enzymes concerned with ATP production, main to multi-organ toxicity.

Nickel (Ni): Released all through combustion, nickel can cause pores and skin irritation, breathing problems, and environmental toxicity.

Zinc (Zn): A usual place stabilizer in plastics, immoderate zinc launch can disrupt aquatic ecosystems.

Copper (Cu): Frequently related to digital waste plastics, copper can result in poisonous ash deposits affecting soil and water.

16.2.2 Acid Gases

Burning plastics, in particular people with halogenated components, releases acid gases along with HCl and HBr. These gases are incredibly corrosive and contribution to acid rain that may harm ecosystems and infrastructure (Teuten et al., 2009). Acid gases contribute to the acidification of soils and water bodies, in addition to degrading biodiversity and agricultural productivity (Lithner et al., 2011).

16.2.3 Particulate Matter (PM)

Incineration produces pleasant particulate remains containing soot, unburned carbon, and hint metals. This debris penetrate deep into the lungs and bloodstream, inflicting cardiovascular and respiration illnesses. PM from plastic incineration frequently consists of adsorbed poisonous chemicals, increasing its fitness dangers (Arena, 2012).

16.3 EARLY EXPOSURE TO PLASTIC ADDITIVES AND DEVELOPMENTAL DELAYS, ADHD, OR AUTISM SPECTRUM DISORDERS IN CHILDREN

Obviously, plastic additives, extensively employed which are supposedly expected to enhance the performance and robustness of plastic materials, have been found to cause serious health problems, especially for children. Developmental delays, attention-deficit/hyperactivity disorder (ADHD), and/or autism spectrum disorders (ASDs) have been linked to exposures to certain additives, e.g., phthalates, bisphenol A (BPA), and BFRs. These results are due to the endocrine-disrupting and neurotoxic actions of these compounds which disrupt normal developmental processes.

The chemical additive is a serious concern for families and children to avoid due, to exposure to these additives by children through multiple pathways:

1. Ingestion: Additives can be ingested via food packages with plastic, baby bottles, and containers. In the case of BPA, it is widely used to coat cans and plastic containers for food (where it "leaks" out of the containers to food) as well as liner of plastic containers for foods (where it mainly comes out of the containers to the food (Muncke, 2011)).
2. Inhalation: VOCs released from plastic materials, such as flooring or furniture, contribute to indoor air pollution. Also because of the highly susceptible position of children living in most indoor activities, they have full risk of being exposed to these chemicals.

   When inhaled, VOCs and other chemicals such as BFRs and phthalates can lead to toxic effects due to their interaction with the respiratory system, brain, and endocrine functions that disrupt physiological processes through oxidative stress and inflammation. Neurotoxic BFRs, e.g., disrupt synaptic plasticity and neuronal communication. However, exposure to such agents during pregnancy has been demonstrated to result in cognitive impairments, behavioral disorders, and increased risk of other disorders, including ADHD and ASD (Nagy Kuti, 2016).

   Inhaled VOCs and BFRs have been associated with the degree of permanent developmental and health problems. In so doing these compounds might be shown to have negative effects on mitochondrial metabolism as well as to induce the formation of ROS. As a consequence, ROS damages neurons and glial cells that are critical for normal brain maturation (Muncke, 2011). Findings of research outcomes support that relationships between indoor air toxic exposures and the loss of cognitive function have been documented among children, such as in learning, memory, and social functioning.
3. Dermal Absorption: Dermal absorption of plastic toys, apparel, and personal care articles with additive substances (i.e., phthalates, PBDD/Fs) might take place from handling these articles or dermal contact with these items. In addition, one questionnaire concluded that children's health is at risk when PBDD/Fs are enclosed in the plastic elements of toys. Based on an estimate of toxicity equivalent quantity (TEQ) daily intake from PBDD/Fs-contaminated plastic toys due to child-mouthing behavior (IPEN, 2020).
4. Prenatal Exposure: Additives can cross the placental barrier and deliver exposure to potentially toxic agents during critical windows of brain and organogenesis (Teuten et al., 2009).

16.3.1 MECHANISMS OF TOXICITY

16.3.1.1 Endocrine Disruption

In children, plastic additives, like BPA and phthalates, act as/interfere with natural hormones thereby interfering with endocrine signaling. These interferences are relayed onto brain development indices resulting in cognitive and behavioral impairment. An example is BPA, which has been demonstrated to bind to putative estrogenic receptors and modulate both neuronal outgrowth and synapse maturation during brain development (Muncke, 2011).

16.3.1.2 Neurotoxicity

BFRs and heavy metals (e.g., lead and cadmium) present neurotoxicity to those used in plastics. These molecules modulate synaptic plasticity and neuronal communication, which are both critical to learning and memory. Early childhood housing has been associated with decreased IQ and elevated risk for neurodevelopmental disorders (Nagy and Kuti 2016).

16.3.1.3 Oxidative Stress

Numerous plastic additives cause the generation of ROS, which triggers oxidative stress and inflammation. During this process neurons and glial cells are significantly damaged via mechanisms that result in both developmental and behavioral abnormalities.

16.3.2 Health Impacts

16.3.2.1 Developmental Delays

Exposure to plastic additives during prenatal and early childhood stages has been associated with delays in speech, motor skills, and cognitive abilities. For instance, research showed that children whose level of phthalate exposure reflects higher exposure performed worse on memory and problem-solving tasks (Muncke, 2011).

16.3.2.2 Attention-Deficit/Hyperactivity Disorder (ADHD)

ADHD is a persistent neuropsychological condition characterized by attention problems and hyperactivity-impulsivity, with current research identifying deficits in brain development and alterations in brain plasticity as fundamental pathophysiological mechanisms of ADHD. Plastic additives likely interact with the brain's plastic mechanisms (particularly neurogenesis and myelin) to potentially modulate alertness, attention, and impulsivity in ADHD patients, such as showing potential enhancer effects of poly(ethylene glycol) and hyaluronic acid on learning and memory, or potentially attenuating the adverse effects of nicotine.

ADHD is one of the most common childhood neurodevelopmental disorders and is defined by continuous inattention, impulsivity, and hyperactivity. It disrupts the child's academic, social, and behavioral functioning. More and more research shows an association between environmental exposure, such as exposure to plastic additives, BPA, and phthalates, and increasing prevalence of ADHD symptoms. These adjuvants act on the developing neural and endocrine systems, consequently causing cognitive and behavioral deficits.

Dopamine Dysregulation

It is already known that plastic agents are able to interfere with the attentional, reward, and impulse control pathways of dopamine. Evidence suggests that BPA alters dopamine receptor signaling via estrogen receptor interaction in the brain to reduce dopamine biosynthesis and release (Rochman, 2013; Teuten et al., 2009). Phthalates also disrupt dopaminergic neuronal function to damage cognitive processes related to attention and behavioral control (Kim et al., 2019; Muncke, 2011). The human research indicated that higher concentrations of phthalates in the urine are associated with greater ADHD symptoms such as impulsivity and attention deficits (Biederman et al., 2012).

Many plastic additives, including BFRs, are neurotoxic. These substances accumulate in the brain, affecting synaptic plasticity, learning, and memory. Nagy and Kuti (2016) noted that children who are exposed to BFRs during sensitive periods of brain development have an increased prevalence of ADHD symptoms. Similarly, imprinting of BPA exposure in utero is associated with impaired development of neural circuits in the prefrontal cortex, a brain region that plays a critical role in attentional regulation and impulsivity (Smith et al., 2021). These findings are corroborated by animal studies showing that prenatal exposure to BPA induces hyperactivity and inattentive-like behavior in offspring (Rochester, 2013; Lithner et al., 2011).

Endocrine-disrupting chemicals (EDCs), such as BPA and phthalates, induce or suppress basic hormones involved in brain development. BPA exerts control over neuronal growth and synapse formation by interacting with estrogen receptors (Muncke, 2011; Rochester, 2013). Dysfunction of hormonal systems in early fetal and childhood periods has been reported to correlate with ADHD-like profiles characterized by hyperactivity and emotional dysregulation (Biederman et al., 2012). Furthermore, the phthalates' effects have also been described to affect thyroid hormone activity, which is crucial for fetal brain development and engages in the development of ADHD symptoms (Kim et al., 2019).

Plastic additives produce ROS leading to oxidative stress and neuroinflammation. (The above) may cause traumatic injury to the developing neuronal and glial cells, in turn, under the context of brain function (Teuten et al., 2009; IPEN, 2020). Oxidative stress is severely linked to ADHD since it can either deteriorate attentional control and/or hyperactivity (Muncke, 2011; Nagy Kuti, 2016). In human studies, children with greater oxidative stress biomarkers in the blood are also at a greater risk of developing ADHD symptoms in response to plastic additives (Kim et al., 2019).

Decreasing plastic additive exposure appears to attenuate ADHD. Household use of plastic is discouraged for food storage, and especially while food is being cooked, and attempts should be made to avoid phthalates- and BPA-containing plastics. On a broader level, stricter policies regulating the use of neurotoxic plastic additives are critical to safeguarding children's neurological health.

16.3.2.3 Autism Spectrum Disorders (ASDs)

ASD is a complex group of developmental disorders characterized by difficulties in social interaction, communication, and restrictive, repetitive behavior. Plastic additives are often used in the production of plastics and are potentially associated with various biological processes, including neurological development.

ASD is a neurodevelopmental disorder group with abnormal social communication, peculiar social interest, and repetitive behavior. Involvement of genetic risk is dominant but environmental exposures, such as plastic additives BPA and BFRs, also increase the risk of ASD. These chemicals exert their effect in key areas of fetal and early cerebral development by disrupting neural connectivity and damaging synaptic and post-synaptic function.

Early exposure to BPA, BFR, and other neurotoxicants is linked to an increased risk of ASD. These chemicals have the effect of disrupting neural communication as well as synaptic function, both of which play critical roles in social communication and social interaction. A further study report on human research evidence has supported that exposure to BPAs (BPA) at high levels during pregnancy increases the risk of autistic-like behavior in children (Lithner et al., 2011). Evidence in support of maternal, in utero BPA exposure also indicates that it is associated with a higher probability of displaying an autistic-like phenotype in children (Lithner et al., 2011; Biederman et al., 2012). Rats treated prenatally with BPA show repetitive behaviors, memory loss, and social play, and have a human equivalent of ASD (Rochester, 2013, Muncke, 2011).

Synapse formation and pruning in the neural network are controlled by the plastic additives and these processes affect the efficiencies of neural communication and so on. In utero exposure to BPA alters synaptic formation and circuits in brain regions that play a role in social processes, e.g., amygdala and prefrontal cortex (Muncke, 2011; Rochester, 2013). Phthalates also interfere with neuronal signaling (i.e., they alter the circuitry involved in cognitive and emotional control of communication) (Kim et al., 2019; Teuten et al., 2009). Epidemiological research found that in-utero maternal exposure to higher BPA levels is associated with higher autism-like traits in children (Lithner et al., 2011; Biederman et al., 2012).

Chronic oxidative damage caused by plasticizers leads to toxic effects on neurons and glia to cause neuroinflammation. Neuroinflammation subsequently offers a plausible mechanism by which

ASD can cause delays in brain development and synaptic tiling (Muncke, 2011; Nagy Kuti, 2016). According to research evidence, prenatal exposure to BPA and BFRs is associated with increased oxidative stress biomarkers in children with ASD (Rochester, 2013; Smith et al., 2021).

Endocrine disruptors (e.g., BPA) disrupt hormonal signaling of great consequence for brain development. BPA mimics estrogen and thyroid hormone agonists by blocking thyroid hormone activity, which is essential for neural growth and differentiation (Teuten et al., 2009; Kim et al., 2019). Hormonal derangements during pregnancy restructure brain networks that mediate social communication and interaction, through to ASD (Muncke, 2011; Rochester, 2013). Evidence found in the research supports the idea that children prenatally exposed to BPA at high concentrations of BPA display social and cognitive deficits (Smith et al., 2021). Regions of greater plastic additive contamination are associated with increased prevalence of ASD (Kim et al., 2019; Smith et al., 2021).

To reduce the potential association between plastic-saturating additives and ASD, exposed women during pregnancy and families need to reduce exposure and not accept products with BPA and take advantage of ventilation and use plastic-free items. Greater action at global scale is needed to remove the pollution from the environment and set stricter constraints for the neurotoxic chemical application to the plastics.

16.4 HOW FOOD PACKAGING LEACHES ADDITIVES INTO FOOD PRODUCTS AND THEIR SUBSEQUENT IMPACT ON COMMUNITY HEALTH

Food packaging plays a crucial role in preserving food quality and extending shelf life. Nevertheless, it has been increasingly attributed to the influx of chemical contamination as a result of the leaching of food additives from container materials into food. Plasticizers, stabilizers, and flame-retardant additives are added to improve the physical properties of packaging materials, but their potential health effects have to be considered when leaching into food stuff. This section explores the mechanisms of additive migration, factors influencing leaching, and the public health implications of exposure to these substances.

Additives migration from food package into food follows through some modes:

1. Diffusion

Additives can migrate from packaging material to food in contact between them, particularly during an extended period of time. Diffusion rates are also influenced by the chemical characteristics of the additive and the food matrix (Muncke, 2011).

The migration of additives from packaging material into food via diffusion is a critical concern in food safety. This process occurs when additives migrate across the material matrix into food due to the concentration gradient between the packaging and the food. The rate of diffusion depends on factors such as the chemical properties of the additive, the composition of the packaging material, and the characteristics of the food matrix (Muncke, 2011).

Polyethylene and PVC are commonly used packaging materials, and additives like plasticizers, stabilizers, and antioxidants are often incorporated into them. These additives, designed to enhance the functionality of the packaging, are not chemically bound to the polymer matrix, making them susceptible to diffusion over time. For instance, phthalates, commonly used as plasticizers, have been shown to migrate into food products stored in plastic containers, particularly under prolonged storage conditions (Teuten et al., 2009).

Diffusion is also influenced by the nature of the food product. Foods with high water content can facilitate the migration of water-soluble additives, while fatty foods enhance the diffusion of lipophilic substances like BPA and phthalates. Studies have documented that cheese and butter wrapped in PVC films exhibit elevated levels of phthalates due to their affinity for fatty substances (Lithner et al., 2011). Extended storage periods further exacerbate the diffusion process, increasing the likelihood of contamination.

2. Solvent Interaction

Food which is high in fat or acid content is able to dissolve packaging additives, allowing them to migrate to the food. Phthalates and BPA for example, are easily leached into fatty foods (cheese and oils) (Lithner et al., 2011).

The solvent properties of the food matrix determine the extent of additive migration. Acidic environments can break down polymer bonds in packaging materials, releasing embedded additives into the food. Lipids, on the other hand, solubilize non-polar compounds, facilitating their transfer from the packaging. Research has shown that food stored in polycarbonate containers exhibits higher concentrations of BPA, particularly in products like mayonnaise and butter, due to the solubility of BPA in fats (Teuten et al., 2009).

The health implications of solvent-induced additive migration are profound. BPA and phthalates are known to interfere with hormonal regulation, causing endocrine disruption, developmental delays, and increased cancer risks. The European Food Safety Authority (EFSA) has established migration limits for these substances, but the widespread use of plastic packaging continues to pose risks. Additionally, the combined effects of multiple additives interacting with diverse food matrices complicate risk assessments, necessitating further research into solvent-mediated migration pathways (Muncke, 2011).

3. Thermal Processes

Food heating in plastic containers, especially during microwave heating, increases the leaching of additives. This is particularly a risk for the mobility of additives, such as BPA, which is known to show increased mobility at higher temperatures (Muncke, 2011).

Heating food in plastic containers significantly accelerates the migration of additives, particularly when exposed to microwave radiation or high-temperature cooking. Heat enhances the mobility of additives within the packaging material, breaking down the polymer structure and facilitating their transfer into food. BPA, a common component of polycarbonate plastics, exhibits increased mobility at elevated temperatures, making it a significant concern in food safety (Muncke, 2011).

Microwave heating poses specific risks due to uneven temperature distribution, which can create localized hotspots in the plastic material. These hotspots amplify the degradation of additives, leading to higher migration rates. Studies have shown that reheating soups and sauces in plastic containers results in detectable levels of BPA and phthalates, exceeding recommended safety thresholds (Teuten et al., 2009). Similarly, single-use plastic containers, not designed for repeated exposure to heat, release greater quantities of additives when reused for heating purposes.

4. Mechanical Stress

Repeated contact or abrasion of packaging materials can result in the leaching of additives from food by surface abrasion.

Mechanical stress, such as repeated handling, folding, or abrasion of packaging materials, can result in the leaching of additives into food. This is particularly true for single-use plastics, which are prone to degradation under repeated mechanical actions. Surface abrasion during packaging or handling compromises the integrity of the material, releasing additives like phthalates and stabilizers into food products (Teuten et al., 2009).

Mechanical stress is often exacerbated by environmental factors such as temperature fluctuations and moisture. For instance, packaging used in frozen foods may experience repeated expansion and contraction, weakening the polymer matrix and facilitating additive release. Similarly, rough handling during transportation and storage increases the likelihood of surface damage and additive migration.

The implications of mechanical stress on food safety are significant. Studies have documented higher levels of phthalates in processed meats stored in plastic films subjected to repeated handling. The presence of microplastics, resulting from the breakdown of packaging materials, further amplifies health risks by serving as carriers for adsorbed additives. These microplastics, ingested along with food, can accumulate in the gastrointestinal tract, releasing additives directly into the body (Lithner et al., 2011).

16.4.1 Common Additives Found in Food Packaging

i. Bisphenol A (BPA)

BPA is extensively used in the manufacture of polycarbonate plastics and epoxy resins, making it a common component in items such as food containers, water bottles, and the inner linings of canned foods. Its ability to leach into food and beverages, especially when exposed to heat or acidic conditions, raises significant concerns. Research has linked BPA exposure to endocrine disruption, as it mimics estrogen and interferes with hormonal balance. Chronic exposure has been associated with reproductive issues such as reduced fertility, metabolic disorders like insulin resistance, and potential links to cardiovascular diseases and obesity. Given its prevalence and impact, BPA contamination remains a critical public health concern.

ii. Phthalates

Phthalates are plasticizers primarily used to enhance the flexibility and durability of plastics like PVC. These chemicals can easily migrate into fatty foods, particularly during storage or heating. Once ingested, phthalates act as endocrine disruptors, interfering with hormonal pathways and potentially leading to developmental and reproductive health issues. Prenatal exposure to phthalates has been linked to adverse outcomes, including reduced male reproductive function and neurodevelopmental issues such as impaired cognitive and motor skills. The widespread use of phthalates in consumer products necessitates stricter regulations and safer alternatives to mitigate health risks (Muncke, 2011).

iii. Flame Retardants

Flame retardants, commonly incorporated into certain packaging materials to reduce flammability, pose a significant health risk due to their potential to migrate into food. These compounds, which include BFRs, are persistent in the environment and bioaccumulate in human tissues. Chronic exposure has been associated with neurotoxicity, affecting cognitive function and behavior, as well as an increased risk of cancers. Studies have also highlighted their potential to disrupt thyroid hormone regulation, which is critical for brain development and metabolism. The health hazards linked to flame retardants underscore the need for reduced usage and safer substitutes in food-related applications (Teuten et al., 2009).

iv. Heavy Metals

Heavy metals such as lead, cadmium, and chromium are often used as stabilizers, pigments, or other additives in plastic packaging materials. These metals can leach into food, particularly in acidic conditions or at elevated temperatures, posing a serious health risk. Lead exposure is associated with neurodevelopmental deficits in children, including lowered IQ and behavioral issues, while cadmium can induce oxidative stress and DNA damage, contributing to kidney damage and cancer risk. Chromium, depending on its oxidation state, can also lead to toxic effects, including respiratory issues and organ damage. The persistence of these metals in the environment and their bioaccumulative nature amplify their impact, calling for stringent control measures in food packaging applications (Lithner et al., 2011).

16.4.2 Factors Influencing Additive Leaching

Food Composition: The composition of the food significantly affects the extent to which additives leach from packaging materials. Foods high in fat, salt, or acid are particularly prone to absorbing packaging additives because of their chemical compatibility with certain plastic components. For example, fatty foods readily interact with plasticizers, while acidic foods can accelerate the migration of heavy metals or other stabilizers. This increased absorption can heighten the risk of chemical exposure, making it crucial to match food types with appropriate packaging materials that minimize leaching.

Storage Conditions: The conditions under which packaged food is stored play a vital role in the rate of additive migration. Elevated temperatures, such as those experienced during prolonged storage or microwaving, can increase the mobility of additives within the packaging. Similarly, high humidity can weaken the structural integrity of certain packaging materials, facilitating leaching. Longer storage durations further compound this issue, allowing more time for additives to migrate into food. These factors underscore the importance of proper storage practices and temperature-resistant packaging materials to limit chemical exposure (Teuten et al., 2009).

Type of Packaging Material: The chemical composition of the packaging material determines the type and amount of additives used, which in turn influences leaching behavior. Flexible plastics, such as those used in wraps or pouches, often contain higher concentrations of plasticizers to enhance their flexibility, making them more prone to leaching. On the other hand, rigid plastics, like those used in containers, are more likely to incorporate stabilizers or pigments, which can also migrate under certain conditions. Understanding the compatibility between packaging materials and their intended food applications is key to minimizing leaching risks.

Recycling Practices: Recycled plastic packaging can contain higher concentrations of additives or contaminants due to the mixing of different plastic types during the recycling process. This blending can introduce unexpected chemical residues or elevate the levels of existing additives, increasing the potential for migration into food. Additionally, the degradation of polymers during repeated recycling cycles can alter the structural properties of the material, making it more susceptible to leaching. Enhancing recycling practices, such as better sorting and purification methods, is critical to ensuring the safety of recycled packaging materials (Lithner et al., 2011).

16.4.3 Public Health Implications

The translocation of packaging additives into food has broad health impacts on populations:

1. Endocrine Disruption

Hormonal signaling is disrupted by the additives BPA and phthalates and the resulting effects include infertility, precocious puberty, and metabolic conditions. These effects are particularly concerning for vulnerable groups such as pregnant women and children (Muncke, 2011).

2. Cancer Risks

Long-term exposure to some additives, such as BPA and flame retardants, is also related to the development of various cancers, like breast and prostate cancer.

3. Neurological Effects

Heavy metals/flame retardants leached from food can compromise cognitive development and raise the incidence of neurodegenerative disorders such as Alzheimer's/Parkinson's disease.

4. Immune System Suppression
Chronic exposure to chemical contaminants from food packaging suppresses the immune system, and as a consequence, people are more likely to contract infectious diseases and suffer from chronic diseases.

5. Reproductive and Developmental Disorders
Maternal exposures to packaging additives may result in the risk of miscarriage, preterm birth, and developmental delays in offspring.

16.4.4 Community-Level Impacts

1. Healthcare Costs
The increase in chemical exposure-related chronic diseases imposes a great strain on healthcare systems, especially in poor-income communities where access to medical care is low.

2. Inequitable Exposure
Marginalized communities often rely on processed and packaged foods due to economic constraints, leading to disproportionate exposure to harmful additives.

3. Environmental Persistence
Mismanagement of food packaging materials causes environmental contamination which, in turn, enhances soil, water, and air contamination risks through the release of food additives.

16.5 MITIGATION STRATEGIES FOR OPEN BURNING OF PLASTICS

Open burning of plastics continues to be a rampant practice, especially in areas where there is a lack of waste management infrastructure. Despite its convenience and low cost in the short term, this practice leads to the emissions of poisonous substances, including dioxins, furans, heavy metals, and PM, which has a significant negative consequence on public health and the environment. The management of the impact of these phenomena requires an overall integrated approach in the areas of technology, regulation, and community initiatives to tackle the causes and effects of open burning.

16.5.1 Promoting Proper Waste Segregation

Efficient waste separation is an important first step in minimizing the amount of plastic that is burned openly. Waste management infrastructures, (and) through the partitioning of the recyclable plastics from the non-recyclable plastics wastes, can transport the recyclable plastics to be recycled to recycling plants and properly dispose of the non-recyclable wastes. For instance, Public Awareness Programmes can be very effective in influencing negative waste disposal behavior by informing communities of the proper ways to sort plastic waste. Furthermore, the creation and deployment of sorting points by both governments and NGOs (Jambeck et al., 2015) which create separate containers for recyclable and non-recyclable waste on the household level can support improved management of waste streams.

16.5.2 Expanding Recycling Programs

Recycling offers an environmentally friendly alternative to burning plastics. While recycling rates are still low in many countries, improving recycling infrastructure can drastically reduce the dependency on open burning. Building recycling facilities such as establishing local recycling plants reduces transportation costs and makes recycling accessible to more communities. Promoting mechanical and chemical recycling is concerned with the recycling of plastics into new products and the breaking down of plastics into plastic raw materials for the production of new plastics. Both approaches can be used to optimize plastic waste management (Hopewell et al., 2009).

16.5.3 Advancing Waste-to-Energy Technologies

WTE plants represent a viable solution for recycling plastic waste into electricity and/or heat. In contrast to open burning, WTE plants are provided with advanced filtration devices in order to stop harmful emissions and mitigate the environmental and health impact. High-temperature combustion is the temperature sufficiently high to break down dioxins and other pollutants. Implementation of emission control systems that utilize scrubbers, filters, and electrostatic precipitators, by which PM and gaseous pollutants can be remediated.

16.5.4 Encouraging the Use of Biodegradable Alternatives

The replacement of these traditional plastics with organic plastics could help minimize the environmental hazard of plastic litter. These materials are broken down more rapidly and have lower amounts of toxic constituents when burnt or disposed of improperly. Examples include *Polylactic Acid (PLA)* (A biodegradable polymer made from renewable resources such as corn starch), *Starch-Based Plastics* (these are quickly lost and at least environmentally inert) (Teuten et al., 2009).

16.5.5 Strengthening Legislation and Policy Frameworks

The adoption of the strictest waste disposal control would be most helpful to the practice of not doing open burning. Governments have actions that they could take to decrease plastic waste and encourage biodegradable alternatives, including plastic bans and levies on non-recyclable packaging, which could restrict plastic use and levy a fee on non-recyclable packaging, respectively, that could reduce the amount of plastic that is made. The compulsion of waste collection services, thus providing regular and low-cost waste collection services, can stop illegal waste dumping and burning. In addition, extended producer responsibility (EPR), responsibility for the life cycle of a plastic product from its birth to its end of life, design for recyclability, and investment in waste handling infrastructure can be achieved (European Commission, 2018).

16.5.6 Improving Community Engagement and Education

Providing educational resources to waste treatment communities about waste treatment, including composting and recycling. Collaboration with Local Leaders by the mobilization of community leaders for spreading awareness and adoption of sustainable actions.

16.5.7 Developing Alternative Waste Management Systems

The provision of accessible and environmentally friendly waste disposal solutions can reduce the use of open burning. These systems include centralized waste collection points which employ the power of a population to demarcate, preferably, sections into which waste is to be placed for safe disposal from which it can be harvested later. Also, composting Organic Waste by Elimination of diversion of the combustible stream results in less waste volume produced, as well as useful secondary products, such as compost.

Significantly, landfill Upgrades where landfill layouts with liners and leachate control systems are successful in minimizing contamination risk.

16.5.8 Supporting Innovation in Waste Management Technology

The research and development process of new technologies with investment in basic research and development potential can enhance waste management. Examples include the following:

i. Plastic-to-Fuel Conversion: Non-recyclable plastics can be converted to valuable fuels through pyrolysis technology, offering a fresh energy path and simultaneously eliminating Waste (Arena, 2012).
ii. Smart Waste Bins: Using sensors these bins can be programmed to alert waste collection services when they are "full" so they can be disposed of accordingly and not be left to burn illegally.

16.5.9 Collaborating with Private Sector and NGOs

Government, industry, and NGO collaboration should be able to improve activities to reduce open burning. For instance, Corporate Responsibility Programs where companies may perform environmentally sustainable waste management tasks and waste management tasks and NGO-Led Recycling Programs which include enterprises (like Plastic Bank) that aim to promote social recycling schemes by offering a "payment" for plastic waste (Lithner et al., 2011).

16.5.10 Monitoring and Enforcement

Stricter enforcement of anti-burning laws coupled with monitoring infrastructures not only ensures compliance with law but also discourages illegal activity. The involvement of the reporting local residents in the reporting of illegal burning can foster the crackdown. Also, the introduction of financial sanctions against illegal open burning practices serves as a disincentive and as a result compliance with waste management legislation is higher.

16.6 CONCLUSION

With growing concerns about plastic waste, there is a renewed focus on adopting circular economy models that prioritize reducing plastic production, promoting recyclable materials, and encouraging the use of biodegradable alternatives. Rethinking plastic usage and waste management is essential for achieving environmental sustainability and safeguarding public health. As Rachel Carson wisely observed, "In nature, nothing exists on its own." This interconnectedness between human activities and ecosystems demands a comprehensive approach to managing plastic waste and its consequences. Burning plastic waste, while effective in significantly reducing waste volume and recovering energy, introduces harmful substances such as dioxins, furans, and heavy metals into the environment. These pollutants pose serious risks to human health, contributing to endocrine disruption, developmental delays, and chronic diseases. The migration of additives from food packaging into consumables further compounds these issues, emphasizing the urgent need for balanced, multi-faceted solutions involving technological innovation, regulatory measures, and community participation.

Incineration remains a critical tool in managing excessive plastic waste, particularly in densely populated urban areas with limited landfill space. However, its environmental footprint must be minimized. Advanced WTE technologies equipped with emission control systems, such as scrubbers and filters, can reduce harmful pollutants, making incineration a viable option with reduced impacts on air quality and human health. Policymakers should ensure that incineration complements recycling and waste reduction efforts, reserving it for non-recyclable waste streams while setting clear guidelines to maximize recycling opportunities. This integrated approach aligns with the goals of circular economy models, balancing waste management efficiency with sustainability. The widespread use of plastic additives in food packaging raises concerns about their migration into food products. EDCs like BPA, phthalates, and certain heavy metals have been linked to adverse health effects, including cancer, hormonal imbalances, and developmental disorders. Mitigating these risks requires a multi-pronged strategy: adopting safer alternatives such as non-toxic or biodegradable polymers, strengthening global and national regulations to restrict hazardous additives, and raising consumer awareness about safe food storage practices, such as avoiding heating food in plastic containers. The success of these measures hinges on active collaboration among governments, industries, and communities. Investments in waste management infrastructure, such as recycling facilities and centralized waste collection systems, are critical to reducing open burning and illegal disposal. Partnerships with NGOs and private organizations can further amplify efforts to combat plastic pollution and drive sustainable practices. Looking ahead, innovative solutions like chemical recycling and plastic-to-fuel technologies hold promise for transforming plastic waste into valuable resources. Increased funding for research and development can accelerate breakthroughs in these areas, reducing the environmental footprint of plastic waste. Given the transboundary nature of plastic pollution, global cooperation is essential to implement consistent policies and practices across regions. By combining technological advancements, robust regulatory frameworks, and active community engagement, societies can tackle the challenges posed by plastic waste and chemical leaching effectively. This integrated approach ensures a sustainable future where waste is managed responsibly, ecosystems are protected, and public health is prioritized.

REFERENCES

Al-Salem, S. M., Lettieri, P., & Baeyens, J. (2009). Recycling and recovery routes of plastic solid waste (PSW): A review. *Waste Management*, 29(10), 2625–2643. https://doi.org/10.1016/j.wasman.2009.06.004

Andrady, A. L., & Neal, M. A. (2009). Applications and societal benefits of plastics. *Philosophical Transactions of the Royal Society B: Biological Sciences*, 364(1526), 1977–1984. https://doi.org/10.1098/rstb.2008.0304

Arena, U. (2012). Process and technological aspects of municipal solid waste gasification: A review. *Waste Management*, 32(4), 625–639. https://doi.org/10.1016/j.wasman.2011.09.025

Biederman, J., Mick, E., Faraone, S. V., & Wozniak, J. (2012). The influence of gender on the course and clinical correlates of attention-deficit hyperactivity disorder from childhood to adulthood. *Journal of Clinical Psychiatry*, 63(10), 1084–1092.

Bullard, R. D. (2000). *Dumping in Dixie: Race, class, and environmental quality*. Westview Press.

Cole, M., Lindeque, P., Halsband, C., & Galloway, T. S. (2011). Microplastics as contaminants in the marine environment: A review. *Marine Pollution Bulletin*, 62(12), 2588–2597. https://doi.org/10.1016/j.marpolbul.2011.09.025

Denison, R. A. (1997). Environmental life-cycle comparisons of recycling, landfilling, and incineration: A review of recent studies. *Annual Review of Energy and the Environment*, 22(1), 191–237. https://doi.org/10.1146/annurev.energy.22.1.191

Ellen MacArthur Foundation. (2017). The new plastics economy: Rethinking the future of plastics & catalysing action. Retrieved from [Ellen MacArthur Foundation] https://ellenmacarthurfoundation.org

European Commission. (2018). A European strategy for plastics in a circular economy. Communication from the Commission to the European Parliament, the Council, the European Economic and Social Committee and the Committee of the Regions. Retrieved from [European Commission] https://eur-lex.europa.eu/legal-content/EN/TXT/?uri=COM:2018:28:FIN

FAO/WHO. (2006). *Evaluation of certain food additives and contaminants: Sixty-seventh report of the Joint FAO/WHO Expert Committee on Food Additives*. WHO Technical Report Series, No. 940. https://iris.who.int/handle/10665/43592

Geyer, R., Jambeck, J. R., & Law, K. L. (2017). Production, use, and fate of all plastics ever made. *Science Advances*, 3(7), e1700782. https://doi.org/10.1126/sciadv.1700782

Gullett, B. K., Lemieux, P. M., Lutes, C. C., Winterrowd, C. K., & Winters, D. L. (2001). Emissions of PCDDs and PCDFs from uncontrolled, domestic waste burning. *Chemosphere*, 43(4–7), 721–725.

Hartley, B. L., Pahl, S., Veiga, J., Vlachogianni, T., Vasconcelos, L., Maes, T., & Thompson, R. C. (2018). Exploring public views on marine litter in Europe: Perceived causes, consequences, and pathways to change. *Marine Pollution Bulletin*, 133, 945–955. https://doi.org/10.1016/j.marpolbul.2018.05.061

Hermabessiere, L., Dehaut, A., Paul-Pont, I., Lacroix, C., Jezequel, R., Soudant, P., & Duflos, G. (2017). Occurrence and effects of plastic additives on marine environments and organisms: A review. *Chemosphere*, 182, 781–793. https://doi.org/10.1016/j.chemosphere.2017.05.096

Hopewell, J., Dvorak, R., & Kosior, E. (2009). Plastics recycling: Challenges and opportunities. *Philosophical Transactions of the Royal Society B: Biological Sciences*, 364(1526), 2115–2126. https://doi.org/10.1098/rstb.2008.0311

International Agency for Research on Cancer. (2012). IARC monographs on the evaluation of carcinogenic risks to humans: Volume 100F. Chemical agents and related occupations. https://publications.iarc.fr/Book-And-Report-Series/Iarc-Monographs-On-The-Identification-Of-Carcinogenic-Hazards-To-Humans/Chemical-Agents-And-Related-Occupations-2012

International Pollutants Elimination Network (IPEN). (2020). Plastic and toxic additives: The hidden threat of plastic pollution. https://ipen.org/site/plastics-toxic-additives

Jambeck, J. R., Geyer, R., Wilcox, C., Siegler, T. R., Perryman, M., Andrady, A., Narayan, R., & Law, K. L. (2015). Plastic waste inputs from land into the ocean. *Science*, 347(6223), 768–771. https://doi.org/10.1126/science.1260352

Kim, H. T., Kim, J. K., Cha, H. G., Kang, M. J., Lee, H. S., Khang, T. U., Yun, E. J., Lee, D. H., Song, B. K., & Park, S. J. (2019). Biological valorization of poly(ethylene terephthalate) monomers for upcycling waste PET. *ACS Sustainable Chemistry & Engineering,* 7(22), 19396–19406.

Kirchherr, J., Reike, D., & Hekkert, M. (2017). Conceptualizing the circular economy: An analysis of 114 definitions. *Resources, Conservation and Recycling*, 127, 221–232. https://doi.org/10.1016/j.resconrec.2017.09.005

Law, K. L., & Thompson, R. C. (2014). Microplastics in the seas. *Science*, 345(6193), 144–145. https://doi.org/10.1126/science.1254065

Lithner, D., Larsson, Å., & Dave, G. (2011). Environmental and health hazards of additives in plastics. *Science of the Total Environment*, 409(18), 3309–3324. https://doi.org/10.1016/j.scitotenv.2011.04.038

Muncke, J. (2011). Endocrine disrupting chemicals and other substances of concern in food contact materials: An updated review of exposure, effect, and risk assessment. *Journal of Steroid Biochemistry and Molecular Biology*, 127(1–2), 118–127. https://doi.org/10.1016/j.jsbmb.2010.10.004

Nagy, Á., & Kuti, R. (2016). The environmental impact of plastic waste incineration. *Academic and Applied Research in Military and Public Management Science*, 15(3), 235–242. Retrieved from [AARMS Journal] www.uni-nke.hu

Plastic Pollution Coalition. (n.d.). *Plastic pollution coalition.* Retrieved June 6, 2025, from https://en.wikipedia.org/wiki/Plastic_Pollution_Coalition

PlasticsEurope. (2022). *Plastics – the Facts 2022: An analysis of European plastics production, demand and waste data.* PlasticsEurope. https://plasticseurope.org/knowledge-hub/plastics-the-facts-2022--2/

Rochman, C. M., Hoh, E., Kurobe, T., & Teh, S. J. (2013). *Ingested plastic transfers hazardous chemicals to fish and induces hepatic stress.* Scientific Reports, 3, 3263.

Smith, M., Love, D. C., Rochman, C. M., & Neff, R. A. (2021). Microplastics in seafood and the implications for human health. *Current Environmental Health Reports*, 5(3), 375–386.

Teuten, E.L., Saquing, J.M., Knappe, D.R.U., Barlaz, M.A., Jonsson, S., Bjorn, A., Rowland, S.J., Thompson, R.C., Galloway, T.S., Yamashita, R., Ochi, D., Watanuki, Y., Moore, C., Viet, P.H., Tana, T.S., Prudente, M., Boonyatumanond, R., Zakaria, M.P., Akkhavong, K., & Takada, H. (2009). Transport and release of chemicals from plastics to the environment and to wildlife. *Philosophical Transactions of the Royal Society B: Biological Sciences*, 364(1526), 2027–2045. https://doi.org/10.1098/rstb.2008.0284

Wiedinmyer, C., Yokelson, R. J., & Gullett, B. K. (2014). Global emissions of trace gases, particulate matter, and hazardous air pollutants from open burning of domestic waste. *Environmental Science & Technology*, 48(16), 9523–9530.

Wilson, D. C., Velis, C., & Cheeseman, C. (2006). Role of informal sector recycling in waste management in developing countries. *Habitat International*, 30(4), 797–808.

World Health Organization. (2002). *Global assessment of the state-of-the-science of endocrine disruptors.* https://www.who.int/publications/i/item/9241562153

World Health Organization. (2023). *Global plastics treaty: Ending plastic pollution.* United Nations Environment Programme. https://www.wired.com/story/plastic-pollution-emergency-united-nations

Zheng, J., & Suh, S. (2019). Strategies to reduce the global carbon footprint of plastics. *Nature Climate Change*, 9(5), 374–378. https://doi.org/10.1038/s41558-019-0459-z

17 Nanotechnology in Bioplastic Wastes Management

Sarah Chioma Nwakanma, Joshua O. Akhigbe, Sesan Abiodun Aransiola, Juliana Okwena Pondei, Ojo Paul Adewoye, Suhasini Bhatnagar, and Naga Raju Maddela

17.1 INTRODUCTION

The increase in plastic usage has raised both environmental and health concerns (Lagaron and Lopez-Rubio, 2011; Musa *et al.*, 2024) globally, leading to accumulation and pollution of plastic waste in marine and terrestrial ecosystems. These plastics tend to persist in the environment as a result of their physical and chemical qualities combined with their resistance to biodegradation (Auta *et al.*, 2022). Even though approaches like recycling and re-use of plastic have been adopted to tackle this issue, the end-of-life of these conventional plastic products still poses a significant challenge; thus, the need for an eco-friendly alternative is a necessity. This need has given rise to the use of green polymers or bioplastics, which are fully biodegradable or partially/non-biodegradable plastics obtained from renewable biomass sources like plant starch, cellulose obtained from plant cell walls, vegetable fats, and oils (Atiwesh *et al.*, 2021; Aransiola *et al.*, 2023a) unlike conventional plastics. These bioplastics are more degradable offering reduced carbon footprint and are quite known for their eco-friendly characteristics (Aransiola *et al.*, 2023b; Lamba and Singh, 2024).

Bioplastics have different applications in different sectors and fields. Starch for example is a type of bioplastic used in biomedical implant devices and also in food packaging industries (Afolalu *et al.*, 2022). However, the challenges that come with the adoption of bioplastic involve low mechanical properties, varying degradation rates and poor barrier properties. According to studies, there are different forms of bioplastics known which are as follows: bio-based biodegradable bioplastics (e.g. starch, cellulose, polylactic acid), petroleum-based biodegradable bioplastics (e.g. polycaprolactone (PCL), polybutylene adipate-co-butylene terephthalate (PBAT), and bio-based non-biodegradable bioplastics (e.g. bio-based polyethylene (PE), polyamide (PA), polyethylene terephthalate (PET) (Fredi and Dorigato, 2021; Auta *et al.*, 2022). A key point to consider is that though bioplastics are an ideal choice in comparison to their non-biodegradable conventional counterparts, they are not all biodegradable or degradable and therefore still pose issues of waste accumulation and pollution just like conventional plastic products. It has been estimated that by 2025 the global production of bioplastic would increase to 2.89 million tons, and if effective waste management strategies are not adopted, it would result in major environmental problems (Fredi and Dorigato, 2021). Issues like CO_2 emission, negative land use change, greenhouse emission that causes ozone layer destruction, and thereby climate change (Atiwesh *et al.*, 2021; Maddela *et al.*, 2024) are problems associated with non-biodegradable bioplastics and are therefore what limits bioplastic in being an excellent choice.

To truly address plastic economy sustainability, there has to be a balance between mechanical properties, biodegradability, and cost of production of the plastic product (Abioye *et al.*, 2021). As much as bioplastics are biodegradable, they have a high cost of production and limited mechanical

DOI: 10.1201/9781003602828-21

strength; hence, techniques like nanotechnology are considered to increase their durability and sustainability as well as their mechanical properties. Recent research has established that nanotechnology is an effective means used in tackling plastic waste accumulation. These advancements suggest that nanomaterials can play a crucial role in optimizing bioplastics for a variety of applications as they are known for their excellent chemical and physical properties (Chausali *et al.*, 2022), from packaging to medical devices by increasing the technical optical, magnetic, and mechanical properties (Carroccio *et al.*, 2022). Enhanced bioplastics by nanomaterials not only reduce environmental pollution but also contribute to resource recovery through biodegradation processes, which is essential for sustainable waste management. This chapter aims to review bioplastic and its mechanism of degradation, nanotechnology, and nanomaterials and how nanotechnology can be harnessed in solving bioplastic waste accumulation.

17.2 NANOTECHNOLOGY IN BIOPLASTIC MANAGEMENT

Plastic pollution is a major danger to ecosystems and all forms of life. Conventional biological techniques, like enzyme degradation, cannot effectively solve the plastic pollution problem because of environmental limitations. As a solution, engineered nanomaterials have become absolutely vital to a variety of industries like infrastructure, aerospace, energy, and medicine. Such nanomaterials have the potential to improve present-day microplastic removal methods while also providing new solutions to nascent issues (Pathak and Kumar, 2017).

17.2.1 Bioplastics – An Overview

Plastics have many benefits, but they are also one of the most destructive inventions ever created by humanity. Plastics do not decompose naturally and that is the main problem with plastics. Traditional methods such as chemical and thermal processes consume a lot of energy to depolymerize polymers and usually leave by-products which can create new environmental issues (Chandran *et al.*, 2023). The increase in the bioplastics industry has occurred over the last decades due to the rising concern about the usage of fossil resources for traditional plastics production and their adverse effects on ecosystems currently (Abioye *et al.*, 2019). At present, approximately 40% of plastic waste is landfilled, 25% incinerated, and another 22% pass through unmanaged or rejected recycling processes (Degan and Shinde, 2019). As a result, in the last decades there have been a great deal of attempts to produce biodegradable polymers such as poly(lactide), polybutylene succinate, polyalkanoates, and many natural biopolymers originated from plants, wood, and seaweeds (Lindström and Österberg, 2020).

Plastics have been traditionally synthesized from petrochemicals via non-reversible processes (Tsang *et al.*, 2019). More lately, bioplastics are being presented as a sustainable alternative. Bioplastics are often considered more environmentally friendly than synthetic plastics and are made from predominantly renewable resources. This type of plastic, called bioplastics, is made by using a large variety of waste materials from organisms, making them a sustainable and green option. Bioplastics include an extensive variety of bio-based materials arising from renewable sources (proteinous materials, wood, potatoes, corn, vegetable oils, cereal harvests, and food waste), incorporating polysaccharides, lipids, and proteins (Zhao *et al.*, 2023).

The substitution of bioplastics is also very favorable from the point of view of global heating, since producing petroleum-based plastic products requires significantly more energy of 77 MJ/kg, compared to bioplastics which require 57 MJ/kg. Considering adverse effects of landfilling biowaste on groundwater and greenhouse gas emissions from landfill sites, the transformation of food products and agricultural residues into bioplastic materials is an optimal method for waste management (Roy, 2022). The bioplastics industry has developed rapidly over the last couple of decades, motivated by innovations in different fields and basic environmental, economic, and social problems associated with conventional plastics. Bioplastics will increasingly replace traditional plastics on an industrial scale (Harrison *et al.*, 2018) in the upcoming years. The transition to bioplastics widens the waste management possibilities, as fossil-based plastics only allow for the more linear paradigm while also favoring renewable resources.

According to European Bioplastics (2021), total bioplastics production globally was found to be about 2.4 million tons in 2021. This number is expected to increase by over 200%, reaching between 6 and 7.5 million tons by 2026, representing approximately 2% of total anticipated global plastic production. Their versatility has allowed bioplastics to enter many different applications, including packaging, consumer goods, electronics, and the automotive sector (Folino *et al.*, 2023), which has contributed to this gradual growth in the production volumes of bioplastics.

17.3 BIOPLASTIC DEGRADATION

Bioplastics consist of various types of materials having different polymers and structures and mainly, the characteristics affecting their environmental degradation (Folino *et al.*, 2023). The ability of bioplastic to disintegrate or biodegrade in a specific environment is not directly determined by its source materials, but rather it's due to the physicochemical characteristics of the very bioplastic, like thickness, hydrophobic feature, molecular weight, crystallinity, and biopolymer's melting temperature (Figure 17.1) (Batori *et al.*, 2018).

The rate of decay of a bioplastic not only relies on particular characteristics inherent to the material but also is dependent on certain environmental conditions such as the existence of microorganisms (bacteria and fungi) which can digest it (Bilo *et al.*, 2018). The state of the organic matter is critical to understanding if it will decompose purely by abiotic factors (as is controlled by heat, sunlight, other dead properties) or whether microbially mediated processes also have a role to play in

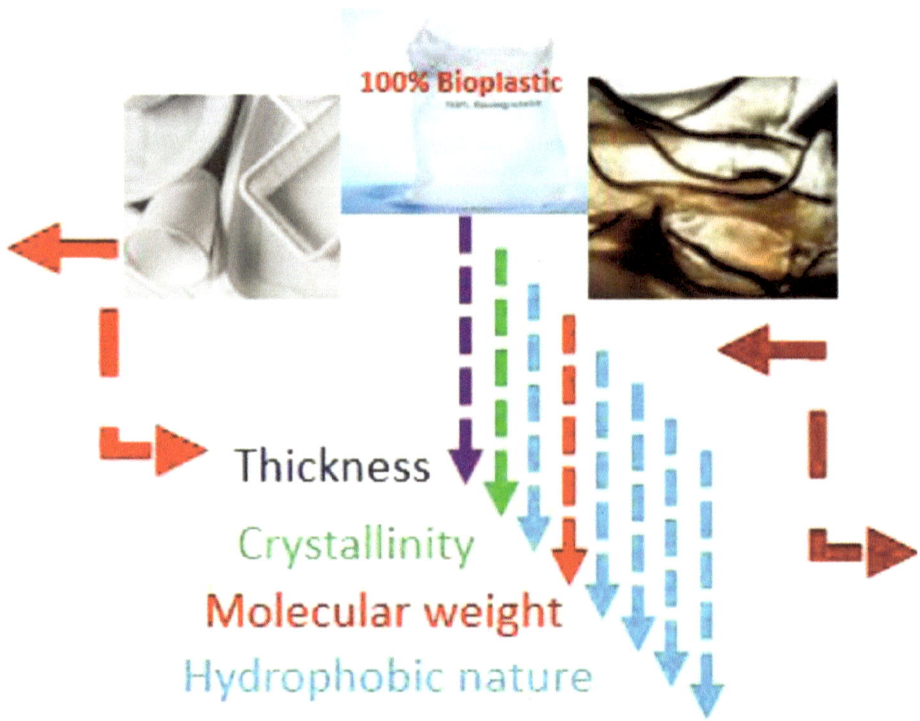

FIGURE 17.1 Degradability pointers of bioplastics.

decomposition (Abe *et al.*, 2022). If degradation is carried out by microorganisms, it is called biodegradation (Bilo *et al.*, 2018), and the bioplastic will be transformed into CO_2, H_2O, NH_4^+, N_2, H_2, and biomasses. In contrast, if decomposition occurs solely via abiotic fragmentation, the process is termed degradation and can result in recalcitrant microplastic particles (Rujnić-Sokele *et al.*, 2017).

In terms of bioplastics, one main benefit is that whatever residues remain after the decomposition or biodegradation process are harmless to other living organisms (Urbanek *et al.*, 2017).

17.4 NANOTECHNOLOGY AND NANO PLASTICS

Iadiz *et al.* (2017) defined nanotechnology as

> the design, characterization, production, and application of materials, devices, and systems by controlling shape and size at the nanoscale; an intermediate length scale between bulk materials as we know them in nature and molecular structures at whose level quantum mechanical effects are frailer.

The US National Nanotechnology Initiative (NNI) (2020) gave their definition as, 'Nanotechnology is the understanding and control of matter at dimensions between approximately 1 and 100 nanometers, where unique phenomena enable novel applications'. Nanotechnology is a multi-disciplinary field spanning physics, chemistry, biology, and different engineering sciences. Nanomaterials are materials with one or more dimensions of less than 100 nanometers and exist as tubes, particles, rods, and fibers. They are used in many scientific disciplines, including agriculture, medicine, environmental science, and energy generation (Hulla *et al.*, 2015). Nanomaterials are unique in their mechanical, optical, electrical, and magnetic properties due to their small building block structure (particles, grains, or phases). These properties depend on their grain size and size distribution (Saad and Elmahjubi, 2018).

17.5 NANOTECHNOLOGY IN BIOPLASTIC MANAGEMENT

Nanomaterials exhibit a unique electron transfer and separation process due to their favorable physicochemical characteristics like low production cost, increased surface activity, minute size with low bandgap energy, and photo-stability properties, hence, play an important role in environmental protection by photocatalytic degradation of plastic debris and other waste (Rani *et al.*, 2023). It is a well-known property about nanoparticles that they are biodegradable. The process of degradation is irreversible and the structural changes consist of a loss of mechanical properties, damage, and depolymerization of the material (Remya *et al.*, 2022).

Table 17.1 shows some nanomaterials and their function as applied to bioplastic degradation.

Chamdran *et al.* (2023) reviewed various experiments conducted in the area of nanotechnology-mediated plastic degradation.

By the amalgamation of hydrogen tetrachloroaurate (III) ($HAuCl_4 \cdot 3H_2O$), gold (Au) nanoparticles were synthesized using an aqueous extract of *Ananas comosus* leaves. These gold nanoparticles showed their photocatalytic properties when a film of low-density polyethylene (LDPE) was exposed, resulting in the degradation of the LDPE film. The LDPE film degraded 51.4% of its weight after 24 hours in sunlight, representing a notable improvement in polymer degradation.

Bimetallic core/shell palladium/gold (Pdcore/Aushell) nanoparticles were consolidated using the reductant, *A. comosus* leaf extract, from Pd(II) and Au(III) solutions. These nanoparticles showed photocatalytic activity when integrated into a solid-phase degradation process of LDPE films under solar irradiation. Within 240 hours of sunlight exposure, the weight of the polymer impregnated with nanoparticles decreased by 55.8±5.9%. Notably, the nanoparticles were also capable of degrading LDPE films in the absence of light, achieving a degradation rate of 1.90±0.3% (Rodríguez-Hernández *et al.*, 2019).

TABLE 17.1
Nanomaterials and Their Functions

S/N	Nanomaterial	Function	Reference
1.	Nanoclay	Improves microbial access	Bandyopadhay and Singha (2022), Manohar and Kumar (2023)
2.	Silver nanoparticles	Exhibit antimicrobial action that supports biofilm breakdown	Sharma and Thakur (2021)
3.	Titanium dioxide (TiO_2)	Exhibit photocatalytic activity	Pathak and Singh (2023), Kim et al. (2023)
4.	Zinc oxide (ZnO)	UV-assisted degradation of bioplastic polymers	Pathak and Singh (2023), Manohar and Kumar (2023)
5.	Iron oxide (Fe_2O_3)	Redox properties facilitate oxidative degradation	Manohar and Kumar (2023)
6.	Silicon dioxide (SiO_2)	Provides surface area for microbial adhesion	Sharma and Thakur (2021), Manohar and Kumar (2023)
7.	Starch nanoparticles	Increases hydrophilicity, enhancing microbial interaction	Sharma and Ghosh (2022)
8.	Cellulose nanofibers	Promotes microbial colonization	Bandyopadhyay and Singha (2022), Sharma and Ghosh (2022)

Raghavendra *et al.* (2017) studied nanoparticle degradation properties using photocatalysis and reported that after 200 hours of irradiation; the rate of degradation of LDPE films increased after the incorporation of zinc oxide nanoparticles (ZnO). The reduction in weight of the polymer further established ZnO nanoparticles photocatalytic activity. In a specific test, ZnO nanoparticles metabolized from *Areca catechu* demonstrated the ability to break down LDPE films due to their photocatalytic properties. As expected, the introduction of nanoparticles of ZnO to the LDPE nanocomposite films contributed to the degradation of the film.

After titanium dioxide's (TiO_2) photocatalytic activity was first demonstrated, this method has been recognized as a sustainable solution to different environmental challenges. It has been useful across multiple research areas, like decomposition of pollutants, water splitting, and conversion of carbon dioxide. Specifically, photocatalytic conversion of plastic waste has gained much attention as a possible solution to the increasing problem of plastic disposal (Kim *et al.*, 2023). During the photocatalytic process, plastics decompose upon absorbing energy from sunlight, producing active free radicals. When energy from photons excite the catalysts, photogenerated electrons (e−) and holes (h+) are created and moved to the surface of the catalyst to begin various reactions. Photoexcited holes can directly oxidize plastic materials, while other free radicals, such as hydroxyl and superoxide radicals (formed when electrons and holes react with O_2 or H_2O), also contribute to plastic degradation (Chu *et al.*, 2022). Oxidation of plastic happens more easily because of its lower potential, reducing the total potential required for conjunction with hydrogen (H_2) evolution. As a result, a variety of catalysts that have lower valence band positions and slimmer bandgaps can be applied in plastic oxidation. In fact, plastics can act as sacrificial reagents, allowing for oxidation by low-energy photogenerated holes (e.g. methanol, lactic acid, and triethanolamine) (Chu *et al.*, 2022). This makes discarded plastic materials a readily available and easily accessible base material for the generation of H_2 and other chemical materials of value, with photocatalytic plastic conversion being a less energy-intensive procedure compared to water splitting. Consequently, considerable advancements are anticipated in using photocatalysis as a method for catabolizing plastic waste.

Research has explored the breakdown of conventional plastics like PET and PE using various catalysts, including microbes and nanomaterials. These catalysts break down polymers into their single units and go a step further to decompose them into simpler molecules, eventually ending in mineralization. This represents a paradigm shift in the understanding of biodegradable plastics. In this context, the breakdown of polymers can be facilitated by nanomaterials under optimal conditions and sunlight irradiation (Harmaen *et al.*, 2016).

Recently, many studies have focused on plastic control practices, such as recycling and degradation, aided by nanomaterials. In this context, TiO_2 nanoparticles have emerged as promising semiconductors or catalytic materials for addressing a diverse array of plastic contaminants due to the stability of their chemical structure, low cost, non-toxicity, and greater photocatalytic activity (David *et al.*, 2020). Their high surface area and pore size make these materials excellent catalysts for plastic degradation. When TiO_2 nanoparticles are integrated on polymer surfaces, effective decomposition occurs, outperforming the standard photocatalysis of polymers under natural conditions. At first, interactions occur between the nanoparticle materials and the chemical bonds of the polymers, leading to bond cleavage. This results in the formation of monomer polymer units, which further degrade into basic molecules such as H_2O and CO_2 in the presence of sunlight irradiation (Rani *et al.*, 2023).

17.6 CONCLUSION

Nanotechnology holds vast promise for the prospect of bioplastic management by addressing many of the limitations associated with bioplastics. Through the integration of nanomaterials, it is possible to enhance the mechanical, electrical, magnetic, optical, thermal, and barrier properties of bioplastics, as well as enhancing biodegradability. These enhancements broaden the bioplastics applications in various fields, contributing to the global shift toward sustainable plastic alternatives. Nanomaterials when incorporated in bioplastics are effective at initiating plastic degradation for pollutant decomposition, water splitting, and CO_2 conversion and are therefore effective in waste management processes. Ultimately, interdisciplinary research and more studies are crucial to fully explore the potential of the role nanotechnology plays in the sustainability of bioplastics and their waste management.

REFERENCES

Abe, M. M., Branciforti, M. C., Montagnolli, R. N., Morales, M. A. M., Jacobus, A. P., Brienzo, M., & Nallin, R. (2022). Production and assessment of the biodegradation and ecotoxicity of xylan- and starch based bioplastics. *Chemosphere, 287*, 132290. https://doi.org/10.1016/J.CHEMOSPHERE.2021.132290

Abioye, A. A., Afolalu, S. A., Yusuf, O. O., Emetere, M. E., Ongbali, S. O., Mfon, U., Samuel, O. D., & Olayinka, A. (2021). Nanotechnology applications in synthesis of bioplastic. *IOP Conference Series: Materials Science and Engineering, 1107*, 012190. https://doi.org/10.1088/1757-899X/1107/1/012190

Abioye, A. A., Rotimi, D. O., Fasanmi, O. O., Abioye, O. P., Afolalu, S. A., Owoeye, T. F., Ajayi, O. O., & Obuekwe, C. C. (2019). Synthesis and characterization of selected starch nanoparticles as matrix reinforcements for low density polyethylene. *Journal of Physics Conference Series, 1378*(4), 042070. https://doi.org/10.1088/1742-6596/1378/4/042070

Afolalu, S. A., Yusuf, O. O., Emetere, M. E., Ongbali, S. O., & Ademuyiwa, F. (2022). Bioplastic: A sustainable remedy to manage environmental waste. In Ayeni, A.O., Oladokun, O., Orodu, O.D. (eds), *Advanced Manufacturing in Biological, Petroleum, and Nanotechnology Processing. Green Energy and Technology*. Springer, Cham. https://doi.org/10.1007/978-3-030-95820-6_22

Aransiola, S. A., Ikhumetse, A. A., Babaniyi, B. R., Abioye O. P., & Oyedele, O. J., & Falade. N. O. (2023a). Phytoaccumulation of micro- and nanoplastics: Root uptake. In Maddela, N.R., Reddy, K.V., Ranjit, P. (eds), *Micro and Nanoplastics in Soil*. Springer, Cham. https://doi.org/10.10007/978-3-031-21195-9_8

Aransiola, S. A., Victor-Ekwebelem, M. O., Ajiboye, A. E., Leh-Togi Zobeashia, S. S., Ijah, U. J. J., & Oyedele, O. J. (2023b). Ecological impacts and toxicity of micro- and nanoplastics in agroecosystem. In Maddela, N.R., Reddy, K.V., Ranjit, P. (eds), *Micro and Nanoplastics in Soil*. Springer, Cham. https://doi.org/10.10007/978-3-031-21195-9_10

Atiwesh, G., Mikhael, A., Parrish, C. C., Banoub, J., & Lee, T. A. T. (2021). Environmental impact of bioplastic use: A review. *Heliyon, 7*(9), e07918.

Auta, S. H., Abioye, O. P., Aransiola, S. A., Bala, J. D., Chukwuemeka, V. I., Hassan, A., Aziz, A., and Fauziah S.H. (2022). Enhanced microbial degradation of PET and PS microplastics under natural conditions in mangrove environment. *Journal of Environmental Management, 304*, 114273. https://doi.org/10.1016/j.jenvman.2021.114273

Bandyopadhyay, S., & Singha, A. (2022). Nanotechnology in bioplastic degradation: Advances and future scope. *Journal of Applied Polymers, 14*(3), 233–241.

Batori, V., Åkesson, D., Zamani, A., Taherzadeh, M. J., & Sárvari Horvath, I. (2018). Anaerobic degradation of bioplastics: A review. *Waste Management, 80*, 406–413. https://doi.org/10.1016/j.wasman.2018.09.040

Bilo, F., Pandini, S., Sartore, L., Depero, L. E., Gargiulo, G., Bonassi, A., Federici, S., & Bontempi, E. (2018). A sustainable bioplastic obtained from rice straw. *Journal of Cleaner Production, 200*, 357–368. https://doi.org/10.1016/j.jclepro.2018.07.252

Carroccio, S. C., Scarfato, P., Bruno, E., Aprea, P., Dintcheva, N. T. & Filippone, G. (2022). Impact of nanoparticles on the environmental sustainability of polymer nanocomposites based on bioplastics or recycled plastics – A review of life-cycle assessment studies. *Journal of Cleaner Production, 335*,130322. https://doi.org/10.1016/j.jclepro.2021.130322

Chandran, R. R., Thomson, B. I., Natishah, A. J., Mary, J., & Nachiyar, V. (2023). Nanotechnology in plastic degradation. *Biosciences Biotechnology Research Asia, 20*(1), 53–68. http://dx.doi.org/10.13005/bbra/3068

Chausali, N., Saxena, J., & Prasad, R. (2022). Recent trends in nanotechnology applications of bio-based packaging. *Journal of Agriculture and Food Research, 7*, 100257. https://doi.org/10.1016/j.jafr.2021.100257

Chu, S., Zhang, B., Zhao, X., Soo, H. S., Wang, F., Xiao, R., & Zhang, H. (2022). Photocatalytic conversion of plastic waste: From photodegradation to photosynthesis. *Advanced Energy Materials, 12*, 2200435.

David, M., Prakash, L., Sangeetha, J., Naik, J., Thangadurai, D., & Thimmappa, S. C. (2020). Degradation of plastics using nanomaterials. In Kharisov, B.I., Martinez L.M.T, Kharissova, O.V. (eds), *Handbook of Nanomaterials and Nanocomposites for Energy and Environmental Applications.* Springer, Germany (pp. 1–13).

Degan, T., & Shinde, S. L. (2019). Waste-plastic processing provides global challenges and opportunities. *Materials Research Society Bulletin, 44*(6), 436–437.

European Bioplastics. (2021). Global bioplastics production will more than triple within the next five years. In *16th EUBP Conference* (pp. 1–2). European Bioplastics.

Folino, A., Pangallo, D., & Calabrò, P. S. (2023). Assessing bioplastics biodegradability by standard and research methods: Current trends and open issues. *Journal of Environmental Chemical Engineering, 11*(2), 109424. https://doi.org/10.1016/j.jece.2023.109424

Fredi, G., & Dorigato, A. (2021). Recycling of bioplastic waste: A review. *Advanced Industrial and Engineering Polymer Research, 4*(3), 159–177. https://doi.org/10.1016/j.aiepr.2021.06.006

Harmaen, A. S., Khalina, A., Ali, H. M., & Azowa, I. (2016). Thermal, morphological, and biodegradability properties of bioplastic fertilizer composites made of oil palm biomass, fertilizer, and poly(hydroxybutyrate-co-valerate). *International Journal of Polymer Sciences, 2016*, 1–8.

Harrison, J. P., Boardman, C., O'Callaghan, K., Delort, A.-M., & Song, J. (2018). Biodegradability standards for carrier bags and plastic films in aquatic environments: A critical review. *Royal Society Open Science, 5*(5), 171792.

Hulla, J. E., Sahu, S. C., & Hayes, A. W. (2015). Nanotechnology: History and future. *Human & Experimental Toxicology, 34*(12), 1318–1321.

Iadiz, M. A. R., Bamedi, M., & Fakour, S. R. (2017). Periodontal diseases and recently applied nanotechnology: A review article. *Health, 9*, 345–351.

Karan, H., Funk, C., Grabert, M., Oey, M., & Hankamer, B. (2019). Green bioplastics as part of a circular bioeconomy. *Trends in Plant Science, 24*, 237–249. https://doi.org/10.1016/j.tplants.2018.11.010

Kawashima, N., Yagi, T., & Kojima, K. (2019). How do bioplastics and fossil-based plastics play in a circular economy? *Macromolecular Materials and Engineering, 304*, 1–14.

Kim, J. Y., & Youn, D. H. (2023). Nanomaterials for advanced photocatalytic plastic conversion. *Molecules, 28*(18), 6502. https://doi.org/10.3390/molecules28186502

Lagaron, J. M., & Lopez-Rubio, A. (2011). Nanotechnology for bioplastics: opportunities, challenges and strategies. *Trends in Food Science & Technology, 22*(11), 611–617. https://doi.org/10.1016/j.tifs.2011.01.007

Lamba, V., & Singh, R. (2024). From waste to resource: Exploring bioplastics and biosorbents as sustainable solutions for plastic waste management. *Nanotechnology Perceptions, 20*(S11), 1204–1211.

Lindström, T., & Österberg, F. (2020). Evolution of biobased and nanotechnology packaging: A review. *Nordic Pulp & Paper Research Journal, 35*(4), 491–515. https://doi.org/10.1515/npprj-2020-0042

Maddela, N. R., Aransiola, S. A., Ezugwu, C. I., Winkelstroter Eller, L.K., Scalvensi, L., & Meng, F. (Eds.). (2024). *Microbial Biotechnology for Bioenergy* (1st ed.). ISBN-9780443141126. https://doi.org/10.1016/C2022-0-02979-2

Manohar, A., & Kumar, P. (2023). Biodegradation of bioplastics with nanomaterials: A review of mechanisms and applications. *Environmental Nanotechnology, Monitoring & Management, 18*, 100695.

Musa, I. O., Auta, H. S., Ilyasu, U. S., Aransiola, S. A., Makun, H. A., Adabara, N. U., Abioye, O. P., Aziz, A., Jayanthi, B., Maddela, N. R., & Prasad, R. (2024). Micro and nanoplastics in environment – Degradation, detection, ecological impact. *International Journal of Environmental Research, 18*(1). https://doi.org/10.1007/s41742-023-00551-9

National Nanotechnology Initiative (NNI). (2020). The National Nanotechnology Initiative: Strategic Plan 2020. U.S. Government. Available at: www.nano.gov

Pathak, D., & Singh, R. (2023). Zinc oxide and TiO_2 nanoparticles in accelerating bioplastic breakdown. *International Journal of Green Nanotechnology, 20*(1), 67–75.

Pathak, V. M., & Kumar, N. (2017). Implications of SiO_2 nanoparticles for in vitro biodegradation of low-density polyethylene with potential isolates of *Bacillus*, *Pseudomonas*, and their synergistic effect on *Vigna mungo* growth. *Energy, Ecology, and Environment, 2*(6), 418–427. https://doi.org/10.1007/s40974-017-0068-5

Raghavendra, G. M., Kim, D., & Seo, J. (2017). One-step synthesis of starch-silver nanoparticle solution and its application to antibacterial paper coating. *International Journal of Biological Macromolecules, 107*, 2285–2290. https://doi.org.10.1016/j.ijbiomac.2017.10.108

Rani, M., Meenu, N., & Shanker, U. (2023). The role of nanomaterials in plastics biodegradability. In *Elsevier eBooks* (pp. 283–308). https://doi.org/10.1016/b978-0-323-89858-4.00012-9

Remya, R. R., Julius, A., Suman, T. Y., Mohanavel, V., Karthick, A., Pazhanimuthu, C., Samrot, A. V., & Muhibbullah, M. (2022). Role of nanoparticles in biodegradation and their importance in environmental and biomedical applications. *Journal of Nanomaterials, 3*, 1–15. http://dx.doi.org/10.1155/2022/6090846

Rodríguez-Hernández, A. G. , Muñoz-Tabares, J. A., Aguilar-Guzmán, J. C., & Vazquez-Duhalt, R. (2019). A novel and simple method for polyethylene terephthalate (PET) nanoparticle production. *Environmental Science Nano, 6*, 2031–2036.

Roy, P. (2022). Nothing is 'waste' in agriculture: From nanotechnology and bioprocesses perspectives In Sharma, A., Choudhury, M., Singh, P, Singh, S. (eds.) (pp. 131–148). *Agriculture Waste Management and Bioresource: The Circular Economy Perspective*. Wiley.https://doi.org/10.1002/9781119808428.ch6

Rujnić-Sokele, M., & Pilipović, A. (2017). Challenges and opportunities of biodegradable plastics: A mini review. *Waste Management Research, 35*, 132–140. https://doi.org/10.1177/0734242X16683272

Saad, S. A., & Elmahjubi, A. (2018). Nanotechnology: Concepts, importance and the current state of scientific research. *CEST-2018, AIJR Proceedings, 4*, 552–562.

Sharma, P., & Thakur, N. (2021). Bioplastic degradation enhanced by nanomaterials. *Materials Today: Proceedings, 45*(2), 1153–1160.

Sharma, S., & Ghosh, R. (2022). Role of cellulose nanofibers in bioplastic biodegradation. *Polymer Science Advances, 15*(5), 303–312.

Tsang, Y. F., Kumar, V., Samadar, P., Yang, Y., Lee, J., Ok, Y. S., Song, H., Kim, K. H., Kwon, E. E., & Jeon, Y. J. (2019). Production of bioplastic through food waste valorization. *Environment international, 127*, 625–644. https://doi.org/10.1016/j.envint.2019.03.076

Urbanek, A. K., Rymowicz, W., Strzelecki, M. C., Kociuba, W., Franczak, Ł., & Mirończuk, A. M. (2017). Isolation and characterization of Arctic microorganisms decomposing bioplastics. *AMB Express, 7*(1). https://doi.org/10.1186/s13568-017-0448-4

Zhao, X., Wang, Y., Chen, X., Yu, X., Li, W., Zhang, S., Meng, X., Zhao, Z. M., Dong, T., Anderson, A., Aiyedun, A., Li, Y., Webb, E., Wu, Z., Kunc, V., Ragauskas, A., Ozcan, S., & Zhu, H. (2023). Sustainable bioplastics derived from renewable natural resources for food packaging. *Matter,* 6(1), 97–127.

18 Social Implications of Bioplastic Waste

Olumide Ayobami Iyiola and Oluwatobi Sunday Akande

18.1 INTRODUCTION

The growing environmental concerns associated with conventional plastics have spurred the development of bioplastics; materials derived from renewable biomass such as corn starch, sugarcane, and other plant-based resources. Bioplastics are positioned as a sustainable alternative to petroleum-based plastics, offering potential solutions to the escalating issues of plastic pollution and its impact on ecosystems. Unlike traditional plastics, many bioplastics are designed to decompose more rapidly under specific conditions, reducing the long-term environmental footprint (Crawford & Quinn, 2017; Khoshtinat , 2023). However, the adoption of bioplastics raises critical questions about their production, use, and disposal, particularly from the perspective of their social and environmental implications.

While bioplastics offer several environmental advantages, such as lower greenhouse gas emissions and reduced dependence on fossil fuels, their benefits are not universally realized. The production of bioplastics often competes with food production by utilizing agricultural land and resources, creating potential conflicts in resource allocation (Rosenboom et al., 2022; Mongkoldhumrongkul & Sukkanta, 2022). Moreover, effective disposal remains a significant challenge. Bioplastics require specific conditions for degradation, such as industrial composting facilities, which are not widely accessible in many regions, particularly rural and low-income areas (Ogwu et al., 2025; Mihai et al., 2021). Improper disposal may lead to environmental contamination, undermining their perceived ecological benefits (Zhu & Wang, 2020).

The challenges surrounding bioplastics extend beyond environmental concerns to their broader socio-economic implications. In particular, rural and vulnerable communities often bear the brunt of inadequate waste management systems, which are ill-equipped to handle the specialized disposal requirements of bioplastics. This disparity exacerbates existing inequalities; as such, regions frequently lack the industrial composting facilities necessary for proper bioplastic degradation (Mihai et al., 2021Mihai, 2017). Furthermore, the lifecycle of bioplastics, from production to disposal, raises ethical concerns about resource allocation and environmental justice, as agricultural land diverted for bioplastic production could otherwise support food security in vulnerable areas (Rosenboom et al., 2022).

18.1.1 Environmental Impact of Bioplastic Waste

The increasing threats of plastics to the natural environment encouraged the production of bioplastics from renewable biomass resources. The premium quality of bioplastics is mainly produced by treating starch with glycerol. Plastics are basically non-biodegradable synthetic or semi-synthetic

DOI: 10.1201/9781003602828-22

products. Plastics are widely used commercial products made of synthetic and semi-synthetic raw materials. Plastics are non-biodegradable or take centuries to degrade under natural climatic conditions (Crawford & Quinn, 2017). These are light, versatile, moisture-resistance, tough, and cost-efficient, which are commercially used for packaging, textiles, and in electronic devices all over the world. Plastics are considered non-biodegradable because it disintegrates in a period beyond the human time frame (Li et al., 2005).

With the advances in technology and the increase in the global population, plastic materials have found wide applications in every aspect of life and industries. However, most conventional plastics such as polyethylene, polypropylene, polystyrene, poly(vinyl chloride), and poly(ethylene terephthalate) are non-biodegradable, and their increasing accumulation in the environment has been a threat to the planet. To overcome all these problems, some steps have been undertaken. The first strategy involved production of plastics with high degree of degradability (Khoshtinat , 2023). The environmental problems caused by discarded synthetic plastics have paved the way for the search for substitutes. Bioplastics, which are both functionally similar to synthetic plastics and environmentally sustainable, are touted as promising new materials to address these problems. The word "bio-plastic" is used confusingly. In our understanding, however, bioplastics consist of either biodegradable plastics (i.e., plastics produced from fossil materials) or bio-based plastics (i.e., plastics synthesized from biomass or renewable resources) (Khoshtinat , 2023). Biodegradable plastics are seen by many researchers as a promising solution to above-described problem because they are "environmentally-friendly".

18.1.2 Degradation Process of Bioplastic

The bioplastics are introduced as an alternative to non-renewable petroleum-based plastics due to their ability to decompose in shorter time as compared to petroleum-based plastics. Bioplastics decompose through microbial activity of soil. In soil, microorganisms produce enzymes under optimal conditions to degrade bioplastics (Liwarska-Bizukojc, 2021). Different factors affecting the decomposition activity of bioplastics, including pretreatment, microorganisms, and characteristics of polymer. The general characteristics of polymers include molecular weight, mobility, and nature of plasticizers, tactility, and crystallinity. The decomposition activity usually takes place in optimal conditions, including humid conditions because hydrophilic enzymes usually occur in moist environment. In soil, various organisms including bacteria, fungi, or worms secrete amylase enzyme in external environment which breaks down starch (Mary et al., 2021).

This enzyme degrades the insoluble starch materials to soluble products which are usually absorbed by microbial cells; these products include maltose and glucose. During decomposition, the large polymers are degraded into monomers by breaking hydrogen bonds through enzymes, thus viscosity is reduced and the product is liquefied (Kunam et al., 2024). Different saccharides are formed in a process known as scarification following the breakage of bonds. The microbial activity of soil is increased in the presence of water, oxygen, temperature, and light intensity.

Thus, the bioplastics readily degrade in presence of water. The bioplastics are usually coated on paper bags to increase the stability of paper bags and to retain oil and moisture (Emadian et al., 2017).

18.1.3 Potential Environmental Benefits and Drawbacks of Bioplastic Waste

Bioplastics are emerging to be highly controversial when it comes to determining their impact on the environment. While bioplastics are often hailed as excellent alternatives to conventional plastics, they are also associated with shortcomings (Licciardello and Piergiovanni, 2020). Let us consider the case of biodegradable bioplastics. Biodegradable bioplastics can decompose into natural materials through microbial mechanisms and blend harmlessly into the soil (Qin et al., 2021; Kaur

& Chauhan , 2024). This decomposition process is aided by water and/or oxygen. For example, when a cornstarch-derived bioplastic is composted, the cornstarch molecules slowly absorb water and swell up when buried. This causes the starch bioplastic to break apart into small fragments that can then be easily digested by bacteria (Qin et al., 2021, Kundu & Payal , 2022).

Biodegradable plastics offer a lot of advantages, such as increased soil fertility, low accumulation of bulky plastic materials in the environment (which invariably will minimize injuries to wild animals), and reduction in the cost of waste management. Furthermore, biodegradable plastics can be recycled to useful metabolites (monomers and oligomers) by microorganisms and enzymes. A second strategy involves degradation of some petroleum-derived plastics by biological processes. A typical example can be seen in the case of some aliphatic polyester, such as polycaprolactone (PCL) (Khoshtinat, 2023) and biodegradable plastics that can be degraded by enzymes and microorganisms (Khoshtinat, 2023). Studies have also shown that polycarbonates (particularly the aliphatic types) possess some degree of biodegradability (Yu et al., 2021).

Two classes of biodegradable plastics, aliphatic polymers and aliphatic–aromatic co-polymers, have become popular in the industry. Aliphatic polymers can be biodegraded with relative ease, however, hold little market value due to their fragile nature and low durability (Jha et al., 2024;). In turn, development of aliphatic–aromatic co-polymers that combine the excellent mechanical properties of conventional aromatic polymers and biodegradability of aliphatic polymers has received a boost (Lim & San Thian , 2022; Savitha et al., 2022). However, proper disposal of biodegradable plastics (i.e., complete neutralization and little contamination with other waste and human food) remains challenging, especially in the existing framework of waste management (Lim & San Thian, 2022; Shah et al., 2014).

Composting plays an important and growing role in sustainable organic waste management and recycling. However, plastics are one of the main contaminants in composts. Biodegradable plastics are meant to address this problem. Composting of these materials also reduces their environmental impact in fact that they will largely be converted to CO_2 and not to CH_4 as if they were in a landfill (Nordahl et al., 2023). Above all, composting represents one of the technologies of processing of biodegradable municipal waste (Ansar et al., 2025).

18.2 IMPACT OF BIOPLASTIC WASTE ON HEALTH AND SAFETY

Plastic waste management is a critical issue. Over 300 million metric tons of plastics are produced in the world annually, and about 50% of this volume is for disposal applications, products that are discarded within a year of their purchase. It is the boon and bane of our times. Although there are multiple uses, its waste and the resultant pollution clog up our rivers, oceans, lands and adversely affect the biodiversity (Kolawole & Iyiola, 2023). Challenges in waste monitoring and strategic management range from environmentally friendly economic sectors exist but face difficulties in terms of capital, unstable output of products, competitive advantages, and low profits. The rapid loss of biodiversity, depletion of natural ecosystems, unsustainable use of natural resources, depletion of non-renewable resources, and over-exploitation of fishing resources, especially in the inshore areas.

Emerging pollution of inland river sediments with microplastics and consequently the implications on rivers as vectors of transport of microplastics into the ocean are worth consideration. The collection of municipal solid waste is a public service that has important impacts on public health and the appearance of towns and cities (Kwun Omang et al., 2021).

18.2.1 Potential Health Concerns from Bioplastic Degradation

In medicine alone, the diversity of plastics' uses is incredible. Prosthetics, engineered tissues, and micro needle patches for drug delivery are all possible with polymers (Mansuri et al., 2025). In many sectors of society, plastics have replaced glass, wood, fibers, and metal in various products, including dishware, clothing, food packaging, personal care products, and others; these uses have already been explored elsewhere (Patti & Acierno, 2022). Polymers have demonstrated their value in pharmaceuticals, as controlled drug delivery systems used by tens of millions of people each year, and in orthopedics: the polymer polymethylmethacrylate is used as bone cement in total hip replacements (ASM). The array of different types of polymers that can be produced has had an enormous effect on the quality of human health through these innovations in medical research.

However, the widespread use of plastics facilitates continuous contact of these materials with the human body and with it daily exposure to ingredients in plastics. Although plastics components do not have significant bioaccumulation potential (except when accidentally ingested and becoming entrapped in the gastrointestinal system), bio monitoring studies (Rodrigues et al., 2023) have demonstrated the presence of steady-state concentration of plastics' components in the human body, thereby reflecting the ongoing balance of constant exposure, metabolism, and excretion of these compounds. This situation implies that in today's plastics-enabled society there are no control groups to be found to analyze the effects on human health from low-level environmental exposures to plastic constituents. Everybody is being exposed to some degree at any given time from gestation through death. Detectable levels of bisphenol A (BPA) have been found in the urine of 95% of the adult population of the United States (Colorado-Yohar et al., 2021).

In recent years there have been several epidemiological studies and controlled animal experiments performed concerning the health effects of plastic components, such as endocrine disrupting BPA and di-(2-ethylhexyl) phthalate (DEHP). Associations were found between exposure to these compounds and destructive effects on health and reproduction, such as early sexual maturation, decreased male fertility, aggressive behavior, and others (Vandenburg, 2007; Rodrigues et al., 2023).

Several rodent and human studies have found correlations between DEHP exposure and harmful health effects, including changes to the female and male reproductive systems, increased waist circumference, and insulin resistance (Wang et al., 2024). The "cocktail effect", i.e., the fact that environmental exposures do not occur in isolation but are cumulative, is an issue requiring further study. Although evidence due to research so far has been sufficient for governmental agencies to take action, more is still being done to demonstrate conclusively that the negative health effects listed above and others are due to the compounds such as BPA and DEHP in plastic. For example, a recent study found a significant association between urinary concentration of BPA and obesity among children and adolescents (Aktağ et al.,2021). However, as noted by the authors, "explanations of the association cannot rule out the possibility that obese children ingest food with higher BPA content or have greater adipose stores of BPA" (Aktağ et al., 2021).

18.2.2 Effects of Bioplastic Waste on Rural Communities

Rural lands and populations serve as essential providers of food and resources for expanding urban areas, yet these communities face significant environmental challenges. The rising prevalence of bioplastic waste presents a novel dimension to the existing environmental crisis in rural settings. Coupled with climate change, biodiversity loss, and general plastic pollution, bioplastic waste complicates efforts to maintain the integrity of rural ecosystems. These challenges exacerbate socio-economic problems such as poverty, depopulation, and marginalization, which are often the outcome of unsustainable linear economy practices like the overreliance on fossil fuels, resource depletion, intensive agriculture, and waste mismanagement (Hariram et al., 2023).

In many low- and middle-income countries, rural areas lack adequate infrastructure for waste management and sanitation services, leaving them particularly vulnerable to bioplastic pollution. Globally, approximately 1.9 billion people do not have access to regular waste collection services in rural areas, highlighting the disparity in infrastructure development between urban and rural environments (Vinti & Vaccari, 2022). As a result, significant amounts of waste, including bioplastic materials, end up leaking into the natural environment. These deficiencies contribute to severe pollution, undermining the livelihoods and health of rural populations.

Pollution from bioplastics compounds existing issues by entering rural ecosystems through interconnected air, water, and soil pathways. Rural communities often rely on open landfills, illegal dumping practices, and the burning of waste to manage their waste streams (Vinti & Vaccari, 2022; Mihai et al., 2021). These activities release hazardous compounds, microplastics, and toxic byproducts into the environment, further contaminating vital natural resources such as soil and freshwater bodies. This pollution not only disrupts agricultural productivity but also compromises the quality of drinking water and air, presenting long-term health risks to rural populations.

Bioplastics, while often marketed as environmentally friendly, may not always break down effectively in natural environments, especially without industrial composting conditions. Instead, they can fragment into microplastics, which infiltrate soil and water systems, affecting both terrestrial and aquatic ecosystems. Rural agricultural activities are particularly vulnerable, as contaminated soil can reduce crop yields and introduce harmful pollutants into the food chain. Furthermore, the microplastic contamination of water bodies affects local fisheries, reducing their productivity and threatening the food security of dependent communities.

The impacts of bioplastic waste on rural areas extend beyond environmental degradation to economic and social dimensions. As rural communities face declining agricultural productivity due to polluted soils and waters, they are often forced to migrate to urban centers, contributing to rural depopulation. This depopulation creates a cycle of abandonment and reduced investment in rural infrastructure, further exacerbating the vulnerability of these areas to environmental challenges. Additionally, the loss of arable land due to pollution can increase reliance on imported food, raising costs and undermining local economies.

Efforts to mitigate bioplastic waste pollution in rural communities are often constrained by a lack of education and awareness about waste management practices. In many cases, rural residents are not adequately informed about the proper disposal and potential hazards of bioplastic materials. This lack of awareness perpetuates harmful practices like open dumping and burning, emphasizing the need for community-focused education campaigns and policy interventions aimed at sustainable waste management. Policy responses to the issue must prioritize the development of affordable and accessible waste management systems tailored to rural contexts. Investments in decentralized waste collection and processing facilities can help reduce pollution while creating local job opportunities.

18.3 SOCIAL AND ENVIRONMENTAL IMPACTS OF BIOPLASTIC WASTE ON COMMUNITIES

Bioplastics are emerging as a promising alternative to traditional plastics, driven by the need for more sustainable options. This chapter stresses the importance of developing robust disposal systems for biodegradable materials and supports recycling initiatives to minimize the need for new resources. Bioplastics are materials made from renewable sources like plants, biomass, or microorganisms, offering a more sustainable alternative to traditional plastics (Rosenboom et al., 2022). They can be bio-based, biodegradable, or both (Nazareth et al., 2022). The goal of bioplastics is to address the environmental issues linked with conventional plastics, which are made from fossil fuels and have harmful effects throughout their life cycle, from extraction and production to use and disposal (Moshood et al., 2022). Bio-based bioplastics often use crops or forestry byproducts, which helps cut carbon emissions and reduce reliance on fossil fuels (Mongkoldhumrongkul & Sukkanta, 2022; Rosenboom et al., 2022). They also offer a way to use fewer finite resources, such as petroleum, which is the base material for traditional plastics (Zhao et al., 2023).

Bioplastics can potentially improve waste management. Being biodegradable, they break down into natural elements, reducing plastic waste in landfills and oceans (Ogwu et al., 2025). Some can be recycled or composted, supporting circular economy practices by promoting reuse and recycling.

Bioplastics generally have a lower environmental impact than traditional plastics and offer sustainable alternatives (Altalhi, 2022). Their adoption drives research and innovation in sustainable materials (Palsra & Chauhan, 2020).

18.3.1 Bioplastic Waste Management

Composting is a common method for bioplastics made from materials like starch, which are specifically designed to be compostable (Ahsan et al., 2023). During composting, organic matter is broken down into nutrient-rich compost in a controlled environment. Certified compostable bioplastics can decompose in industrial composting facilities, where conditions like temperature, humidity, and microbial activity help them break down into water, carbon dioxide, and biomass within a set period (Nizamuddin et al., 2024). Composting is a sustainable disposal method as it creates valuable compost that enriches the soil (Manea et al., 2024). Another method is anaerobic digestion, where microorganisms break down organic materials without oxygen. Some bioplastics, such as certain bio polyesters, can be processed this way to produce biogas (Vardar et al., 2022).

The disposal of bioplastics involves several methods, each with its own environmental implications. Anaerobic digestion uses microorganisms to break down bioplastics without oxygen, producing biogas (mainly methane and carbon dioxide) for energy (Yu et al., 2023). Mechanical recycling involves sorting, cleaning, shredding, and reprocessing bioplastic waste into granules or pellets, which helps reduce the need for new raw materials. Its effectiveness depends on the type of bioplastic, contamination levels, and recycling infrastructure (Kumar et al., 2023). The choice of disposal method for bioplastics depends on factors like the specific type of bioplastic, local waste management infrastructure, and environmental considerations (Mousavi et al., 2024). Effective management also requires proper collection systems, clear labeling, and consumer education.

18.3.2 Effects of Bioplastic Waste on Rural Communities

Rural lands and rural populations are the main providers of food and resources for the ever-expanding urban areas. At the same time, rural communities around the world are facing serious environmental challenges related to climate change, biodiversity loss, and plastic pollution of the natural environment. Rural communities are often lacking basic services such as waste and sanitation management services, particularly in low- and middle-income countries (Vinti & Vaccari, 2022). At the global level, 1.9 billion people lack access to regular waste collection schemes in rural areas (Vinti & Vaccari, 2022) and waste management infrastructure is underdeveloped (Kumari & Raghubanshi, 2023), resulting in domestic plastic pollution leakage into the natural environment. Cheap plastic packaging materials are accessible to rural communities and even to remote villages (Mihai et al., 2021). Plastic pollution covers a wide spectrum of natural and semi-natural habitats that are found in rural areas, including coastal and island communities, threatening SDGs 14 (Life below Water) and 15 (Life on Land).

18.3.2.1 Water Resources Pollution

Plastic pollution in rural water resources remains an understudied and often overlooked issue. Despite the lower population density in rural areas, macroplastic (>5 mm) and microplastic (<5 mm) pollution have become persistent problems, largely due to inadequate or nonexistent waste management systems (Akindele & Alimba, 2021). Without formal waste collection or recycling programs, rural communities frequently resort to unsanctioned disposal methods such as open dumping and burning, which significantly contribute to the accumulation of plastics in water bodies (Mihai, 2019). These practices lead to the leaching of toxic substances into soil and water, thereby threatening the safety and usability of water resources. Bioplastic waste, while marketed as a more sustainable alternative, may also contribute to water resource pollution if not properly managed. Misconceptions about their biodegradability can lead to improper disposal, as many bioplastics require industrial composting conditions to decompose effectively. In water systems, bioplastic fragments may behave similarly to conventional plastics, fragmenting into microplastics that persist in aquatic environments (Zhu & Wang, 2020). These fragments can adsorb harmful substances such as heavy metals and hydrophobic organic contaminants further exacerbating water pollution.

Bioplastics can serve as substrates for microbial colonization, leading to the formation of a "plastisphere" in aquatic ecosystems (Wright et al., 2020). This microbial ecosystem may alter natural water systems by introducing new pathogens or disrupting existing microbial communities. The presence of such microbial activity can degrade water quality, making it unsafe for consumption and agricultural use. Furthermore, rural communities often depend on these water sources for irrigation and daily needs, amplifying the health risks posed by polluted water. In regions with inadequate waste management infrastructure, water pollution from plastics, including bioplastics, poses a direct threat to sustainable development. Polluted water resources impact agricultural productivity, aquatic biodiversity, and human health, especially in areas where livelihoods depend on the availability of clean water. Addressing this issue requires targeted interventions such as improving waste management systems, educating communities about bioplastic disposal, and implementing strict regulations for plastic and bioplastic waste handling. A collaborative effort among governments, local communities, and non-governmental organizations is essential to mitigate the effects of plastic and bioplastic waste on rural water resources.

18.3.2.2 Soil Pollution and Farming Activities

Rural land, particularly agricultural soil, is highly vulnerable to microplastic pollution due to its direct exposure to agricultural practices and waste mismanagement. Agro-ecosystems play an essential role in ensuring food security and maintaining biodiversity (Gairola et al., 2025). However, these systems face mounting pressure from climate change, population growth, and the introduction of microplastics into the soil. Microplastics, often originating from degraded plastics, synthetic fertilizers, and plastic mulching, can alter soil structure, reduce fertility, and impact water retention, thereby compromising agricultural productivity.

The infiltration of microplastics into soil presents a multifaceted challenge. These particles can disrupt soil microbial communities, which are essential for nutrient cycling and organic matter decomposition (Zhu et al., 2020). Also, plastic particles can adsorb and transport harmful chemicals such as pesticides and heavy metals, further exacerbating soil contamination and introducing these pollutants into the food chain (Okoye et al., 2022). Farmers in rural communities are particularly affected by these impacts, as they depend on healthy soil for their livelihoods. Reduced soil productivity can lead to lower crop yields, jeopardizing food security in areas where agriculture is already under strain. Furthermore, the persistence of microplastics in soil can lead to long-term degradation of arable land, making it less viable for future farming activities. Efforts to address soil pollution from microplastics require a comprehensive approach, including stricter regulations on plastic use in agriculture, the promotion of biodegradable alternatives, and improved waste management practices to prevent plastic leakage into rural environments. Also, educating farmers about the risks associated with plastic usage and encouraging sustainable farming practices can help mitigate the impact of soil pollution. With coordinated efforts, it is possible to protect rural agricultural systems from the growing threat of microplastic contamination and ensure the sustainability of food production for future generations.

18.3.2.3 Food Contamination and Public Health Threats

Food packaging is one of the largest sectors for plastic use globally. In 2020, food packaging accounted for 39.6% of total plastic production in Europe, highlighting the scale of its prevalence (Stasiškienė et al., 2022). These packaging materials are not only composed of plastics but also include combinations of other materials such as metals, glass, paper, adhesives, and coatings, all of which may introduce a variety of harmful substances into the food supply. While plastics themselves are designed to protect and preserve food, they can also pose significant health risks when they degrade or leach into food products.

Plastic packaging often contains residues of solvents used in the manufacturing process, as well as unintentional substances like impurities, oligomers, and byproducts from the degradation of plastics (Mihali et al., 2021). These substances can migrate from the packaging into food, especially when exposed to heat, moisture, or acidic conditions. This migration can lead to the contamination of food with potentially harmful chemicals, such as BPA, phthalates, and other endocrine-disrupting compounds. Over time, prolonged exposure to these chemicals has been linked to various health issues, including reproductive problems, hormonal imbalances, and increased risks of cancer (Vandenburg, 2007; Trasande et al., 2012).

The presence of microplastics in food, particularly from the consumption of packaged or processed goods, further compounds the risk. Studies have shown that microplastics can enter the human body through direct ingestion of contaminated food, water, or through the inhalation of particles in the air (Ageel et al., 2022). Once ingested, microplastics may accumulate in the gastrointestinal tract and other organs, potentially causing physical harm or contributing to inflammatory responses (Browne et al., 2011). Although the long-term effects of microplastic exposure on human health remain unclear, the growing body of evidence raises serious concerns, particularly in vulnerable populations such as children and pregnant women.

In rural communities, where food security often relies on locally produced crops and limited access to healthcare, the contamination of food by plastics represents a significant public health threat. The widespread use of plastic packaging, combined with inadequate waste management infrastructure, increases the likelihood of plastics contaminating both the environment and the food chain.

18.3.2.4 Illegal Plastic Waste Trade

Plastics, with their unique properties of durability, flexibility, and low cost, have become an essential global commodity, deeply embedded in nearly every aspect of modern life. From packaging to electronics, construction materials, and agricultural applications, the widespread use of plastic has made it an integral part of industrial and consumer products. However, this ubiquity also means that plastic waste has proliferated at an alarming rate, especially in rural communities where waste management systems are often inadequate or nonexistent. As a result, large quantities of plastic waste accumulate in the environment, contributing to pollution and posing long-term environmental challenges.

In many rural areas, the lack of formal recycling infrastructure and proper waste disposal systems has led to a rise in the illegal plastic waste trade. In this illicit trade, plastic waste is often illegally dumped or shipped to countries with less stringent environmental regulations. These regions, typically low- or middle-income countries, become destinations for plastic waste that is not processed or recycled properly in the originating countries. This trade is driven by the desire to bypass expensive waste management systems, enabling the exporting countries to avoid costly recycling processes or landfill fees (Rossi & Morone, 2023).

Illegal plastic waste trade has far-reaching social and environmental consequences. The uncontrolled importation of plastic waste often leads to unsafe disposal practices, such as open burning or dumping in landfills, both of which result in toxic emissions and contamination of local water and soil resources. Moreover, plastic waste that is improperly recycled or dumped may eventually find its way into nearby rivers and oceans, exacerbating pollution on a global scale (Kibria et al., 2023). Rural communities that receive plastic waste often lack the resources and technical capacity to manage these materials safely, increasing the risk of exposure to hazardous chemicals and microplastics that can enter the food chain (Akindele & Alimba, 2021).

This trade in illegal plastic waste also highlights broader issues of environmental justice and inequality. Rural communities, which are already vulnerable due to limited infrastructure and economic opportunities, bear the burden of this waste trade. The influx of plastic waste into these areas not only harms local ecosystems but also places a significant strain on community health and local economies.

18.4 CONCLUSION

The shift from conventional plastics to bioplastics represents a critical milestone in addressing the global environmental crises caused by plastic waste. Derived from renewable resources such as corn, sugarcane, and other biomaterials, bioplastics provide a sustainable alternative to traditional petroleum-based plastics. Their potential environmental benefits are significant, including reduced dependence on finite fossil fuels, lower greenhouse gas emissions, and the possibility of minimizing the ecological footprint associated with plastic products (Sultan et al., 2020). Despite their promise, bioplastics are not without challenges, and their adoption must be approached thoughtfully to ensure they achieve their intended environmental objectives.

One of the most pressing concerns surrounding bioplastics is their reliance on agricultural resources. The feedstocks used in their production, such as corn and sugarcane, often compete with crops meant for food and feed, raising critical issues about land use and food security (Brizga et al., 2021). In regions already grappling with food shortages, diverting arable land to cultivate crops for bioplastic production exacerbates existing pressures on agricultural systems. This competition for resources disproportionately impacts rural and low-income communities that depend on agriculture for their livelihoods, intensifying socio-economic inequalities (Zhang & Zhao, 2024). A more sustainable approach would involve prioritizing alternative feedstocks such as agricultural residues, non-food biomass, or algae, which do not compete with food production and reduce the strain on arable land (Meyers et al., 2022).

Another significant challenge lies in the end-of-life management of bioplastics. While many bioplastics are marketed as biodegradable or compostable, the conditions required for their proper decomposition are often stringent, necessitating industrial composting facilities (Rujnić-Sokele & Pilipović, 2017). These facilities are scarce, particularly in developing and underdeveloped regions, which lead to improper disposal of bioplastics. When discarded inappropriately, bioplastics may persist in the environment, breaking down into microplastics that contaminate soil and water systems (Narancic et al., 2018). This not only undermines the ecological advantages of bioplastics but also exacerbates the growing burden of plastic waste management, particularly in rural and under-resourced areas.

The lack of awareness regarding bioplastics further complicates their disposal. A common misconception is that bioplastics are inherently environmentally friendly and will degrade naturally in any setting. This misunderstanding has led to inappropriate disposal practices such as open dumping and burning, which release harmful pollutants into the air and soil (European Bioplastics, 2021). Such practices negate the environmental benefits of bioplastics and highlight the need for robust public education campaigns. These campaigns should emphasize proper disposal methods, the importance of separating bioplastics from conventional plastics, and the environmental implications of improper waste management (Hottle et al., 2017). The socio-economic implications of bioplastic production and disposal are equally significant. While the bioplastics industry holds the potential to generate economic opportunities, including jobs in research, production, and waste management, the benefits are not evenly distributed. Rural and low-income communities, already burdened by limited resources and infrastructure, often bear the environmental and economic costs of bioplastic production and waste mismanagement (Shen et al., 2020). Policymakers must ensure that the transition to bioplastics includes measures to equitably distribute these benefits, particularly by supporting communities that are most vulnerable to the associated challenges.

Efforts to make the bioplastics industry sustainable must involve an integrated approach. Investing in research and development is critical to improve the efficiency of bioplastic production processes and identify innovative feedstocks that do not compete with food resources (Nanda & Berruti, 2021). Additionally, advancing technologies to enable bioplastics to degrade effectively in natural environments could help reduce their environmental impact (Narancic et al., 2018). Simultaneously, governments and industry stakeholders must prioritize the development of infrastructure for composting and recycling bioplastics to prevent environmental contamination and promote a circular economy. Regulations and standards are also essential to guide the responsible adoption of bioplastics. Setting clear criteria for biodegradability and compostability, incentivizing the use of sustainable feedstocks, and enforcing waste management guidelines are important steps in mitigating the environmental and socio-economic challenges associated with bioplastics (Karan et al., 2019). Collaborative efforts at the global level will be necessary to harmonize policies, share best practices, and ensure that all countries, particularly those with limited resources, can benefit from the transition to bioplastics.

REFERENCES

Ageel, H. K., Harrad, S., & Abdallah, M. A. E. (2022). Occurrence, human exposure, and risk of microplastics in the indoor environment. *Environmental Science: Processes & Impacts*, *24*(1), 17-31.

Ahsan, W. A., Hussain, A., Lin, C., & Nguyen, M. K. (2023). Biodegradation of different types of bioplastics through composting—a recent trend in green recycling. *Catalysts*, *13*(2), 294.

Akindele, E. O., & Alimba, C. G. (2021). Plastic pollution in rural water resources: A growing concern. *Environmental Science Journal*, 28(3), 456–472.

Aktağ, E., Yurdakök, K., Yalçın, S. S., & Kandemir, N. (2021). Urinary bisphenol A levels in prepubertal children with exogenous obesity according to presence of metabolic syndrome. *Journal of Pediatric Endocrinology and Metabolism*, *34*(4), 495-502.

Altalhi, A. (2022). Bio plastics and their role in sustainable waste management. *Journal of Environmental Management*, 14(2), 150–160.

Andrady, A. L., & Neal, M. A. (2009). Applications and societal benefits of plastics. *Philosophical Transactions of the Royal Society B: Biological Sciences*, 364(1526), 1977–1984. https://doi.org/10.1098/rstb.2008.0304

Ansar, A., Du, J., Javed, Q., Adnan, M., & Javaid, I. (2025). Biodegradable Waste in Compost Production: A Review of Its Economic Potential. *Nitrogen*, *6*(2), 24.

Brizga, J., Hubacek, K., & Feng, K. (2021). Environmental impacts of bioplastics: A review. *Sustainability*.

Browne, M. A., Dissanayake, A., Galloway, T. S., Lowe, D. M., & Thompson, R. C. (2011). Ingested microplastic translocates to the circulatory system of the mussel, *Mytilus edulis*. *Environmental Science & Technology*, 42(4), 5026–5031. https://doi.org/10.1021/es800249a

Calafat, A. M., Kuklenyik, Z., Reidy, J. A., Caudill, S. P., Ekong, J., & Needham, L. L. (2005). Urinary concentrations of bisphenol A and 4-nonylphenol in a human reference population. *Environmental Health Perspectives*, 113(4), 391–395. https://doi.org/10.1289/ehp.7534

Colorado-Yohar, S. M., Castillo-González, A. C., Sánchez-Meca, J., Rubio-Aparicio, M., Sánchez-Rodríguez, D., Salamanca-Fernández, E., ... & Chirlaque, M. D. (2021). Concentrations of bisphenol-A in adults from the general population: A systematic review and meta-analysis. *Science of the Total Environment*, *775*, 145755.

Crawford, R. J., & Quinn, J. (2017). *Plastics engineering*. Elsevier.

Emadian, S. M., Onay, T. T., & Demirel, B. (2017). Biodegradation of bioplastics in natural environments. *Waste Management*, 59, 526–536.

European Bioplastics. (2021). Understanding the Environmental Benefits and Challenges of Bioplastics.

Gairola, A., Kumar, V., Sharma, D., Kumar, S., & Rathore, S. S. (2025). Ecosystem Services in Agro-Ecosystems: A Step Towards Sustainability. In *Ecological Solutions to Agricultural Land Degradation* (pp. 401-429). Springer, Singapore.

Geyer, R., Jambeck, J. R., & Law, K. L. (2017). Production, use, and fate of all plastics ever made. *Science Advances*, 3(7), e1700782. https://doi.org/10.1126/sciadv.1700782

Hariram, N. P., Mekha, K. B., Suganthan, V., & Sudhakar, K. (2023). Sustainalism: An integrated socio-economic-environmental model to address sustainable development and sustainability. *Sustainability*, *15*(13), 10682.

Hottle, T. A., Bilec, M. M., & Landis, A. E. (2017). Sustainability assessments of bioplastics: A review. *Sustainable Materials and Technologies*.

Jha, S., Akula, B., Enyioma, H., Novak, M., Amin, V., & Liang, H. (2024). Biodegradable biobased polymers: A review of the State of the Art, Challenges, and Future Directions. *Polymers*, *16*(16), 2262.

Karan, H., Funk, C., & Grabert, M. (2019). Green bioplastics as part of a circular bioeconomy. *Trends in Plant Science*.

Kaur, R., & Chauhan, I. (2024). Biodegradable plastics: Mechanisms of degradation and generated bio microplastic impact on soil health. *Biodegradation*, *35*(6), 863-892.

Khoshtinat, S. (2023). State-of-the-art review of aliphatic polyesters and polyolefins biodeterioration by microorganisms: from mechanism to characterization. *Corrosion and Materials Degradation*, *4*(4), 542-572.

Kibria, M. G., Masuk, N. I., Safayet, R., Nguyen, H. Q., & Mourshed, M. (2023). Plastic waste: Challenges and opportunities to mitigate pollution and effective management. *International Journal of Environmental Research*, *17*(1), 20.

Kolawole, A. S., & Iyiola, A. O. (2023). Environmental pollution: Threats, impact on biodiversity, and protection strategies. In *Sustainable utilization and conservation of Africa's biological resources and environment* (pp. 377-409). Singapore: Springer Nature Singapore.

Kumar, R., Sadeghi, K., Jang, J., & Seo, J. (2023). Mechanical, chemical, and bio-recycling of biodegradable plastics: A review. *Science of the Total Environment*, *882*, 163446.

Kumari, T., & Raghubanshi, A. S. (2023). Waste management practices in the developing nations: Challenges and opportunities. *Waste management and resource recycling in the developing world*, 773-797.

Kunam, P. K., Ramakanth, D., Akhila, K., & Gaikwad, K. K. (2024). Bio-based materials for barrier coatings on paper packaging. *Biomass Conversion and Biorefinery*, *14*(12), 12637-12652.

Kundu, R., & Payal, P. (2022). Biodegradation study of potato starch-based bioplastic. *Current Chinese Chemistry*, *2*(2), e190421192895.

Kwun Omang, D. I., John, G. E., Inah, S. A., & Bisong, J. O. (2021). Public health implication of solid waste generated by households in Bekwarra Local Government area. *African Health Sciences*, *21*(3), 1467-1473.

Li, H., Li, L., Schwartz, D. J., Metz, M. V., Marks, T. J., Liable-Sands, L., & Rheingold, A. L. (2005). Coordination copolymerization of severely encumbered isoalkenes with ethylene: Enhanced enchainment mediated by binuclear catalysts and cocatalysts. *Journal of the American Chemical Society*, *127*(42), 14756-14768.

Licciardello, F., & Piergiovanni, L. (2020). Packaging and food sustainability. In *The interaction of food Industry and environment* (pp. 191-222). Academic Press.

Lim, B. K. H., & San Thian, E. (2022). Biodegradation of polymers in managing plastic waste—A review. Science of the Total Environment, *813*, 151880.

Liwarska-Bizukojc, E. (2021). Effect of (bio) plastics on soil environment: A review. *Science of the Total Environment*, *795*, 148889.

Manea, E. E., Bumbac, C., Dinu, L. R., Bumbac, M., & Nicolescu, C. M. (2024). Composting as a sustainable solution for organic solid waste management: Current practices and potential improvements. *Sustainability*, *16*(15), 6329.

Mansuri, S. S., Patil, H. S., Nunse, D. K., Jumde, S. M., Gupta, P. R., Talele, S. G., & Borse, L. B. (2025). Polymer and Plastic Fundamentals and Its Connecting with Current Environmental, Biomedical, and Pharmaceutical Engineering. In *Sustainability in Polymer Technology and Plastic Engineering* (pp. 117-255). Apple Academic Press.

Mary, C., Vaj, E. A., Subhash, A., & Paul, N. M. (2021). Comparative study on the extracellular amylase activity of amylolytic bacteria isolated from different soil types and earthworms gut. *International Journal of Botany Studies*, *6*(2), 115-121.

Meyers, J. K., Narancic, T., & O'Connor, K. E. (2022). Bioplastics and their role in waste management systems. *Waste Management*.

Mihai, F. C. (2017). Rural waste management: Challenges and opportunities. *Waste Management Journal*, 68, 389–400.

Mihai, F. C., Gündoğdu, S., Markley, L. A., Olivelli, A., Khan, F. R., Gwinnett, C., ... & Molinos-Senante, M. (2021). Plastic pollution, waste management issues, and circular economy opportunities in rural communities. *Sustainability*, *14*(1), 20.

Mongkoldhumrongkul, K., & Sukkanta, J. (2022). Sustainable bioplastics: Advances and environmental challenges. *Renewable Resources Journal*, *34*(2), 113–121.

Moshood, T. D., Nawanir, G., Mahmud, F., Mohamad, F., Ahmad, M. H., & AbdulGhani, A. (2022). Sustainability of biodegradable plastics: New problem or solution to solve the global plastic pollution?. *Current Research in Green and Sustainable Chemistry*, *5*, 100273.

Mousavi, M., Kowsari, E., Gheibi, M., Cheshmeh, Z. A., Teymoorian, T., & Ramakrishna, S. (2024). Assessing bioplastics' economic, commercial, political, and energy potential with circular economy modeling: A sustainable solution to plastic waste management. *Materials Circular Economy*, *6*(1), 6.

Nanda, S., & Berruti, F. (2021). A review on bioplastics and their sustainable production. *Sustainable Chemistry and Engineering*.

Nazareth, M. C., Marques, M. R., Pinheiro, L. M., & Castro, Í. B. (2022). Key issues for bio-based, biodegradable and compostable plastics governance. *Journal of Environmental Management*, *322*, 116074.

Nizamuddin, S., Baloch, A. J., Chen, C., Arif, M., & Mubarak, N. M. (2024). Bio-based plastics, biodegradable plastics, and compostable plastics: Biodegradation mechanism, biodegradability standards and environmental stratagem. *International Biodeterioration & Biodegradation*, *195*, 105887.

Nordahl, S. L., Preble, C. V., Kirchstetter, T. W., & Scown, C. D. (2023). Greenhouse gas and air pollutant emissions from composting. *Environmental science & technology*, *57*(6), 2235-2247.

Ogwu, M. C., Iyiola, A. O., Fawole, W. O., & Izah, S. C. (2025). Biobased Resource Recovery Techniques in the Global South. In *Sustainable Bioeconomy Development in the Global South: Volume II Bioeconomy Techniques* (pp. 455-484). Singapore: Springer Nature Singapore.

Okoye, C. O., Addey, C. I., Oderinde, O., Okoro, J. O., Uwamungu, J. Y., Ikechukwu, C. K., ... & Odii, E. C. (2022). Toxic chemicals and persistent organic pollutants associated with micro-and nanoplastics pollution. *Chemical Engineering Journal Advances*, *11*, 100310.

Palsra, P., & Chauhan, G. S. (2020). Bio plastics in waste management: Innovations and sustainability perspectives. *Waste Resources Management Journal*, *22*(4), 345–356.

Patti, A., & Acierno, D. (2022). Towards the sustainability of the plastic industry through biopolymers: properties and potential applications to the textiles world. *Polymers*, *14*(4), 692.

Qin, M., Chen, C., Song, B., Shen, M., Cao, W., Yang, H., & Gong, J. (2021). A review of biodegradable plastics to biodegradable microplastics: Another ecological threat to soil environments?. *Journal of Cleaner Production*, *312*, 127816.

Rodrigues, C. C., Salla, R. F., & Rocha, T. L. (2023). Bioaccumulation and ecotoxicological impact of micro (nano) plastics in aquatic and land snails: Historical review, current research and emerging trends. *Journal of hazardous materials*, *444*, 130382.

Rosenboom, J., Langer, R., & Traverso, G. (2022). Bioplastics for a circular economy: Current challenges and perspectives. *Advanced Materials*, *34*(5), 2109178. https://doi.org/10.1002/adma.202109178

Rossi, F., & Morone, P. (2023). North–South waste trade: Prime Example of the circular economy or Major environmental Threat?. *Circular Economy and Sustainability*, *3*(4), 2159-2182.

Rujnić-Sokele, M., & Pilipović, A. (2017). Challenges and Opportunities of Bioplastics. Tehnički Vjesnik.

Savitha, K. S., Paghadar, B. R., Kumar, M. S., & Jagadish, R. L. (2022). Polybutylene succinate, a potential bio-degradable polymer: Synthesis, copolymerization and bio-degradation. *Polymer Chemistry*, *13*(24), 3562-3612.

Shen, L., Worrell, E., & Patel, M. (2020). Environmental impacts and market prospects of bioplastics. *Renewable and Sustainable Energy Reviews*.

Singh, B., & Li, Z. (2012). Endocrine-disrupting chemicals in plastics: A review. *Journal of Environmental Monitoring*, *14*(2), 598–614.

Stasiškienė, Ž., Barbir, J., Draudvilienė, L., Chong, Z. K., Kuchta, K., Voronova, V., & Leal Filho, W. (2022). Challenges and strategies for bio-based and biodegradable plastic waste management in Europe. *Sustainability*, *14*(24), 16476.

Sultan, M., Tabassum, R., & Tahir, A. (2020). Bioplastics: A Sustainable Alternative to Plastics. *Environmental Progress & Sustainable Energy*.

Trasande, L., Attina, T. M., & Blustein, J. (2012). Association between urinary bisphenol A concentration and obesity prevalence in children and adolescents. *JAMA*, 308(11), 1113–1121. https://doi.org/10.1001/jama.2012.936

Vardar, S., Demirel, B., & Onay, T. T. (2022). Degradability of bioplastics in anaerobic digestion systems and their effects on biogas production: A review. *Reviews in Environmental Science and Bio/Technology*, 1-19.

Vert, M. (2005). Aliphatic polyesters: Great degradable polymers that cannot do everything. *Biomacromolecules*, *6*(2), 538–546.

Vinti, G., & Vaccari, M. (2022). Solid waste management in rural communities of developing countries: An overview of challenges and opportunities. *Clean Technologies*, *4*(4), 1138-1151.

Wang, S., Xu, K., Du, W., Gao, X., Ma, P., Yang, X., & Chen, M. (2024). Exposure to environmental doses of DEHP causes phenotypes of polycystic ovary syndrome. *Toxicology*, *509*, 153952.

Wright, R. J., Erni-Cassola, G., Zadjelovic, V., Latva, M., & Christie-Oleza, J. A. (2020). Marine plastisphere: Developing a global research agenda. *Environmental Science & Technology*, *54*(19), 11611–11625.

Yu, C., Dongsu, B., Tao, Z., Xintong, J., Ming, C., Siqi, W., ... & Yalei, Z. (2023). Anaerobic co-digestion of three commercial bio-plastic bags with food waste: Effects on methane production and microbial community structure. *Science of The Total Environment*, *859*, 159967.

Yu, W., Maynard, E., Chiaradia, V., Arno, M. C., & Dove, A. P. (2021). Aliphatic polycarbonates from cyclic carbonate monomers and their application as biomaterials. *Chemical Reviews*, *121*(18), 10865-10907.

Zhang, Y., & Zhao, W. (2024). Social capital's role in mitigating economic vulnerability: Understanding the impact of income disparities on farmers' livelihoods. *World Development*, *177*, 106515.

Zhao, J., Ren, J., Wang, L., & Zhao, L. (2023). Advancing circular economy with bioplastics: A systems perspective. *Journal of Cleaner Production*, 380, 135085. https://doi.org/10.1016/j.jclepro.2022.135085

Zhou, C., Ma, J., & Shi, Q. (2019). Factors affecting bioplastic degradation in soil environments. *Agricultural and Environmental Science Journal*, *45*(7), 657–672.

Zhu, H., & Wang, L. (2020). Bioplastic degradation and environmental impact. *Journal of Environmental Chemistry*, *27*(3), 315–328.

19 Politics and Policies of Bioplastic Regulations

J.O. Otedo, Saheed Ibrahim Musa, Comfort Nneka Okoji, Moses Nathan, A.O. Isoje, I. Onosakponome, J.B. Austen, P. Ebrahimbabaie, and E.C. Unoka

19.1 HISTORICAL CONTEXT OF BIOPLASTIC REGULATIONS

19.1.1 Early Efforts in Bioplastics Policy

Polyhydroxybutyrate (PHB), the very first bioplastic, was identified in 1926 by Maurice Lemoigne, a French researcher, during his work with the bacterium *Bacillus megaterium*. Nevertheless, this discovery was overlooked for many years until the petroleum crisis of the mid-1970s sparked a renewed focus on finding alternatives to traditional petroleum-based plastics. This was thus the driving force behind early attempts in bioplastics policy and led to the creation of frameworks, production methods, and various applications for a wide range of bioplastics by the start of the 21st century (Musa & Ikhajiagbe, 2023).

Initial bioplastic policies focused on marine pollution and waste management, with both governments and organizations keen on reducing plastic waste in water bodies (Musa *et al.*, 2022a). Also, stakeholders such as governments, industries, and non-governmental organizations (NGOs) voluntarily promoted the use of bioplastics (Barrowclough & Birkbeck, 2022). However, over time, national and international policies began to focus on regulating both the production and use of bioplastics, while also setting standards for their biodegradability and compostability (Filiciotto & Rothenberg, 2020; Barrowclough & Birkbeck, 2022). Furthermore, incentives like tax breaks, research and development grants, and subsidies were introduced to promote the production and utilization of bioplastics (Shah & Gangadeen, 2023).

19.1.2 Significant Legislative Milestones

There are a number of legislative milestones in bioplastics regulation that have contributed to the adaptation, growth, and continued development of bioplastics in the European Union (EU), the United Kingdom (UK), the United States of America (US), and globally. Some of these notable policy milestones are briefly described below (Musa & Ikhajiagbe, 2024). The policy of Extended Producer Responsibility (EPR) was first introduced in Sweden in 1990, by Thomas Lindhqvist, in a report to the Swedish Ministry of Environment (Liu & Ramakrishna, 2021; Gaur *et al.*, 2022). It is a regulatory framework that mandates producers to be responsible for the environmental impact of their products throughout their entire lifecycle, including after consumers deem the products no longer useful; with particular emphasis on the recycling and disposal phases of the products (Gaur *et al.*, 2022). Its aim is to help countries achieve their recycling and recovery targets by shifting some of the waste management costs from local governments to producers (Musa *et al.*, 2022b). A major advantage of this approach is that companies are motivated to adopt more sustainable practices (Insights, 2023).

DOI: 10.1201/9781003602828-23

It was first introduced in Europe in the early 1990s by the EU, then later launched in the UK in 1997 where it has evolved over the years across various phases (with significant updates made with the Environment Act of 2021). A revised and improved EPR framework is currently being phased in, in the UK, with full adoption expected by 2025. It was launched in Nigeria in 2014 by the National Environmental Standards and Regulations Enforcement Agency (NESREA) (Yakubu *et al.*, 2024), and numerous countries have now adopted it to mitigate the environmental impact of products throughout their entire lifecycle (EXPRA, 2024).

The **EU Plastics Strategy** is another significant milestone in bioplastic regulations, which was introduced by the EU in **2018**. It is a comprehensive strategy whose major objective is to cut down on plastic waste while encouraging the use of sustainable materials. Part of this policy includes proposals to support bio-based, biodegradable, and compostable plastics (Stoica *et al.*, 2022; Beatriz Barbosa Moreno *et al.*, 2023; Insights, 2023). The **Single-Use Plastics** (SUPs) **Directive** is also a notable bioplastics policy milestone introduced by the EU in **2019**. It banned the sale of SUP items like plastic plates, plastic cutlery, cotton swabs, food containers, polystyrene cups, and more. Similarly, to the EU Plastics Strategy, its objective is to cut down on plastic waste and promote the use of sustainable materials (Stoica *et al.*, 2022; Insights, 2023; Alok & John, 2024).

The revision of the existing **Packaging and Packaging Waste Directive (PPWD)** (Tin Sin Lee & Soo Tueen Bee, 2024) to include bio-based, biodegradable, and compostable plastics (Torkelis *et al.*, 2024) is another notable milestone in bioplastics regulation, as it focuses on the EU's approach to deal with waste, over packaging, sustainable production and consumption practices. This revision has made it possible for the PPWD to align with the European Green Deal (*European Bioplastics Documents*, 2022).

In the UK, the aforementioned **Environment Act of 2021**, and the **UK Plastics Pact of 2018** are worthy of note. The former brought about a new and improved EPR framework for packaging, which is scheduled for full implementation by 2021 (Stoica *et al.*, 2022); while the latter is a voluntary initiative to stop the use of harmful plastic packaging, and to make sure only reusable, recyclable, and compostable plastic packaging are in use by 2025. While in the US, **California's Senate Bills 270 (SB 270) of 2014 and 54 (SB 54) of 2016** are also notable. SB 270 barred stores from offering SUP carryout bags to customers, but permitted the sale and distribution of recycled paper bags and reusable grocery bags that meet specific standards and are sold at a regulated price (*California's Law Banning Plastic Bags*, 2014). SB 54, also referred to as the Plastic Pollution Prevention and Packaging Producer Responsibility Act, aimed to reduce packaging waste in California by transferring the plastic pollution burden from consumers to producers, thereby requiring manufacturers to achieve recycling and climate targets for plastic products.

These pioneering policies all put together have thus laid the foundation for both the integration and the current development of bioplastics into the global economy (Musa *et al.*, 2023). It should be noted that recent policies focus on a more comprehensive life-cycle approach by addressing plastic pollution from production to disposal stages, which is inclusive of promoting circular economies and lowering plastics' carbon footprint (Barrowclough and Birkbeck, 2022).

19.2 GLOBAL REGULATORY FRAMEWORKS

Compostable bioplastics are becoming increasingly popular worldwide, but stronger growth is being impeded by substantial market failures and a lack of government action to remedy them. According to Joseph and Paul (2008), each of the policy tools is explained in terms of how it is intended to solve fail commercial market along the value chain for biodegradable bioplastics, along with some of the implementation-related opportunities and challenges (Musa *et al.*, 2022c).

The market circumstances that foster innovation and growth in the biodegradable bioplastics industry are mostly determined by government laws. Upholding product integrity across the value chain is crucial for the biodegradable bioplastics markets to operate as intended under a circular economy strategy. This calls for technical laws that outline the product's approved applications in the economy as well as standards that attest to the legitimacy of the product's features, if needed. For products to accurately tell consumers about their features, labels must be applied. To keep these waste streams out of landfills, regulations mandating trash segregation and the composting of appropriately biodegradable bioplastics are required. Procurement policies are a useful tool for governments to create demand in emerging economies and allocate funds to encourage investment. For markets to operate effectively. It is necessary to have policy vehicles that make it easier for standards and regulations from different jurisdictions to converge. When combined, these interrelated government initiatives can create a supportive environment for biodegradable bioplastics that advances consumer welfare, eases commerce, gives businesses access to markets and spurs innovation, and helps local communities' economies and environments.

The WEF is supporting evidence-based policy initiatives for a circular plastics economy in partnership with the Ellen MacArthur Foundation and McKinsey & Company (WEF, 2016). The Open-ended Working Group recommends that nations develop universal technoeconomic analysis methodologies to measure the environmental benefit of bioplastics, as well as clearly define and standardize the identification of bio (degradable) plastics. Additionally, bioplastic production processes should be improved to make bioplastics economically and environmentally competitive with plastics derived from fossil fuels. Strategies to reduce plastic leakage into seas, on 80% by 2040 are provided by a recent report. These recommendations call for a 90% reduction in waste exports to nations with high leakage rates, a doubling of the world's mechanical recycling capacity, better recycling design to increase the amount of plastic that can be recycled globally from 21% to 54%, and the application of tried-and-true methods to eradicate the main sources of microplastics. By January 2021, all biodegradable plastics and several SUPs, such as straws, silverware, food and drink containers made of polystyrene, and cotton buds, will no longer be allowed for sale in the EU; exports of low-grade plastic waste outside the EU will be restricted in compliance with the Basel agreements; and the EU has implemented several plastics initiatives as part of the European Green Deal and its Circular Economy Action Plan, with the goal of 50% recycling for plastic packaging by 2030(Coady *et al.*, 2019). According to TCB 2020, a tax of €800 per ton was imposed on non-recycled plastic to encourage the manufacturing sector to switch to recyclable, reusable, or compostable materials. EPR programs, which are often restricted to packaging materials, have been adopted in many EU nations and have shown to be helpful in moving EOL expenses from local governments to producers. Their application has to be enhanced, notably by unifying definitions of what is termed "recyclable" from a local and chemical standpoint, and their scope needs to be extended to other plastic-intensive industries (such as agriculture, textiles, medicine, and construction) (Watkin, 2017). The EU is creating regulations for bioplastics to establish which uses are suitable for them in order to stop businesses from making fraudulent sustainability claims, or "greenwashing." The EU intends to update and unify current standards (such EN 13432) as part of the regulatory framework in order to take more realistic biodegradation tests into account (Coady *et al.*, 2019). China, the biggest manufacturer of SUPs worldwide, said recently that, by 2025, it will only allow biodegradable polymers and not recyclable materials. Chinese producers therefore intend to significantly boost PLA production to 700,000 tons annually and the combined output of PBAT and PBS to 1.24 million tons annually by 2023. Plans for the controlled disposal of these enhanced volumes are still unknown, and these capacities may have an impact on the prices of these polymers on the global market. A number of other nations have established financial subsidies for bioplastic, including South Korea, Malaysia, Singapore, Japan, and Malaysia. In the United States, the current administration has committed to more environment-focused and climate-focused policies. Among its objectives include tackling ocean trash and enticing private sector investment into indigenous renewable energy technology and materials (Biden, 2020). In February 2020, the Break Free from Plastic Pollution Act was presented to the House of Representatives. This law proposes to tax carryout bags and hold plastic manufacturers financially accountable for collecting and recycling their products in accordance with EPR principles, thereby limiting the amount of single-use items and non-recyclables in markets after 2022. Additionally, the measure forbids the export of plastic garbage to nations outside of the Organization for Economic Cooperation and Development (OECD). These initiatives, however, conflict with laws that have supported the manufacturing of plastics generated from fossil fuels. Incentives and lax restrictions have made shale gas and fracking possible in the United States, allowing plastic made from these fossil fuels. Despite estimates that incentives for sustainable technologies could save 28% of carbon emissions, reduce deaths from air pollution by 46%, and increase government revenues by 3.8% of gross domestic product (GDP), the International Monetary Fund estimates that 85% of global subsidies benefit fossil fuels (OECD, 2009).

Future advancements in biorefining and industrial biotechnology will present the chance to launch brand-new bio-based businesses and revolutionize already-existing ones. It is anticipated that this would lead to a notable surge in economic activity and the creation of new and larger markets for bio-based products (OECD, 2009). But in order to do this within the framework of a larger, more competitive bioeconomy, a conducive environment is required. Examples of this include developing and sustaining markets for environmentally sustainable products, providing funding for basic and applied research, and making investments in infrastructure and education that serve multiple purposes. Furthermore, this would need to be paired with more immediate measures including encouraging public discussion and raising support for the adoption and application of globally recognized life cycle analysis (LCA) standards, in addition to a variety of additional incentives intended to encourage environmentally friendly behavior. A rising number of governments throughout the world have been creating strategies and policy frameworks in recent years to aid in the growth of a competitive and sustainable bioeconomy. A number of these policies promote bio-based products or the bioeconomy in general by providing general assistance for the advancement of biochemicals, biomaterials, and bioplastics. The majority of them prioritize innovation and research. A small number of nations have created policies specifically aimed at advancing the development of bioplastics. In 2018, Hundertmark *et al.* There is also a broader bioeconomy framework given. As with any product, the development of bioplastics can be divided into multiple stages, and for each stage, specific policy instruments may be applicable, for example:

- Research and innovation include promoting technological transfer, sponsoring demonstrator projects, boosting market introduction, offering investment-related subsidies, and supporting research and development through public money.
- Early Market Support: Defending against well-established (conventional) options, such as fiscal measures and (temporary) tax incentives; assisting in the creation of governance and communication tools (standards, labels); putting production cost-cutting measures into place; offering encouraging the use of process by-products; imposing quotas; and imposing tariffs to safeguard comparable domestic products (like the now-canceled 54-cent-per-gallon tariff imposed by the US on imported ethanol).
- Large Commercial Market: Policies that impact the prices and availability of feedstocks, as well as market-pulling strategies like public procurement; waste and packaging laws, which are essential for guaranteeing proper disposal of waste at the end of its useful life and permitting bioplastics to utilize the current waste collection infrastructure. Even if the market for bioplastics is expanding currently, its percentage of the total market for plastic production is very modest (0.5%).

In 2022, the European Commission issued a set of regulations governing the use of bio-based plastics, their source, labeling, and disposal, as well as the use of biodegradable and compostable plastics. The European Green Deal, Circular Economy Action Plan, and Plastics Strategy all introduced the policy framework to promote a sustainable plastics economy. In particular, it aims to increase our understanding of bioplastics. The framework, which holds novel materials to the same exacting standards as conventional materials, makes clear when, how, and for what purposes these materials can assist the environment (EU, 2022).

Although some laws, including the EU Taxonomy, the Single Use Plastics Directive, the EU Taxonomy, the Plastic Carrier Bags Directive, the PPWD, and the Waste Framework Directive, address certain aspects and applications of bio-based, biodegradable, and compostable plastics, there is currently no EU law that specifically addresses bioplastics. The Commission's Communication for the EU policy framework also represents the Commission's views and objectives on these issues and will guide the creation of future EU policies, such as those concerning carbon removal, microplastics, eco-design for sustainable products, and green claims, even though it is not legally binding. Although there isn't now a single EU regulation addressing bioplastics, there are different rules that are utilized in specific aspects and applications of bio-based, biodegradable, and compostable plastics such are addressed by the Plastics Directive, EU Taxonomy, the Plastic Carrier Bags Directive, the PPWD, and the Waste Framework Directive. Furthermore, even though the Commission's Communication for the EU policy framework is not legally binding, it does represent the Commission's opinions and goals regarding these topics and will direct the development of future EU policies, including those pertaining to microplastics, carbon removal, eco-design for sustainable products, and green claims. European Bioplastics (EUBP) and its members applaud the Commission's initiative to develop the first comprehensive policy framework on bioplastic materials. The Communication offers a thorough overview of our industry in Europe and recognizes the benefits and potential that our materials have to offer real environmental benefits. However, a few enduring myths about land use, the assessment of environmental benefits, purported dangers of cross-contamination of waste streams, and biodegradability in various settings persisted in the policy framework in spite of substantial and reliable scientific evidence.

The current legal framework does not recognize or support the main advantages of bio-based plastics, which include the use of renewable resources to create plastic products and their contribution to the transition to a circular and carbon-neutral bioeconomy. Additionally, the framework takes an extremely narrow and cautious stance regarding the unique quality of biodegradability in particular settings, ignoring the enormous potential for the circular economy—particularly with regard to industrially compostable plastics—and generally impeding innovation in this nascent but promising field.

It is feared that these Communication flaws would prevent the Commission from fully embracing the shift to bio-based products, which is necessary to help Europe reduce its dependency on fossil fuels and accomplish its ambitious climate change and circularity ambitions. It is also worried that the current strategy would hinder rather than advance innovation and the European market for sustainable material solutions.

19.3 STAKEHOLDERS IN BIOPLASTIC REGULATION

Stakeholders in bioplastic regulation are groups of people or organizations from different sectors that shape the use or are affected by the policies of bioplastics in a particular region. These stakeholders try to determine where bioplastic products are most effective, where challenges are, and the causes and they also balance the economic, ecological, and social considerations. These stakeholders encompass government agencies, investors, consumers, manufacturers, researchers, industry associations, etc.

19.3.1 Government Agencies

Government agencies are responsible for creating and enforcing regulations to ensure the safety and environmental compliance of bioplastics. The government agencies involved are as follows:

1. Environmental Commission: The European Commission has launched a comprehensive policy strategy, for bio-based, biodegradable, and compostable plastics as a component of the European Green Deal and Circular Economy Action Plan, and Plastic Strategy. This initiative seeks to enhance awareness and encourage the environmental advantages of bioplastics (European Bioplastics, 2022).
2. EUBP: Although it is not a government body, EUBP collaborates closely with EU institutions to advocate for the interests of the bioplastics value chain and participate in current policy debates (European Bioplastics, 2023).
3. United Nations Environment Programme (UNEP): The UNEP offers support and tools to help countries to manage plastic waste, including bioplastics, by providing legislative frameworks and policy strategies.
4. European Chemicals Agency (ECHA): In the EU, the ECHA reviews the safety and ecological implications and consequences of chemical substances in bioplastics, implementing the REACH regulations which are the registration, evaluation, authorization, and restriction of chemicals (European Chemicals Agency, 2023).
5. National Institute of Standards and Technology (NIST): NIST designs benchmarks supporting manufacturers in quality assurance and regulating compliance.

19.3.2 Industry Associations

Industry associations are organizations focused on the interests of stakeholders and businesses involved in bioplastic regulation, with a goal to modify, develop, and promote policies and strategies that guide the manufacture, usability, and discarding of bioplastics. These associations primarily engage in advocacy, offer guidance on legal compliance, support research initiatives, and enhance best practices within the industry.
Some key industry associations include the following:

1. Bioplastic Product Institute: The Bioplastic Product Institute promotes organic products and establishes regulations for biodegradable materials.
2. EUBP: They enhance rules that govern sustainability and circular economy.
3. American Bioplastics Association: They improve the growth of bioplastics and collaborate with law makers to build a regulatory environment that is advantageous to the bioplastic sector.
4. International Biodegradable Products: They enhance the growth of regulations suitable for biodegradable and organic products, enhancing collaboration on bioplastic regulations internationally.

19.3.3 Manufacturers

Manufacturers of bioplastics play a vital importance in the transition toward more sustainable materials. Some of these manufacturers include the following (Emergen Research, 2023):

1. BASF SE: It is the largest chemical producer based in Germany. BASF EF deals on compostable bioplastics derived from renewable sources like sugar, it is utilized in biodegradable disposable bags, farm films, and compostable coffee capsules.
2. Eastman Chemical Company: It is based in the US and produces cellulose-based bioplastics that are environmentally friendly, resistant to chemicals, BPA-free, and very stable.
3. Plantic Technologies: It is a company based in Australia focused on producing eco-friendly plastics sourced from sustainable resources.

4. Futamura Group: It is a Japan-based bioplastic industry that deals on a variety of bioplastic materials like polymer films, synthetic films, fibrous casting, and purified carbon, and they are also ethical and sustainable.
5. Natureworks LLC: NatureWorks is a US-based bioplastic company that focuses solely on the manufacture of bioplastics obtained from plants like cornstarch. They are also into injection molding, foam, 3D printing, cards, fibers, etc.

According to Mererub, 2023, some of the roles played by manufacturers in bioplastic regulation include the following:

1. Innovation and Development: Manufacturers develop new bioplastic materials to improve on existing ones. This includes, through research of alternative feedstock, enhancing biodegradability and the reduction of production costs.
2. Sustainable Production: Manufacturers aim at manufacturing bioplastics through the utilization of sustainable resources like; corn, sugarcane, and other plant-based products. This decreases dependence on hydrocarbon fuels and minimizes the carbon emissions of plastic production.
3. Market Expansion: Manufacturers help expand the market for sustainable alternatives by creating bioplastic products that meet consumer needs. This entails the production of biodegradable packaging, disposable utensils, and other daily items.
4. Collaboration with Stakeholders: Manufacturers work with law makers, researchers, and various other stakeholders that help contribute to the promotion of bioplastics. Such collaborations include joining industry associations, contributing to governing discussions, and supporting public symposiums and campaigns.
5. Economic Incentives: Manufacturers often gain from government allocations and reductions that promote sustainable materials. This in turn helps drive research and development that helps increase the production of bioplastics.

19.3.4 Non-Governmental Organizations and Advocacy Groups in Bioplastic Regulation

19.3.4.1 Non-Governmental Organizations

NGOs make a substantial impact on bioplastic regulations by campaigning for sustainable practices, raising public awareness, and collaborating with law makers and industry stakeholders. Some of these NGOs include the following:

1. Greenpeace: They advocate for more rigid laws on the manufacture of bioplastics and waste management through the incorporation of degradable and compostable materials (Alice, 2018).
2. World Wildlife Fund (WWF): WWF works in terms of reducing pollution caused by plastics, thereby enhancing sustainable alternatives through public awareness and symposiums (WWF Positions, 2020).
3. Plastic Pollution Coalition: It aims at the disintegration of plastic pollution and enhancing the usability of biodegradable and compostable plastics through advocacy and community engagement.
4. Break Free from Plastics: Their focus is to educate on plastic pollution by campaigning for laws that should be utilized for sustainable materials (Gaia, 2022).

19.3.4.2 Advocacy Groups

Advocacy groups contribute greatly to the engagement of bioplastics regulations by creating enlightenment on the ecological impact of bioplastics. Some of these advocacy groups include the following:

1. Plastics Industry Association (PLASTICS): This association promotes the development of bioplastics by educating the industry, government, and public about the usefulness of bioplastics.
2. CITEO: This is a French company that is responsible for household packaging and graphic papers. They campaign for clear definitions and regulations for bio-based biodegradable and compostable plastics (Axel and Romane, 2021).

19.4 POLICY FRAMEWORK AND APPROACHES

The global market for compostable bioplastics is witnessing significant expansion; however, market inefficiencies and insufficient policy interventions impede optimal growth. Effective government support through strategic policy is instrumental to establishing conducive environments that facilitate the sustainable growth and development of compostable bioplastics markets (USAID, 2024).

Londo and Meeusen (2010) identified policymaking for the bio-based economy as a complex task that cuts across several policy areas. Globally, there has been an increasing interest in the number of countries crafting strategies and policies to foster a competitive and sustainable bioeconomy. While many focus on research and innovation, few nations have specific policies supporting biochemicals, biomaterials, and bioplastics development. To bridge this gap, targeted policies are needed to promote bio-based products and drive bioplastics growth (OECD 2013). The development of bio-based products benefits from policy-driven efforts to reduce market barriers (Uyarra *et al.*, 2014). The US Department of Agriculture's (USDA) BioPreferred program represents a notable model for integrated demand-side innovation procurement policy, specifically targeting compostable bioplastics. This initiative was established under the 2018 Farm Bill and leverages federal procurement and certification to boost bio-based product adoption. By providing a framework for federal agencies to meet statutory requirements, the program promotes the purchase and specification of bio-based products, including compostable bioplastics, where relevant (USAID, 2024).

19.4.1 Incentives for Bioplastics Development

In response to the pressing global issue of climate change, numerous legislative acts have been enacted worldwide. These laws either introduce novel solutions to emerging challenges or update existing frameworks to address the evolving needs of environmentally conscious societies and policymakers (Stasiškienė *et al.*, 2022). Governments worldwide employ diverse direct and indirect measures to foster bioplastics development, primarily leveraging financial incentives to stimulate private sector investment. The United Nations emphasizes the need for incentives to transform consumption habits, production, and research, focusing on sustainable materials (UNEP, 2018). In an attempt to address the global plastic crisis, governments worldwide have intensified efforts through legislative acts. The EU's plastics and plastic waste management policy draws heavily from the Basel Convention, a global treaty focused on hazardous waste management, with a strong emphasis on preventing and minimizing plastic waste (UNEP, 2011). This treaty is the foundation for the EU's efforts to tackle plastic pollution. In addition, the Sustainable Development Goals (SDGs) play a crucial role in addressing plastic pollution through Agenda 2030. Specifically, SDG 12 promotes sustainable consumption and production patterns, providing a framework for reducing plastic waste (United Nation (UN), 2015).

According to the OECD, various measures and incentives can impact the development of bioplastics. These include the following:

1. Fiscal incentives through tax legislation
2. Preferential treatment in public procurement programs
3. Streamlined regulations for waste management
4. Access to community composting facilities for biowaste
5. Government-funding development programs for collaborative industry-university projects
6. Initiatives promoting market introduction and awareness
7. Supportive agricultural policies

Countries like Malaysia, Japan, South Korea, Singapore, and China have made significant strides in encouraging bioplastics research and investment through incentives, including; Tax waivers, grants, loans, and subsidies (USAID, 2024). These measures support companies exploring sustainable alternatives, driving innovation and advances in bioplastics (UNEP, 2018).

19.4.2 Restrictions on Conventional Plastics

According to European Bioplastics (2021), replacing all traditional plastics with bio-based or biodegradable plastics is not feasible in the near term. At the moment, bioplastic production represents a small percentage of around 1% of the total 368 million tons of plastic produced annually, however, efforts have been made by various Governments worldwide to restrict the use of conventional plastics, especially SUPs. Plastic pollution arises from market and policy failures, requiring effective policy interventions to address negative externalities (UNEP, 2021; Watkins et al., 2019). To control plastic pollution, governments employ policy instruments targeting the plastic lifecycle. According to available data, more than 60 countries have implemented bans, taxes, and levies to curb plastic waste (World Bank, 2022). China took the giant stride by banning plastic waste leading to a significant reduction in plastic waste (Brooks *et al.*, 2018). China's waste import ban has prompted several countries to explore alternative solutions for managing plastic waste. In Europe, Europe's EPR imposes fees on plastic production, but these fees are low and not globally enforced (Monier *et al.*, 2014). The fight against SUP pollution is gaining momentum in Africa, with Rwanda's innovative law banning plastic bags and other SUP items paving the way for other countries to adopt similar measures (Adam *et al.*, 2020; Xie and Martin 2022). In Nigeria, The Lagos State Ministry of Environment and the Lagos Waste Management Authority (LAWMA) have developed a plastic waste management policy aligned with the national policy to overcome challenges faced in addressing plastic waste management. Despite notable achievements, several difficulties continue to pose a challenge including enforcement, infrastructure, compliance and integration of the informal sector, limited resources, and public awareness. The Challenges in Nigeria include a lack of national-level policy standardizing the definition of SUPs, voluntary and rudimentary implementation of EPR schemes, inconsistent efforts to promote behavioral change, and failure to implement important measures such as levies on SUPs and taxes. Additionally, the private sector lacks incentives to invest in sustainable packaging business models. While the Plastic Bags Prohibition Bill (2019) represents a trend, it could benefit from including affordable alternatives to plastic bags and market-based instruments (World Bank, 2024). Kenya's SUP regulation began in 2005, addressing concerns over waste management and environmental harm. A government study revealed widespread plastic ingestion by cattle and infrastructure damage. The 2005 law banned thin plastic bags and levied thicker ones. However, subsequent attempts to implement a complete ban have failed due to resistance from powerful business stakeholders (Uche, 2023). Current policies worldwide emphasize eliminating SUP bags and foam products while addressing plastic waste prevention, management, and reduction. Additionally, many policies encourage sustainable plastic production by incorporating renewable materials and recycled content (Xie and Martin 2022).

19.4.3 Certification and Labeling Standards

Standardization sets industry-wide criteria for products, services, and processes, promoting clarity, quality, and fairness. Compliance is usually voluntary, encouraging market participants to adopt best practices. Adherence to standards is typically voluntary, relying on individual market participants to choose compliance (Asia Pacific Economic Cooperation (APEC), 2011). The Scientists' Coalition (2023) identifies a critical limitation in the bioplastics industry, arising from inconsistent usage of key terms, including bioplastic, bio-based, biodegradable, and compostable. This lack of standardization results in unclear product labeling, ambiguity surrounding material characteristics, and uncertainty regarding disposal and benefits (Sijtsema *et al.*, 2016; Napper and Thompson, 2019; Purkiss, *et al.*, 2022; Theobald *et al.*, 2024).

The leading European certification organizations DIN CERTCO (Germany) and Vinçotte (Belgian certification organization) verify bioplastics' composability and biodegradability. Certification benefits include differentiation, conformity, consumer assurance, simplified waste sorting, and quality guarantees. These standards enable safe composting and ensure bioplastics meet ASTM D6866 requirements (USAID, 2024).

19.5 CONFLICTS OF INTEREST AMONG STAKEHOLDERS

Fletcher *et al.* (2021) explored the challenges and issues associated with bio-based and biodegradable plastics, presenting them as eco-friendly alternatives to traditional plastics. They discuss the perspectives of a variety of stakeholders, including manufacturers, consumers, waste management professionals, and policymakers, highlighting the hurdles involved in integrating these materials into mainstream use.

Bio-based plastics, sourced from renewable resources, and biodegradable plastics, which decompose under specific environmental conditions, are both seen as environmentally friendly alternatives to fossil fuel-based plastics. However, their widespread adoption faces several obstacles, including high production costs, environmental impacts tied to raw material sourcing, and challenges related to waste management and disposal. Misunderstandings among consumers and inconsistent regulatory frameworks also contribute to the slow integration of these plastics into the market.

A key issue raised in the chapter is the confusion surrounding the terms "bio-based" and "biodegradable." Many consumers incorrectly believe these plastics will break down easily in all environments. In reality, they require controlled conditions, such as industrial composting facilities, to decompose effectively. This misunderstanding often leads to improper disposal, undermining the environmental advantages of these materials and putting additional strain on waste management systems. The authors stress the importance of clearer labeling and public education to resolve this issue.

On the policy side, Fletcher *et al.* (2021) advocate for more consistent and robust regulations governing the production, labeling, and disposal of bio-based and biodegradable plastics. Harmonizing standards across regions would foster market confidence and encourage innovation. The authors also call for increased investment in waste management infrastructure, such as composting facilities, and for incentives that promote sustainable manufacturing practices, which are necessary to facilitate the transition toward bio-based plastics.

Ultimately, the chapter emphasizes that a collaborative approach, involving transparent communication among all stakeholders, along with well-defined policies and standards, is essential to overcome the challenges associated with these materials. Such efforts are necessary to realize the potential of bio-based and biodegradable plastics to reduce environmental impacts and support more sustainable practices within the plastics industry.

19.5.1 Politics of the Plastics Life Cycle

Nielsen *et al.* explored the political dimensions of the global plastic crisis, focusing on how political decisions influence the production, consumption, and waste management of plastics throughout their life cycle. They argue that solving the plastic crisis requires not only technological and environmental solutions but also a deep understanding of political structures and decision-making processes that shape the way plastics are produced, consumed, and disposed of.

The authors examine the entire plastic life cycle—production, consumption, waste management, and disposal—highlighting how political actions and the influence of multinational corporations affect each stage. The global plastic industry is dominated by powerful corporations that often prioritize economic growth over environmental concerns, lobbying against regulations that could reduce plastic production or increase operational costs. Despite mounting evidence of the environmental and health risks of plastics, these industries resist change to maintain the status quo.

The chapter also delves into the political challenges associated with plastic production, particularly the industry's reliance on fossil fuels and the petrochemical sector's role in sustaining current practices. Regarding consumption, the authors note that demand for plastic products is driven by convenience, which is reinforced through advertising and market strategies that often downplay the environmental consequences of plastic use.

On waste management and disposal, the chapter addresses the challenges associated with plastic recycling programs and the global movement of plastic waste. Recycling efforts are often hampered by inconsistent infrastructure and fluctuating markets for recycled materials. The authors also discuss the political struggles over plastic waste exports, particularly as countries like China have banned the importation of plastic waste, forcing developing countries to shoulder the burden of waste disposal, and exacerbating global inequality. Nielsen *et al.* (2019) argue that solving the plastic crisis requires significant political change at all levels, from local to global. They call for stronger international cooperation to create binding agreements that promote circular economies, reduce plastic production, and improve waste management practices. The authors emphasize that engaging diverse stakeholders—governments, industries, civil society, and consumers—is crucial in developing and enforcing effective policies. In summary, the chapter emphasizes that significant political commitment and action are essential to reform the production, usage, and disposal of plastics; otherwise, the social and environmental impacts of plastic pollution will persist and intensify. A comprehensive and coordinated approach is necessary to address the political factors that contribute to the plastic crisis.

19.5.2 Future Directions in Bioplastic Policy

Due to the increasing global approaches for eliminating plastic pollution, transitioning toward a circular economy, and reducing environmental harm from materials, much of the anticipated regulatory changes on bioplastics are driven by this. Although bioplastics offer a promising alternative to traditional plastics, regulation of these materials is still under development. In the coming years, expect the following major regulatory changes to impact bioplastics:

19.5.3 Increased Standards for Compostable and Biodegradable

That means governments may impose stricter rules on bioplastics to ensure they truly live up to the marketed aspects of being biodegradable and compostable. Look for more stringent specifications on definitions such as "biodegradable" to ensure that bioplastics can degrade safely, through non-toxic by-products, in natural conditions (soil or water). It teased expanded regulations enforcing not just industrial but home composting, as well as certifying products that biodegrade properly in decentralized systems. Such policies could range from third-party testing to life-cycle assessments and potentially clearer labeling laws for consumers on how to dispose of a product. Stronger enforcement and the possibility of fines for deception will also likely spur more responsibility. There are also likely to be international standards to ensure consistent markets around the world.

19.5.4 Extended Producer Responsibility (EPR)

EPR policies can be a game changer for curbing plastics by requiring brands to take responsibility for their products from manufacturing to disposal. In the case of bioplastics, EPR would mandate producers to ensure that their products are collected and recycled or properly disposed of in order to prevent pollution and minimize landfill waste. It would prompt manufacturers to create bioplastics that are recycled- or compostable-ready and invest in systems to ensure product returns. EPR may save local governments the costs associated with disposing of bioplastics and further reinforce a circular economy in these materials by ensuring they are reused, rather than discarded.

19.5.5 Carbon Taxes and Incentives for Bioplastics

Bioplastic production and consumption are likely to, in one way or the other, be influenced by the introduction of carbon taxes as well as environmental incentives. Carbon taxes could be introduced for bioplastics across their complete lifecycle from production to landfilling where taxation is imposed on the carbon released into the environment. While there are many benefits in using bioplastics as compared to traditional plastics, the use of fossil fuels or energy-intensive processes in their production makes it carbon-intensive. To counter this, subsidies can be provided for bioplastics derived from renewable materials or produced with low-carbon processes such as manufacturing practices that are more environmentally friendly.

19.5.6 Strategies for Infrastructure Development through the Use of Public–Private Partnerships

To effectively recycle and compost bioplastics, the required infrastructure needs to be established and implemented, which is an expensive venture. Funding and establishment of the infrastructure required to manage bioplastic waste can be provided through public–private partnerships (PPPs). These synergetic initiatives between governments and industry stakeholders are key to developing most of the waste management solutions aiming to recycle bioplastics and thus ensure their responsible post-consumer use.

The Role of Innovation in Shaping Bioplastic Policies

Innovation will be key to shaping bioplastic policies, as technological advancements continue to improve the sustainability, performance, and scalability of bioplastics. Policymakers must stay ahead of these innovations to ensure that regulations support progress while reducing plastic waste and promoting a circular, low-carbon economy. The key roles of innovation in shaping bioplastic policies are as follows:

1. **Material Innovations and Performance:** Research is helping to achieve improved strength and durability while enhancing the biodegradability of bioplastics. Materials like PLA and PHA are promising, although the issue of scale-up and cost control still presents challenges. Meanwhile, the innovations in genetic engineering and fermentation processes are assisting in making production more efficient which may prompt encouraging policies for adoption.
2. **Likely Biodegradability and Circular Economy:** Production and use of bioplastics that can undergo biodegradation in waste facilities will lead to a reduction in environmental efforts. The focus of much of the development work, however, is in fact on how to improve the rate and safety of biodegradation of such materials especially in the oceans. This may cause the

introduction of some restrictions in Latin America on sectors such as packaging or how it is allowed to "dispose" of bioplastics in order to reduce waste accumulation.

3. **Feedstocks Carbon Footprint Reducing:** The use of renewable non-food feedstocks such as algae agricultural waste or cellulose will decrease the tension on food production and enhance the sustainability of bioplastics. Trends in the development of low-carbon feedstocks may impact the reduction of emissions policies including lifecycle-based and carbon footprint assessment or reporting becoming a component of the existing legislation.
4. **End-of-Life and Waste Management Options:** The rise of bioplastics will require improvements in waste management, particularly in recycling and composting. It is possible that new policies that promote the circular economy will be developed along with a bioplastics regulation acting as the framework for how waste by biomaterials is dealt with.

Recommendations for Policymakers and Stakeholders

To harness the potential of bioplastics while addressing their challenges, both policymakers and industry stakeholders must collaborate. The following recommendations can help ensure bioplastics contribute positively to environmental sustainability:

1. **Establish Clear and Consistent Standards:** By handling bioplastics in the governance, countries should come up with harmonized and open standards, which span from sourcing to biodegradability. Precise, convincing labeling will make it easy for consumers to make the right decision and elude the "greenwashing" concept. Industry stakeholders should partner with the government in the conservation of their assertions about bioplastics, which are defended with scientific data.
2. **Promote Research and Innovation:** Policymakers should motivate bioplastics research and development, by giving grants and tax incentives for projects aimed at optimizing performance and scalability. Industry players should give a premium to the innovation of bioplastics that are cheap and no-pollution, with the major part of the circular economy. Alliances with academic bodies can shorten the way to superior bioplastic solutions.
3. **Strengthen EPR:** Policymakers should ensure that EPR programs work, thus making sure the producers are responsible for their products from the very beginning to recycling or composting. This could provoke more ecological designs of products. At the same time, sector partners should drive to easier-to-recycle or compost bioplastic products and provide garbage collection and waste separation systems to help customers to properly dispose of them.
4. **Support a Circular Economy for Bioplastics Bioplastics:** Governments should be made to fit into the existing recycling and composting infrastructures for the circular economy. Leaders in the industry should join forces with municipalities and waste management firms to build infrastructure that facilitates the recycling and composting of bioplastics, thus ensuring these materials do not end up as waste.
5. **Mandatory Ecological Footprint Transparency:** The governments should demand life cycle assessments (LCAs) of the bioplastics to follow their environmental impact from production to disposal. This information will enable the regulators and the users to get a full picture of bioplastics in terms of their environmental costs. The industry players should carry out and spread the independent LCAs disclosing their products' true environmental impact.
6. **Increase Public Awareness and Consumer Education:** Governments should launch public education campaigns to inform consumers about the benefits and limitations of bioplastics, including how to dispose of them properly. Similarly, industry stakeholders should lead in educating the public, making it clear how bioplastics contribute to environmental sustainability and the importance of proper disposal practices.

By working together, policymakers and industry players can guide the development of bioplastics in a way that maximizes their environmental benefits, fosters innovations, and supports a circular economy.

19.6 CONCLUSION

The politics and policies surrounding bioplastics regulations have evolved significantly over the years, reflecting the complex interplay between economic, environmental, and social interests. Governments around the world have begun to recognize the potential of bioplastics to reduce greenhouse gas emissions, conserve non-renewable resources, and decrease plastic waste. As a result, many countries have implemented policies and regulations to promote the development and use of bioplastics. However, the regulatory landscape for bioplastics remains fragmented and inconsistent, with different countries and regions adopting varying approaches to regulating bioplastics. This has created challenges for companies seeking to develop and market bioplastic products, as they must navigate a complex web of regulations and standards. Despite these challenges, there is a growing recognition of the need for more comprehensive and coordinated regulations to support the development of a sustainable bioplastics industry. This includes the need for clearer standards and labeling requirements, as well as more robust policies to support the recycling and composting of bioplastics.

Furthermore, the politics of bioplastics regulations are also influenced by the interests of various stakeholders, including the fossil fuel industry, agricultural producers, and environmental groups. These stakeholders often have competing interests and priorities, which can shape the regulatory landscape for bioplastics. The politics and policies of bioplastics regulations are complex and multifaceted, reflecting the diverse interests and priorities of various stakeholders. While there have been significant advances in promoting the development and use of bioplastics, there is still a need for more comprehensive and coordinated regulations to support the growth of a sustainable bioplastics industry. As the bioplastics industry continues to evolve, it is likely that the politics and policies surrounding bioplastics regulations will remain a critical area of focus and debate.

REFERENCES

Abdo, S. M., and Ali, G. H. (2019). Analysis of polyhydroxybutrate and bioplastic production from microalgae. *Bulletin of the National Research Centre*, 43(1), 23–45.

Adam, I., Walker, T. R., Bezerra, C. J., and Clayton, A. (2020). Policies to reduce single-use plastic marine. *Marine Policy.*

Alice, H. (2018). *What's the Answer to Our Plastic Problem?* GreenPeace.

Alves, L., Silva, M., and Lima, M. (2023). Sustainable feedstocks for bioplastics: A review of non-food raw materials. *Renewable and Sustainable Energy Reviews*, 163, 112349.

Asia Pacific Economic Cooperation (APEC) (2011). "2011 Leaders' Declaration." Web Resource. November.

Axel, D., and Roxanne, C. (2021). Citeo's Feedback on the Policy Framework on Biobased, Biodegradable, and Compostable Plastics. *CITEO*, E-002997/2021

Barbosa, L., Ferreira, P., and Garcia, S. (2020). Recent innovations in bioplastics and their potential impact on sustainability policies. *Journal of Cleaner Production*, 253, 119–875.

Barrowclough, D., and Birkbeck, C. D. (2022). Transforming the global plastics economy: The role of economic policies in the global governance of plastic pollution. *Social Sciences*, 11(1), 26–39.

Biden, H. (2020). The Biden Plan for a Clean Energy Revolution and Environmental Justice. Biden Harris.

Brooks, A.L., Wang, S., and Jambeck, J.R. (2018). The Chinese import ban and its impact on global plastic waste trade. *Science Advances*, 4(6), 20.

Coady, D., Parry, I., Le, N.-P., and Shang, B. (2019). *Global Fossil Fuel Subsidies Remain Large: An Update Based on Country-Level Estimates*. IMF Working Papers.

Das, A. P., and John, A. (2024). *Sustainable Management of Urban Plastic Waste Through Circular Economic Approaches*. CRC Press.

Eerhart, A. J. J. E., Faaij, A. P. C., and Patel, M. K. (2012). Replacing fossil-based PET with biobased PEF; process analysis, energy and GHG balance. *Energy & Environmental Science*, 5, 6407–6422.

Emergen Research (2023). Top 8 Bioplastic Companies Transforming the World in an Eco-friendly Way. *Emergen Research* www.emergenressearch.com

European Bioplastics (2021). Bioplastics facts and figures. https://docs.european-bioplastics.org/publications/EUBP_Facts_and_figures.pdf

European Bioplastics (2022). EU Policy Framework on Biobased, Biodegradable and Compostable Plastics. *European Bioplastics*, COM(2022), 682, https://environment.ec.europa.eu/system/files/2022

European Bioplastics (2023). Overview of Relevant EU legislation for Bioplastics. European bioplastics. www.european-bioplastics.org.

European Chemical Agency (2023). REACH: Registration, Evaluation, Authorization, and Restriction of Chemicals. https://echa.europa.eu/

European Commission (2015). Closing the loop—An EU Action Plan for the Circular Economy. 2015. Available online: https: //eur-lex.europa.eu/legal-content/EN/TXT/?uri=CELEX:52015DC0614

European Commission (2019a). Directive (EU) 2019/904 on the reduction of the impact of certain plastic products on the environment. *Official Journal of the European Union*, 12(2), 12–23.

European Commission (2022). The European Green Deal. European Commission.

Expra (2024). EXPRA summarized the over 30 years of extended producer responsibility implementation and best practices in a new document: "30 YEARS OF EPR: HOW TO MAKE THE BEST OUT OF IT". Expra.eu. https://expra.eu/2024/09/18/expra-summarized-the-over-30-years-of-extended-producer-responsibility-implementation-and-best-practices-in-a-new-document-30-years-of-epr-how-to-make-the-best-out-of-it/

Filiciotto, L., and Rothenberg, G. (2020). Biodegradable plastics: Standards, policies, and impacts. *ChemSusChem*, 14(1). https://doi.org/10.1002/cssc.202002044

Fletcher, C. A., Niemenoja, K., Hunt, R., Adams, J., Dempsey, A., and Banks, C. E. (2021). Addressing stakeholder concerns regarding the effective use of bio-based and biodegradable plastics. *Resources*, 10(10), 95. https://doi.org/10.3390/resources10100095

Gaur, A., Gurjar, S. K., and Chaudhary, S. (2022). Circular System of Resource Recovery and Reverse Logistics Approach: Key to Zero Waste and Zero Landfill. In *Advanced Organic Waste Management* (pp. 365–381). https://doi.org/10.1016/b978-0-323-85792-5.00008-3

Hundertmark, T., McNally, C., Simons, T. J., and Vanthournout, H. (2018). *No Time to Waste: What Plastics Recycling Could Offer*. McKinsey on Chemicals. McKinsey & Company.

Insights, W. B. M. (2023, July 5). *EU Bioplastics Policy: Progress and Prospects*. World Bio Market Insights. https://worldbiomarketinsights.com/eu-bioplastics-policy-progress-and-prospects/

Kerry, J. and Butler, P. (2008). *Smart Packaging Technologies for Fast Moving Consumer Goods*. John Wiley & Sons Ltd.

Kleiner, M. (2020). The future of bioplastics: Challenges in standardization and environmental claims. *Environmental Science & Technology*, 54(3), 1234–1242.

Lee, Tin Sin, and Bee, Soo Tueen. (2024). *Microplastics Pollution and Worldwide Policies on Plastic Use*. CRC Press.

Liu, L., & Ramakrishna, S. (2021). *An Introduction to Circular Economy*. Springer.

Londo, M., & Meeusen, M. J. G. (2010). Policy Making for the Biobased Economy. In Langeveld, H., Sanders, J., & M. J. G. Meeusen (Eds.), *The Biobased Economy: Biofuels, Materials and Chemicals in the Post-Oil Era* (pp. 203–213). Earthscan. www.taylorfrancis.com/books/e/9781849774802/chapters/10.4324/9781849774802-23

Meerub, A. (2023). Bioplastics: A Sustainable Choice for a Sustainable Future. www.finance.yahoo.com

Monier, V., Hestin, M., Cavé, J., Laureysens, I., Watkins, E., Reisinger, H., and Porsch, L. (2014). *Development of Guidance on Extended Producer Responsibility*, 23(2), 23–42.

Moreno, B. B., Rodrigues, B., Afonso, L., Jimenez, P. C., and Castro, Ítalo B. (2023). High incidence of false biodegradability claims related to single-use plastic utensils sold in Brazil. *Sustainable Production and Consumption*, 41, 1–8.

Moses, N., Ezenwata, I. S., and Musa, S. I. (2023). Some physicochemical properties and heavy metals in water from Oboshi River, Ibusa in Delta State, Nigeria. *Journal of Applied Sciences and Environmental Management*, 27(6), 1291–1294.

Musa, S. I., and Ikhajiagbe, B. (2023). Seed bio-priming with phosphate solubilizing bacteria strains to improve rice (*Oryza sativa* L. var. FARO 44) growth under ferruginous ultisol conditions. *Journal of Biotechnology, Computational Biology and Biotechnology*, 104(1), 33–51.

Musa, S. I., and Ikhajiagbe, B. (2024). Effects of plant growth promoting bacteria (PGPB) rhizo-inoculation on soil physico-chemical, bacterial community structure and root colonization of rice (*Oryza sativa* L. var. FARO 44) grown in ferruginous ultisol conditions. *Science World Journal*, 19(3), 894–899.

Musa, S. I., Isitua, C. C., Moses, N., and Dioha, I. J. (2022b). Challenges and future scope of bio-nanotechnology in algal biofuel and life cycle analysis. In *Bionanotechnology towards Green Energy* (p. 567). Taylor and Francis, CRC Press.

Musa, S. I., Moses, N., and Ikhajiagbe, B. (2022c). Seed Priming with Phytohormones. Plant Hormones-Recent Advances, New Perspectives and Applications. IntechOpen, London. Page 456.

Musa, S. I., Ogala, H., Okoji, C., Moses, N., Nathaniel, I., and Ojei, S. (2022a). Applications of bio-nanoenabled materials and coatings for affordable, reliable and sustainable green energy production. In *Bionanotechnology towards Green Energy* (p. 567). Taylor and Francis, CRC Press.

Musa, S.I., Shehu, M. S., Ogala, H., Onukwu, I. S., Ebrahimbabaie, P., Okoji, C., Chodize, E., Moses, N., and Ezinne, O. M. (2023). Global advances in bioenergy production technologies. In *Microbial Biotechnology for Bioenergy*. Elsevier.

Napper, I. E., and R.C. Thompson (2019). Environmental deterioration of biodegradable, oxo-biodegradable, compostable, and conventional plastic carrier bags in the sea, soil, and open-air over a 3-year period. *Environmental Science and Technology*, 53(9), 4775–4783.

National Institute of Standards and Technology (2022). Bioplastics and Standards Development: Supporting the U.S. Bioplastics Industry. www.nist.gov/

Nielsen, Tobias D., Hasselbalch, J., Holmberg, K., and Stripple, J. (2020). Politics and the plastic crisis: A review throughout the plastic life cycle. *WIREs Energy and Environment*, 9, e360. https://doi.org/10.1002/wene.360

OECD (2009). Organisation for Economic CoOperation and Development.

OECD (2013). "Policies for Bioplastics in the Context of a Bioeconomy", OECD Science, Technology and Industry Policy Papers, No. 10, OECD Publishing, Paris. www.oecd-ilibrary.org/docserver/5k3xpf9rrw6den.pdf?expires=1730458152&id=id&accname=guest&checksum=511D553B3DDFEFCFDC54240CAF3A5A97. (accessed on 01 November 2024).

Pilla, S. (2011). *Handbook of Bioplastics and Biocomposites Engineering Applications*. Wiley-ScrivenerPublishingLLC. Richard Coles and MarkJ.Kirwan, Food and Beverage Packaging Technology.

Purkiss, D., Allison, A. L., Lorecatto, F., Michie, S., and Miodownik, M. (2022). The Big Compost Experiment: Using citizen science to assess the impact and effectiveness of biodegradable and compostable plastics in UK home composting. *Frontiers in Sustainability*, 3. https://doi.org/10.3389/frsus.2022.942724

Scientists' Coalition for an Effective Plastics Treaty (2023). Policy Brief: The Global Plastics Treaty: What Is the Role of Bio-Based Plastic, Biodegradable Plastic, and Bioplastic? (Possible Core Obligation 8).

Shah, K. U., and Gangadeen, I. (2023). Integrating bioplastics into the US plastics supply chain: Towards a policy research agenda for the bioplastic transition. *Frontiers in Environmental Science*, 11. https://doi.org/10.3389/fenvs.2023.1245846

Sijtsema, S. J., Onwezen, M. C., Reinders, M. J., Dagevos, H., Partanen, A., and Meeusen, M. (2016). Consumer perception of bio-based products—An exploratory study in 5 European countries. *NJAS: Wageningen Journal of Life Sciences*, 77(1), 61–69.

Stasiškienė, Ž., Barbir, J., Draudvilienė, L., Chong, Z. K., Kuchta, K., Voronova, V., and Leal, F. W. (2022). Challenges and strategies for bio-based and biodegradable plastic waste management in Europe. *Sustainability*, 14(24),16476.

Stoica, M., Stoica, D., Ivan, A. S., and Dragomir, C. M. B. (2022). Biopolymers: Regulatory and Legislative Issues. In Nadda, A. K., Sharma, S., & Bhat, R. (Eds.), *Biopolymers*. Springer Series on Polymer and Composite Materials. Springer, Cham. https://doi.org/10.1007/978-3-030-98392-5_4

The Pew Charitable Trusts and SYSTEMIQ (2020). Breaking the Plastic Wave – A Comprehensive Assessment of Pathways towards Stopping Ocean Plastic Pollution. The Pew Charitable Trusts and SYSTEMIQ.

The Times Editorial Board (2020) Editorial: China Is Putting the U.S. to Shame in the Fight against Plastic Trash (*Los Angeles Times*, 2020).

Theobald, B., Risani, R., Donaldson, L., Bridson, J. H. Kingsbury, J. M. Pantos, O., Weaver, L., Lear, G., Pochon, X., Zaiko, A., Smith, D. A., Anderson, R., Davy, B., Davy, S., Doake, F., Masterton, H., Audrezet, F., Maday, S. D. M., Wallbank, J. A., Barbier, M., Greene, A. F., Parker, K., Harris, J., Northcott, G. L., and Abbel, L. (2024). An investigation into the stability and degradation of plastics in aquatic environments using a large-scale field-deployment study. *Science of the Total Environment*, 917, 170301, ISSN 0048-9697, https://doi.org/10.1016/j.scitotenv.2024.170301

Thompson, R. C., Moore, C. J., and Vom Saal, F. S. (2022). Biodegradable plastics and their environmental impact. *Environmental Research Letters*, 17(9), 093003.

Torkelis, A., Dvarionienė, J., and Denafas, G. (2024). The factors influencing the recycling of plastic and composite packaging waste. *Sustainability*, 16(21), 9515. https://doi.org/10.3390/su16219515

Uche, O. L (2023). Plastic waste regime in Rwanda, Kenya and South Africa: A comparative case study. *American Journal of Law*, 5(2), 54–85.

UNEP (2011). Basel Convention on the Control of Transboundary Movement of Hazardous Wastes. Available online: https://wedocs.unep.org/20.500.11822/8385

UNEP (2018). Plastics: Single-Use Plastics: A Roadmap for Sustainability. Available online: www.unep.org/resources/ report/single-use-plastics-roadmap-sustainability (accessed on 28 October 2024).

United Nation (UN). (2015) Transforming Our World: The 2030 Agenda for Sustainable Development. Available online: https://sustainabledevelopment.un.org/post2015/transformingourworld/publication

United Nations Environment Programme (UNEP) (2021a). Tackling Plastic Pollution. Legislative Guide for the Regulation of Single-Use Plastic Products.

United Nations Environment Programme (UNEP) (2021b). From Pollution to Solution: A Global Assessment of Marine Litter and Plastic Pollution. Available https://wedocs.unep.org/bitstream/handle/20.500.11822/36963/POLSOL.pdf. Accessed on 27 October 2021e pollution in West Africa.

USAID. (2024). Promoting Compostable Bioplastics in the APEC Region. Policy Frameworks to Enable Trade, Investment and Innovation. www.apec.org/docs/default-source/publications/2024/3/224_cti_promoting-compostable-bioplastics-in-the-apec-region.pdf?sfvrsn=9dbe0010_2 (accessed on 01 November 2024).

Uyarra, E., Edler, J., García-Estévez, J., Georghiou, L., and Yeow, J. (2014). Barriers to innovation through public procurement: A supplier perspective. *Technovation*, 34(10), 631–645. https://doi.org/10.1016/j.technovation.2014.04.003

Watkins, E. (2017). *EPR in the EU Plastics Strategy and the Circular Economy: A Focus on Plastic Packaging*. Institute for European Environmental Policy.

Watkins, E., Schweitzer, J., Leinala, E., and Börkey, P. (2019) Policy Approaches to Incentivise Sustainable Plastic Design. Environment Working Paper. ENV/WKP, 8, 1–61.

Williams, J. H., DeBenedictis, A., Ghanadan, R., Mahone, A., Moore, J., Morrow III, William R., Price, S. and Torn, M. S. (2012). *Science*, 335, 53–59.

World Bank (2022). Tackling Plastic Pollution: Toward Experience-Based Policy Guidance. Washington, DC: The World Bank. https://documents1.worldbank.org/curated/en/099100104052238589/pdf/P170869059c4e100409c2706e4d1b0b25e1.pdf. (accessed on 01 November 2024).

World Bank (2024). Improving Solid Waste and Plastics Management in Lagos State: A Way Forward. file:///C:/Users/hp/Downloads/P1761781eb744507f184b01f525451f4014.pdf (accessed on 01 November 2024).

World Economic Forum, Ellen MacArthur Foundation & McKinsey & Company (2016). The New Plastics Economy: Rethinking the future of plastics. Report on the envisioned shift from linear to circular plastic economies, addressing strategies such as bioplastics, advanced recycling and extended producer responsibility.

WWF Position. (2020). Biobased and Biodegradable Plastic. Position Paper, No Plastic in Nature www.files.worldwidelife.org

Xie, J. and Martin. J. (2022). *Plastic Waste Management in Rwanda: An Ex-Post Policy Analysis*. The World Bank. https://documents1.worldbank.org/curated/en/099530006222226992/pdf/P1791430174c860e0a9af0cb4b442deb33.pdf (accessed on 01 November 2024).

Yakubu, S., Miao, B., Hou, M., and Zhao, Y. (2024). A review of the ecotoxicological status of microplastic pollution in African freshwater systems. *Science of the Total Environment*, 946, 174092. https://doi.org/10.1016/j.scitotenv.2024.174092

20 Future Perspectives of Bioplastics

E.C. Unoka, J.O. Otedo, Sheila O. Michael, A.O. Isoje, I. Onosakponome, Comfort Nneka Okoji, P. Ebrahimbabaie, and Saheed Ibrahim Musa

20.1 INTRODUCTION

The extensive application of traditional plastics has contributed to significant environmental problems, particularly the emission of heat-trapping gases and the drawbacks closely linked to plastic waste management (Geyer et al., 2017). This plastic waste, largely obtained from petochemicals, is a substantial source of carbon emissions. Bioplastics are polymeric materials that are either bio-based/non-biodegradable, bio-based/biodegradable, or both (Nanda et al., 2022). Bioplastics production from sustainable sources such as plants and microorganisms has emerged as an ecofriendly alternative to the traditional plastics produced from petrochemicals. Unlike the non-decomposable plastics derived from petrochemicals, which are a significant precursor to ecological pollution and carbon footprint, bioplastics are produced from biomass raw materials and serve as carbon sink. While the biodegradable products decompose into carbon dioxide by the action of microorganisms, the bio-based products help to drive the circular economy, making them non-hazardous and non-toxic to the ecosystem. This has led to an increased demand for bioplastic products. Progressive research on the development of advanced bioplastic production technologies is pivotal to meeting the large-scale demand and diverse applications of bioplastics (Pulungan et al., 2020; Arikan et al., 2021). Technological advancements in bioplastic will arouse the interest of end-users and lead to increased availability of innovative bioplastic products for biomedicals, packaging, and vital industrial usage, and the satisfaction of consumers.

Before the discovery of petrochemicals as raw materials for producing plastics, plastic materials were produced from renewable sources in the olden days. Historically, the development and application of bioplastics existed as far back as the 19th century with the production of the first cellulose-based bioplastic known as celluloid by an American scientist, John Wesley Hyatt Jr. in 1869 (El-Kadi, 2010), which had its primary application as eyeglass frames, films, and toys. This was followed by cellophane which was basically used as packaging materials. Other significant historical achievements in bioplastic production from natural materials such as milk, soya beans, microorganism (*Bacillus megaterium*), lactic acid, carob, and cotton waste were recorded between the year 1897 and 2000. In recent decades, advanced bioplastic characterization and production technologies such as surface response methodology (Pulungan et al., 2020), thermogravimetric analysis (TGA), attenuated total reflectance (AFM), X-ray diffraction (XRD), scanning electron microscopy (SEM), and Fourier transform infrared spectroscopy (FTIR), and processing methods such as blown film extrusion, fermentation, evaporation, polymerization, injection molding, fiber spinning, compression molding, film casting, blow molding, casting, and thermoforming have been explored by different researchers for innovative bioplastic production from diverse biomass raw materials including *Morus* sp. (mulberry) leaves, banana peels, perennial grass (*Miscanthus* fiber),

 DOI: 10.1201/9781003602828-24

gum arabic, potatoes, yellow pumpkin, rice straw, algae (*Spirulina platensis*), algal biomass, cassava peel (*Manihot utilisima*), neem, mango seed, groundnut, giant red grass, mulberry, mollasses, feather quil, and waste newspaper have been adopted which has tremendously transformed the bioplastic industry and global market (Hempel et al., 2011; Quil et al., 2011; Cifriadi et al., 2017; Moura et al., 2017; Author et al., 2018; Bilo et al., 2018; Maulida, et al., 2018; Fathanah et al., 2018; Starch et al., 2018; Shellikeri et al., 2018; Aziz et al., 2019; Revadhi, et al., 2019; Arikan et al., 2021; Suresh et al., 2022). Polymers like polyhydroxybutyrate (PHB), cellulose, polyvinyl chloride (PVC), starch, protein-based polymers, polyhydroxyalkanoates (PHA), and polyethylene (PE) are currently utilized to make bioplastics from microalgae (Abdo & Ali, 2019).

Bioplastics are widely applied in diverse sectors such as medical, agriculture, automotive, food, packaging, and construction. Widely utilized bioplastic products include boxes, bags, films, packaging materials in various industries, shopping bags, bioplastic cups, medical tools, food wrappers, drug capsules, raincoats, and fruit and vegetable wrappers.

Rapid expansion is predicted in the bioplastic sector with an expected fiscal rise from USD 9.2 billion to USD 20 billion by 2026 (Intrado GlobeNewswire 2021; Grand View Research 2021; Nanda et al., 2022;). In 2020, bioplastic synthesis exceeded 2.1 million tons and a 38.1% rise is expected in 2025. Asia has the highest global bioplastic production, next to Europe, then South and North America being the least (European Bioplastics 2021). Global manufacturing capability of non-biodegradable and biodegradable bioplastic was estimated as 47.9% and 52.1%, respectively, in 2023; and 38.0% and 62.0% in 2028 (Grand View Research 2021).

Significant growth rate has been recorded for decomposable bioplastics, which includes polybutylene succinate (PBS), polylactic acid (PLA), polybutylene adipate (PBA) terephthalate (PBAT), PHA, and starch-composites, with the production volume estimated at 55.5% (1.16 Mt) of the global bioplastic sector while bio-based bioplastics such as bio-PE terephthalate (bio-PET), and bio-polyamides (bio-PA) accounts for about 40% (Brizga et al., 2020, European Bioplastics, 2021).

Different industries have made remarkable advances in different bioplastic production. The first ever reused bio-PET bottle was unveiled by Coca-Cola company. Solvay, Nova Chemicals, and Petrobras are the leading bio-PE industries, which manufactures about one-fifth of the global commercial portion of the drop-in bioplastics. A yearly commercial production capacity of 20,000 tons of polybutylene succinate-co-terephthalate (PBST) was realized by Sinopec Yizheng Chemical Fiber Co. Ltd. in 2020.

The development of bioplastic composite is another major breakthrough recorded in bioplastic production. This has led to a great transformation in the mechanical, structural, and functional features of bioplastics, and its processability (Díez-Pascual, 2019). Ecofriendly bio-additives have also been developed for bioplastic formulations to replace the non-ecofriendly synthetic ones; however, their application in bioplastic production is selective (Suresh et al., 2022).

The synthesis and utilization of bioplastics will benefit the world economy if there are significant adjustments in consumer preferences and supportive regulatory framework to advance their applications (Ghasemlou et al., 2024).

20.2 ECONOMIC VIABILITY OF BIOPLASTICS IN A CHANGING MARKET

In an always evolving market, bioplastics, which are environmentally friendly plastics, are gaining massive grounds (Porto et al., 2023; Mousavi et al., 2024), with its production capacity predicted to continue increasing significantly. The worldwide industrial capacity of bioplastics market size is predicted to reach USD 44.7 billion by 2030, representing 18.4% yearly growth (Bioplastics Market Size and Outlook, 2030, 2024) from 2023 to 2024. This is helped by the fact that they are produced from easily available natural resources like corn or potato starch, milk protein, gelatin, sugarcane, wheat, algae, and other microorganisms (Abdo & Ali, 2019), thus, helping us to reduce dependence on fossil fuels. However, the procedure for making bioplastics is quite difficult and those made from other sources apart from microalgae are unstable, have reduced mechanical and water resistance properties, and contribute to the depletion of food reserves. Microalgae are thus the preferred source of bioplastics due to their fast development, nutrient concentration, and the high quality of bioplastics gotten from them.

The drive toward less hazardous and sustainable packaging, government regulations restricting the usage of petrochemicals in production of items like medical devices and food packages, growing landfill issues, and other environmental concerns across the world are some of the driving forces behind the increased quest for bioplastics in the packaging industry (Bioplastics Market Size and Outlook, 2030, 2024). The European Union's *Green Deal*, which seeks to achieve a 2050 zero carbon footprint goal and thus deal with the global ecological crisis (Di Bartolo et al., 2021; Awasthi et al., 2022), is a good example of government's policy support for bioplastics. Other forms of support include subsidies, tax incentives, and favorable regulations toward bioplastics production. As a result of well-founded concerns over the application of conventional plastics, the clamor for sustainable plastic alternatives has led to the emergence of bioplastics as a credible option, seeing that they have lower greenhouse gas (GHG) emission, a reduced carbon footprint, and fit perfectly within the circular economy model – a model which promotes the recycling and reuse of waste materials at the end of their life span thus reducing the need for new resources (Muyiwa & Oluwaseun, 2023; Mousavi et al., 2024). This development will most remarkably contribute in promoting more sustainable commercial plastic life cycles and mitigating the Earth's environmental and sustainability issues (Di Bartolo et al., 2021) and consequently generate massive opportunities for the global bioplastics market in the coming years.

This increased demand for bio-based plastics, coupled with new product development and rapid pace of innovation (Anli & Kishore, 2024), have led manufacturers to proactively turn their attention toward the development of bioplastics. And to mitigate the concerns arising from the hazardous effects of plastic waste, these companies are constantly creating product lines using renewable and recycled bioplastics. Also, some businesses are determined to use only recycled or sustainably sourced materials by a specific timeframe (Bioplastics Market Size & Outlook, 2030, 2024).

The bioplastics market is divided based on product into biodegradable and non-biodegradable segments – with the former accounting for a higher revenue share. Based on usage, the market is divided into agriculture, packaging, automotive and transportation, textiles, electronics, etc. Of these segments, the packaging sector accounts for the highest revenue share as bioplastics are widely used in various industries. Starch blends, PET, PE, PBAT, PLA, and PBS, are among the most popular bioplastics used as packaging materials. The bioplastics market is a strongly competitive one, with several key players dominating the market and driving innovation and market growth. Some of these big players include Toyota Tsusho Corporation, TEIJIN Limited, PTT MCC Biochem Company Limited, Total Energies Corbion, Toray Industries Incorporated, and NatureWorks LLC (Bioplastics Market Size & Outlook, 2030, 2024). These companies do a lot of research and development to ensure the performance, sustainability, and cost-effectiveness of their products.

However, the economic viability of bioplastics is greatly impacted by a number of challenges. One of the challenges is their higher costs of production when compared to traditional plastics, as a result of their small production volume (Anli & Kishore, 2024). But recent technological advancements and economies of scale are now helping with this. Another challenge is the still nascent stage of bioplastics, recycling infrastructure, which has led to sorting and processing issues (Mousavi et al., 2024).

The availability of renewable resources (which is most crucial for bioplastics production), poor performance of some bioplastics when compared to traditional plastics (Anli & Kishore, 2024), fierce competition from other more favored bio-based products such as biofuels (OECD, 2013), and some unfavorable government policies toward both development and usage of bioplastics (Di Bartolo et al., 2021; Ali et al., 2022) are numerous more challenges besetting the economic viability of bioplastics. These challenges notwithstanding, bioplastics are seeing a marked increase in demand, especially in the packaging and disposable products sectors. This continuous global expansion of the bioplastic market is driven by aforesaid factors, such as research developments, investment, and growing awareness of the dangers of plastics derived from fossil fuels (Yap et al., 2022). To buttress this point, in a research by Grand View, analysts stated that in 2015, less than 2% market share of the plastic sector was dominated by bioplastics, but by 2030, bioplastics are expected to dominate above 40% by 2030, which would make bioplastics a billion-dollar venture.

Thus, while the bioplastics global market is indeed making massive gains, it is imperative to sustain research and development in its technology with the hope that this will lead to improved production efficiency, final products of the highest quality, and relatively cost-friendly products. Also, there is a need for further research to see how to incorporate more natural resources as raw materials for production purposes, so as to increase production volume. Furthermore, collaborative efforts from the government in the form of friendly policies and incentives will help in the growth of this still seemingly nascent industry. Robust biodegradation and recycling strategies and practices must be implemented for the effective management of bioplastic products at the expiration of their lifespan (Fredi & Dorigato, 2021).

Therefore, in order for bioplastics to become economically viable and have a truly sustainable impact, these issues of production costs, scalability, and technological constraints must be wholly addressed while taking advantage of the ever-increasing market opportunities, customer demand, and government support.

20.3 IMPROVED RECYCLING TECHNIQUES OF BIOPLASTICS

Improved recycling techniques for bioplastics focus on enhancing the circular economy and reducing environmental impact. Here are some advancements and methods.

i Chemical recycling

Chemical recycling is an innovative method for transforming byproduct materials into useful chemicals, such as building blocks and polymer fragments, which can be reintegrated into the polymer production process (Merchan et al., 2022). This method is especially important for biopolymers, particularly aliphatic polyesters, focusing on controlled depolymerization to conserve resources rather than just reducing waste from slow degradation. Depolymerization typically involves solvolytic chain cleavage and heat, utilizing various techniques to break down polymers into their constituent monomers or partially into oligomers. Common methods include glycolysis, alcoholysis, aminolysis, hydrolysis, and ammonolysis. Hydrolysis involves both water permeation and the hydrolysis process: if water diffuses slowly, degradation occurs mainly on the surface (heterogeneous erosion), while faster diffusion leads to bulk degradation (homogeneous erosion). In alcoholysis, alcohol acts as a nucleophile, enabling a transesterification reaction that cleaves polyester ester bonds, resulting in monomers or valuable products. Glycolysis inserts glycol into polymer chains, cleaving ester bonds and substituting them with hydroxyl termini (Merchan et al., 2022).

ii Mechanical recycling

Mechanical recycling, the primary technique for plastic recovery, is recognized for its cost-effectiveness, use of simpler innovation, and reduced ecological footprint compared to chemical recycling (Cosate de Andrade et al., 2016). Mechanical recycling consists of key and auxiliary recycling processes. The key recycling process is a self-sustaining method applicable only to premium plastic waste with a defined history, allowing the recycled material to be used in applications similar to virgin plastics. It typically involves reprocessing uncontaminated waste, like production scraps, back into pellets or resin without sorting or cleaning. In contrast, secondary recycling deals with post-consumer plastics and often results in materials with lower mechanical properties owing to diminished purity and breakdown (Cosate de Andrade et al., 2016). The mechanical recycling process involves several steps: waste collection, screening, sorting, grinding, washing, drying, compounding/extrusion, and granulation. The steps may be performed in varying order based on the characteristics of the plastic waste. While mechanical recycling is a long-standing technique for conventional plastics, caution is necessary when recycling bioplastics, especially biodegradable ones, like PLA, PHAs, and PGA (Lamberti et al., 2020). These materials are sensitive to heat and can undergo thermal degradation, which diminishes their properties. Their high hygroscopicity also promotes hydrolytic breakdown at elevated temperatures, making it essential to thoroughly dry these bioplastics before reprocessing (Lamberti et al., 2020).

iii Enzymatic and microbial recycling

Enzymatic and microbial depolymerizations are cutting-edge practices that use enzymes and microbial activity to selectively disintegrate decomposable bioplastics, aiming to retrieve building blocks and valuable chemicals (Fredi & Dorigato, 2021). This distinguishes them as true recycling methods, unlike biodegradation and composting, which are primarily disposal processes. While they fall under the umbrella of chemical recycling, their novelty justifies separate discussion. Research and patents have investigated the depolymerization of decomposable aliphatic polyesters like PLA, PCL, PBS, PTT, and PBA, often utilizing enzymes such as lipase or proteinase-K to break down oligomers. However, this technology is still developing and tends to be lagging, particularly for incredibly crystalline polymers, as increasing temperatures to accelerate reactions can degrade the enzymes (Fredi & Dorigato, 2021). Microbial enzymes represent a promising pathway for enhancing bioplastic recycling due to their mild processing conditions and ability to release monomers. Many hydrolyzing enzymes have been identified, with bacteria producing various extracellular polyester-hydrolyzing enzymes, including proteases, lipases, cutinases, and esterases (Fredi & Dorigato, 2021). Some of the most effective polyester hydrolases come from fungi, with ongoing discoveries of new candidates (Niaounakis, 2019). For example, the cutinase-like enzyme (CLE1) from *Cryptococcus* sp. S-2 exhibits significant activity against various bioplastic polymers, positioning it as a promising candidate for advancing bioplastic recycling.

iv Improved sorting technologies

Improved sorting technologies, such as optical sorting systems, infrared technology, and robotics, are increasingly utilized in bioplastic recycling (Sohrab, 2004). These methods employ sensors, cameras, Artificial Intelligence (AI) algorithms, and near-infrared (NIR) technology to differentiate various materials for recycling. Optical sorting systems improve material separation accuracy, enhancing the efficiency of recycling facilities by analyzing heat maps of items (Cesar et al., 2024). For example, they can accurately identify and separate plastic bottles of different colors, ensuring appropriate recycling processes and reducing contamination. Robotics has also revolutionized sorting in recycling facilities, enabling efficient management of large volumes of incoming materials (Pučnik et al., 2024). AI algorithms optimize material flow, energy use, and waste reduction. They can predict machine maintenance needs, preventing breakdowns. Additionally, machine learning enhances the speed and accuracy of waste sorting, allowing for better identification of materials like plastics, metals, and paper. NIR technology analyzes reflected light spectra to accurately identify different polymers in plastics, facilitating accurate sorting according to the polymer composition (Pučnik et al., 2024). This advancement is crucial for improving material purity in recycling, ensuring that only high-quality recyclable materials are processed while effectively removing non-recyclable contaminants.

20.4 CHALLENGES OF BIOPLASTICS

Scalability and commercialization of bioplastics, high production cost, limited raw materials supply, and recycling problems Parisa Ebrahimbabaie

The future of bioplastics promises significant environmental benefits, yet it is fraught with challenges, particularly in terms of scalability, commercialization, high production costs, and the sustainable availability of raw materials. Addressing these hurdles is key to the broader adoption of bioplastics as a promising option to conventional petrochemical-derived plastics.

20.5 SCALABILITY AND COMMERCIALIZATION

Scaling up the production of bioplastics to meet global demand is one of the most pressing challenges. Traditional plastics have well-established infrastructure and processes that have been optimized over decades, making production highly cost-efficient. Bioplastics, on the other hand, are still relatively new and require significant investment in infrastructure to be produced on a comparable scale (Negrete-Bolagay & Guerrero, 2024). Processes like microbial fermentation, used to produce some bioplastics, are more energy- and resource-intensive than petrochemical methods, which impacts their scalability. Large-scale production requires optimization in various production technologies, such as extrusion, polymerization, and fermentation (Akinsemolu et al., 2024).

Moreover, commercialization faces several obstacles. While there is growing environmental awareness, consumer demand for bioplastics remains inconsistent, and there is stiff competition from fossil fuel-based plastics, which are cheaper and more accessible. Regulatory support, such as subsidies for bioplastics or penalties for conventional plastics, could drive commercialization forward. Policies that create economic incentives, as well as more stringent environmental regulations, would further accelerate bioplastic adoption by industries and consumers (Negrete-Bolagay & Guerrero, 2024).

New technologies and innovations in biotechnology could address these scalability issues. Advancements in synthetic biology and metabolic engineering have the potential to streamline the manufacturing process, making it more efficient and reducing costs. By optimizing fermentation processes and developing new, higher-yielding microbial strains, it could be possible to scale up production to meet market demands (Akinsemolu et al., 2024). Governments and industries need to foster partnerships that facilitate large-scale production and commercialization, as seen with earlier technological shifts in energy and materials industries (Walker & Rothman, 2020).

20.6 HIGH PRODUCTION COSTS

Bioplastics currently face a significant cost disadvantage compared to conventional plastics, primarily due to the more complex and resource-intensive production processes. Petrochemical plastics benefit from a mature infrastructure that allows for mass production at a lower cost. Conversely, bioplastics rely on plant-based feedstocks, which are often more expensive to grow, harvest, and process into commercial polymers (Akinsemolu et al., 2024). Also, the processing of biomass into polymers typically requires more energy than the production of petrochemical plastics, adding to the overall cost (Negrete-Bolagay & Guerrero, 2024).

Efforts are underway to reduce these costs through technological advancements. For instance, improvements in microbial fermentation, enzyme efficiency, and biotechnological processes could lower the energy consumption associated with bioplastic production (Philp et al., 2013). Integrating bioplastic production into existing waste-management systems, such as using agricultural residues as feedstocks, could also significantly lower raw material costs and reduce reliance on virgin plant materials (Negrete-Bolagay & Guerrero, 2024). Despite these advances, however, bioplastics remain 20–50% more expensive than conventional plastics, making it difficult for them to compete in price-sensitive markets like packaging and consumer goods (Akinsemolu et al., 2024).

In addition to improving production efficiencies, market conditions and government incentives could help reduce costs. Tax incentives for companies investing in bioplastics and subsidies for research and development are potential policy tools to offset the current cost barriers. As demand for bioplastics increases and production scales up, economies of scale could eventually bring costs down to a level where bioplastics can compete directly with petrochemical plastics (Negrete-Bolagay & Guerrero, 2024).

20.7 LIMITED RAW MATERIALS SUPPLY

Bioplastics are primarily produced from sustainable biomass sources like corn, sugarcane, or starch. However, the reliance on agricultural feedstocks presents a significant challenge due to the limited availability of land and the potential for competition with edible crops. As demand for bioplastics grows, there is concern that using food crops for plastic production could exacerbate food security issues and put additional pressure on land and water resources (Negrete-Bolagay & Guerrero, 2024).

To address these concerns, researchers are focusing on developing alternative raw materials that do not interfere or limit with food production. Lignocellulosic biomass, derived from non-food crops like grasses, wood, and agricultural residues, offers a promising solution. These materials are abundant and do not require the same level of resource input as food crops, making them more sustainable in the long term (Akinsemolu et al., 2024). Additionally, waste materials from industrial processes, such as food and paper waste, could be repurposed as raw materials for bioplastic production, further reducing reliance on agricultural resources (Philp et al., 2013).

However, using lignocellulosic biomass and waste materials introduces new technical challenges. These materials are more difficult to break down into monomers that can be used to produce plastics, requiring more advanced chemical and enzymatic processes (Negrete-Bolagay & Guerrero, 2024). Ongoing research is exploring ways to make these processes more efficient, such as through the use of genetically modified enzymes or microorganisms that can more easily degrade lignocellulose into its constituent parts (Akinsemolu et al., 2024).

20.8 DISPOSAL PROBLEM

The disposal of biodegradable bioplastic materials in conventional disposal facilities doesn't promote biodegradation. Many a times, these materials end up in waste streams, landfills, incinerators, etc., where they hardly decompose because of the absence of the required conditions for decomposition. Effective biodegradation is achieved when the industrial composting conditions, such as type of environment, aeration, temperature, time duration, presence of microorganisms, and chemical balance applicable to a particular bioplastic, are met. To address this problem, the right biodegradable infrastructures and systems required to successfully sort and collect these bioplastics products at the end of their life span without mixing them up with the conventional ones should be created; and the end-users of these products should be properly educated on the best disposal approach needed for different biodegradable bioplastic products. Disposal policies and laws for biodegradable bioplastics should be enacted by the relevant authorities and implemented. For instance, in California, there is a policy that compels bioplastic producers to account for the lifecycle of their products, from production stage to final disposal (https://rethinkwaste.org/2023/06/15/the-problem-with-bioplastics/). Slow biodegradation of some compostable bioplastics is also an issue that needs to be tacked. However, researches are ongoing to develop and adopt recycling techniques for biodegradable bioplastics.

20.9 LACK OF PROPER IDENTIFICATION MEANS FOR BIOPLASTIC PRODUCTS IN THE MARKET

Actions such as proper labeling and recognition strategies, targeted toward effectively distinguishing between biodegradable bioplastic products and conventional ones, should be implemented. This will help to forestall the wrong identification of compostable bioplastics which, often leads to disposal of these bioplastic materials along with conventional ones (Dolci et al., 2023).

20.10 GLOBAL REGULATORY BARRIERS

The chemical industry is heavily regulated because of the operational and environmental hazards that arise from industrial products and their production pathways (Watkins et al., 2019). In 2017, the United Nations introduced 17 Sustainable Development Goals (SDGs) which prompted companies to adopt sustainable practices globally (United Nations (UN), 2020). However, the shift to ecofriendly plastics remains sluggish, requiring country-specific policies (Filiciotto & Rothenberg, 2021).

There are ambiguities in the waste management of bio-based plastics; no identified guidelines have been established for the specific collection of bioplastics. This deficiency has resulted in the indiscriminate disposal of bio-based plastics, together with petrochemical-derived plastics and other municipal waste. Standardization of systems of waste collection as well as establishing a uniform waste collection structure, which would lead to the effective sorting of bio-based plastic waste, is required. Local and regional authorities must thus be put in place to implement, administer, and oversee such a setup (Stasiškienė et al., 2022). In 2017, China took the giant stride by banning plastic waste, leading to a substantial reduction in plastic waste (Brooks et al., 2018). Although the China model has served as an eye-opener to some countries to seek alternative solutions on how to dispose of their plastic waste, such as implementing recycling processes and developing biodegradable plastics, the wind of change has not been that successful. Europe's extended producer responsibility (EPR) imposes fees on plastic production, but these fees are low and not globally enforced (Monier et al., 2014). Another discouraging factor in the relatively slow production and market penetration of bioplastics is that the cost of developing and producing bioplastic products that will be completely biodegradable is quite expensive; thus, companies will tend to stick with the cheaper conventional plastics (Filiciotto & Rothenberg, 2021).

There is a need for bioplastics companies that produce more than 1 ton to maintain maximum compliance with registration and regulations involving chemicals. Although there are guidelines established globally to lessen the unrestrained release of toxic substances, sadly they are not stringently followed (European Chemicals Agency (ECHA), 2019a). The REACH (Registration, Evaluation, Authorization, and Restriction of Chemicals) regulation at the EU level was introduced to serve as a form of protection for the well-being of humans and the environment from harmful substances (ECHA, 2019b). This regulation prohibits substances of very high concern (SVHCs) that are capable of causing cancer, acting as mutagens, harmful to reproduction, and persistent substances capable of bioaccumulating (Kubicki, 2011). It is surprising to see that despite the long-term deleterious effects of microplastics, they are not listed as part of the SVHCs.

Another major drawback is that the emphasis on sustainability is somewhat streamlined to international rules instead of domesticating them locally. For instance, the US policies on environment places a lot of emphasis on fuels produced from CO_2 and biodegradable plastics (House, 2012). Even though California has introduced legislation to phase out non-compostable plastics, this law has faced resistance (Singh, 2019). Other countries, such as China, also possess inadequate guidelines on plastics that are biodegradable (OECD, 2013). In Africa, the compliance level is insignificant, only Rwanda is setting the pace for others with its policy that bans imports as well as single-use plastics (Babayemi et al., 2019).

The inconsistencies in globally approved labeling for bioplastic represent another barrier to the use of bioplastic. According to Stasiškienė et al. (2022), the vagueness in labeling bio-based plastics and their products, as well as the absence of a robust and identification system backed by law results in their inappropriate handling and ultimately leads to landfill waste. This can be addressed by standardized obligatory labeling of bioplastics, hence clearly providing a distinctive feature that could be used to differentiate them from other materials, thereby driving consumers' choice of packaging. This is also important in allowing consumer and waste-management bodies to reduce plastic waste pollution as suitable management of the used bioplastics will be adhered to.

The lack of international harmonization and inconsistent regulations has resulted in a fragmented policy framework in bioplastic production, use, labeling, and disposal (Watkins et al., 2019). This fragmentation poses significant challenges for the bioplastics industry, consumers, and the environment. Addressing these challenges would result in a more cohesive and effective bioplastic regulatory framework, thus promoting sustainable development and environmental stewardship.

20.11 IMPACT OF BIOPLASTIC ON CLIMATE CHANGE

Recently, bioplastics have drawn interest as an ecofriendlier substitute with conventional plastics, with promises to reduce GHG emissions by sourcing raw materials from renewable sources. However, while bioplastics hold potential for climate change mitigation, there are challenges in production and lifecycle that need to be carefully assessed (Ali et al., 2023).

Bioplastics which are broadly classified into two main categories include bio-based plastics, derived partially or wholly from renewable biomass sources. Some of which are biodegradable, while others are chemically identical to traditional plastics and thus non-biodegradable and biodegradable plastics, which can break down naturally, either with oxygen present (aerobic conditions) or without (anaerobic conditions). Not all polymers that decompose naturally are bio-based, as some petroleum-based plastics are engineered to biodegrade (Diksha et al., 2022). Examples of bioplastics include PLA and PHAs, which are both biological and compostable, while bio-based PE (bio-PE) is chemically similar to conventional PE and does not biodegrade (Costa et al., 2023).

20.12 CLIMATE CHANGE MITIGATION: POTENTIAL BENEFITS OF BIOPLASTICS

a. Decrease in GHG Pollutions: One of the primary advantages of bioplastics is that they rely on renewable biomass, which absorbs carbon dioxide (CO_2) from the atmosphere during its growth. This process can partially offset emissions from production via carbon sequestration in biomass in which the crops used in bioplastic production capture atmospheric CO_2, creating a temporary carbon sink (Rosenboom et al., 2022). Although this carbon is eventually released upon decomposition, the initial sequestration reduces the net CO_2 emitted and lowering of fossil fuel dependence since bioplastics are derived from plants rather than petroleum, they reduce reliance on fossil fuels, potentially lowering emissions from fossil fuel extraction, processing, and transportation (Ghada et al., 2021).
b. Biodegradability and End-of-Life Benefits: Many biological plastics, such as PLA and PHA, are made to break down under certain conditions, which can reduce emissions associated with waste disposal. Their ability to decompose in composting facilities reduces methane emissions that occur when organic waste decomposes anaerobically in landfills. Also, there is reduction in plastic waste and pollution, where in marine and terrestrial environments, certain biodegradable bioplastics reduce long-term pollution, which has implications for ecosystem health and the atmospheric carbon balance (Narancic et al., 2020).

20.13 CHALLENGES ASSOCIATED WITH BIOPLASTICS THAT CAN CONTRIBUTE NEGATIVELY TO CLIMATE CHANGE

While biological plastics have the capacity to lessen reliance on fossil fuels and offer some environmental benefits, their impact on climate change is mixed. In many cases, the production and end-of-life stages of bioplastics can create emissions that offset or even exceed their environmental benefits (Emadian et al., 2017). Some of these challenges include the following:

1. Land Use and Deforestation: The production of bioplastics often requires crops like corn, sugarcane, and cassava. Large-scale bioplastic production can lead to deforestation to create

more agricultural land, which releases stored carbon from trees and soil, contributing to GHG emissions (Nanda & Berruti, 2020). There is also soil carbon loss in which the clearing of natural land for crop cultivation not only releases carbon from trees but also disturbs soil, which can release additional stored carbon into the atmosphere.

2. Agricultural Emissions via Use of Fertilizer and Pesticide: Crops grown for bioplastics require water, fertilizers, and pesticides, which have their own carbon footprints (Brizga et al., 2020). Synthetic fertilizers, in particular, emit nitrous oxide, a potent GHG. Also, in areas with limited water supply, the intensive water needs of crops used for bioplastics can stress water resources, indirectly impacting ecosystems that help regulate carbon emissions.
3. Energy-Intensive Production Processes: Some bioplastics require significant energy during production, and if this energy comes from fossil fuels, it can result in high emissions (Hottle et al., 2013). For example, PLA bioplastics involve fermentation processes that can be energy-intensive. There is also the high lifecycle emission for certain bioplastics, since not all bioplastics are created equal, bio-based PE (PE) and PE terephthalate (PET), which mimic traditional plastics, can have high lifecycle emissions similar to their petroleum-based counterparts.
4. End-of-Life Issues Such As Methane Emissions in Landfills: If bioplastics end up in landfills, they can produce methane, a potent GHG, as they decompose under anaerobic (oxygen-free) conditions (Kale et al., 2007). This is particularly a concern if the waste is not properly managed or if industrial composting facilities are unavailable.
5. Limited Degradability: Not all bioplastics are biodegradable or compostable. Some may persist in the environment like traditional plastics, leading to microplastic pollution and potential ecosystem impacts.
6. Competition with Food Supply: Using agricultural land for bioplastic production can compete with food production, especially in areas where arable land is limited. This can indirectly affect food prices and lead to agricultural expansion, increasing emissions due to land-use change (Piemonte & Gironi, 2011).
7. Waste Management and Infrastructure Challenges: Bioplastics often require specialized recycling or composting infrastructure, which is not available everywhere. If they are incorrectly sorted, bioplastics can contaminate conventional plastic recycling streams, increasing waste and emissions from disposal (Islam et al., 2024).

20.14 COMPARATIVE ANALYSIS AND LIFECYCLE ASSESSMENT

Lifecycle assessments (LCA) provide a comprehensive view of the effect of bioplastics on climate compared to traditional plastics (International Organization for Standardization (ISO), 2006). LCAs consider components such as extraction, manufacturing, transportation, raw material, usage, and disposal. Findings indicate variation in climate impact: not all bioplastics offer the same environmental benefits. For example, bio-based PLA might have less carbon footprint than petroleum-based PE but still generates emissions due to high energy inputs and room for improvement through innovations in feedstock sourcing, energy efficiency in manufacturing, and effective end-of-life solutions can further reduce the carbon footprint of bioplastics (Islam et al., 2024).

20.15 CONCLUSION

The future of bioplastics depends on overcoming key challenges in scalability, commercialization, production costs, and the availability of sustainable raw materials. While these challenges are significant, they are not insurmountable. Advances in biotechnology, material science, and process engineering are pivotal in addressing these concerns, also continued investment in research and development is highly essential. Additionally, policy interventions, such as subsidies and tax incentives, can facilitate the global adoption of bioplastics. With these efforts, bioplastics could eventually become a viable, sustainable substitute to traditional petrochemical-derived plastics, helping to reduce plastic pollution and mitigate environmental degradation.

REFERENCES

Abdo, S. M. & Ali, G. H. (2019). Analysis of polyhydroxybutrate and bioplastic production from microalgae. *Bulletin of the National Research Centre*, 43(1). https://doi.org/10.1186/s42269-019-0135-5

Akinsemolu, A. A., et al. (2024). Recycling technologies for biopolymers: Current challenges and future directions. *Polymers*. https://doi.org/10.3390/polym16192770

Ali, S. S., Abdelkarim, E. A., Elsamahy, T., Al-Tohamy, R., Li, F., Kornaros, M., Zuorro, A., Zhu, D. and Sun, J. (2023). Bioplastic production in terms of life cycle assessment: A state-of-the-art review. *Environmental Science Ecotechnology*, 15: 100254.

Anli, D. A. & Kishore, G. (2024). Microalgae: An emerging source of bioplastics production. *Discover Environment*, 2(1). https://doi.org/10.1007/s44274-024-00038-0

Arikan, E., Bouchareb, E., Bouchareb, R., Yagci, N. and Dizge, N. (2021). Innovative Technologies Adopted for the Production of Bioplastics at Industrial Level. https://doi.org/10.1007/978-981-16-1823-9_3

Author, S. J., Sharma, U. and Goswami G. (2018). Bio-plastic from waste newspaper. *International Journal of Research in Engineering and Technology*, 2(30): 1–12.

Awasthi, S. K., Kumar, M., Kumar, V., Sarsaiya, S., Anerao, P., Ghosh, P., Singh, L., Liu, H., Zhang, Z. and Awasthi, M. K. (2022). A comprehensive review on recent advancements in biodegradation and sustainable management of biopolymers. *Environmental Pollution*, 307: 119600. https://doi.org/10.1016/j.envpol.2022.119600

Aziz, I. A., Wan, C., Robiah, S. and Adollah, R. (2019). Fibre based bioplastic film from *Morus* sp. (mulberry) leaves for medical purpose. *Journal of Physics Conference Series*, 1372(1): 012069.

Babayemi, J. O., Nnorom, I. C., Osibanjo, O. and Weber, R. (2019). *Environmental Sciences Europe*, 31, 23–52. Article No. 60.

Bilo, F., Pandini, S., Sartore, L., Depero, L. E., Gargiulo, G. and Bonassi, A. A. (2018). Sustainable bioplastic obtained from rice straw. *Journal of Cleaner Production*, 200: 357–368.

Bioplastics Market Size & Outlook, 2030. (2024). Grandviewresearch.com. www.grandviewresearch.com/horizon/outlook/bioplastics-market-size/global

Brizga, J., Hubacek, K. and Feng, K. (2020). The unintended side effects of bioplastics: Carbon, land, and water footprints. *One Earth*, 3: 45–53.

Brooks, A. L., Wang, S. and Jambeck, J. R. (2018). The Chinese import ban and its impact on global plastic waste trade. *Science Advances*, 4(6): 20.

Cesar, L., Bin Daej, M. A. A. and Alexandridis, P. (2024). Recent developments in technology for sorting plastic for recycling: The emergence of artificial intelligence and the rise of the robots. *Recycling* 9(4): 59. https://doi.org/10.3390/recycling9040059

Che Ab Aziz, N. I., Zakaria, Y., Md Muslim, N. Z. and Nik Hassan, N. F. (2023). Emerging and advanced technologies in biodegradable plastics for sustainability. In Ali, G.A.M., Makhlouf, A.S.H. (eds) *Handbook of Biodegradable Materials*. Springer, Cham. https://doi.org/10.1007/978-3-031-09710-2_21

Cifriadi, A., Panji, T., Wibowo, N. A. and Syamsu, K. (2017). Bioplastic production from cellulose of oil palm empty fruit bunch. *Earth and Environmental Science*, 65(1): 012011.

Cosate de Andrade, M. F., Souza, P. M. S., Cavalett, O. and Morales, A. R. (2016). Life cycle assessment of poly(lactic acid) (PLA): Comparison between chemical recycling, mechanical recycling and composting. *Journal of Polymers and the Environment*, 24: 372–384. https://doi.org/10.1007/ s10924-016-0787-2

Costa, A., Encarnação, T., Tavares, R., Todo, B. T. and Mateus, A. (2023). Bioplastics: Innovation for green transition. *Polymers*, 15(3): 517.

de Moura, I. G., de Sa, A. V., Sofia, A., Machado, L., Vera, A. and Machado, A. (2017). Bioplastics from agro-wastes for food packaging applications. *Food Packaging*, 1: 223–267

Di Bartolo, A., Infurna, G. and Dintcheva, N. T. (2021). A review of bioplastics and their adoption in the circular economy. *Polymers*, 13(8): 1229. https://doi.org/10.3390/polym13081229

Díez-Pascual, A. M. (2019). Synthesis and applications of biopolymer composites. *International Journal of Molecular Sciences*, 20: 2321. https://doi.org/10.3390/ijms20092321

Diksha, Y., Anamika, S., Anjali, Y., Bharti, M. and Manish, S. (2022). Development and sustainability of bioplastics: A review. *Asian Journal of Green Chemistry*, 6(2): 112–128.

Dolci, G., Rigamonti, L. and Grosso, M. (2023). The challenges of bioplastics in waste management. *Waste Management & Research*, 41(8): 1281–1282. https://doi.org/10.1177/0734242X231181999

Ekawardhani, Y. A., Pasaribu, C. Y., Rohmah, A. N. and Salsabila, O. (2021). Bioplastic technology as packaging innovation. *IOP Conference Series: Materials Science and Engineering*. https://doi.org/10.1088/1757-899X/1158/1/012008

El-Kadi, S. (2010) *Bioplastic Production from Inexpensive Sources: Bacterial Biosynthesis, Cultivation System, Production and Biodegradability*. VDM Publishing House, New York.

Emadian, S. M., Onay, T. T. and Demirel, B. (2017). Biodegradation of bioplastics in natural environments. *Waste Management*, 59: 526–536.

European Bioplastics (2021) Bioplastics market data. www.european- bioplastics. org/ market/ (Accessed on 2 November 2024).

European Chemicals Agency (ECHA) (2019a). Annex XV Restriction Report: Proposal of Restriction for Intentionally Added Microplastics. https://echa.europa.eu/documents/10162/05bd96e3-b969-0a7c-c6d0-441182893720

European Chemicals Agency (ECHA) (2019b). Authorization-ECHA can be found under https://echa.europa.eu/substances-of-very-high-concern-identification-explained

Fathanah, U., Lubis, M. R., Nasution, F. and Masyawi, M. S. (2018). Characterization of bioplastic based from cassava crisp home industrial waste incorporated with chitosan and liquid smoke. *Energy & Environmental Science*, 334(1): 012073.

Filiciotto, L. and Rothenberg, G. (2021). Biodegradable plastics: Standards, policies and impacts. *ChemSusChem*, 14(1): 56–72.

Fredi, G. and Dorigato, A. (2021). Recycling of bioplastic waste: A review. *Advanced Industrial and Engineering Polymer Research*, 4(3): 159–177. https://doi.org/10.1016/j.aiepr.2021.06.006

Geyer, R., Jambeck, J. and Law, K. (2017). Production, use, and fate of all plastics ever made. *Science Advances*, 3: e1700782.

Ghada, A., Abanoub, M., Christopher, C., Parrish, J., Banoub, T. and Le, T. (2021). Environmental impact of bioplastic use: A review. *Heliyon*, 7(9): e07918.

Ghasemlou, M., Barrow, C. J. and Adhikari, B. (2024). The future of bioplastics in food packaging: An industrial perspective. *Food Packaging and Shelf Life*, 43: 101279. https://doi.org/10.1016/j.fpsl.2024.101279

Grand View Research (2021). Bioplastics Market Size, Share & Trends Analysis Report by Product (Biodegradable, Non-biodegradable), by Application (Packaging, Automotive & Transportation, Textile), by Region, and Segment Forecasts, 2021–2028. www. grand viewr esear ch. com/ indus try- analy sis/ bioplastics- indus try#: ~: text= The% 20glo bal% 20bio plast ics% 20market% 20size% 20was% 20est imated% 20at% 20USD% 208. 3,USD%209.2% 20bil lion% 20in% 202020. & text= The% 20glo bal% 20bioplast ics% 20mar ket% 20is,USD% 2026.0% 20bil lion% 20by202027. (Accessed 2 November, 2024).

Hempel, F., Bozarth, A. S., Lindenkamp N., Klingl, A., Zauner, S. and Linne, U. (2011). Microalgae as bioreactors for bioplastic production. *Agris*, 12: 2–7.

Hottle, T., Bilec, M. and Landis, A. (2013). Sustainability assessments of bio-based polymers. *Polymer Degradation and Stability*, 98: 1898–1907.

House T. W. (2012). *Industrial Biotechnology*, 8: 97–102.

International Organization for Standardization (ISO) (2006). Environmental management-Life cycle assessment-Principles and framework. International Organization for Standardization: Geneva, 14040.

Intrado GlobeNewswire (2021). Bioplastics Market Size Expected to Reach USD 19.93 Billion by 2026 | Fortune Business Insights™. www. globe newsw ire. com/ news- relea se/ 2021/ 03/ 30/22017 86/0/ en/ Biopl astics- Market- Size- Expec ted- to- Reach-USD- 19- 93- Billi on- by- 2026- Fortu ne- Busin ess- Insig hts html (Accessed 2 November 2024).

Islam, M., Xayachak, T., Haque, N., Lau, D., Bhuiyan, M. and Pramanik, M. B. (2024). Impact of bioplastics on environment from its production to end-of-life. *Process Safety and Environmental Protection*, 188: 151–166.

Lamberti, F. M., Román-Ramírezm, L. A. and Wood, J. (2020). Recycling of bioplastics: Routes and benefits. *Journal of Polymers and the Environment*, 28: 2551–2571.

Maulida, K. T., Harahap, M. B. and Ginting, M. H. S. (2018). *IOP Conference Series: Materials Science and Engineering*, 309: 012068. https://doi.org/10.1088/1757-899X/309/1/012068

Merchan, A. L., Fischöder, T., Hee, J., Lehnertz, M. S., Osterthun, O. and Pielsticker, S. (2022). Chemical recycling of bioplastics: Technical opportunities to preserve chemical functionality as path towards a circular economy. *Green Chemistry*, 24: 9428–9449.

Monier, V., Hestin, M., Cavé, J., Laureysens, I., Watkins, E., Reisinger, H. and Porsch, L. (2014). Development of Guidance on Extended Producer Responsibility (EPR). https://ec.europa.eu/environment/archives/waste/eu_guidance/pdf/Guidance%20on%20EPR%20-%20Final%20Report.pdf

Mousavi, M., Elaheh, K., Gheibi, M., Zahra, A. C., Termeh, T., and Ramakrishna, S. (2024). Assessing bioplastics' economic, commercial, political, and energy potential with circular economy modeling: A sustainable solution to plastic waste management. *Materials Circular Economy*, 6(1). https://doi.org/10.1007/s42824-023-00098-2

Muyiwa, O. and Oluwaseun, K. (2023). *Digital Innovations for a Circular Plastic Economy in Africa*. Routledge eBooks. Informa. https://doi.org/10.4324/9781003278443

Nanda, S. and Berruti, F. (2020). Bioplastics from renewable sources: A critical review. *Materials Today Chemistry*, 16: 100248.

Nanda, S., Patra, B. R., Patel, R., Bakos, J. and Dalai, A. K. (2022). Innovations in applications and prospects of bioplastics and biopolymers: A review *Environmental Chemistry Letters* 20: 379–395. https://doi.org/10.1007/s10311-021-01334-4

Narancic, T., Cerrone, F., Beagan, N. and Kevin, O. (2020). Recent advances in bioplastics: Application and biodegradation. *Polymers*, 12: 920. https://doi.org/10.3390/polym12040920

Negrete-Bolagay, D. and Guerrero, V. H. (2024). Opportunities and challenges in the application of bioplastics: Perspectives from formulation, processing, and performance. *Polymers (Basel)*, 16(18): 2561. https://doi.org/10.3390/polym16182561. PMID: 39339026; PMCID: PMC11434805.

Niaounakis, M. (2019). Recycling of biopolymers e the patent perspective. *European Polymer Journal*, 114: 464e475.

Kale, G., Auras, R. and Singh, S. P. (2007). Degradation of PLA in anaerobic environments. *Journal of Polymers and the Environment*, 15: 242–249.

Kubicki M. (2011). *Guidance on the Preparation of an Application for Authorisation*. European Chemicals Agency, Helsinki.

OECD (2013). Policies for Bioplastics in the Context of a Bioeconomy, can be found at www.oecd-ilibrary.org/science-and-technology/policies-for-bioplastics-in-the-context-of-a-bioeconomy_5k3xpf9rrw6d-en

Philp, J. C., et al. (2013). Bioplastics science from a policy perspective. *New Biotechnology*, 30(6): 635–646. https://doi.org/10.1016/j.nbt.2013.03.010

Piemonte, V. and Gironi, F. (2011). Land use change and environmental impact of bioplastics. *Sustainable Chemical Processes*, 3(1): 22–29.

Pučnik, R., Dokl, M., Fan, Y.V., Vujanović, A., Pintarič, Z. N., Aviso, K. B. (2024). A waste separation system based on sensor technology and deep learning: A simple approach applied to a case study of plastic packaging waste. *Journal of Cleaner Production*, 540: 141762.

Pulungan, M. H., Kapita, R. A. D. and Dewi, I. A. (2020). Optimisation on the production of biodegradable plastic from starch and cassava peel flour using response surface methodology. *IOP Conference Series: Earth and Environmental Science*, 475(1): 012019.

Quill, F., Ullah, A., Vasanthan, T., Bressler, D., Elias, A. L and Wu, J. (2011). Bioplastics from feather quill. *Biomacromolecules*, 12(10): 3826–3832.

Revadhi, T. and Nanthini, R. (2019). Biosynthesis of polyhydroxybutyrate from giant reed grass hydrolysate and evaluation of its drug releasing profiles. *Journal of Drug Delivery and Therapeutics*, 9: 8–15.

Rosenboom, J. G., Langer, R. and Traverso, G. (2022). Bioplastics for a circular economy. *Nature Reviews Materials*, 7: 117–137.

Shellikeri, A., Kaulgud, V., Yaradoddi, J., Ganachari, S., Banapurmath, N. and Shettar, A. (2018). Development of neem based bioplastic for food packaging application. *Materials Science and Engineering*, 376(1): 012052.

Singh, M. (2019). The most ambitious US law to tackle single-use plastics faces make-or-break moment, can be found at www.theguardian.com/us-news/2019/sep/13/california-plastics-legislation-single-use

Sohrab (2004). Innovations in Recycling: Advanced Sorting Technologies & Circular Economy Models.https://medium.com/@sohrabjam/innovations-in-recycling-advanced-sorting-technologies-circular-economy-models-8f8d55f8357a

Starch, G., Plasticizer, S.A.S., Sariningsih, N., Putra, Y.P., Pamungkas, W. P. and Amri, A. (2018). Bioplastic from chitosan and yellow pumpkin starch with castor oil as plasticizer. *Materials Science and Engineering*, 333(1): 1–8.

Stasiškienė, Z., Barbir, J., Draudviliene, L., Chong, Z. K., Kuchta, K., Voronova, V. and Leal, F. W. (2022). Challenges and strategies for bio-based and biodegradable plastic waste management in Europe. *Sustainability*, 14(24): 16476. https://doi.org/10.3390/su142416476

Suresh, S. S., Abhinay, T., Saumya, S. and Kumar, A. (2022). Recent advances in bioplastics: Synthesis and emerging perspective Iran. *Journal of Chemistry and Chemical Engineering*, 41: 8.

The Problems with Bioplastics. (2010). https://rethinkwaste.org/2023/06/15/the-problem-with-bioplastics/. (Accessed 04 November 2024).

United Nations (UN) (2020). *The Sustainable Development Goals Report*, UN.

Walker, S. and Rothman, R. (2020). Life cycle assessment of bio-based and fossil-based plastic: A review. *Journal of Cleaner Production*, 261: 121–157. https://doi.org/10.1016/j.jclepro.2020.121157

Watkins, E., Schweitzer, J., Leinala, E. and Börkey, P. (2019). Policy Approaches to Incentivise Sustainable Plastic Design Environment Working Paper. ENV/WKP. 8: 1–61

Yap, X. Y., Gew, L. T., Khalid, M. and Yow, Y.-Y. (2022). Algae-based bioplastic for packaging: A decade of development and challenges (2010–2020). *Journal of Polymers and the Environment*. https://doi.org/10.1007/s10924-022-02620-0

Index